Contents

Preface

This text is about MATLAB®. If you use MATLAB or are considering using it, this book is for you. This text represents an alternative to learning MATLAB on your own with or without the help of the documentation that comes with the software. The informal style of this book makes it easy to read and, as the title suggests, it provides the tools you need to master MATLAB. As a programming language and data visualization tool, MATLAB offers a rich set of capabilities for solving problems in engineering, scientific, computing, and mathematical disciplines. The fundamental goal of this text is to help you increase your productivity by showing you how to use these capabilities efficiently. Because of the interactive nature of MATLAB, the material is generally presented in the form of examples that you can duplicate by running MATLAB as you read this book.

This text covers only topics that are of use to a general audience. The material presented generally applies to all computer platforms. None of the *Toolboxes,* *Blocksets,* and other *Libraries* that are available for additional cost are discussed, although some are referred to in appropriate places. There are simply too many additional product items to consider in one book. However, based on user feedback, this book does discuss linking MATLAB to programs written in C, FORTRAN, and Java. Furthermore, it demonstrates how MATLAB can be dynamically linked to other applications on the PC platform.

Since MATLAB continues to evolve as a software tool, this text focuses on MATLAB Version 7. For the most part, the material applies to 6.x and 5.x versions of MATLAB as well. When appropriate, distinctions between versions are made.

Many thanks to the folks at the MathWorks, specifically Penny Anderson, Rich Ellis, Tim Farajian, Paul Fricker, Steve Lord, Bob Gilmore, and Peter Webb. They carefully reviewed and offered many helpful suggestions that have improved the presentation of this book.

We the authors encourage you to give us feedback on this book. What are the best features of the text? What areas need more work? What topics should be left out? What topics should be added? We can be reached at the e-mail address *mm@eece.maine.edu.* In addition, errata, all examples in the text, and other related material can be found at *http://www.eece.maine.edu/mm.*

DUANE HANSELMAN
BRUCE LITTLEFIELD

1

Getting Started

1.1 INTRODUCTION

This text assumes that you have some familiarity with matrices and computer programming. Matrices and arrays, in general, are at the heart of MATLAB, since all data in MATLAB are stored as arrays. Besides offering common matrix algebra operations, MATLAB offers array operations that allow you to quickly manipulate sets of data in a wide variety of ways. In addition to its matrix orientation, MATLAB offers programming features that are similar to those of other computer programming languages. Finally, MATLAB offers graphical user interface (GUI) tools that enable you to use MATLAB as an application development tool. This combination of array data structures, programming features, and GUI tools makes MATLAB an extremely powerful tool for solving problems in many fields. In this text, each of these aspects of MATLAB is discussed in detail. To facilitate learning, detailed examples are presented.

1.2 TYPOGRAPHICAL CONVENTIONS

The following conventions are used throughout this book:

Bold italics	New terms or important facts
Boxed text	Important terms and facts
Bold Initial Caps	Keyboard key names, menu names, and menu items
`Constant width`	Computer user input, function and file names, commands, and screen displays

1

| Boxed constant width | Contents of a script, function, or data file |

Constant width italics   User input that is to be replaced and not taken
                                literally, such as » `help` *functionname*

Italics   Window names, object names, book titles,
                                toolbox names, company names, example text,
                                and mathematical notations

1.3 WHAT'S NEW IN MATLAB 7

MATLAB 7 is yet another evolutionary change in MATLAB. The *Command* window remains the primary user interface, *Figure* windows are used to display graphical information and to create graphical user interfaces (GUIs), and a text editor is provided for writing and editing MATLAB code. The *MATLAB Desktop* coordinates the position and visibility of a number of other windows, such as *Workspace*, *Editor, Help, Command History*, etc.

Under the surface, several numerical and operational changes have been made to the software. In addition to having added or modified several internal numerical algorithms, MATLAB now supports mathematical operations on data types other than double-precision arrays, which have always been at the heart of MATLAB. Perhaps even more importantly, the MATLAB command interpreter now includes acceleration features, collectively called *The MATLAB JIT-Accelerator*. This accelerator, which first appeared in MATLAB 6.5, dramatically increases the speed of loop operations by interpreting and executing code within a loop as a whole, rather than line by line, as was done previously. This acceleration eliminates the need to *vectorize* some code, which requires you to create and maximize the use of arrays and array mathematics to achieve optimal performance. To make use of the JIT-Accelerator, loop operation code must follow specific guidelines, which are covered in this text. When these guidelines are not followed, loop operation code is interpreted at the much slower line-by-line rate.

In sum, MATLAB 7 represents an evolutionary change in MATLAB. The basic operation of MATLAB and its capabilities haven't changed in any dramatic way. In almost all cases, MATLAB code written for MATLAB 6 will run without change in MATLAB 7. For the most part, the new and changed features of MATLAB 7 can increase your productivity in solving problems with MATLAB.

1.4 WHAT'S IN *MASTERING MATLAB 7*

MATLAB documentation, in both hard copy and electronic format, exceeds 5000 pages of information and help text. Given this exhaustive documentation and help, *Mastering MATLAB 7* does not attempt to be a comprehensive tutorial or reference, which just isn't possible in a single book containing less than 1000 pages. *Mastering MATLAB 7* does not even attempt to document all functions within the base product of MATLAB. This too is impossible, as there are more than 300 built-in functions in MATLAB, and approximately 1000 more M-file functions that are part of the base product of MATLAB. In light of this documentation set, the goals of *Mastering MATLAB 7* include the following: (1) introduce MATLAB to the novice user, (2) illustrate all key features and capabilities of MATLAB, and (3) demonstrate by example how to write efficient MATLAB code.

This text is intended to be a valuable resource when the MATLAB documentation set is unavailable. It follows the rule of providing 80 percent of the information needed in 20 percent of the space required to cover everything. The book is also intended to be a valuable resource when the MATLAB documentation set is available. In this case, it supplies numerous examples of efficient MATLAB coding that demonstrate how the many features of MATLAB come together to solve real problems.

Equally important is what's not included in *Mastering MATLAB 7*. The book does not discuss all of the windows, menus, menu items, submenus, and dialog boxes of the MATLAB user interface. While these user-interface aspects of MATLAB are very important, there is no room for them in this text. Moreover, they are difficult to cover effectively in a text and are best learned by hands-on exploration. Rather, this text focuses on the mathematical, programming, and graphical features that facilitate problem solution.

This book was written using information about MATLAB Version 7.0. As MATLAB evolves between Versions 7.0 and 8.0, some things are bound to change. As a result, there may be isolated areas in this book where information about new features is missing and, worse yet, isolated areas where information is incorrect. The authors have no control over MATLAB. We also cannot rewrite the text to reflect minor MATLAB releases. We are thankful that the makers of MATLAB are very careful when introducing new features and when changing old features. Historically, old features are grandfathered for one major release, and sometimes for additional releases. As a result, even though the book reflects MATLAB Version 7.0, it will undoubtedly be useful for all future 7.x versions as well.

To support this text, the authors maintain the Mastering MATLAB website at *http://www.eece.maine.edu/mm*. At the site, you can find errata for the text, as well as MATLAB script M-files for creating all of the figures in the book. The authors also encourage constructive feedback about the text at the e-mail address *mm@eece.maine.edu*.

2

Basic Features

Running MATLAB creates one or more windows on your computer monitor. One of these windows, entitled MATLAB, is commonly called the ***MATLAB desktop***. This window is the primary graphical user interface for MATLAB. Within the *MATLAB* window, there is a window called the ***Command*** window, which is the primary place wherein you interact with MATLAB. The prompt >> is displayed in the *Command* window, and when the *Command* window is active, a blinking cursor appears to the right of the prompt. This cursor and prompt signify that MATLAB is waiting to perform a mathematical operation.

2.1 SIMPLE MATH

Just like a calculator, MATLAB can do basic math. Consider the following simple example: Mary goes to the office supply store and buys four erasers at 25 cents each, six memo pads at 52 cents each, and two rolls of tape at 99 cents each. How many items did Mary buy, and how much did they cost?

To solve this problem with a calculator, you enter

```
4 + 6 + 2 = 12 items
4 × 25 + 6 × 52 + 2 × 99 = 610 cents
```

In MATLAB, this problem can be solved in a number of different ways. First, the calculator approach can be taken:

```
>> 4+6+2
ans =
    12
```

```
>> 4*25 + 6*52 + 2*99
ans =
   610
```

Note that MATLAB doesn't care about spaces, for the most part, and that multiplication takes precedence over addition. Note also that MATLAB calls the result ans, which is short for *answer* for both computations.

As an alternative, the problem can be solved by storing information in *MATLAB variables*:

```
>> erasers = 4
erasers =
     4
>> pads = 6
pads =
     6
>> tape = 2;
>> items = erasers + pads + tape
items =
    12
>> cost = erasers*25 + pads*52 + tape*99
cost =
   610
```

Here, we created three MATLAB variables—erasers, pads, and tape—to store the number of each item. After entering each statement, MATLAB displayed the results, except in the case of tape. The semicolon at the end of the line tells MATLAB to evaluate the line, but not to display the answer. Finally, rather than calling the results ans, we told MATLAB to call the number of items purchased items, and the total price paid cost. At each step, MATLAB remembered past information. Because MATLAB remembers things, let's ask what the average cost per item was:

```
>> average_cost = cost/items
average_cost =
    50.833
```

Since the term "average cost," is two words and MATLAB variable names must be one word, an underscore was used to create the single MATLAB variable average_cost.

In all, MATLAB offers the following basic arithmetic operations:

Operation	Symbol	Example
Addition	+	3 + 22
Subtraction	−	54.4 − 16.5
Multiplication	*	3.14 * 6
Division	/ or \	19.54/7 or 7\19.54
Exponentiation	∧	2∧8

The order in which these operations are evaluated in a given expression is determined by the usual rules of precedence, summarized as follows:

> Expressions are evaluated from left to right, with the exponentiation operation having the highest precedence, followed by multiplication and division, having equal precedence, and then by addition and subtraction, having equal precedence.
>   Parentheses can be used to alter this ordering, in which case these rules of precedence are applied within each set of parentheses, by starting with the innermost set and proceeding outward.

More information about precedence rules can be displayed by typing `help precendence` at the MATLAB prompt.

2.2 THE MATLAB WORKSPACE

As you work in the *Command* window, MATLAB remembers the commands you enter, as well as the values of any variables you create. These commands and variables are said to reside in the ***MATLAB workspace*** or ***Base workspace***, and they can be recalled whenever you wish. For example, to check the value of `tape`, all you have to do is ask MATLAB for it by entering its name at the prompt:

```
>> tape
tape =
      2
```

If you can't remember the name of a variable, you can ask MATLAB for a list of the variables it knows by using the MATLAB command `who`:

```
>> who
Your variables are:
```

```
ans              cost      items    tape
average_cost     erasers   pads
```

Note that MATLAB doesn't tell you the value of all of the variables; it merely gives you their names. To find their values, you must enter their names individually at the MATLAB prompt.

To recall previous commands, MATLAB responds to the **Cursor** keys on your keyboard. For example, pressing the ↑ key once recalls the most recent command to the MATLAB prompt. Repeatedly pressing ↑ scrolls back through prior commands, one at a time. In a similar manner, pressing the ↓ key scrolls forward through commands. By pressing the → or ← keys, you can maneuver within a given command at the MATLAB prompt, thereby editing the command in much the same way that you edit text in a word-processing program. Other standard editing keys, such as **Delete** or **Backspace**, **Home**, and **End**, perform their commonly assigned tasks. The **Tab** key is useful for variable-name completion. Once a scrolled command is acceptable, pressing the **Return** key with the cursor *anywhere* in the command tells MATLAB to process it. Finally, and perhaps most useful, the **Escape** key erases the current command at the prompt. For those of you familiar with the EMACS editor, MATLAB also accepts common EMACS editing control-character sequences, such as **Control-U** to erase the current command.

2.3 ABOUT VARIABLES

Like any other computer language, MATLAB has rules about variable names. Earlier, it was noted that variable names must be a single word containing no spaces. More specifically, MATLAB variable-naming rules are listed, as follows:

Variable-Naming Rules	Comments/Examples
Variable names are case sensitive.	Cost, cost, CoSt, and COST are all different MATLAB variables.
Variable names can contain up to 63 characters. Any characters beyond the 63rd are ignored.	Howaboutthisvariablename
Variable names must start with a letter, followed by any number of letters, digits, or underscores.	how_about_this_variable_name X51483
Punctuation characters are not allowed, because many of them have special meanings in MATLAB.	a_b_c_d_e

There are some specific exceptions to these naming rules. MATLAB has several names that cannot be used for variables. These names are *keywords* and form a *reserved word list* for MATLAB:

Reserved Word List
for end if while function return elseif case otherwise switch continue else try catch global persistent break

This list is returned as an output of the iskeyword function. MATLAB will report an error if you try to use a reserved word as a variable. However, you can use words similar to keywords by capitalizing one or more letters. The function isvarname('teststring') returns True (1) if the character string argument 'teststring' is a valid variable name; otherwise, it returns False (0).

In addition, just as your calculator stores constants such as π, MATLAB has a number of special variables:

Special Variables	Description
ans	Default variable name used for results
beep	Makes computer sound a beep
pi	Ratio of the circumference of a circle to its diameter
eps	Smallest number that, when added to 1, creates a number greater than 1 on the computer
inf	Stands for infinity (e.g., 1/0)
NaN or nan	Stands for Not-a-Number (e.g., 0/0)
i or j	Stands for $\sqrt{-1}$
nargin	Number of function input arguments
nargout	Number of function output arguments
realmin	Smallest usable positive real number
realmax	Largest usable positive real number
bitmax	Largest usable positive integer stored in double-precision format
varargin	Variable number of function input arguments
varargout	Variable number of function output arguments

If you reuse a variable (such as tape, in the earlier example) or assign a value to one of the special variables from the preceding list, the prior value of the variable is overwritten and lost. However, any other expressions computed with the prior values do not change. Consider the following example:

```
>> erasers = 4;
>> pads = 6;
>> tape = 2;
```

```
>> items = erasers + pads + tape
items =
    12
>> erasers = 6
erasers =
     6
>> items
items =
    12
```

Here, using the first example again, we found the number of items Mary purchased. Afterward, we changed the number of erasers to six, overwriting its prior value of 4. In doing so, the value of items has not changed. Unlike a common spreadsheet program, MATLAB does not recalculate the number of items based on the new value of erasers. ***When MATLAB performs a calculation, it does so using the values that it knows at the time the requested command is evaluated.*** In the preceding example, if you wish to recalculate the number of items, the total cost, and the average cost, it is necessary to recall the appropriate MATLAB commands and ask MATLAB to evaluate them again.

The special variables given earlier follow this guideline, also, with the exception of the fact that the special values can be restored. When you start MATLAB, the variables have their original values; when you change their values, the original special values are lost. To restore a special value, all you have to do is ***clear*** the overwritten value. For example,

```
>> pi
ans =
        3.1416
>> pi = 1.23e-4
pi =
      0.000123
>> clear pi
>> pi
ans =
        3.1416
```

shows that pi has the special value of 3.1416, to five significant digits; is overwritten with the value of 1.23e-4; and then, after being cleared using the clear function, has its original special value once again.

2.4 COMMENTS, PUNCTUATION, AND ABORTING EXECUTION

As we saw earlier, placing a semicolon at the end of a command suppresses printing of the computed results. This feature is especially useful for suppressing the results of intermediate calculations. For instance,

```
>> erasers
erasers =
      6
>> items = erasers + pads + tape;
>> cost = erasers*25 + pads*52 + tape*99;
>> average_cost = cost/items
average_cost =
        47.143
```

displays the average cost of the items that Mary bought when she purchased six erasers, rather than the original four. The intermediate results items and cost were not printed, because semicolons appear at the ends of the commands defining them.

In addition to semicolons, MATLAB uses other punctuation symbols. For example, all text after a percent sign (%) is taken as a comment statement:

```
>> tape = 2   % number of rolls of tape purchased
```

The variable tape is given the value of 2, and MATLAB simply ignores the percent sign and all text following it.

If they are separated by commas or semicolons, multiple commands can be placed on one line:

```
>> erasers = 6, pads = 6; tape = 2
erasers =
      6
tape =
      2
```

Commas tell MATLAB to display results; semicolons suppress printing.

Sometimes expressions or commands are so long, that it is convenient to continue them onto additional lines. In MATLAB, statement continuation is denoted by three periods in succession, as shown in the following code:

```
>> average_cost = cost/items   % command as done earlier
average_cost =
        47.143
```

```
>> average_cost = cost/... % command with valid continuation
items
average_cost =
      47.143
>>average_cost = cost... % command with valid continuation
/items
average_cost =
      47.143
>>average_cost = cost... command with valid continuation (no % needed)
/items
average_cost =
      47.143
>>average_cost = cost/it... % command with INvalid continuation
ems
??? ems
   |
Error: Missing MATLAB operator.
```

Note that statement continuation works if the three periods appear between variable names and mathematical operators, but not in the middle of a variable name. That is, variable names cannot be split between two lines. Furthermore, all text after the three periods is considered to be a comment, so no percent symbol is needed. In addition, since comment lines are ignored, they cannot be continued either, as the following example shows:

```
>> % Comments cannot be continued...
>> either
??? Undefined function or variable 'either'.
```

In this case, the (...) in the comment line is part of the comment and is not processed by MATLAB.

Finally, MATLAB processing can be interrupted at any time by pressing **Control-C** (that is, pressing the **Ctrl** and **C** keys simultaneously).

2.5 COMPLEX NUMBERS

One of the most powerful features of MATLAB is that it does not require any special handling for complex numbers. Complex numbers are formed in MATLAB in several ways. Examples of complex numbers include the following:

```
>> c1 = 1-2i  % the appended i signifies the imaginary part
c1 =
   1.0000 - 2.0000i
>> c1 = 1-2j  % j also works
c1 =
   1.0000 - 2.0000i
>> c1 = complex(1,-2) % a function that creates complex numbers
c1 =
   1.0000 - 2.0000i
>> c2 = 3*(2-sqrt(-1)*3)
c2 =
   6.0000 - 9.0000i
>> c3 = sqrt(-2)
c3 =
        0 + 1.4142i
>> c4 = 6+sin(.5)*1i
c4 =
   6.0000 + 0.4794i
>> c5 = 6+sin(.5)*1j
c5 =
   6.0000 + 0.4794i
```

In the last two examples, multiplication by 1i and 1j are used to form the imaginary part. Multiplication by 1i or 1j is required in these cases, since sin(.5)i and sin(.5)j have no meaning in MATLAB. Termination with the characters i and j, as shown in the first two examples, works only with numbers and not with expressions.

Some programming languages require special handling for complex numbers, wherever they appear. In MATLAB, no special handling is required. Mathematical operations on complex numbers are written the same way as those on real numbers:

```
>> c6 = (c1+c2)/c3  % from the above data
c6 =
  -7.7782 - 4.9497i
>> c6r = real(c6)
c6r =
       -7.7782
```

```
>> c6i = imag(c6)
c6i =
     -4.9497
>> check_it_out = 1i^2  % sqrt(-1) squared must be -1!
check_it_out =
    -1
```

In general, operations on complex numbers lead to complex number results. However, in the last case, MATLAB is smart enough to drop the zero imaginary part of the result. In addition, the foregoing shows that the functions real and imag extract the real and imaginary parts of a complex number, respectively.

As a final example of complex arithmetic, consider the Euler identity that relates the polar form of a complex number to its rectangular form $M \angle \theta = M e^{j\theta} = a + bj$, where the polar form is given by a magnitude M and an angle θ and the rectangular form is given by $a + bj$. The relationships among these forms are $M = \sqrt{a^2 + b^2}$, $\theta = \tan^{-1}(b/a)$, $a = M \cos(\theta)$, and $b = M \sin(\theta)$.

In MATLAB, the conversion between polar and rectangular forms makes use of the functions real, imag, abs, and angle:

```
>> c1
c1 =
   1.0000 - 2.0000i
>> mag_c1 = abs(c1) % magnitude
mag_c1 =
    2.2361
>> angle_c1 = angle(c1) % angle in radians
angle_c1 =
   -1.1071
>> deg_c1 = angle_c1*180/pi % angle in degrees
deg_c1 =
  -63.4349
>> real_c1 = real(c1) % real part
real_c1 =
    1
>> imag_c1 = imag(c1) % imaginary part
imag_c1 =
    -2
```

The MATLAB function abs computes the magnitude of complex numbers or the absolute value of real numbers, depending on which one you assign it to compute. Likewise, the MATLAB function angle computes the angle of a complex number in radians. *MATLAB does not natively perform trigonometric operations with units of degrees; however, basic trigonometric functions supporting angles in degrees are provided in MATLAB 7.*

2.6 FLOATING-POINT ARITHMETIC

In almost all cases, numerical values in MATLAB are represented in double-precision arithmetic, using a binary (base 2) representation internally. Such representation is the most common representation used by computers and is a native format for numerical coprocessors. Because of this representation, not all numbers can be represented exactly. There are limiting values that can be represented, and there is a recognizable lower limit for addition.

The largest positive real number that can be represented is

```
>> format long  % tell MATLAB to display more precision
>> realmax
ans =
    1.797693134862316e+308
```

The smallest positive number that can be represented is

```
>> realmin
ans =
    2.225073858507201e-308
```

The smallest number that can be added to 1 to produce a number larger than 1 in double precision is

```
>> eps
ans =
    2.220446049250313e-016
```

Generalizing this concept, eps(x) is the smallest increment that can be added to x to produce a number larger than x, as the following examples show:

```
>> eps(1) % same as eps by itself
ans =
    2.220446049250313e-016
```

```
>> eps(10)
ans =
     1.776356839400251e-015

>> eps(1e10)
ans =
     1.907348632812500e-006
```

As the magnitude of a number increases, the distance between values that can be represented in finite precision (which is what eps(x) returns) increases as well.

The consequences of the limitations of finite-precision arithmetic are sometimes strange. For example, as shown here, addition is not *exactly* commutative:

```
>> 0.42 - 0.5 + 0.08
ans =
    -1.387778780781446e-017

>> 0.08 - 0.5 + 0.42   % rearrange order
ans =
     0

>> 0.08 + 0.42 - 0.5   % rearrange order again
ans =
     0
```

All three of these results should be zero, but they are not. In each case, the arithmetic was performed from left to right. The issue here is that not all of the numbers can be represented exactly in double-precision arithmetic. In fact, only 0.5 has an exact representation. When numbers cannot be represented exactly, they are approximated with as much precision as possible—leading to inevitable errors in computed results. For the most part, these errors are minor; otherwise, double-precision arithmetic wouldn't be used in modern computers. In practice, the problems with double-precision arithmetic occur most often when asking MATLAB to compare two numbers for equality or inequality. Clearly, within MATLAB, $0.42 - 0.5 + 0.08$ does not equal $0.08 - 0.5 + 0.42$, even though our brains can do the exact arithmetic to show that it does.

A second drawback of finite-precision arithmetic appears in function evaluation. Not only is finite-precision arithmetic unable to always represent function arguments exactly, but also most functions cannot themselves be represented exactly as the following code shows:

```
>> sin(0)
ans =
    0
>> sin(pi)
ans =
    1.224646799147353e-016
```

Here, both results should be zero, but $\sin(\pi)$ is not. It is interesting to note that the error here and the error in the previous example are both less than eps.

Finally, when using double-precision, floating-point arithmetic to represent integers, only integers up to a limit can be represented *exactly*. The limiting value is $2^{53} - 1$, which is returned by MATLAB as follows:

```
>> bitmax
ans =
    9.007199254740991e+015
```

2.7 MATHEMATICAL FUNCTIONS

Lists of the common functions that MATLAB supports are shown in the tables on page 17. Most of these functions are used in the same way you would write them mathematically:

```
>> x = sqrt(2)/2
x =
    0.7071
>> y = asin(x)
y =
    0.7854
>> y_deg = y*180/pi
y_deg =
    45.0000
```

These commands find the angle where the sine function has a value of $\sqrt{2}/2$. Note, again, that MATLAB uses radians, not degrees, in trigonometric functions. Other examples include the following:

```
>> y = sqrt(3^2 + 4^2)   % show 3-4-5 right triangle relationship
y =
    5
```

```
>> y = rem(23,4)  % remainder function, 23/4 has a remainder of 3
y =
    3
>> x = 2.6, y1 = fix(x), y2 = floor(x), y3 = ceil(x), y4 = round(x)
x =
    2.6000
y1 =
    2
y2 =
    2
y3 =
    3
y4 =
    3
```

Trigonometric Function	Description
acos	Inverse cosine
acosd	Inverse cosine returning degrees
acosh	Inverse hyperbolic cosine
acot	Inverse cotangent
acotd	Inverse cotangent returning degrees
acoth	Inverse hyperbolic cotangent
acsc	Inverse cosecant
acscd	Inverse cosecant returning degrees
acsch	Inverse hyperbolic cosecant
asec	Inverse secant
asecd	Inverse secant returning degrees
asech	Inverse hyperbolic secant
asin	Inverse sine
asind	Inverse sine returning degrees
asinh	Inverse hyperbolic sine
atan	Inverse tangent
atand	Inverse tangent returning degrees
atanh	Inverse hyperbolic tangent
atan2	Four-quadrant inverse tangent

Trigonometric Function	Description
cos	Cosine
cos	Cosine with argument in degrees
cosh	Hyperbolic cosine
cot	Cotangent
cot	Cotangent with argument in degrees
coth	Hyperbolic cotangent
csc	Cosecant
csc	Cosecant with argument in degrees
csch	Hyperbolic cosecant
sec	Secant
sec	Secant with argument in degrees
sech	Hyperbolic secant
sin	Sine
sin	Sine with argument in degrees
sinh	Hyperbolic sine
tan	Tangent
tan	Tangent with argument in degrees
tanh	Hyperbolic tangent

Exponential Function	Description
^	Power
exp	Exponential
expm1	Exponential minus 1 [i.e., $\exp(x) - 1$]
log	Natural logarithm
log10	Base 10 logarithm
log1p	Natural logarithm of $x + 1$ [i.e., $\log(x + 1)$]
log2	Base 2 logarithm and floating-point number dissection
nthroot	n^{th} real root of real numbers
pow2	Base 2 power and floating-point number scaling
reallog	Natural logarithm limited to real nonnegative values
realpow	Power limited to real-valued arguments
realsqrt	Square root limited to real-valued values
sqrt	Square root
nextpow2	Next-higher power of 2

Complex Function	Description
abs	Absolute value or magnitude
angle	Phase angle in radians
conj	Complex conjugate
imag	Imaginary part
real	Real part
unwrap	Unwrap phase angle
isreal	True for real values
cplxpair	Sort vector into complex conjugate pairs
complex	Form complex number from real and imaginary parts

Rounding and Remainder Function	Description
fix	Round toward zero
floor	Round toward negative infinity
ceil	Round toward positive infinity
round	Round toward nearest integer
mod	Modulus or signed remainder
rem	Remainder after division
sign	Signum function

Coordinate Transformation Function	Description
cart2sph	Cartesian to spherical
cartpol	Cartesian to cylindrical or polar
pol2cart	Cylindrical or polar to Cartesian
sph2cart	Spherical to Cartesian

Number Theoretic Function	Description
factor	Prime factors
isprime	True for prime numbers
primes	Generate list of prime numbers
gcd	Greatest common divisor
lcm	Least common multiple

Number Theoretic Function	Description
rat	Rational approximation
rats	Rational output
perms	All possible combinations
nchoosek	All combinations of N elements taken K at a time

Specialized Function	Description
airy	Airy function
besselj	Bessel function of the first kind
bessely	Bessel function of the second kind
besselh	Bessel function of the third kind
besseli	Modified Bessel function of the first kind
besselk	Modified Bessel function of the second kind
beta	Beta function
betainc	Incomplete beta function
betaln	Logarithm of beta function
ellipj	Jacobi elliptic function
ellipke	Complete elliptic integral
erf	Error function
erfc	Complementary error function
erfcx	Scaled complementary error function
erfinv	Inverse error function
expint	Exponential error function
gamma	Gamma function
gammainc	Incomplete gamma function
gammaln	Logarithm of gamma function
legendre	Associated Legendre function
cross	Vector cross product
dot	Vector dot product

3

The MATLAB Desktop

As stated in the preceding chapter, running MATLAB creates one or more windows on your computer monitor. One of these windows, entitled *MATLAB*, is commonly called the *MATLAB* desktop. This window contains or manages all other windows that are part of MATLAB. Depending on how you set up MATLAB, some windows associated with the desktop may or may not be visible, and some may or may not reside within (i.e., be ***docked*** in) the *MATLAB* window. Management of the *MATLAB* desktop window and its associated windows is not discussed in this text. If you are familiar with other window-based programs, manipulation of windows in MATLAB will be familiar to you. The *MATLAB* desktop menu items change, depending on which window is active. In addition, there are many helpful contextual menus (which are accessed by pressing the right mouse button over an item). If you are not familiar with window-based programs, it may be beneficial to seek general assistance in this area from other sources. In any case, to find out which windows are associated with the desktop, investigate the items on the **Desktop** menu within the *MATLAB* desktop window.

3.1 MATLAB WINDOWS

The windows used in MATLAB include the (1) *Command*, (2) *Command History*, (3) *Current Directory* (browser), (4) *Workspace* (browser), (5) *Help* (browser), (6) *Editor*, and (7) *Profiler*. The following table gives an overview of the purpose of each of these windows:

Window	Description
Command	Issues commands to MATLAB for processing
Command History	Running history of prior commands issued in the *Command* window
Current Directory	GUI for directory and file manipulation in MATLAB
Workspace	GUI for viewing, editing, loading, and saving MATLAB variables
Help	GUI for finding and viewing on-line documentation
Editor	Text editor for creating M-files
Profiler	Tool for optimizing M-file performance

3.2 MANAGING THE MATLAB WORKSPACE

Within the *MATLAB* desktop, actions taken in all windows support computations performed in the *Command* window. As a result, the rest of this chapter provides more detailed information about the *Command* window.

The data and variables created in the *Command* window reside in what is called the **MATLAB workspace** or **Base workspace**. In addition to viewing variables in the *Workspace* window, you can see what variable names exist in the MATLAB workspace by issuing the command who:

```
>> who

Your variables are:

angle_c1       c4             cost           pads
ans            c5             deg_c1         real_c1
average_cost   c6             erasers        tape
c1             c6i            imag_c1
c2             c6r            items
c3             check_it_out   mag_c1
```

The variables you see may differ from those just listed, depending on what you've asked MATLAB to do since you opened the program. For more detailed information, use the command whos:

```
>> whos
  Name             Size           Bytes  Class
    angle_c1       1x1                8  double array
    ans            1x1                8  double array
```

average_cost	1x1	8	double array
c1	1x1	16	double array (complex)
c2	1x1	16	double array (complex)
c3	1x1	16	double array (complex)
c4	1x1	16	double array (complex)
c5	1x1	16	double array (complex)
c6	1x1	16	double array (complex)
c6i	1x1	8	double array
c6r	1x1	8	double array
check_it_out	1x1	8	double array
cost	1x1	8	double array
deg_c1	1x1	8	double array
erasers	1x1	8	double array
imag_c1	1x1	8	double array
items	1x1	8	double array
mag_c1	1x1	8	double array
pads	1x1	8	double array
real_c1	1x1	8	double array
tape	1x1	8	double array

Grand total is 21 elements using 216 bytes

Here, each variable is listed, along with its size, the number of bytes used, and its class. Since MATLAB is array oriented, all of the variables belong to the class of double-precision arrays, even though all of the above variables are scalars. Later, as other data types or classes are introduced, the information in this last column will become more useful. The preceding list is also displayed in the *Workspace* window, which can be viewed by typing workspace at the MATLAB prompt or by choosing the **Workspace** menu item on the **Desktop** menu of the *MATLAB* desktop.

The command clear deletes variables from the MATLAB workspace. For example,

```
>> clear real_c1 imag_c1 c*
>> who
Your variables are:
angle_c1          deg_c1       mag_c1
ans               erasers      pads
average_cost      items        tape
```

deletes the variables real_c1, imag_c1, and all variables starting with the letter c. Other options for the clear function can be identified by asking for information with the command help or helpwin:

```
>> help clear
 CLEAR   Clear variables and functions from memory.
    CLEAR removes all variables from the workspace.
    CLEAR VARIABLES does the same thing.
    CLEAR GLOBAL removes all global variables.
    CLEAR FUNCTIONS removes all compiled M- and MEX-functions.
    CLEAR ALL removes all variables, globals, functions and MEX links.
    CLEAR ALL at the command prompt also removes the Java packages import
    list.

    CLEAR IMPORT removes the Java packages import list at the command
    prompt. It cannot be used in a function.
    CLEAR CLASSES is the same as CLEAR ALL except that class definitions
    are also cleared. If any objects exist outside the workspace (say in
    userdata or persistent in a locked m-file) a warning will be issued and
    the class definition will not be cleared. CLEAR CLASSES must be used if
    the number or names of fields in a class are changed.
    CLEAR JAVA is the same as CLEAR ALL except that java classes on the
    dynamic java path (defined using JAVACLASSPATH) are also cleared.

    CLEAR VAR1 VAR2 ... clears the variables specified. The wildcard
    character '*' can be used to clear variables that match a pattern. For
    instance, CLEAR X* clears all the variables in the current workspace
    that start with X.

    CLEAR -REGEXP PAT1 PAT2 can be used to match all patterns using regular
    expressions. This option only clears variables. For more information on
    using regular expressions, type "doc regexp" at the command prompt.
    If X is global, CLEAR X removes X from the current workspace, but
    leaves it accessible to any functions declaring it global.
    CLEAR GLOBAL X completely removes the global variable X.
```

CLEAR GLOBAL -REGEXP PAT removes global variables that match regular
expression patterns.
Note that to clear specific global variables, the GLOBAL option must
come first. Otherwise, all global variables will be cleared.

CLEAR FUN clears the function specified. If FUN has been locked by
MLOCK it will remain in memory. Use a partial path (see PARTIALPATH) to
distinguish between different overloaded versions of FUN.   For
instance, 'clear inline/display' clears only the INLINE method for
DISPLAY, leaving any other implementations in memory.

CLEAR ALL, CLEAR FUN, or CLEAR FUNCTIONS also have the side effect of
removing debugging breakpoints and reinitializing persistent variables
since the breakpoints for a function and persistent variables are
cleared whenever the m-file changes or is cleared.

Use the functional form of CLEAR, such as CLEAR('name'), when the
variable name or function name is stored in a string.

Examples for pattern matching:
```
    clear a*               % Clear variables starting with "a"
    clear -regexp ^b\d{3}$ % Clear variables starting with "b" and
                           %    followed by 3 digits
    clear -regexp \d       % Clear variables containing any digits
```

See also who, whos, mlock, munlock, persistent.

Obviously, the clear command does more than just delete variables. Its other uses
will become apparent as you become familiar with more of MATLAB's features.

3.3 MEMORY MANAGEMENT

MATLAB allocates memory for variables as they are created and for M-file functions as they are used. Depending on the computer on which the program is
installed, it is possible for MATLAB to run out of memory, making it impossible to
do any further work. When you eliminate variables by using the clear command,
MATLAB frees up the memory used by the variables cleared. Over time, however,
it is possible for memory to become fragmented, leaving MATLAB's memory space

populated by variables surrounded by numerous small fragments of free memory. Since MATLAB always stores variables in contiguous chunks of memory, these fragments of free memory may not be reusable. To alleviate this problem, the `pack` command performs memory garbage collection. The command saves all MATLAB workspace variables to disk, clears all variables from the workspace, and then reloads the variables back into the workspace. On completion, all fragments of free memory are consolidated into one large, usable block. Depending on how much memory is allocated to MATLAB on your computer, how long you've been running a particular MATLAB session, and how many variables you've created, you may or may not ever need to use the `pack` command.

3.4 NUMBER DISPLAY FORMATS

When MATLAB displays numerical results, it follows several rules. By default, if a result is an integer, MATLAB displays it as an integer. Likewise, when a result is a real number, MATLAB displays it with approximately four digits to the right of the decimal point. If the significant digits in the result are outside of this range, MATLAB displays the result in scientific notation, similar to the display of a scientific calculator. You can override this default behavior by specifying a different numerical format. From the *Command* window **File** menu, choose the **Preferences** menu. Alternatively, type the appropriate MATLAB `format` command at the prompt. With the special variable `pi`, the numerical display formats produced by different `format` selections are as follows:

MATLAB Command	pi	Comments
`format short`	3.1416	5 digits
`format long`	3.14159265358979	16 digits
`format short e`	3.1416e+000	5 digits plus exponent
`format long e`	3.14159265358979e+000	16 digits plus exponent
`format short g`	3.1416	Best of `format short` or `format short e`
`format long g`	3.14159265358979	Best of `format long` or `format long e`
`format hex`	400921fb54442d18	Hexadecimal, floating point
`format bank`	3.14	2 decimal digits
`format +`	+	Positive (+), negative (−), or zero (0)
`format rat`	355/113	Rational approximation

format debug	Structure address = 1214830 m = 1 n = 1 pr = 11d60d0 pi = 0 3.1416	Internal storage information in addition to short g

> Note: In MATLAB, the internal representation of a number does not change when different display formats are chosen; only the display changes. All calculations using double-precision numbers are performed using double-precision arithmetic.

3.5 KEEPING A SESSION LOG

Sometimes it is useful to keep a log of all work performed during a MATLAB session. The *Command History* window keeps a chronological log of the functions and commands executed during current and past MATLAB work sessions, but does not show the results. MATLAB provides a command diary, which saves all *Command* window activity to a text file in the current directory. The help text for diary describes its use:

```
>> help diary
 DIARY Save text of MATLAB session.
    DIARY filename causes a copy of all subsequent command window input
    and most of the resulting command window output to be appended to the
    named file.  If no file is specified, the file 'diary' is used.
    DIARY OFF suspends it.
 DIARY ON turns it back on.
 DIARY, by itself, toggles the diary state.
```

3.6 SYSTEM INFORMATION

MATLAB provides a number of commands that provide information about the computer in use, as well as about the MATLAB version in use. The command computer returns a character string identifying the computer in use:

```
>> computer
ans =
PCWIN
```

In this case, the computer is a PC that is running *Windows XP*. The command version returns a character string identifying the MATLAB version:

>> version

ans =

7.0.0.151483 (R14)

Note that the command ver returns information about MATLAB, as well as installed toolboxes:

> ver

MATLAB Version 7.0.0.051483 (R14)

MATLAB License Number: 9754

Operating System: Microsoft Windows XP Version 5.2 (Build 2600: Service Pack 1)

Java VM Version: Java 1.4.2 with Sun Microsystems Inc. Java HotSpot(TM) Client VM

MATLAB   Version 7.0   (R14)

Mastering MATLAB Toolbox   Version 6.0

MATLAB licensing information can be found by using the commands license and hostid:

>> hostid

    '9754'

>> license

ans =

9754

Of course, your results from entering these commands will most likely be different from those shown because your computer and MATLAB version will probably differ from those used to produce this text.

3.7 THE MATLAB SEARCH PATH

MATLAB uses a *search path* to find information stored in files on disk. MATLAB's files are organized into numerous directories and subdirectories.

> The list of all directories where MATLAB's files are found is called the MATLAB search path or simply MATLAB path.

Use of the MATLAB search path is described next. When you enter cow at the MATLAB prompt by typing >>cow, MATLAB does the following: (1) It checks to see if cow is a *variable* in the MATLAB workspace. (2) If not, it checks to see if cow is a built-in function. (3) If not, it searches for a file named cow.m in the *current directory*. (4) If none exists, it checks to see if cow.m exists anywhere on the *MATLAB search path* by searching in the order in which the path is specified. (5) If cow isn't found at this point, MATLAB reports an error.

MATLAB takes appropriate action according to this search strategy. If cow is a variable, MATLAB uses the variable. If cow is a built-in function, it is called. If cow.m is a file (either in the current directory or anywhere on the MATLAB search path) the file cow.m is opened, and MATLAB acts on what is found in the file. As is documented in Chapters 4 and 12, MATLAB has two basic file types that are of common use; both are simple text files containing MATLAB commands. (See Chapters 4 and 12 for further information regarding these *M-files*).

In reality, the search procedure is more complicated than that just described, because of advanced features in MATLAB. For the most part, however, this search procedure provides sufficient detail for basic MATLAB work. (More detailed information regarding the MATLAB search path can be found in Chapter 12.)

When MATLAB starts up, it defines a default MATLAB search path that points to all of the directories in which MATLAB stores its files. This search path can be displayed and modified in several ways. The easiest way is to use the *Path Browser*, which is a graphical user interface designed for viewing and modifying the MATLAB search path. The path browser is made available by choosing **Set Path ...** from the **File** menu on the *MATLAB* desktop window. Since the MATLAB search path already points to all directories where MATLAB stores its files, the primary purpose for accessing the path browser is to add your own MATLAB file storage directories to the search path.

To display the MATLAB search path in the *Command* window, MATLAB provides the function matlabpath. In addition, the features of the path browser can be duplicated in the *Command* window by using the functions path, addpath, and rmpath. For more information regarding these functions, see the on-line documentation.

4

Script M-Files

For simple problems, entering your requests at the MATLAB prompt in the *Command* window is fast and efficient. However, as the number of commands increases, or when you wish to change the value of one or more variables and reevaluate a number of commands, typing at the MATLAB prompt quickly becomes tedious. MATLAB provides a logical solution to this problem. It allows you to place MATLAB commands in a simple text file and then tell MATLAB to open the file and evaluate the commands exactly as it would have if you had typed the commands at the MATLAB prompt. These files are called **script files** or **M-files**. The term "script" signifies that MATLAB simply reads from the script found in the file. The term *M-file* means that script filenames must end with the extension '.m', as in, for example, example1.m.

4.1 SCRIPT M-FILE USE

To create a script M-file, click on the blank page icon on the MATLAB desktop toolbar, or choose **New** from the **File** menu and select **M-file**. This procedure brings up a text editor window wherein you can enter MATLAB commands. The following script M-file shows the commands from an example considered earlier:

```
% script M-file example1.m
   erasers = 4; % number of each item
   pads = 6;
   tape = 2;
```

```
items = erasers + pads + tape
cost = erasers*25 + pads*52 + tape*99
average_cost = cost/items
```

This file can be saved to disk and executed immediately by (1) choosing **Save and Run** from the **Debug** menu, (2) pressing the **Save and Run** button on the *Editor* toolbar, or (3) simply pressing the function key **F5**. Alternatively, you can save this file as the M-file `example1.m` on your disk by choosing **Save** from the **File** menu; then, at the MATLAB prompt, it's just a matter of typing the name of the script file without the `.m` extension:

```
>> example1
items =
    12
cost =
   610
average_cost =
      50.833
```

When MATLAB interprets the `example1` statement, it follows the search path described in Chapter 3. In brief, MATLAB prioritizes current MATLAB variables ahead of M-file names. If `example1` is not a current MATLAB variable or built in function name, MATLAB opens the file `example1.m` (if it can find it) and evaluates the commands found there just as if they had been entered directly at the *Command* window prompt. As a result, commands within the M-file have access to all of the variables in the MATLAB workspace, and all of the variables created by the M-file become part of the workspace. Normally, the M-file commands are not displayed as they are evaluated. The `echo on` command tells MATLAB to display, or echo, commands to the *Command* window as they are read and evaluated. You can probably guess what the `echo off` command does. Similarly, the command `echo` by itself toggles the echo state.

This ability to create script M-files makes it simple to answer "what if?" questions. For example, you could repeatedly open the `example1.m` M-file, change the number of `erasers`, `pads`, or `tape`, and then save and execute the file.

The utility of MATLAB comments is readily apparent when you use script files, as shown in `example1.m`. Comments allow you to document the commands found in a script file, so that you do not forget the commands when you view them in the future. In addition, the use of semicolons at the ends of lines to suppress the display of results allows you to control script-file output, so that only important results are shown. Because of the utility of script files, MATLAB provides several functions that are particularly helpful when used in M-files:

Function	Description
beep	Makes computer beep
disp(variablename)	Displays results without identifying variable names
echo	Controls *Command* window echoing of script file contents as they are executed
input	Prompts user for input
keyboard	Temporarily gives control to keyboard. (Type return to return control to the executing script M-file.)
pause or pause(n)	Pauses until user presses any keyboard key, or pauses for n seconds and then continues
waitforbuttonpress	Pauses until user presses mouse button or any keyboard key

When a MATLAB command is not terminated by a semicolon, the results are displayed in the *Command* window, with the variable name identified. For a prettier display, it is sometimes convenient to suppress the variable name. In MATLAB, this is accomplished with the command disp:

```
>> items
items =
     12
>> disp(items)
     12
```

Rather than repeatedly edit a script file for computations for a variety of cases, you can employ the input command to prompt for input as a script file is executed. For example, reconsider the example1.m script file, with the following modifications:

```
% script M-file example1.m

erasers = 4;   % Number of each item
pads = 6;
tape = input('Enter the number of rolls of tape purchased > ');
items = erasers + pads + tape
cost = erasers*25 + pads*52 + tape*99
average_cost = cost/items
```

Running this script M-file produces this result:

```
>> example1
Enter the number of rolls of tape purchased > 3
items =
     13
cost =
   709
average_cost =
      54.538
```

In response to the prompt, the number 3 was entered and the **Return** or **Enter** key was pressed. The remaining commands were evaluated as before. The function input accepts any valid MATLAB expression for input. For example, running the script file again and providing different input gives the following result:

```
>> example1
Enter the number of rolls of tape purchased > round(sqrt(13))-1
items =
     13
cost =
   709
average_cost =
      54.538
```

In this case, the number of rolls of tape was set equal to the result of evaluating the expression

```
round (sqrt(13))-1.
```

To see the effect of the echo command, add it to the script file and execute it:

```
% script M-file example1.m
echo on
erasers = 4;   % Number of each item
pads = 6;
tape = input('Enter the number of rolls of tape purchased > ');
items = erasers + pads + tape
```

```
cost = erasers*25 + pads*52 + tape*99
average_cost = cost/items
echo off
```

```
>> example1
erasers = 4;   % Number of each item
pads = 6;
tape = input('Enter the number of rolls of tape purchased > ');
Enter the number of rolls of tape purchased > 2
items = erasers + pads + tape
items =
    12
cost = erasers*25 + pads*52 + tape*99
cost =
   610
average_cost = cost/items
average_cost =
      50.833
echo off
```

As you can see, the echo command made the result much harder to read. On the other hand, the echo command can be very helpful when debugging more complicated script files.

4.2 BLOCK COMMENTS AND CODE CELLS

In prior versions of MATLAB, comments were line oriented. That is, comments began with an initial percent sign and continued to the end of the current line. To continue the comments on the next line required another initial percent sign. Therefore, a block of comments would all start with initial percent signs, as shown here:

```
% This is an example of multiple line comments
% in an M-file. Each line requires an initial % sign or MATLAB
% assumes the line contains code to be executed.
```

While the block of comments is visually simple, it can become cumbersome to manage later, when comment text is augmented or edited. In this case, the percent signs must remain at the beginnings of the lines, and the comment text must flow after the initial percent signs on each line. In the past, to make it less cumbersome, the MATLAB editor included commands for adding or removing the initial percent signs on a highlighted block of lines. Now, MATLAB 7 supports *block comments* through the use of the syntax %{ and %}. These symbols mark the beginning and end, respectively, of a block of text to be treated as comments by MATLAB. For example, using block comment syntax on the previous example produces

```
%{
This is an example of multiple line comments
in an M-file. Each line requires an initial % sign or MATLAB
assumes the line contains code to be executed.
(Now lines can be added and edited as desired without having to
place percent signs at the beginning of each line.)
%}
```

In addition to their utility in composing multiline comments, block comments will allow you to rapidly turn on and off the interpretation and execution of any number of lines of code in an M-file. This feature is particularly helpful in creating and debugging large M-files. Simply adding %{ before and %} after a block of MATLAB code turns the enclosed code into comments that are not processed by MATLAB. When this feature is used, different sections of a script file can be executed at different times during the editing and debugging processes.

   In the past, the MATLAB editor offered commands for executing a block of highlighted code in an editor window. In MATLAB 7, the editor now supports the selective execution of M-file code through the use of **code cells**. A code cell is simply a block of M-file code that starts with a comment line containing two percent signs followed by a space (i.e., %%). The code cell continues to the end of the M-file or to the beginning of another code cell. From within the MATLAB editor, cells can be created, individually executed, and sequentially executed, thereby enabling effective M-file debugging. The **Cell** menu in the *Editor* window facilitates these operations. It is important to note that the syntax for code cells is interpreted by the editor, not by the MATLAB command interpreter. As a result, when an M-file is executed after its name has been entered in the *Command* window, code-cell syntax is ignored, and all executable lines in the file are processed.

4.3 SETTING EXECUTION TIME

When the name of an M-file is typed at the *Command* window prompt, the M-file is executed immediately. In some circumstances (such as large programming projects, or situations wherein the data to be manipulated becomes available only over a long period of time while MATLAB is running), it is convenient to specify execution times for M-files. In MATLAB, this capability is accomplished by using ***timer objects***. A timer object is created with the function `timer`. For example,

```
>> my_timer = timer('TimerFcn','MfileName','StartDelay',100)
```

creates the timer object stored in the variable `my_timer`, which executes the M-file *MfileName* 100 seconds after the timer has been started by using the `start` function:

```
>> start(my_timer) % start the timer in the variable my_timer
```

In general terms, timer function syntax has the form

```
>> t=timer('PropertyName1',PropertyValue1,'PropertyName2',PropertyValue2,...)
```

where the arguments identify property names as character strings paired with corresponding property values.

Neither MATLAB nor the computer operating system is held up while a timer is running, but not executing code. MATLAB and other programs can be used during this idle time. When the timer object initiates code execution, the timer takes control of MATLAB and executes the code, just as if the code had been typed at the MATLAB prompt. When this execution ends, MATLAB returns control to the *Command* window prompt.

The timer object has many more features. For example, '*MfileName*' can be any statement executable at the MATLAB prompt. It can be a script M-file, a function handle or function M-file, or a set of MATLAB commands. You also can specify that timer code be executed on a periodic basis or executed a specified number of times. You can specify four different M-files or code sequences to be executed under different conditions. For example,

```
>> my_timer = timer('TimerFcn','Mfile1',...
                    'StartFcn','Mfile2',...
                    'StopFcn,, 'Mfile3',...
                    'ErrorFcn','Mfile4');
```

creates a timer function that executes (1) `'Mfile1'` as the primary timer code that may be repeatedly executed, (2) `'Mfile2'` when the timer is started with the `start` function, (3) `'Mfile3'` when the timer is stopped with the `stop` function, and (4) `'Mfile4'` if a MATLAB error occurs while any of these functions are executing. (See the MATLAB documentation for more complete information about timer objects.)

4.4 STARTUP AND FINISH

When MATLAB starts up, it executes two script M-files, `matlabrc.m` and `startup.m`. The first, `matlabrc.m`, comes with MATLAB and generally should not be modified. The commands in this M-file set the default *Figure* window size and placement, as well as a number of other default features. The default MATLAB search path is set by retrieving the script file `pathdef.m` from `matlabrc.m`. The *Path Browser* and *Command* window functions for editing the MATLAB search path maintain the file `pathdef.m`, so there is no need to edit it with a text editor.

Commands in `matlabrc.m` check for the existence of the script M-file `startup.m` on the MATLAB search path. If the M-file exists, the commands in it are executed. This optional M-file `startup.m` typically contains commands that add personal default features to MATLAB. For example, it is common to put one or more `addpath` or `path` commands in `startup.m` to append additional directories to the MATLAB search path. Similarly, the default number-display format can be changed (e.g., `format compact`). Since `startup.m` is a standard script M-file, there are no restrictions as to what commands can be placed in it. (However, it's probably not wise to include the command `quit` in `startup.m`!) On single-user installations, `startup.m` is commonly stored in the `toolbox/local` subdirectory on the MATLAB path. On network installations, a convenient location for your `startup.m` file is the default directory where you start MATLAB sessions.

When you terminate MATLAB via the **Exit MATLAB** item on the **File** menu in the *MATLAB* desktop window (or by typing `exit` or `quit` at the MATLAB prompt) MATLAB searches the MATLAB path for a script M-file named `finish.m`. If one is found, the commands in it are executed before MATLAB terminates. For example, the following `finish.m` prompts the user for confirmation before quitting, and the command `quit cancel` provides a way to cancel quitting:

```
%FINISH Confirm Desire for Quitting MATLAB

question = 'Are You Sure You Want To Quit?';
button = questdlg(question,'Exit Request','Yes','No','No');

switch button
case 'No'
   quit cancel;  % how to cancel quitting!
end
% 'Yes' lets script and MATLAB end.
```

5

Arrays and Array Operations

All of the computations considered to this point have involved single numbers called scalars. Operations involving scalars are the basis of mathematics. At the same time, when we wish to perform the same operation on more than one number at a time, performing repeated scalar operations is time consuming and cumbersome. To solve this problem, MATLAB defines operations on data arrays.

5.1 SIMPLE ARRAYS

Consider the problem of computing values of the sine function over one half of its period, namely, $y = \sin(x)$ over $0 \le x \le \pi$. Since it is impossible to compute $\sin(x)$ at all points over this range (there are an infinite number of them), we must choose a finite number of points. In doing so, we sample the function. To pick a number, let's evaluate $\sin(x)$ every 0.1π in this range; that is, let $x = 0, 0.1\pi, 0.2\pi, \dots, 1.0\pi$. If you were using a scientific calculator to compute these values, you would start by making a list, or an array, of the values of x. Then, you would enter each value of x into your calculator, find its sine, and write down the result as the second array y. Perhaps you would write the arrays in an organized fashion, as follows:

x	0	0.1π	0.2π	0.3π	0.4π	0.5π	0.6π	0.7π	0.8π	0.9π	π
y	0	0.31	0.59	0.81	0.95	1	0.95	0.81	0.59	0.31	0

As shown, x and y are ordered lists of numbers; that is, the first value or element in y is associated with the first value or element in x, the second element in y is associated with the second element in x, and so on. Because of this ordering, it is common to refer to individual values or elements in x and y by using subscripts; for example, x_1 is the first element in x, y_5 is the fifth element in y, and x_n is the nth element in x.

MATLAB handles arrays in a straightforward, intuitive way. Creating arrays is easy—just follow the preceding visual organization to create the following arrays:

```
>> x = [0 .1*pi .2*pi .3*pi .4*pi .5*pi .6*pi .7*pi .8*pi .9*pi pi]
x =
  Columns 1 through 7
        0    0.3142    0.6283    0.9425    1.2566    1.5708    1.8850
  Columns 8 through 11
   2.1991    2.5133    2.8274    3.1416
>> y = sin(x)
y =
  Columns 1 through 7
        0    0.3090    0.5878    0.8090    0.9511    1.0000    0.9511
  Columns 8 through 11
   0.8090    0.5878    0.3090    0.0000
```

To create an array in MATLAB, all you have to do is start with a left bracket, enter the desired values separated by spaces (or commas), and then close the array with a right bracket. Notice that finding the sine of the values in x follows naturally. MATLAB understands that you want to find the sine of each element in x and place the results in an associated array called y. This fundamental capability makes MATLAB different from other computer languages.

Since spaces separate array values, complex numbers entered as array values cannot have embedded spaces, unless the expressions are enclosed in parentheses. For example, [1 -2i 3 4 5+6i] contains five elements, whereas the identical arrays [(1 - 2i) 3 4 5+6i] and [1-2i 3 4 5+6i] contain four.

5.2 ARRAY ADDRESSING OR INDEXING

In the previous example, since x has more than one element, (it has 11 values separated into columns), MATLAB gives you the result with the columns identified. As shown, x is an array having one row and 11 columns; or, in mathematical jargon, it is a row vector, a 1-by-11 array, or simply an array of length 11.

In MATLAB, individual array elements are accessed by using subscripts; for example, x(1) is the first element in x, x(2) is the second element in x, and so on. The following code is illustrative:

```
>> x(3)  % The third element of x
ans =
   0.6283
```

```
>> y(5)  % The fifth element of y
ans =
    0.9511
```

To access a block of elements at one time, MATLAB provides ***colon notation***:

```
>> x(1:5)
ans =
         0    0.3142    0.6283    0.9425    1.2566
```

These are the first through fifth elements in x. The notation 1:5 says, start with 1 and count up to 5. The code

```
>> x(7:end)
ans =
        1.885    2.1991    2.5133    2.8274    3.1416
```

starts with the seventh element and continues to the last element. Here, the word end signifies the last element in the array x. In the code

```
>> y(3:-1:1)
ans =
    0.5878    0.3090         0
```

the results contain the third, second, and first elements in reverse order. The notation 3:-1:1 says, start with 3, count down by 1, and stop at 1. Similarly, the results in the code

```
>> x(2:2:7)
ans =
    0.3142    0.9425    1.5708
```

consists of the second, fourth, and sixth elements in x. The notation 2:2:7 says, start with 2, count up by 2, and stop when you get to 7. (In this case, adding 2 to 6 gives 8, which is greater than 7, and so the eighth element is not included.) The code

```
>> y([8 2 9 1])
ans =
    0.8090    0.3090    0.5878         0
```

uses another array, [8 2 9 1], to extract the elements of the array y in the order we wanted them! The first element taken is the eighth, the second is the second, the third is the ninth, and the fourth is the first. In reality, [8 2 9 1] itself is an array that addresses the desired elements of y. Note that in the code

```
>> y([1 1 3 4 2 2])
ans =
         0          0     0.5878     0.8090     0.3090     0.3090
```

there is no requirement that the array used as an index contain unique elements. This allows you to rearrange and duplicate array elements arbitrarily. Using this feature leads to efficient MATLAB coding.

Addressing one array with another works, as long as the addressing array contains integers between 1 and the length of the array. In contrast, consider the following code:

```
>> y(3.2)
??? Subscript indices must either be real positive integers or logicals.
>> y(11.6)
??? Subscript indices must either be real positive integers or logicals.
>> y(11)
??? Index exceeds matrix dimensions.
```

In these examples, MATLAB simply returns an error when the index is noninteger and returns a statement that no such value exists when the index exceeds the dimensions of the variable. In all cases, no numerical output appears.

5.3 ARRAY CONSTRUCTION

Earlier, we entered the values of x by typing each individual element in x. While this is fine when there are only 11 values in x, what if there were 111 values? Two other ways of entering x are as follows:

```
>> x = (0:0.1:1)*pi
x =
  Columns 1 through 7
         0     0.3142     0.6283     0.9425     1.2566     1.5708     1.8850
  Columns 8 through 11
    2.1991     2.5133     2.8274     3.1416
>> x = linspace(0,pi,11)
x =
  Columns 1 through 7
```

```
    0       0.3142      0.6283      0.9425      1.2566      1.5708      1.8850
Columns 8 through 11
  2.1991      2.5133      2.8274      3.1416
```

In the first case, the colon notation (0:0.1:1) creates an array that starts at 0, increments (or counts) by 0.1, and ends at 1. Each element in the array is then multiplied by π to create the desired values in x. In the second case, the MATLAB function linspace is used to create x. This function's arguments are described by

```
linspace(first_value,last_value,number_of_values)
```

Both of these array creation forms are common in MATLAB. The colon notation form allows you to directly specify the increment between data points, but not the number of data points. Using linspace, on the other hand, allows you to directly specify the number of data points, but not the increment between the data points.
Both of these array creation forms result in arrays in which the individual elements are linearly spaced with respect to each other. For the special case where a logarithmically spaced array is desired, MATLAB provides the logspace function:

```
>> logspace(0,2,11)
ans =
  Columns 1 through 7
    1.0000      1.5849      2.5119      3.9811      6.3096     10.0000     15.8489
  Columns 8 through 11
   25.1189     39.8107     63.0957    100.0000
```

Here, we created an array starting at 10^0, ending at 10^2, and containing 11 values. The function arguments are described by

```
logspace(first_exponent,last_exponent,number_of_values)
```

Although it is common to begin and end at integer powers of 10, logspace works equally well when nonintegers are used as the first two input arguments.
When using colon notation, or the functions linspace or logspace, there is often a temptation to enclose expressions in brackets:

```
>> a = [1:7]
a =
     1    2    3    4    5    6    7
>> b = [linspace(1,7,5)]
b =
     1          2.5          4          5.5          7
```

Although using brackets does not change the results and may add clarity to the statements, the added brackets force MATLAB to do more work and take more time, because brackets signify a concatenation operation. In the preceding examples, no concatenation is performed, and so there's no need ask MATLAB to take the time to consider that possibility.

Parentheses do not signify concatenation and therefore do not slow MATLAB down. As a result, parentheses can be used as needed:

```
>> a = (1:7)' % change row to column
a =
    1
    2
    3
    4
    5
    6
    7
```

Sometimes, an array is required that is not conveniently described by a linearly or logarithmically spaced element relationship. There is no uniform way to create these arrays. However, array addressing and the ability to combine expressions can help eliminate the need to enter individual elements, one at a time. For example,

```
>> a = 1:5, b = 1:2:9
a =
    1    2    3    4    5
b =
    1    3    5    7    9
```

creates two arrays. Remember that multiple statements can appear on a single line if they are separated by commas or semicolons. The code

```
>> c = [b a]
c =
    1    3    5    7    9    1    2    3    4    5
```

creates an array c composed of the elements of b, followed by those of a, while the code

```
>> d = [a(1:2:5) 1 0 1]
d =
    1    3    5    1    0    1
```

creates an array d composed of the first, third, and fifth elements of a, followed by three additional elements.

The simple array construction features of MATLAB are summarized in the following table:

Array Construction Technique	Description
x=[2 2*pi sqrt(2) 2-3j]	Creates row vector x containing arbitrary elements
x=first:last	Creates row vector x starting with first, counting by 1, and ending at or before last (Note that x=[first:last] produces the same result, but takes longer, since MATLAB considers both bracket and colon array-creation forms.)
x=first:increment:last	Creates row vector x starting with first, counting by increment, and ending at or before last
x=linspace(first,last,n)	Creates linearly spaced row vector x starting with first, ending at last, having n elements
x=logspace(first,last,n)	Creates logarithmically spaced row vector x starting with 10^{first}, ending at 10^{last}, and having n elements

5.4 ARRAY ORIENTATION

In the preceding examples, arrays contained one row and multiple columns. As a result of this row orientation, the arrays are commonly called row vectors. It is also possible for an array to be a column vector, having one column and multiple rows. In this case, all of the previous array manipulation and mathematics apply without change. The only difference is that results are displayed as columns, rather than as rows.

Since the array creation functions previously illustrated all create row vectors, there must be some way to create column vectors. The most straightforward way to create a column vector is to specify it, element by element, by using *semicolons* to separate values:

```
>> c = [1;2;3;4;5]
c =
     1
     2
     3
     4
     5
```

According to this example, separating elements by spaces or commas specifies elements in different columns, whereas separating elements by semicolons specifies elements in different rows.

To create a column vector using the colon notation `start:increment:end` or the functions `linspace` and `logspace`, you must ***transpose*** the resulting row into a column by using the MATLAB transpose operator (`'`). For example,

```
>> a = 1:5
a =
     1     2     3     4     5
```

creates a row vector using the colon notation format. The code

```
>> b = a'
b =
     1
     2
     3
     4
     5
```

uses the transpose operator to change the row vector a into the column vector b:

```
>> w = b'
w =
     1     2     3     4     5
```

This statement applies the transpose again and changes the column back into a row.

In addition to performing the simple transpose operations just described, MATLAB also offers a transpose operator with a preceding dot. In this case, the ***dot-transpose operator*** is interpreted as the noncomplex conjugate transpose. When an array is complex, the transpose (`'`) gives the complex-conjugate transpose; that is, the sign on the imaginary part is changed, as part of the transpose operation. On the other hand, the dot-transpose (`.'`) transposes the array, but does not conjugate it. The code

```
>> c = a.'
c =
     1
     2
     3
     4
     5
```

shows that `.'` and `'` are identical for real data, while the code

```
>> d = complex(a,a)
d =
  Columns 1 through 4
   1.0000 + 1.0000i   2.0000 + 2.0000i   3.0000 + 3.0000i   4.0000 + 4.0000i
  Column 5
   5.0000 + 5.0000i
```

creates a simple complex row vector from the array a by using the function complex. The code

```
>> e = d'
e =
   1.0000 - 1.0000i
   2.0000 - 2.0000i
   3.0000 - 3.0000i
   4.0000 - 4.0000i
   5.0000 - 5.0000i
```

creates a column vector e that is the complex conjugate transpose of d, while the code

```
>> f = d.'
f =
   1.0000 + 1.0000i
   2.0000 + 2.0000i
   3.0000 + 3.0000i
   4.0000 + 4.0000i
   5.0000 + 5.0000i
```

creates a column vector f that is the transpose of d.

If an array can be a row vector or a column vector, it makes intuitive sense that arrays can have both multiple rows and multiple columns. That is, arrays can also be in the form of matrices. The creation of matrices follows the creation of row and column vectors. ***Commas or spaces are used to separate elements in a specific row, and semicolons are used to separate individual rows:***

```
>> g = [1 2 3 4;5 6 7 8]
g =
     1     2     3     4
     5     6     7     8
```

Here, g is an array or matrix having 2 rows and 4 columns; that is, it is a 2-by-4 matrix, or it is a matrix of dimension 2 by 4. The semicolon tells MATLAB to start a new row between the 4 and the 5. Note the use of line breaks in the following code:

```
>> g = [1 2 3 4
5 6 7 8
9 10 11 12]
g =
      1      2      3      4
      5      6      7      8
      9     10     11     12
```

Thus, in addition to using semicolons, pressing the **Return** or **Enter** key while entering an array tells MATLAB to start a new row. MATLAB strictly enforces the fact that all rows must contain the same number of columns:

```
>> h = [1 2 3;4 5 6 7]
??? Error using ==> vertcat
All rows in the bracketed expression must have the same
number of columns.
```

5.5 SCALAR–ARRAY MATHEMATICS

In the first array example given, the array x is multiplied by the scalar π. Other simple mathematical operations between scalars and arrays follow the same natural interpretation. Addition, subtraction, multiplication, and division by a scalar simply apply the operation to all elements of the array. For example, the code

```
>> g-2
ans =
     -1      0      1      2
      3      4      5      6
      7      8      9     10
```

subtracts 2 from each element in g, while the code

```
>> 2*g - 1
ans =
      1      3      5      7
      9     11     13     15
     17     19     21     23
```

multiplies each element in g by 2 and subtracts 1 from each element of the result. Finally, the code

```
>> 2*g/5 + 1
ans =
      1.4        1.8        2.2        2.6
      3          3.4        3.8        4.2
      4.6        5          5.4        5.8
```

multiplies each element of g by 2, then divides each element of the result by 5, and finally adds 1 to each element.

Note that scalar–array mathematics uses the same order of precedence used in scalar expressions to determine the order of evaluation.

5.6 ARRAY–ARRAY MATHEMATICS

Mathematical operations between arrays are not quite as simple as those between scalars and arrays. Clearly, array operations between arrays of different sizes or dimensions are difficult to define and are of dubious value. However, when two arrays have the same dimensions, addition, subtraction, multiplication, and division apply on an element-by-element basis in MATLAB, as in the following example:

```
>> g  % recall previous array
g =
    1     2     3     4
    5     6     7     8
    9    10    11    12
>> h = [1 1 1 1;2 2 2 2;3 3 3 3]  % create new array
h =
    1     1     1     1
    2     2     2     2
    3     3     3     3
>> g + h  % add h to g on an element-by-element basis
ans =
    2     3     4     5
    7     8     9    10
   12    13    14    15
>> ans - h  % subtract h from the previous answer to get g back
```

```
ans =
      1      2      3      4
      5      6      7      8
      9     10     11     12
>> 2*g - h  % multiplies g by 2 and subtracts h from the result
ans =
      1      3      5      7
      8     10     12     14
     15     17     19     21
>> 2*(g-h)  % use parentheses to change order of operation
ans =
      0      2      4      6
      6      8     10     12
     12     14     16     18
```

Note that array–array mathematics uses the same order of precedence used in scalar expressions to determine the order of evaluation. Note also that parentheses can be used as desired to change the order of operation.

Element-by-element multiplication and division work similarly, but use slightly unconventional notation:

```
>> g.*h
ans =
      1      2      3      4
     10     12     14     16
     27     30     33     36
```

Here, we multiplied the arrays g and h, element by element by using the dot multiplication symbol .*.

> The dot preceding the standard asterisk multiplication symbol tells MATLAB to perform element-by-element array multiplication. Multiplication without the dot signifies matrix multiplication, which is discussed later.

For this particular example, matrix multiplication is not defined:

```
>> g*h
??? Error using ==> *
Inner matrix dimensions must agree.
```

Element-by-element division, or dot division, also requires use of the dot symbol as follows:

```
>> g./h
ans =
       1.0000     2.0000     3.0000     4.0000
       2.5000     3.0000     3.5000     4.0000
       3.0000     3.3333     3.6667     4.0000
>> h.\g
ans =
       1.0000     2.0000     3.0000     4.0000
       2.5000     3.0000     3.5000     4.0000
       3.0000     3.3333     3.6667     4.0000
```

As with scalars, division is defined by using both forward and backward slashes. In both cases, the array *below* the slash is divided into the array *above* the slash.

The dot preceding the forward or backward slash symbol tells MATLAB to perform element-by-element array division. Division without the dot signifies matrix inversion, which is discussed later.

Array or dot division also applies if the numerator is a scalar. Consider the following example:

```
>> 1./g
ans =
            1            0.5       0.33333          0.25
          0.2        0.16667       0.14286         0.125
      0.11111            0.1      0.090909      0.083333
```

In this case, the scalar 1 in the numerator is expanded to an array the same size as the denominator, and then element-by-element division is performed. That is, the preceding code represents a shorthand way of computing

```
>> f=[1 1 1 1; 1 1 1 1; 1 1 1 1] % create numerator
f =
     1     1     1     1
     1     1     1     1
     1     1     1     1
>> f./g
```

ans =

1	0.5	0.33333	0.25
0.2	0.16667	0.14286	0.125
0.11111	0.1	0.090909	0.083333

```
>> f./h
```

ans =

1	1	1	1
0.5	0.5	0.5	0.5
0.33333	0.33333	0.33333	0.33333

This process of automatically expanding scalar values so that element-by-element arithmetic applies is called *scalar expansion*. Scalar expansion is used extensively in MATLAB.

Division without the dot is the matrix division or matrix inversion operation, which is an entirely different operation, as the following code shows:

```
>> g/h
```

Warning: Rank deficient, rank = 1 tol =   5.3291e-015.

ans =

0	0	0.83333
0	0	2.1667
0	0	3.5

```
>> h/g
```

Warning: Rank deficient, rank = 2 tol =   1.8757e-014.

ans =

-0.125	0	0.125
-0.25	0	0.25
-0.375	0	0.375

Matrix division gives results that are not necessarily the same size as g and h. (Matrix operations are discussed in Chapter 17.)

Array exponentiation is defined in several ways. As used with multiplication and division, ^ is reserved for matrix exponentiation, and .^ is used to denote element-by-element exponentiation. When the exponent is a scalar, the scalar is applied to each element of the array. For example,

```
>> g, h  % recalls the arrays used earlier
g =
     1    2    3    4
```

```
       5       6       7       8
       9      10      11      12
h  =
       1       1       1       1
       2       2       2       2
       3       3       3       3

>> g.^2
ans =
       1       4       9      16
      25      36      49      64
      81     100     121     144
```

squares the individual elements of g, whereas

```
>> g^2
??? Error using ==> mpower
Matrix must be square.
```

is matrix exponentiation, which is defined only for square matrices—that is, matrices with equal row and column counts. The code

```
>> g.^-1
ans =
               1             0.5         0.33333            0.25
             0.2         0.16667         0.14286           0.125
         0.11111             0.1        0.090909        0.083333
```

finds the reciprocal of each element in g. The code

```
>> 1./g
ans =
               1             0.5         0.33333            0.25
             0.2         0.16667         0.14286           0.125
         0.11111             0.1        0.090909        0.083333
```

produces the same result as the scalar expansion approach seen earlier.

    When the exponent is an array operating on a scalar, each element of the array is applied to the scalar. For example, the code

```
>> 2.^g
ans =
```

2	4	8	16
32	64	128	256
512	1024	2048	4096

raises 2 to the power of each element in the array g.

If both components are arrays of the same size, exponentiation is applied element by element. Thus,

```
>> g.^h
ans =
```

1	2	3	4
25	36	49	64
729	1000	1331	1728

raises the elements of g to the corresponding elements in h. In this case, the first row is unchanged, since the first row of h contains ones, the second row is squared, and the third row is cubed.

The following example shows that scalar and array operations can be combined:

```
>> g.^(h-1)
ans =
```

1	1	1	1
5	6	7	8
81	100	121	144

The two forms of exponentiation that have scalar parts are further examples of scalar expansion. The results make intuitive sense if the scalars involved are first expanded to the size of the array and element-by-element exponentiation is then applied.

The following table summarizes basic array operations:

Element-by-Element Operation	**Representative Data**
	$A = [a_1\ a_2\ \ldots\ a_n]$, $B = [b_1\ b_2\ \ldots\ b_n]$, $c = $ <a *scalar*>
Scalar addition	$A+c = [a_1+c\ a_2+c\ \ldots\ a_n+c]$
Scalar subtraction	$A-c = [a_1-c\ a_2-c\ \ldots\ a_n-c]$
Scalar multiplication	$A*c = [a_1*c\ a_2*c\ \ldots\ a_n*c]$
Scalar division	$A/c = c\backslash A = [a_1/c\ a_2/c\ \ldots\ a_n/c]$

Array addition	$A+B = [a_1+b_1 \; a_2+b_2 \; \ldots \; a_n+b_n]$
Array multiplication	$A.*B = [a_1*b_1 \; a_2*b_2 \; \ldots \; a_n*b_n]$
Array right division	$A./B = [a_1/b_1 \; a_2/b_2 \; \ldots \; a_n/b_n]$
Array left division	$A.\backslash B = [a_1 \backslash b_1 \; a_2 \backslash b_2 \; \ldots \; a_n \backslash b_n]$
Array exponentiation	$A.^{\wedge}c = [a_1^{\wedge}c \; a_2^{\wedge}c \; \ldots \; a_n^{\wedge}c]$
	$c.^{\wedge}A = [c^{\wedge}a_1 \; c^{\wedge}a_2 \; \ldots \; c^{\wedge}a_n]$
	$A.^{\wedge}B = [a_1^{\wedge}b_1 \; a_2^{\wedge}b_2 \; \ldots \; a_n^{\wedge}b_n]$

5.7 STANDARD ARRAYS

Because of the general utility of standard arrays, MATLAB provides functions for creating a number of them. Standard arrays include those containing all ones or all zeros, identity matrices, arrays of random numbers, diagonal arrays, and arrays whose elements are a given constant. The following are examples:

```
>> ones(3)

ans =

        1       1       1
        1       1       1
        1       1       1
>> zeros(2,5)

ans =

        0       0       0       0       0
        0       0       0       0       0
>> size(g)

ans =

        3       4
>> ones(size(g))

ans =

        1       1       1       1
        1       1       1       1
        1       1       1       1
```

When called with a single input argument, ones(n) or zeros(n), MATLAB creates an n-by-n array containing ones or zeros, respectively. When called with two input arguments, ones(r,c) or zeros(r,c), MATLAB creates an array having r rows and c columns. To create an array of ones or zeros that is the same size as another array, use the size function (discussed later in this chapter) in the argument of ones or zeros.

In the code

```
>> eye(4)
ans =
    1    0    0    0
    0    1    0    0
    0    0    1    0
    0    0    0    1
>> eye(2,4)
ans =
    1    0    0    0
    0    1    0    0
>> eye(4,2)
ans =
    1    0
    0    1
    0    0
    0    0
```

the function eye produces identity matrices by using the same syntax style as that used to produce arrays of zeros and ones. An identity matrix or array is all zeros, except for the elements $A(i,i)$, where $i = 1:min(r,c)$, in which $min(r,c)$ is the minimum of the number of rows and columns in A: In the code

```
>> rand(3)
ans =
    0.9501    0.4860    0.4565
    0.2311    0.8913    0.0185
    0.6068    0.7621    0.8214
>> rand(1,5)
ans =
    0.4447    0.6154    0.7919    0.9218    0.7382
>> b = eye(3)
b =
    1    0    0
```

```
     0     1     0
     0     0     1
>> rand(size(b))
ans =
      0.1763          0.9169          0.0579
      0.4057          0.4103          0.3529
      0.9355          0.8937          0.8132
```

the function rand produces uniformly distributed random arrays whose elements lie between 0 and 1. On the other hand, the function randn produces arrays whose elements are samples from a *zero-mean, unit-variance* normal distribution:

```
>> randn(2)
ans =
      -0.4326         0.1253
      -1.6656         0.2877
>> randn(2,5)
ans =
      -1.1465      1.1892       0.3273      -0.1867      -0.5883
       1.1909     -0.0376       0.1746       0.7258       2.183
```

The function diag creates diagonal arrays in which a vector can be placed at any location parallel to the main diagonal of an array:

```
>> a = 1:4  % start with a simple vector
a =
     1     2     3     4
>> diag(a)   % place elements on the main diagonal
ans =
     1     0     0     0
     0     2     0     0
     0     0     3     0
     0     0     0     4
>> diag(a,1)   % place elements 1 place up from diagonal
ans =
     0     1     0     0     0
     0     0     2     0     0
```

```
    0      0      0      3      0
    0      0      0      0      4
    0      0      0      0      0
>> diag(a,-2)  % place elements 2 places down from diagonal
ans =
    0      0      0      0      0      0
    0      0      0      0      0      0
    1      0      0      0      0      0
    0      2      0      0      0      0
    0      0      3      0      0      0
    0      0      0      4      0      0
```

With the preceding standard arrays, there are several ways to create an array whose elements all have the same value. Some of them are as follows:

```
>> d = pi;  % choose pi for this example
>> d*ones(3,4)  % slowest method (scalar-array multiplication)
ans =
        3.1416         3.1416         3.1416         3.1416
        3.1416         3.1416         3.1416         3.1416
        3.1416         3.1416         3.1416         3.1416
>> d+zeros(3,4) % slower method (scalar-array addition)
ans =
        3.1416         3.1416         3.1416         3.1416
        3.1416         3.1416         3.1416         3.1416
        3.1416         3.1416         3.1416         3.1416
>> d(ones(3,4))  % fast method (array addressing)
ans =
        3.1416         3.1416         3.1416         3.1416
        3.1416         3.1416         3.1416         3.1416
        3.1416         3.1416         3.1416         3.1416
>> repmat(d,3,4)  % fastest method (optimum array addressing)
ans =
        3.1416         3.1416         3.1416         3.1416
        3.1416         3.1416         3.1416         3.1416
        3.1416         3.1416         3.1416         3.1416
```

For small arrays, all of the methods used in this example are fine. However, as the array grows in size, the multiplications required in the scalar multiplication approach slow the procedure down. Since addition is often faster than multiplication, the next-best approach is to add the desired scalar to an array of zeros. Although they are not intuitive, the last two methods explained are the fastest for large arrays. They both involve array indexing, as described earlier.

The solution d(ones(r,c)) creates an r-by-c array of ones and then uses this array to index and duplicate the scalar d. Creating the temporary array of ones takes time and uses memory, thereby slowing down this approach, despite the fact that no floating-point mathematics are used. The solution repmat(d,r,c) calls the function repmat, which stands for replicate matrix. For scalar d, this function performs the following steps:

```
D(r*c) = d;          % a row vector whose (r*c)-th element is d
D(:) = d;            % scalar expansion to fill all elements of D with d
D = reshape(D,r,c); % reshape the vector into the desired r-by-c shape
```

This MATLAB code uses scalar expansion to create a vector having r*c elements all equal to d. The vector is then reshaped, by using the function reshape, into an r-by-c array. (The functions repmat and reshape are discussed in more detail later.)

5.8 ARRAY MANIPULATION

Because arrays are fundamental to MATLAB, there are many ways to manipulate them in MATLAB. Once arrays are formed, MATLAB provides powerful ways to insert, extract, and rearrange subsets of arrays by identifying subscripts of interest. Knowledge of these features is a key to using MATLAB efficiently. To illustrate the array manipulation features of MATLAB, consider the following examples:

```
>> A = [1 2 3;4 5 6;7 8 9]
A =
      1     2     3
      4     5     6
      7     8     9
>> A(3,3) = 0  % set element in 3rd row, 3rd column to zero
A =
      1     2     3
```

```
     4     5     6
     7     8     0
```

The preceding code changes the element in the third row and third column to zero. The code

```
>> A(2,6) = 1  % set element in 2nd row, 6th column to one
A =
     1     2     3     0     0     0
     4     5     6     0     0     1
     7     8     0     0     0     0
```

places the number one in the second row, sixth column. Since A does not have six columns, the size of A is increased as necessary, and filled with zeros, so that the array remains rectangular.

The code

```
>> A(:,4) = 4
A =
     1     2     3     4     0     0
     4     5     6     4     0     1
     7     8     0     4     0     0
```

sets the fourth column of A equal to 4. Since 4 is a scalar, it is expanded to fill all of the elements specified. This is another example of scalar expansion. MATLAB performs scalar expansion to simplify statements that can be interpreted unambiguously. For example, the preceding statement is equivalent to the following, more cumbersome, statement:

```
>> A(:,4) = [4;4;4]
A =
     1     2     3     4     0     0
     4     5     6     4     0     1
     7     8     0     4     0     0
>> A(:,4) = [4 4 4]   % but a row can't be squeezed into a column!

??? In an assignment  A(:,matrix) = B, the number of elements in
the subscript of A and the number of columns in B must be the same.
```

Let's start over and look at other array manipulations. The code

```
>> A = [1 2 3;4 5 6;7 8 9];  % restore original data

>> B = A(3:-1:1,1:3)
B =
       7       8       9
       4       5       6
       1       2       3

>> B = A(end:-1:1,1:3)   % same as above
B =
       7       8       9
       4       5       6
       1       2       3
```

creates an array B by taking the rows of A in reverse order. The word end automatically denotes the final or largest index for a given dimension. In this example, end signifies the largest row index 3. The code

```
>> B = A(3:-1:1,:)
B =
       7       8       9
       4       5       6
       1       2       3
```

does the same as the previous example. The final single colon means take all columns. That is, : is equivalent to 1:3 in this example, because A has three columns. The code

```
>> C = [A B(:,[1 3])]
C =
       1       2       3       7       9
       4       5       6       4       6
       7       8       9       1       3
```

creates C by appending, or concatenating, all rows in the first and third columns of B to the right of A. Similarly, the code

```
>> B = A(1:2,2:3)
```

```
B =
     2     3
     5     6

>> B = A(1:2,2:end)   % same as above
B =
     2     3
     5     6
```

creates B by extracting the first two rows and last two columns of A. Once again, colon notation is used to create index vectors that identify the array elements to extract. In the second instance just shown, end is used to denote the final or largest column index. The statement

```
>> C = [1 3]
C =
     1     3
>> B = A(C,C)
B =
     1     3
     7     9
```

uses the array C to index the array A, rather than specify them directly by using the colon notation start:increment:end or start:end. In this example, B is formed from the first and third rows and first and third columns of A. The code

```
>> B = A(:)
B =
     1
     4
     7
     2
     5
     8
     3
     6
     9
```

builds B by stretching A into a column vector and taking its columns one at a time, in order, which is the simplest form of reshaping an array into an array that has different dimensions, but the same number of total elements.

In MATLAB, the statement

```
>> B = B.'
B =
    1    4    7    2    5    8    3    6    9

>> B = reshape(A,1,9)   % reshape A into 1-by-9
B =
    1    4    7    2    5    8    3    6    9

>> B = reshape(A,[1 9])
B =
    1    4    7    2    5    8    3    6    9
```

illustrates the dot-transpose operation introduced earlier, as well as the function reshape. In this case, reshape works with the indices, supplied as separate function arguments, or supplied as a single vector argument. The code

```
>> B = A  % copy A into B
B =
    1    2    3
    4    5    6
    7    8    9
>> B(:,2) = []
B =
    1    3
    4    6
    7    9
```

redefines B by throwing away all rows in the second column of the original B. When you set something equal to the empty matrix or empty array [], it is deleted, causing the array to collapse into what remains. Note that you must delete whole rows or columns, so that the result remains rectangular.
The statement

```
>> C = B.'
C =
    1    4    7
    3    6    9
```

```
>> reshape(B,2,3)  % reshape is not equivalent to transpose
ans =
     1     7     6
     4     3     9
```

illustrates the transpose of an array and demonstrates that reshape is not the same as transpose. The transpose operator converts the *i*th row to the *i*th column of the result, and so the original 3-by-2 array becomes a 2-by-3 array.
The code

```
>> C(2,:) = []
C =
     1     4     7
```

throws out the second row of C, leaving a row vector, while the code

```
>> A(2,:) = C
A =
     1     2     3
     1     4     7
     7     8     9
```

replaces the second row of A with C. The MATLAB code

```
>> B = A(:,[2 2 2 2])   % create new B array
B =
     2     2     2     2
     4     4     4     4
     8     8     8     8
>> B = A(:,2+zeros(1,4))   % [2 2 2 2]=2+zeros(1,4)
B =
     2     2     2     2
     4     4     4     4
     8     8     8     8
>> B = repmat(A(:,2),1,4)   % replicate 2nd column into 4 columns
B =
     2     2     2     2
     4     4     4     4
     8     8     8     8
```

creates B three ways by duplicating all rows in the second column of A four times. The last approach is fastest for large arrays. The statement

```
>> A, C  % show A and C again
A =
       1        2        3
       1        4        7
       7        8        9
C =
       1        4        7
>> A(2,2) = []
???  Indexed empty matrix assignment is not allowed.
```

shows that you can throw out only entire rows or columns. MATLAB simply does not know how to collapse an array when partial rows or columns are thrown out. Finally

```
>> C = A(4,:)
???  Index exceeds matrix dimensions.
```

produces an error since A does not have a fourth row. According to this result, indices must adhere to the following guidelines:

If A(r,c) appears on the left-hand side of an equal sign and one or more elements specified by (r,c) do not exist, then zeros are added to A as needed, so that A(r,c) addresses known elements. However, on the right-hand side of an equal sign, all elements addressed by A(r,c) must exist or an error is returned.

Continuing on, consider the following:

```
>> C(1:2,:) = A
??? Subscripted assignment dimension mismatch.
```

This example shows that you can't squeeze one array into another array of a different size.

However, the statement

```
>> C(3:4,:) = A(2:3,:)
C =
       1        4        7
       0        0        0
```

```
              1        4       7
              7        8       9
```

reveals that you can place the second and third columns of A into the same-size area of C. Since the second through fourth rows of C did not exist, they are created as necessary. Moreover, the second row of C is unspecified, and so it is filled with zeros.
   The code

```
>> A = [1 2 3;4 5 6;7 8 9]  % fresh data
A =
      1      2      3
      4      5      6
      7      8      9
>> A(:,2:3)              % a peek at what's addressed next
ans =
      2      3
      5      6
      8      9
>> G(1:6) = A(:,2:3)
G =
      2     5     8     3     6     9
```

creates a row vector G by extracting all rows in the second and third columns of A. Note that the shapes of the matrices are different on both sides of the equal sign. The elements of A are inserted into the elements of G by going down the rows of the first column and then down the rows of the second column.
   When (:) appears on the left-hand side of the equal sign, it signifies that elements will be taken from the right-hand side and placed into the array on the left-hand side without changing its shape. In the following example, this process extracts the second and third columns of A and inserts them into the column vector H:

```
>> H = ones(6,1);    % create a column array

>> H(:) = A(:,2:3)   % fill H without changing its shape
H =
      2
      5
      8
      3
```

```
        6
        9
```

Of course, for this code to work, both sides must address the same number of elements.
      When the right-hand side of an assignment is a scalar and the left-hand side is
an array, *scalar expansion* is used. For example, the code

```
>> A(2,:) = 0
A =
        1       2       3
        0       0       0
        7       8       9
```

replaces the second row of A with zeros. The single zero on the right-hand side is
expanded to fill all the indices specified on the left. This example is equivalent to

```
>> A(2,:) = [0 0 0]
A =
        1       2       3
        0       0       0
        7       8       9
```

Scalar expansion occurs whenever a scalar is used in a location calling for an array.
MATLAB automatically expands the scalar to fill all requested locations and then
performs the operation dictated. The code

```
>> A(1,[1 3]) = pi
A =
      3.1416       2.0000       3.1416
           0            0            0
      7.0000       8.0000       9.0000
```

is yet another example in which the scalar π is expanded to fill two locations. Consider again what the function reshape does with scalar input. Let's create a 2-by-4
array containing the number 2:

```
>> D(2*4) = 2  % create array with 8 elements
D =
        0     0     0     0     0     0     0     2
>> D(:) = 2  % scalar expansion
```

```
D =

     2      2      2      2      2      2      2      2
>> D = reshape(D,2,4)   % reshape
D =

     2      2      2      2
     2      2      2      2
```

The first line D(2*4) = 2 causes a row vector of length 8 to be created and places 2 in the last column. Next, D(:) = 2 uses scalar expansion to fill all elements of D with 2. Finally, the result is reshaped into the desired dimensions.

Sometimes, it is desirable to perform some mathematical operation between a vector and a two-dimensional (2-D) array. For example, consider these arrays:

```
>> A = reshape(1:12,3,4)'
A =

      1      2      3
      4      5      6
      7      8      9
     10     11     12
>> r = [3 2 1]

r =

      3      2      1
```

Suppose that we wish to subtract r(i) from the ith column of A. One way of accomplishing it is as follows:

```
>> Ar = [A(:,1)-r(1) A(:,2)-r(2) A(:,3)-r(3)]
Ar =

     -2      0      2
      1      3      5
      4      6      8
      7      9     11
```

Alternatively, we can use indexing:

```
>> R = r([1 1 1 1],:) % duplicate r to have 4 rows
R =

      3      2      1
      3      2      1
```

```
        3      2      1
        3      2      1
>> Ar = A - R % now use element by element subtraction
Ar =
       -2      0      2
        1      3      5
        4      6      8
        7      9     11
```

The array R can also be computed faster and more generally by using the functions ones and size or by using the function repmat, all of which are discussed later. Consider the following example:

```
>> R = r(ones(size(A,1),1),:) % historically this is Tony's trick
R =
        3      2      1
        3      2      1
        3      2      1
        3      2      1
>> R = repmat(r,size(A,1),1)  % often faster than Tony's trick
R =
        3      2      1
        3      2      1
        3      2      1
        3      2      1
```

In this example, size(A,1) returns the number of rows in A.

Sometimes, it is more convenient to address array elements with a single *index*. **When a single index is used in MATLAB, the index counts down the rows, proceeding column by column, starting with the first column**. Here is an example:

```
>> D = reshape(1:12,3,4)  % new data
D =
        1      4      7     10
        2      5      8     11
        3      6      9     12
```

```
>> D(2)
ans =
     2
>> D(5)
ans =
     5

>> D(end)
ans =
    12
>> D(4:7)
ans =
     4     5     6     7
```

The MATLAB functions sub2ind and ind2sub perform the arithmetic to convert to and from a single index to row and column subscripts:

```
>> sub2ind(size(D),2,4)   % find single index from row and column
ans =
    11
>> [r,c] = ind2sub(size(D),11)   % find row and column from single index
r =
     2
c =
     4
```

The element in the second row, fourth column is the 11th indexed element. Note that these two functions need to know the size of the array to search (i.e., size(D), rather than the array D itself).

Subscripts refer to the row and column locations of elements of an array; for instance, A(2,3) refers to the element in the second row and third column of A. The *index* of an element in an array refers to its position relative to the element in the first row, first column, having index 1. Indices count down the rows first and then proceed across the columns of an array. That is, the indices count in the order in which the subscripts appear; for example, if D has three rows, D(8) is the second element in the third column of D.

In addition to addressing arrays on the basis of their subscripts, *logical arrays* that result from logical operations (to be discussed more thoroughly later) can also be used if the size of the array is equal to that of the array it is addressing.

In this case, True (1) elements are retained, and False (0) elements are discarded. For example,

```
>> x = -3:3  % Create data
x =
    -3    -2    -1     0     1     2     3
>> abs(x)>1
ans =
     1     1     0     0     0     1     1
```

returns a logical array with ones (True), where the absolute value of x is greater than one and zeros (False) elsewhere. (The chapter on logical expressions contains more detailed information.)

The statement

```
>> y = x(abs(x)>1)
y =
    -3    -2     2     3
```

creates y by taking those values of x where its absolute value is greater than 1.

Note, however, that even though abs(x)>1 produces the array [1 1 0 0 0 1 1], it is not equivalent to a numerical array containing those values; that is, the code

```
>> y = x([1 1 0 0 0 1 1])
??? Subscript indices must either be real positive integers or logicals.
```

gives an error, even though the abs(x)>1 and [1 1 0 0 0 1 1] appear to be the same vector. In the second case, [1 1 0 0 0 1 1] is a **numeric array** as opposed to a **logical array**. The difference between these two visually equal arrays can be identified by using the function class:

```
>> class(abs(x)>1)    % logical result from logical comparison
ans =
logical
>> class([1 1 0 0 0 1 1]) % double precision array
ans =
double
```

Alternatively, the functions islogical and isnumeric return logical True (1) or False (0) to identify data types:

```
>> islogical(abs(x)>1)
ans =
    1
>> islogical([1 1 0 0 0 1 1])
ans =
    0
>> isnumeric(abs(x)>1)
ans =
    0
>> isnumeric([1 1 0 0 0 1 1])
ans =
    1
```

Because of the differences between logical and numeric arrays, MATLAB generates an error when executing x([1 1 0 0 0 1 1]), because MATLAB arrays do not have indices at zero.

Quite naturally, MATLAB provides the function logical for converting numeric arrays to logical arrays:

```
>> y = x(logical([1 1 0 0 0 1 1]))
y =
    -3    -2    2    3
```

Once again, we have the desired result.

To summarize:

> Specifying array subscripts with numerical arrays extracts the elements having the given numerical indices. On the other hand, specifying array subscripts with logical arrays, which are returned by logical expressions and the function logical, extracts elements that are logical True (1).

Just as the functions ones and zeros are useful for creating numeric arrays, the functions true and false are useful for creating arrays of logical arrays containing True and False values, respectively:

```
>> true
ans =
    1
>> true(2,3)
```

```
ans =
     1      1      1
     1      1      1
>> false
ans =
     0
>> false(1,6)
ans =
     0      0      0      0      0      0
```

Although these results appear to be equal to results returned by equivalent statements that use ones and zeros, they are *logical* arrays, not numeric.

Logical arrays work on two-dimensional arrays, as well as on vectors:

```
>> B = [5 -3;2 -4]   % new data
B =
     5      -3
     2      -4
>> x = abs(B)>2   % logical result
x =
     1      1
     0      1
>> y = B(x)   % grab True values
y =
     5
    -3
     4
```

However, the final result is returned as a column vector, since there is no way to define a two-dimensional array having only three elements. No matter how many elements are extracted, MATLAB extracts all of the true elements by using single-index order and then forms or reshapes the result into a column vector.

The preceding array-addressing techniques are summarized in the following table:

Array Addressing	Description
A(r,c)	Addresses a subarray within A defined by the index vector of desired rows in r and an index vector of desired columns in c

Array Addressing	Description
A(r,:)	Addresses a subarray within A defined by the index vector of desired rows in r and all columns
A(:,c)	Addresses a subarray within A defined by all rows and the index vector of desired columns in c
A(:)	Addresses all elements of A as a column vector taken column by column. If A(:) appears on the left-hand side of the equal sign, it means to fill A with elements from the right hand-side of the equal sign without changing A's shape.
A(k)	Addresses a subarray within A defined by the single index vector k, as if A were the column vector A(:)
A(x)	Addresses a subarray within A defined by the logical array x. Note that x should be the same size as A. If x is shorter than A, the missing values in x are assumed to be False. If x is longer than A, all extra elements in x must be False.

5.9 ARRAY SORTING

Given a data vector, a common task required in numerous applications is sorting. In MATLAB, the function sort performs this task:

```
>> x = randperm(8)  % new data
x =
     7    5    2    1    3    6    4    8
>> xs = sort(x)  % sort ascending by default
xs =
     1    2    3    4    5    6    7    8
>> xs = sort(x,'ascend')  % sort ascending
xs =
     1    2    3    4    5    6    7    8
>> [xs,idx] = sort(x)  % return sort index as well
xs =
     1    2    3    4    5    6    7    8
idx =
     4    3    5    7    2    6    1    8
```

As shown, the sort function returns one or two outputs. The first is an ascending sort of the input argument, and the second is the sort index—for example, xs(k) = x(idx(k)).

> Note that, when a MATLAB function returns two or more variables, they are enclosed by square brackets on the left-hand side of the equal sign. This syntax is different from the array manipulation syntax discussed previously, in which [a,b] on the right-hand side of the equal sign builds a new array, with b appended to the right of a.

In prior versions of MATLAB, sorting in the descending direction required turning around the sort function output by using array-indexing techniques. For example, consider the following code:

```
>> xsd = xs(end:-1:1)
xsd =
     8     7     6     5     4     3     2     1
>> idxd = idx(end:-1:1)
idxd =
   8    1    6    2    7    5    3    4
```

With MATLAB Version 7, the sort direction can be specified when calling the sort function:

```
>> xs = sort(x,'descend')   % sort descending
xs =
     8     7     6     5     4     3     2     1
```

When presented with a two-dimensional array, sort acts differently. For example, in the code

```
>> A = [randperm(6);randperm(6);randperm(6);randperm(6)] % new data
A =
     1     2     5     6     4     3
     4     2     6     5     3     1
     2     3     6     1     4     5
     3     5     1     2     4     6
>> [As,idx] = sort(A)
As =
     1     2     1     1     3     1
     2     2     5     2     4     3
     3     3     6     5     4     5
     4     5     6     6     4     6
```

```
idx =
      1       1       4       3       2       2
      3       2       1       4       1       1
      4       3       2       2       3       3
      2       4       3       1       4       4
```

the sort function sorts, in ascending order, each column independently of the others, and the indices returned are those for each column. In many cases, we are more interested in sorting an array on the basis of the sort of a specific column. In MAT-LAB, this task is easy. For example, in the code

```
>> [tmp,idx] =  sort(A(:,4)); % sort 4-th column only

>> As = A(idx,:) % rearrange rows in all columns using idx
As =
      2       3       6       1       4       5
      3       5       1       2       4       6
      4       2       6       5       3       1
      1       2       5       6       4       3
```

the rows of As are just the rearranged rows of A, in which the fourth column of As is sorted in ascending order.

MATLAB is a flexible program. Rather than sort each column, it is also possible to sort each row. By using a second argument to sort, you can specify in which direction to sort:

```
>> As = sort(A,2) % sort across 2-nd dimension
As =
      1       2       3       4       5       6
      1       2       3       4       5       6
      1       2       3       4       5       6
      1       2       3       4       5       6
>> As = sort(A,1)  % same as sort(A)
As =
      1       2       1       1       3       1
      2       2       5       2       4       3
      3       3       6       5       4       5
      4       5       6       6       4       6
```

Since in A(r,c), the row dimension appears first, sort(A,1) means to sort down the rows. Because the column dimension appears second, sort(A,2) means to sort across the columns. Although not shown, 'descend' can be appended to the function calls (e.g., sort(A,1,'descend')), to return a descending sort. (The chapter on data analysis contains much more information about the sort function.)

5.10 SUBARRAY SEARCHING

Many times it is desirable to know the indices or subscripts of the elements of an array that satisfy some relational expression. In MATLAB, this task is performed by the function find, which returns the subscripts when a relational expression is True, as for example, in the following code:

```
>> x = -3:3
x =
    -3    -2    -1    0    1    2    3
>> k = find(abs(x)>1)   % finds those subscripts where abs(x)>1
k =
    1    2    6    7
>> y = x(k)   % creates y using the indices in k.
y =
    -3    -2    2    3
>> y = x(abs(x)>1) % creates the same y vector by logical addressing
y =
    -3    -2    2    3
```

The find function also works for two-dimensional arrays:

```
>> A = [1 2 3;4 5 6;7 8 9] % new data
A =
    1    2    3
    4    5    6
    7    8    9
>> [i,j] = find(A>5) % i and j are not equal to sqrt(-1) here
i =
    3
    3
    2
    3
```

```
j =

     1

     2

     3

     3
```

Here, the indices stored in i and j are the associated row and column indices, respectively, where the relational expression is True. That is, A(i(1),j(1)) is the first element of A, where A>5, and so on.

Alternatively, find returns single indices for two-dimensional arrays:

```
>> k = find(A>5)
k =

     3

     6

     8

     9
```

Of the two index sets returned for two-dimensional arrays, this latter single-index form is often more useful:

```
>> A(k)  % look at elements greater than 5
ans =

     7

     8

     6

     9

>> A(k) = 0  % set elements addressed by k to zero
A =

     1     2     3
     4     5     0
     0     0     0
>> A = [1 2 3;4 5 6;7 8 9]   % restore data
A =

     1     2     3
     4     5     6
     7     8     9
```

```
>> A(i,j)  % this is A([3 3 2 3],[1 2 3 3])
ans =
      7        8        9        9
      7        8        9        9
      4        5        6        6
      7        8        9        9
>> A(i,j) = 0  % this is A([3 3 2 3],[1 2 3 3]) also
A =
      1        2        3
      0        0        0
      0        0        0
```

These A(i,j) cases are not as easy to understand as the preceding single-index cases. To assume that A(k) is equivalent to A(i,j) is a common MATLAB indexing mistake. Rather, A(i,j) in the previous example is equivalent to

```
>>  [A(3,1) A(3,2) A(3,3) A(3,3)
     A(3,1) A(3,2) A(3,3) A(3,3)
     A(2,1) A(2,2) A(2,3) A(2,3)
     A(3,1) A(3,2) A(3,3) A(3,3)]
ans =
      7        8        9        9
      7        8        9        9
      4        5        6        6
      7        8        9        9
```

where the first-row index in i is coupled with all of the column indices in j to form the first row in the result. Then, the second-row index in i is coupled with the column indices in j to form the second row in the result, and so on. Given the form in this example, the diagonal elements of A(i,j) are equal to those of A(k). Therefore, the two approaches are equal if the diagonal elements of A(i,j) are retained. In the code

```
>> diag(A(i,j))
ans =
      7
      8
      6
      9
```

while diag $(A(i,j))$ = $A(k)$, simply entering $A(k)$ is preferred, as it does not create an intermediate square array.

Similarly, $A(i,j)$ = 0 is equivalent to

```
>> A(3,1)=0; A(3,2)=0; A(3,3)=0; A(3,3)=0; % i(1) with all j
>> A(3,1)=0; A(3,2)=0; A(3,3)=0; A(3,3)=0; % i(2) with all j
>> A(2,1)=0; A(2,2)=0; A(2,3)=0; A(2,3)=0; % i(3) with all j
>> A(3,1)=0; A(3,2)=0; A(3,3)=0; A(3,3)=0  % i(4) with all j
A =
     1     2     3
     0     0     0
     0     0     0
```

Based on the above equivalents, it is clear that $A(i,j)$ is not generally as useful as $A(k)$ for subarray searching using find.

When you only need to find a few indices, the find function offers an alternative syntax that avoids searching the entire array. For example, in the code

```
>> x = randperm(8) % new data
x =
     8     2     7     4     3     6     5     1
>> find(x>4) % find all values greater than 4
ans =
     1     3     6     7
>> find(x>4,1) % find first value greater than 4
ans =
     1
>> find(x>4,1,'first') % same as above
ans =
     1
>> find(x>4,2) % find first two values greater than 4
ans =
     1     3
>> find(x>4,2,'last') % find last two values greater than 4
ans =
     6     7
```

find(expr,n), find(expr,n,'first') and find(expr,n,'last'), where the expression expr is True, return up to n indices. That is, n defines the maximum number of indices returned. If fewer indices satisfying expr exist, fewer are returned.

The preceding concepts are summarized in the following table:

Array Searching	Description
i=find(X)	Return single *indices* of the array X, where its elements are nonzero or True
[r,c]=find(X)	Return row and column *subscripts* of the array X, where its elements are nonzero or True
find(X,n) or find(X,n,'first')	Starting at the beginning of the array X, returns up to n indices, where X is nonzero or True
find(X,n,'last')	Starting at the end of the array X, returns up to n indices, where X is nonzero or True

In addition to using the function find to identify specific values within an array, using the maximum and minimum values and their locations within an array is often helpful. MATLAB provides the functions max and min to accomplish these tasks:

```
>> v = rand(1,6)  % new data
v =
      0.3046    0.1897    0.1934    0.6822    0.3028    0.5417

>> max(v)  % return maximum value
ans =
      0.6822
>> [mx,i] = max(v)  % maximum value and its index
mx =
      0.6822
i =
       4

>> min(v)  % return minimum value
ans =
      0.1897

>> [mn,i] = min(v)  % minimum value and its index
```

```
mn =

    0.1897

i =

    2
```

For two-dimensional arrays, min and max behave a little differently. For example, in the code

```
>> A = rand(4,6)   % new data
A =
    0.1509     0.8537     0.8216     0.3420     0.7271     0.3704
    0.6979     0.5936     0.6449     0.2897     0.3093     0.7027
    0.3784     0.4966     0.8180     0.3412     0.8385     0.5466
    0.8600     0.8998     0.6602     0.5341     0.5681     0.4449
>> [mx,r] = max(A)
mx =
    0.8600     0.8998     0.8216     0.5341     0.8385     0.7027
r =
    4      4      1      4      3      2
>> [mn,r] = min(A)
mn =
    0.1509     0.4966     0.6449     0.2897     0.3093     0.3704
r =
    1      3      2      2      2      1
```

mx is a vector containing the maximum of each column of A, and r is the row index where the maximum appears. The same principle applies to mn and r, relative to the function min. To find the overall minimum or maximum of a 2-D array, one can take one of two approaches:

```
>> mmx = max(mx)   % apply max again to prior result
mmx =
    0.8998
>> [mmx,i] = max(A(:)) % reshape A as a column vector first
mmx =
    0.8998
i =
    8
```

The first of these is essentially max(max(A)), which requires two function calls. The second is preferred in many situations, because it also returns the single index where the maximum occurs (i.e., mmx = A(i)). (The latter approach also works for multi-dimensional arrays, which are discussed in the next chapter.)

When an array has duplicate minima or maxima, the indices returned by min and max are the first ones encountered. To find all minima and maxima requires the use of the find function:

```
>> x = [1 4 6 3 2 1 6]
x =
      1     4     6     3     2     1     6
>> mx = max(x)
mx =
      6
>> i = find(x==mx) % indices of values equal to mx
i =
      3     7
```

(The chapter on data analysis contains more information about the functions min and max.)

5.11 ARRAY MANIPULATION FUNCTIONS

In addition to the arbitrary array-addressing and manipulation capabilities described in the preceding sections, MATLAB provides several functions that implement common array manipulations. Many of these manipulations are easy to follow:

```
>> A = [1 2 3;4 5 6;7 8 9]  % fresh data
A =
      1     2     3
      4     5     6
      7     8     9
>> flipud(A)  % flip array in up-down direction
ans =
      7     8     9
      4     5     6
      1     2     3
>> fliplr(A)  % flip array in the left-right direction
```

```
ans =

     3     2     1
     6     5     4
     9     8     7
>> rot90(A)  % rotate array 90 degrees counterclockwise
ans =

     3     6     9
     2     5     8
     1     4     7
>> rot90(A,2)  % rotate array 2*90 degrees counterclockwise
ans =

     9     8     7
     6     5     4
     3     2     1


>> A = [1 2 3;4 5 6;7 8 9] % recall data
A =

     1     2     3
     4     5     6
     7     8     9
>> circshift(A,1) % circularly shift rows down by 1
ans =

     7     8     9
     1     2     3
     4     5     6
>> circshift(A,[0 1]) % circularly shift columns right by 1
ans =

     3     1     2
     6     4     5
     9     7     8
>> circshift(A,[-1 1]) % shift rows up by 1 and columns right by 1
ans =

     6     4     5
     9     7     8
     3     1     2
```

```
>> B = 1:12   % more data
B =
     1     2     3     4     5     6     7     8     9    10    11    12
>> reshape(B,2,6)  % reshape to 2 rows, 6 columns, fill by columns
ans =
     1     3     5     7     9    11
     2     4     6     8    10    12

>> reshape(B,[2 6]) % equivalent to above
ans =
     1     3     5     7     9    11
     2     4     6     8    10    12
>> reshape(B,3,4)  % reshape to 3 rows, 4 columns, fill by columns
ans =
     1     4     7    10
     2     5     8    11
     3     6     9    12
>> reshape(B,3,[]) % MATLAB figures out how many columns are needed
ans =
     1     4     7    10
     2     5     8    11
     3     6     9    12

>> reshape(B,[],6) % MATLAB figures out how many rows are needed
ans =
     1     3     5     7     9    11
     2     4     6     8    10    12
>> reshape(A,3,2)  % A has more than 3*2 elements, OOPS!
??? Error using ==> reshape
To RESHAPE the number of elements must not change.
>> reshape(A,1,9)  % stretch A into a row vector
ans =
     1     4     7     2     5     8     3     6     9

>> A(:)' % convert to column and transpose; same as the above
```

```
ans =
    1    4    7    2    5    8    3    6    9

>> reshape(A,[],3) % MATLAB figures out how many rows are needed
ans =
    1    2    3
    4    5    6
    7    8    9
```

The following functions extract parts of an array to create another array:

```
>> A   % remember what A is
A =
    1    2    3
    4    5    6
    7    8    9
>> diag(A)   % extract diagonal using diag
ans =
    1
    5
    9
>> diag(ans)   % remember this? same function, different action
ans =
    1    0    0
    0    5    0
    0    0    9
>> triu(A)   % extract upper triangular part
ans =
    1    2    3
    0    5    6
    0    0    9
>> tril(A)   % extract lower triangular part
ans =
    1    0    0
    4    5    0
    7    8    9
```

```
>> tril(A) - diag(diag(A)) % lower triangular part with no diagonal
ans =
     0     0     0
     4     0     0
     7     8     0
```

The following functions create arrays from other arrays:

```
>> a = [1 2;3 4]  % a smaller data array
a =
     1     2
     3     4
>> b = [0 1;-1 0]  % another smaller data array
b =
     0     1
    -1     0
>> kron(a,b)  % the Kronecker tensor product of a and b
ans =
     0     1     0     2
    -1     0    -2     0
     0     3     0     4
    -3     0    -4     0
```

The preceding kron(a,b) is equivalent to

```
>> [1*b 2*b
    3*b 4*b]
ans =
     0     1     0     2
    -1     0    -2     0
     0     3     0     4
    -3     0    -4     0
```

Now consider

```
>> kron(b,a)  % the Kronecker tensor product of b and a
ans =
     0     0     1     2
```

```
    0      0      3      4
   -1     -2      0      0
   -3     -4      0      0
```

The preceding kron(b,a) is equivalent to

```
>> [0*a 1*a
    -1*a 0*a]
ans =
    0      0      1      2
    0      0      3      4
   -1     -2      0      0
   -3     -4      0      0
```

So, kron(a,b) takes each element of its first argument, multiplies it by the second argument, and creates a block array.

One of the most useful array manipulation functions is repmat, which was introduced earlier:

```
>> a  % recall data
a =
    1      2
    3      4
>> repmat(a,1,3) % replicate a once down, 3 across
ans =
    1      2      1      2      1      2
    3      4      3      4      3      4
>> repmat(a,[1 3]) % equivalent to above
ans =
    1      2      1      2      1      2
    3      4      3      4      3      4

>> [a a a] % equivalent to above
ans =
    1      2      1      2      1      2
    3      4      3      4      3      4

>> repmat(a,2,2) % replicate a twice down, twice across
```

```
ans =
        1       2       1       2
        3       4       3       4
        1       2       1       2
        3       4       3       4
>> repmat(a,2) % same as repmat(a,2,2) and repmat(a,[2 2])
ans =
        1       2       1       2
        3       4       3       4
        1       2       1       2
        3       4       3       4
>> [a a; a a] % equivalent to above
ans =
        1       2       1       2
        3       4       3       4
        1       2       1       2
        3       4       3       4
```

As illustrated, the functions `repmat` and `reshape` accept indexing arguments in two ways. The indexing arguments can be passed as separate input arguments, or they can be passed as individual elements in a single-vector argument. In addition, for `repmat`, a single second-input argument `repmat(A,n)` is interpreted as `repmat(A,[n n])`.

Finally, to replicate a scalar to create an array that is the same size as another array, one can simply use `repmat(d,size(A))`:

```
>> A = reshape(1:12,[3 4]) % new data
A =
        1       4       7      10
        2       5       8      11
        3       6       9      12
>> repmat(pi,size(A)) % pi replicated to be the size of A
ans =
        3.1416        3.1416        3.1416        3.1416
        3.1416        3.1416        3.1416        3.1416
        3.1416        3.1416        3.1416        3.1416
```

(Note that the function `size` is discussed in the next section.)

5.12 ARRAY SIZE

In cases where the size of an array or vector is unknown, and is needed for some mathematical manipulation, MATLAB provides the utility functions size, length, and numel:

```
>> A = [1 2 3 4;5 6 7 8]
A =
        1       2       3       4
        5       6       7       8
>> s = size(A)
s =
        2       4
```

With one output argument, the size function returns a row vector whose first element is the number of rows, and whose second element is the number of columns:

```
>> [r,c] = size(A)
r =
        2
c =
        4
```

With two output arguments, size returns the number of rows in the first variable and the number of columns in the second variable:

```
>> r = size(A,1)   % number of rows
r =
        2
>> c = size(A,2)   % number of columns
c =
        4
```

Called with two input arguments, size returns either the number of rows or columns. The correspondence between the second argument to size and the number returned follows the order in which array elements are indexed. That is, A(r,c) has its row index r specified first and hence size(A,1) returns the number of rows. Likewise, A(r,c) has its column index c specified second and so size(A,2) returns the number of columns.

The function `numel` returns the total number of elements in an array:

```
>> numel(A)
ans =
     8
```

The function `length` returns the number of elements along the largest dimension. For example, the code

```
>> length(A)
ans =
     4
```

returns the number of rows or the number of columns, whichever is larger. For vectors, `length` returns the vector length:

```
>> B = -3:3
B =
    -3    -2    -1     0     1     2     3
>> length(B) % length of a row vector
ans =
     7
>> length(B') % length of a column vector
ans =
     7
```

The functions `size` and `length` also work for an array of zero dimension:

```
>> c = [] % you can create an empty variable!
C =
     []
>> size(c)
ans =
     0     0
>> d = zeros(3,0) % an array with one dimension nonzero!
d =
   Empty matrix: 3-by-0
>> size(d)
ans =
     3     0
```

```
>> length(d)
ans =
     0
>> max(size(d)) % maximum of elements of size(d)
ans =
     3
```

As shown, MATLAB allows arrays to have one zero and one nonzero dimension. For these arrays, length and maximum dimension are not the same.

These array size concepts are summarized in the following table:

Array Size	Description
s=size(A)	Returns a row vector s whose first element is the number of rows in A, and whose second element is the number of columns in A
[r,c]=size(A)	Returns two scalars, r and c, containing the number of rows and columns, respectively
r=size(A,1)	Returns the number of rows in A
c=size(A,2)	Returns the number of columns in A
n=length(A)	Returns max(size(A)) for nonempty A, 0 when A has either zero rows or zero columns and the length of A if A is a vector
n=max(size(A))	Returns length(A) for nonempty A, and for empty A returns the length of any nonzero dimension of A
n-numel(A)	Returns the total number of elements in A

5.13 ARRAYS AND MEMORY UTILIZATION

In most modern computers, data transfer to and from memory is more time consuming than floating-point arithmetic, which has been fully integrated into most processors. Memory speed just hasn't kept pace with processor speed, and this has forced computer manufacturers to incorporate multiple levels of memory cache into their products in an attempt to keep processors supplied with data. In addition, computer users work with increasingly larger data sets (variables), which can easily exceed cache capacity. Consequently, efficient memory utilization is critical to effective computing.

MATLAB itself does not perform any explicit memory management. Memory allocation and deallocation within MATLAB use calls to standard C functions (malloc, calloc, free). Therefore, MATLAB relies on the compiler's implementation of these library functions to take appropriate, efficient, system-specific steps for memory allocation and deallocation. Given that there is an inherent tradeoff between memory use and execution speed, MATLAB purposely chooses to use more memory when doing so increases execution speed. Since MATLAB uses memory to gain performance, it is beneficial to consider how memory is allocated in the program and what can be done to minimize memory overuse and fragmentation.

When a variable is created with an assignment statement such as

```
>> P = zeros(100);
```

MATLAB requests a contiguous chunk of memory to store the variable, letting the compiler and the operating system determine where that chunk is allocated. If a variable is reassigned, as in

```
>> P = rand(5,6);
```

the original memory is deallocated, and a new allocation request is made for the new variable in its new size. Once again, it is up to the compiler and the operating system to figure out how to implement these tasks. Clearly, when the reassigned variable is larger than the original, a different chunk of contiguous memory is required to store the new data.

In a case where the variable reassignment just so happens to use exactly the same amount of memory, MATLAB still goes through the memory allocation and deallocation process:

```
>> P = zeros(5,6); % same size as earlier
>> P = ones(6,5);  % same number of elements as earlier
```

As a result, reusing variable names does not eliminate memory allocation or deallocation overhead. However, it does clear the memory used by prior data that are no longer needed. The exception to this reallocation is with scalars. For example, a = 1 followed by a = 2 simply copies the new value into the old memory location.

On the other hand, if an assignment statement addresses indices that already exist in a variable, the associated memory locations are updated, and no allocation or deallocation overhead is incurred. For example,

```
>> P(3,3) = 1;
```

does not force any memory-allocation calls, because P(3,3) already exists. The variable size remains the same, and hence, there's no need to find a different contiguous chunk of memory to store it. However, if an assignment statement increases the size of a variable, as in the code

```
>> P(8,1) = 1;
>> size(P)
ans =
     8     5
```

which makes P grow from 6-by-5 or 30 elements to 8-by-5 or 40 elements, MATLAB requests memory for the revised variable, copies the old variable into the new memory, and then deallocates the old memory chunk, thereby incurring allocation or deallocation overhead.

When all the indices of a variable are addressed in an assignment statement, such as

```
>> P(:) = ones(1,40);
```

the number of elements on the left- and right-hand sides must be equal. When this is true, MATLAB simply copies the data on the right-hand side into the memory that already exists on the left-hand side. No allocation or deallocation overhead is incurred. Furthermore, the dimensions of P remain unchanged:

```
>> size(P)
ans =
    8    5
```

In MATLAB, variables can be declared global in scope by using the global command. (This is discussed elsewhere in the text.) From a memory-management point of view, global variables do not behave any differently than ordinary variables. So there is neither benefit from, nor penalty for, using them.

One interesting thing that MATLAB does to improve performance is a feature called ***delayed copy***.

For example, in the code

```
>> A = zeros(10);
>> B = zeros(10);
>> C = B;
```

the variables A and B each have memory for 100 elements allocated to them. However, the variable C does not have any memory allocated to it. It shares the memory allocated to B. Copying of the data in B to a chunk of memory allocated to C does not take place until either B or C is modified. That is, a simple assignment, such as that shown does *not* immediately create a copy of the right-hand-side array in the left-hand-side variable. When an array is large, it is advantageous to delay the copy. That is, future references to C simply access the associated contents of B, and so to the user it appears that C is equal to B. Time is taken to copy the array B into the variable C only if the contents of B are about to change, or if the contents of C are about to be assigned new values by some MATLAB statement. While the time saved by this delayed-copy feature is insignificant for smaller arrays, it can lead to significant performance improvements for very large arrays.

The delayed-copy scheme applies to functions as well. When a function is called, for example, myfunc(a,b,c), the arrays a, b, and c are not copied into the workspace of the function, unless the function modifies them in some way. By implementing a delayed copy, memory-allocation overhead is avoided, unless the function modifies the variable. As a result, if that function only reads data from the array, there is no performance penalty for passing a large array to a function. (This feature is covered in the chapter on M-file functions as well.)

When you call a function, it is common to pass the result of a computation or a function output directly to another function. For example, in the code

```
>> prod(size(A))
ans =
   100
```

the results from the function size(A) are passed directly to the function prod, without explicitly storing the result in a named variable. The memory used to execute this statement is identical to

```
>> tmp=size(A);
>> prod(tmp)
ans =
   100
```

In other words, even though the first statement did not explicitly create a variable to store the results from size(A), MATLAB created an implicit variable and then passed it to the function prod. The only benefit gained by the first approach is that the implicit variable is automatically cleared after the function call.

In many applications, it is convenient to increase the size of an array as part of a computational procedure. Usually, this procedure is part of some looping structure, but for now, let's consider a simpler example:

```
>> A=1:5
A =
   1   2   3   4   5
>> B=6:10;

>> A=[A;B]
A =
   1   2   3   4   5
   6   7   8   9   10
>> C=11:15;

>> A=[A;C]
A =
```

1	2	3	4	5
6	7	8	9	10
11	12	13	14	15

Every time the variable A is reassigned, new memory is allocated, old data is copied, and old memory is deallocated. If the arrays involved are large or the reassignment occurs numerous times, then memory overhead can significantly reduce algorithm speed. To alleviate the problem, you should **preallocate** all the memory required and then fill it as required. Doing so for the preceding example gives

```
>> A=zeros(3,5)  % grab all required memory up front
A =
     0    0    0    0    0
     0    0    0    0    0
     0    0    0    0    0
>> A(1,:)=1:5;   % no memory allocation here

>> B=6:10;
>> A(2,:)=B;     % no memory allocation here
>> C=11:15;
>> A(3,:)=C      % no memory allocation here
A =
     1    2    3    4    5
     6    7    8    9   10
    11   12   13   14   15
```

While this example is somewhat silly, it serves to illustrate that the memory-allocation process for the variable A is performed only once, rather than once at every reassignment of A.

The following table summarizes the facts discussed in this section:

Syntax	Description
P=zeros(100); P=rand(5,6);	Reassignment of a variable incurs memory allocation and deallocation overhead.
P(3,3)=1;	If the indices on the left exist, no memory allocation and deallocation is performed.
P(8,1)=1;	If the indices on the left do not exist, memory allocation and deallocation occurs.

`P=zeros(5,6);`	Reassignment of a variable incurs memory allocation and deallocation overhead, even if the reassignment does not change the number of elements involved.
`P(:)=rand(1,30)`	If P exists, contents from the right-hand side are copied into the memory allocated on the left. No memory allocation and deallocation is performed.
`B=zeros(10);` `C=B;`	Delayed copy. C and B share the memory allocated to B until either B or C is modified.

`prod(size(A))`
 	
`tmp=size(A);`
`prod(tmp)`
	Implicit and explicit variables require the same memory allocation. However, the implicit variable is automatically cleared.

`myfunc(a,b,c)`	Delayed copy. The variables a, b, and c are not copied into the function workspace, unless the function modifies them. No memory allocation and deallocation overhead is incurred if the variables are not modified within the function.

`A=zeros(3,5);`
`A(1,:)=1:5;`
`A(2,:)=6:10;`
`A(3,:)=11:15;`
	Preallocate all memory for a variable that grows as an algorithm progresses. Memory allocation and deallocation overhead is incurred only once, rather than once per iteration or reassignment.

6

Multidimensional Arrays

In the previous chapter, 1- and 2-D arrays and their manipulation were illustrated. Since MATLAB Version 5 appeared several years ago, MATLAB has added support for arrays of arbitrary dimensions. For the most part, MATLAB supports multidimensional arrays (i.e., *n*-D arrays) by using the same functions and addressing techniques that apply to 1- and 2-D arrays. In general, the third dimension is numbered by **pages**, while higher dimensions have no generic name. Thus, a 3-D array has rows, columns, and pages. Each page contains a 2-D array of rows and columns. In addition, just as all of the columns of a 2-D array must have the same number of rows and vice versa, all of the pages of a 3-D array must have the same number of rows and columns. One way to visualize 3-D arrays is to think of the residential listings (white pages) in a phone book. Each page has the same number of columns and the same number of names (rows) in each column. The stack of all of the pages forms a 3-D array of names and phone numbers.

Even though there is no limit to the number of dimensions, 3-D arrays are predominately used as examples in this chapter, because they are easily visualized and displayed.

6.1 ARRAY CONSTRUCTION

Multidimensional arrays can be created in several ways:

```
>> A = zeros(4,3,2)
A(:,:,1) =
  0    0    0
```

```
     0       0       0
     0       0       0
     0       0       0
A(:,:,2) =
     0       0       0
     0       0       0
     0       0       0
     0       0       0
```

This is an array of zeros, having four rows, three columns, and two pages. The first page is displayed first, followed by the second page. The other common array-generation functions ones, rand, and randn work the same way, simply by adding dimensions to the input arguments.

Direct indexing also works, as in the following example:

```
>> A = zeros(2,3)  % start with a 2-D array
A =
     0       0       0
     0       0       0
>> A(:,:,2) = ones(2,3)   % add a second page to go 3-D!
A(:,:,1) =
     0       0       0
     0       0       0
A(:,:,2) =
     1       1       1
     1       1       1
>> A(:,:,3) = 4   % add a third page by scalar expansion
A(:,:,1) =
     0       0       0
     0       0       0
A(:,:,2) =
     1       1       1
     1       1       1
A(:,:,3) =
     4       4       4
     4       4       4
```

This approach starts with a 2-D array, which is the first page of a 3-D array. Then, additional pages are added by straightforward array addressing.

The functions reshape and repmat can also be used to create *n*-D arrays. For example, in the code

```
>> B = reshape(A,2,9) % 2-D data, stack pages side-by-side
B =

     0    0    0    1    1    1    4    4    4
     0    0    0    1    1    1    4    4    4

>> B = [A(:,:,1) A(:,:,2) A(:,:,3)] % equivalent to above
B =

     0    0    0    1    1    1    4    4    4
     0    0    0    1    1    1    4    4    4

>> reshape(B,2,3,3) % recreate A
ans(:,:,1) =

     0    0    0
     0    0    0
ans(:,:,2) =

     1    1    1
     1    1    1
ans(:,:,3) =

     4    4    4
     4    4    4

>> reshape(B,[2 3 3]) % alternative to reshape(B,2,3,3)
ans(:,:,1) =

     0    0    0
     0    0    0
ans(:,:,2) =

     1    1    1
     1    1    1
ans(:,:,3) =

     4    4    4
     4    4    4
```

reshape can change any dimensional array into any other dimensional array.

The code

```
>> C = ones(2,3) % new data
C =
     1     1     1
     1     1     1
```

```
>> repmat(C,1,1,3) % this form not allowed above 2-D!
??? Error using ==> repmat
Too many input arguments.
```

```
>> repmat(C,[1 1 3]) % this is how to do it
ans(:,:,1) =
     1     1     1
     1     1     1
ans(:,:,2) =
     1     1     1
     1     1     1
ans(:,:,3) =
     1     1     1
     1     1     1
```

replicates C once in the *row* dimension, once in the *column* dimension, and three times in the *page* dimension.

The cat function creates *n*-D arrays from lower dimensional arrays:

```
>> a = zeros(2); % new data
>> b = ones(2);
>> c = repmat(2,2,2);
>> D = cat(3,a,b,c)  % conCATenate a,b,c along the 3rd dimension
D(:,:,1) =
     0     0
     0     0
D(:,:,2) =
     1     1
     1     1
```

```
D(:,:,3) =
      2     2
      2     2

>> D = cat(4,a,b,c)  % try the 4th dimension!
D(:,:,1,1) =
      0     0
      0     0
D(:,:,1,2) =
      1     1
      1     1
D(:,:,1,3) =
      2     2
      2     2

>> D(:,1,:,:)  % look at elements in column 1
ans(:,:,1,1) =
      0
      0
ans(:,:,1,2) =
      1
      1
ans(:,:,1,3) =
      2
      2

>> size(D)
ans =
      2     2     1     3
```

Note that D has two rows, two columns, one page, and three fourth-dimension parts.

6.2 ARRAY MATHEMATICS AND MANIPULATION

As additional dimensions are created, array mathematics and manipulation become more cumbersome. Scalar-array arithmetic remains straightforward, but array–array arithmetic requires that the two arrays have the same size in all dimensions. Since scalar-array and array–array arithmetic remains unchanged from the 2-D case presented in the previous chapter, further illustrations are not presented here.

MATLAB provides several functions for the manipulation of n-D arrays. The function squeeze eliminates *singleton dimensions*; that is, it eliminates dimensions of size 1. For example, in the code

```
>> E = squeeze(D)   % squeeze dimension 4 down to dimension 3
E(:,:,1) =
        0      0
        0      0
E(:,:,2) =
        1      1
        1      1
E(:,:,3) =
        2      2
        2      2
>> size(E)
ans =
        2      2      3
```

E contains the same data as D, but has two rows, two columns, and three pages.

How about a 3-D vector? This case is shown as follows:

```
>> v(1,1,:) = 1:6   % a vector along the page dimension
v(:,:,1) =
        1
v(:,:,2) =
        2
v(:,:,3) =
        3
v(:,:,4) =
        4
v(:,:,5) =
        5
v(:,:,6) =
        6
```

```
>> squeeze(v) % squeeze it into a column vector
ans =
     1
     2
     3
     4
     5
     6

>> v(:) % this always creates a column vector
ans =
     1
     2
     3
     4
     5
     6
```

The function reshape allows you to change the row, column, page, and higher order dimensions, without changing the total number of elements:

```
>> F = cat(3,2+zeros(2,4),ones(2,4),zeros(2,4)) % new 3-D array
F(:,:,1) =
     2     2     2     2
     2     2     2     2
F(:,:,2) =
     1     1     1     1
     1     1     1     1
F(:,:,3) =
     0     0     0     0
     0     0     0     0

>> G = reshape(F,[3 2 4]) % change it to 3 rows, 2 columns, 4 pages
G(:,:,1) =
     2     2
     2     2
     2     2
```

```
G(:,:,2) =
       2       1
       2       1
       1       1
G(:,:,3) =
       1       1
       1       0
       1       0
G(:,:,4) =
       0       0
       0       0
       0       0

>> H = reshape(F,[4 3 2]) % or 4 rows, 3 columns, 2 pages
H(:,:,1) =
       2       2       1
       2       2       1
       2       2       1
       2       2       1
H(:,:,2) =
       1       0       0
       1       0       0
       1       0       0
       1       0       0

>> K = reshape(F,2,12) % 2 rows, 12 columns, 1 page
K =
   2    2    2    2    1    1    1    1    0    0    0    0
   2    2    2    2    1    1    1    1    0    0    0    0
```

Reshaping can be confusing until you become comfortable with visualizing arrays in *n*-D space. In addition, some reshaping requests make more practical sense than others. For example, in the previously displayed code, G has little practical value, whereas K is much more practical, because it stacks the pages of F side by side as additional columns.

The reshaping process follows the same pattern as that for 2-D arrays. Data are gathered first by rows, followed by columns, then by pages, and so on into higher dimensions. That is, all the rows in the first column are gathered, then all the rows in the second column, and so on. Thus, when the first page has been gathered, MATLAB moves on to the second page and starts over with all the rows in the first column.

The order in which array elements are gathered is the order in which the functions sub2ind and ind2sub consider single-index addressing:

```
>> sub2ind(size(F),1,1,1) % 1st row, 1st column, 1st page is element 1
ans =
     1
>> sub2ind(size(F),1,2,1) % 1st row, 2nd column, 1st page is element 3
ans =
     3
>> sub2ind(size(F),1,2,3) % 1st element, 2nd column, 3rd page is element 19
ans =
    19

>> [r,c,p] = ind2sub(size(F),19) % inverse of above
r =
     1
c =
     2
p =
     3
```

The *n*-D equivalent to flipud and fliplr is flipdim:

```
>> M = reshape(1:18,2,3,3) % new data
M(:,:,1) =
     1     3     5
     2     4     6
M(:,:,2) =
     7     9    11
     8    10    12
M(:,:,3) =
    13    15    17
    14    16    18
```

```
>> flipdim(M,1) % flip row order
ans(:,:,1)
      2      4      6
      1      3      5
ans(:,:,2) =
      8     10     12
      7      9     11
ans(:,:,3) =
     14     16     18
     13     15     17
>> flipdim(M,2) % flip column order
ans(:,:,1) =
      5      3      1
      6      4      2
ans(:,:,2) =
     11      9      7
     12     10      8
ans(:,:,3) =
     17     15     13
     18     16     14


>> flipdim(M,3) % flip page order
ans(:,:,1) =
     13     15     17
     14     16     18
ans(:,:,2) =
      7      9     11
      8     10     12
ans(:,:,3) =
      1      3      5
      2      4      6
```

The function shiftdim shifts the dimensions of an array. That is, if an array has r rows, c columns, and p pages, a shift by one dimension creates an array with c rows, p columns, and r pages, as shown in the following example:

```
>> M  % recall data
M(:,:,1) =
      1      3      5
      2      4      6
M(:,:,2) =
      7      9     11
      8     10     12
M(:,:,3) =
     13     15     17
     14     16     18

>> shiftdim(M,1) % shift one dimension
ans(:,:,1) =
      1      7     13
      3      9     15
      5     11     17
ans(:,:,2) =
      2      8     14
      4     10     16
      6     12     18
```

Shifting dimensions by 1 causes the first row on page 1 to become the first column on page 1, the second row on page 1 to become the first column on page 2, and so on.
     In the code

```
>> shiftdim(M,2) % shift two dimensions
ans(:,:,1) =
      1      2
      7      8
     13     14
ans(:,:,2) =
      3      4
      9     10
     15     16
ans(:,:,3) =
      5      6
     11     12
     17     18
```

the first column on page 1 of M becomes the first row on page 1, the first column on page 2 becomes the second row on page 1, and so on. If you are like the authors of this text, shifting dimensions is not immediately intuitive. For the 3-D case, it helps if you visualize M forming a rectangular box with page 1 on the front, followed by page 2 behind page 1, and then page 3 forming the back of the box. Hence, shifting dimensions is equivalent to rotating the box so that a different side faces you.

The function `shiftdim` also accepts negative shifts. In this case, the array is pushed into higher dimensions, leaving singleton dimensions behind, as in this example:

```
>> M % recall data
M(:,:,1) =
        1       3       5
        2       4       6
M(:,:,2) =
        7       9      11
        8      10      12
M(:,:,3) =
       13      15      17
       14      16      18
>> size(M) % M has 2 rows, 3 columns, and 3 pages
ans =
        2       3       3

>> shiftdim(M,-1) % shift dimensions out by 1
ans(:,:,1,1) =
        1       2
ans(:,:,2,1) =
        3       4
ans(:,:,3,1) =
        5       6
ans(:,:,1,2) =
        7       8
ans(:,:,2,2) =
        9      10
ans(:,:,3,2) =
       11      12
```

```
ans(:,:,1,3) =
   13      14
ans(:,:,2,3) =
   15      16
ans(:,:,3,3) =
   17      18
>> size(ans)
ans =
    1     2     3     3
```

The result now has four dimensions. The reader is left to figure out the correspondence between the original data and the shifted result.

In 2-D arrays, the transpose operator swapped rows and columns, converting an r-by-c array into a c-by-r array. The functions permute and ipermute are the *n*-D equivalents of the transpose operator. By itself, permute is a generalization of the function shiftdim. In the following code example:

```
>> M  % recall data
M(:,:,1) =
     1      3      5
     2      4      6
M(:,:,2) =
     7      9     11
     8     10     12
M(:,:,3) =
    13     15     17
    14     16     18

>> permute(M,[2 3 1]) % same as shiftdim(M,1)
ans(:,:,1) =
     1      7     13
     3      9     15
     5     11     17
ans(:,:,2) =
     2      8     14
     4     10     16
     6     12     18
```

```
>> shiftdim(M,1)
ans(:,:,1) =
       1        7       13
       3        9       15
       5       11       17
ans(:,:,2) =
       2        8       14
       4       10       16
       6       12       18
```

[2 3 1] instructs the function to make the second dimension the first, the third dimension the second, and the first dimension the third.

Now, in the simpler code

```
>> permute(M,[2 1 3])
ans(:,:,1) =
       1        2
       3        4
       5        6
ans(:,:,2) =
       7        8
       9       10
      11       12
ans(:,:,3) =
      13       14
      15       16
      17       18
```

[2 1 3] instructs permute to transpose the rows and columns, but to leave the third dimension alone. As a result, each page in the result is a conventional transpose of the original data.

The second argument to permute, called ORDER, must be a permutation of the dimensions of the array, passed as the first argument; otherwise, the requested permutation doesn't make sense, as shown in the following example:

```
>> permute(M,[2 1 1])
??? Error using ==> permute
ORDER cannot contain repeated permutation indices.
```

```
>> permute(M,[2 1 4])
??? Error using ==> permute
ORDER contains an invalid permutation index.
```

The function `permute` can also be used to push an array into higher dimensions. For example, `shiftdim(M,-1)`, shown earlier, is equivalent to

```
>> permute(M,[4 1 2 3])
ans(:,:,1,1) =
      1     2
ans(:,:,2,1) =
      3     4
ans(:,:,3,1) =
      5     6
ans(:,:,1,2) =
      7     8
ans(:,:,2,2) =
      9    10
ans(:,:,3,2) =
     11    12
ans(:,:,1,3) =
     13    14
ans(:,:,2,3) =
     15    16
ans(:,:,3,3) =
     17    18
```

An array always has a unit dimension beyond its size; for example, a 2-D array has one page. That is, all dimensions past the nonunity size of an array are singletons. As a result, in the preceding example, the singleton fourth dimension in M is made the first dimension of the result shown.

For 2-D arrays, issuing the transpose operator a second time returns the array to its original form. Because of the added generality of *n*-D arrays, the function `ipermute` is used to undo the actions performed by `permute`:

```
>> M  % recall data
M(:,:,1) =
```

```
              1        3       5
              2        4       6
M(:,:,2) =
              7        9      11
              8       10      12
M(:,:,3) =
             13       15      17
             14       16      18
>> permute(M,[3 2 1]) % sample permutation
ans(:,:,1) =
              1        3       5
              7        9      11
             13       15      17
ans(:,:,2) =
              2        4       6
              8       10      12
             14       16      18
>> ipermute(ans,[3 2 1]) % back to original data
ans(:,:,1) =
              1        3       5
              2        4       6
ans(:,:,2) =
              7        9      11
              8       10      12
ans(:,:,3) =
             13       15      17
             14       16      18
```

6.3 ARRAY SIZE

As demonstrated in the prior chapter and earlier in this chapter, the function size returns the size of an array along each of its dimensions. The functionality of size is unchanged from the features demonstrated in the last chapter. In addition, the function numel remains unchanged. Consider the following example:

```
>> size(M)  % return array of dimensions
ans =
    2    3    3
>> numel(M) % number of elements
ans =
    18
>> [r,c,p] = size(M) % return individual variables
r =
    2
c =
    3
p =
    3
>> r = size(M,1) % return just rows
ans =
    2
>> c = size(M,2) % return just columns
c =
    3
>> p = size(M,3) % return just pages
p =
    3
>> v = size(M,4) % default for all higher dimensions
v =
    1
```

When the number of dimensions is unknown or variable, the function ndims is useful:

```
>> ndims(M)
ans =
    3
>> ndims(M(:,:,1)) % just the 2-D first page of M
ans =
    2
```

In this last example, M(:,:,1) is a 2-D array, because it has only one page. That is, it has a singleton third dimension. The function ndims is equivalent to the following simple code fragment:

```
>> length(size(M))
ans =
    3
```

The following table summarizes the functions illustrated in this chapter:

n-D Function	Description
ones(r,c, ...) zeros(r,c, ...) rand(r,c, ...) randn(r,c, ...)	Basic *n*-D array creation
reshape(B,2,3,3) reshape(B,[2 3 3])	Reshape array into arbitrary dimensions
repmat(C,[1 1 3])	Replicate array into arbitrary dimensions
cat(3,a,b,c)	Concatenate array along a specified dimension
squeeze(D)	Eliminate dimensions of size equal to 1 (i.e., singleton) dimensions
sub2ind(size(F),1,1,1) [r,c,p]=ind2sub(size(F),19)	Subscript to single-index conversion, and single-index to subscript conversion
flipdim(M,1)	Flip order along a given dimension; *n*-D equivalent to flipud and fliplr
shiftdim(M,2)	Shift dimensions; circular shift for positive second argument; push-out for negative second argument
permute(M,[2 1 3]) ipermute(M,[2 1 3])	Arbitrary permutation and inverse of dimensions; generalization of transpose operator to *n*-D arrays
size(M) [r,c,p]=size(M) r=size(M,1) c=size(M,2) p=size(M,3)	Size of *n*-D array along its dimensions
ndims(M)	Number of dimensions in an array
numel(M)	Number of elements in an array

7

Numeric Data Types

In the preceding chapters, numeric variables were real or complex arrays containing values stored in double precision. Historically, all data in MATLAB was stored in double-precision format. Character strings and logical data were stored as double-precision, 8 byte, real arrays. Needless to say, this resulted in very inefficient memory usage. Character strings need, at most, two bytes per character, and logical arrays require only one bit per element to distinguish between True and False.

Over time, these storage inefficiencies were eliminated. First, character strings became a separate *data type* or *variable class* and were changed to a two-bytes-per-character representation. More recently, logical arrays became a separate data type and were changed to a one-byte-per-value representation. Most recently, single-precision data types and a variety of signed and unsigned integer data types were introduced.

Prior to MATLAB 7, arithmetic operations on single-precision and integer data were undefined. However, sorting, searching, logical comparisons, and array manipulation were supported. To perform arithmetic operations on these data types, it was necessary to convert the data to double precision before performing the desired operation. Then, if desired, the result could be converted back to the original data type. With the release of MATLAB Version 7, most operations on these data types are performed natively, without explicit conversions.

7.1 INTEGER DATA TYPES

MATLAB supports signed and unsigned integer data types having 8-, 16-, 32-, and 64-bit lengths. These data types are summarized in the following table:

Data Type	Description
uint8	Unsigned 8-bit integer in the range 0 to 255 (or 0 to 2^8)
int8	Signed 8-bit integer in the range -128 to 127 (or -2^7 to $2^7 - 1$)
uint16	Unsigned 16-bit integer in the range 0 to 65,535 (or 0 to 2^{16})
int16	Signed 16-bit integer in the range $-32,768$ to 32,767 (or -2^{15} to $2^{15} - 1$)
uint32	Unsigned 32-bit integer in the range 0 to 4,294,967,295 (or 0 to 2^{32})
int32	Signed 32-bit integer in the range $-2,147,483,648$ to 2,147,483,647 (or -2^{31} to $2^{31} - 1$)
uint64	Unsigned 64-bit integer in the range 0 to 18,446,744,073,707,551,615 (or 0 to 2^{64})
int64	Signed 64-bit integer in the range $-9,223,372,036,854,775,808$ to 9,223,372,036,854,775,807, (or -2^{63} to $2^{63} - 1$)

With the exception of the range of definition, each of these integer data types has the same properties. The upper and lower limits of their ranges are given by the intmax and intmin functions, as shown in the following example:

```
>> intmax('int8')
ans =
   127
>> intmin('uint32')
ans =
   0
```

Variables containing integer data can be created in a number of ways. When an array of zeros or ones is desired, the functions zeros and ones can be used:

```
>> m = zeros(1,6,'int8') % specify data type as last argument
m =
   0   0   0   0   0   0
>> class(m) % confirm class of result
ans =
int8
>> n = ones(4,'uint16') % again specify data type as last argument
n =
   1      1      1      1
   1      1      1      1
```

```
    1       1       1       1
    1       1       1       1
```

```
>> class(n) % confirm class of result
ans =
uint16
```

For other values, one must convert or cast results into the desired data type:

```
>> k = 1:7 % create default double precision
k =
    1    2    3    4    5    6    7
>> class(k)
ans =
double
>> kk = uint8(k) % convert using uint8 function
kk =
    1    2    3    4    5    6    7
>> class(kk)
ans =
uint8
>> kkk = cast(k,'uint8') % use more general cast function
kkk =
    1    2    3    4    5    6    7
>> class(kkk)
ans =
uint8
```

Once a variable of a given data type exists, it retains that data type even if other data are inserted into it:

```
>> kkk(3:5) = ones(1,3) % insert double precision values
kkk =
    1    2    1    1    1    6    7

>> class(kkk) % class remains unchanged
ans =
uint8
```

```
>> kkk(5:7) = zeros(1,3,'uint16') % insert uint16 data
kkk =
    1    2    1    1    0    0    0

>> class(kkk) % class remains unchanged
ans =
uint8

>> kk(1:2:end) = pi % insert a noninteger value!
Kkk =
    3    2    3    1    3    0    3
>> class(kkk) % class remains unchanged
ans =
uint8
```

Note that, when a noninteger is inserted, it is first rounded to the nearest integer and then inserted into the array.

Mathematical operations on integer data types of the same kind are defined as in the following code:

```
>> k = int8(1:7) % create new data
k =
    1    2    3    4    5    6    7
>> m = int8(randperm(7)) % more new data
m =
    7    2    3    6    4    1    5

>> k+m % addition
ans =
    8    4    6   10    9    7   12
>> k-m % subtraction
ans =
   -6    0    0   -2    1    5    2
>> k.*m % element by element multiplication
ans =
    7    4    9   24   20    6   35
```

```
>> k./m % element by element division
ans =
     0    1    1    1    1    6    1

>> k % recall data
k =
     1    2    3    4    5    6    7

>> k/k(2)
ans =
     1    1    2    2    3    3    4
```

Addition, subtraction, and multiplication are straightforward. However, in many cases, integer division does not produce an integer result. As shown, MATLAB performs the integer division as if the arrays were double precision and then rounds the result to the nearest integer.

> When dividing integers, MATLAB effectively performs the operation in double precision, rounds the result to the nearest integer, and converts the result back to the integer data type involved.

Mathematical operations between variables of different integer data types are not defined. However, mathematical operations between a double-precision scalar and an integer data type implicitly convert the double-precision scalar to the corresponding integer type and then perform the requested operation, as the following code illustrates:

```
>> m % recall data and data type
m =
     7    2    3    6    4    1    5
>> class(m)
ans =
int8
>> n = cast(k,'uint16') % new data of type uint16
n =
     1    2    3    4    5    6    7
>> m+n % try mixed type addition
??? Error using ==> plus
```

Integers can only be combined with integers of the same class, or scalar doubles.

```
>> n+3 % try adding default double precision constant 3
ans =
     4     5     6     7     8     9    10
>> class(ans)
ans =
uint16

>> n-(1:7) % try nonscalar double precision subtraction
??? Error using ==> minus
```

Integers can only be combined with integers of the same class, or scalar doubles.

> MATLAB supports mixed mathematical operations between a **scalar** double-precision value and an integer data type, but does not support operations between an array of double-precision values and an integer array.

Given the limited range of each integer data type, mathematical operations may produce results that exceed the data type's range. In this case, MATLAB implements saturation. That is, when the result of an operation exceeds that specified by intmin and intmax, the result becomes either intmin or intmax, depending on which limit is exceeded:

```
>> k = cast('hellothere','uint8') % convert a string to uint8
k =
  104   101   108   108   111   116   104   101   114   101
>> double(k)+150 % perform addition in double precision
ans =
  254   251   258   258   261   266   254   251   264   251
>> k+150 % perform addition in uint8, saturate at intmax('uint8')=255
ans =
  254   251   255   255   255   255   254   251   255   251
>> k-110 % perform subtraction in uint8, saturation at intmin('uint8')=0
ans =
    0     0     0     0     1     6     0     0     4     0
```

In sum, MATLAB supports a variety of integer data types. Except for the 64-bit types, these data types are more storage efficient than double-precision data.

Mathematical operations on identical integer data types produce results in the same data type. Operations between mixed data types are defined only between a double-precision scalar and the integer data types. Although not illustrated previously, the double-precision values inf and NaN do not exist for integer data types. In addition, mathematical operations on 64-bit integers are not defined in MATLAB 7, but may be in future versions.

7.2 FLOATING-POINT DATA TYPES

The default data type in MATLAB is double precision, or simply double. This float-ing-point data type conforms to the IEEE standard for double-precision arithmetic. MATLAB supports arrays containing single-precision data as a storage-saving alter-native. Mathematical operations on single-precision data are defined and perform similarly to the integer data types illustrated in the preceding section. As shown in the following example, the values of realmin, realmax, and eps reflect the reduced range and precision of single-precision data:

```
>> realmin('single')
ans =
    1.175494350822288e-038
>> realmax('single')
ans =
    3.402823466385289e+038
>> eps('single')
ans =
    1.192092895507813e-007

>> realmin('double') % compare to corresponding double values
ans =
    2.225073858507201e-308
>> realmax('double')
ans =
    1.797693134862316e+308
>> eps % same as eps(1) and eps('double')
ans =
    2.220446049250313e-016
```

The creation of single-precision data follows the approach used for integer data types:

```
>> a = zeros(1,5,'single') % specify data type as last argument
a =
       0     0     0     0     0
>> b = eye(3,'single') % specify data type as last argument
b =
       1     0     0
       0     1     0
       0     0     1
>> c = single(1:7) % convert default double precision to single
c =
       1     2     3     4     5     6     7
>> d = cast(6:-1:0,'single') % use more general cast function
d =
       6     5     4     3     2     1     0
```

Mathematical operations between single-precision data and between single- and double-precision data produce single-precision results:

```
>> c.^d % element by element exponentiation of singles
ans =
     1    32    81    64    25     6     1
>> c*pi % multiplication by a scalar double
ans =
    3.1416    6.2832    9.4248    12.566    15.708    18.85    21.991
>> d.*rand(size(d)) % element by element multiplication by a double array
ans =
    4.879    0.049307    0.55556    0.6083    0.39744    0.60379    0
>> class(ans)
ans =
single
```

> MATLAB supports mathematical operations between arrays of double- and single-precision arrays, returning single-precision results.

Single-precision data share the special floating-point values of inf and NaN that are well known in double precision:

```
>> c % recall data
c =
    1    2    3    4    5    6    7
>> c(1:2:end) = 0 % inserting double precision does not change data type
c =
    0    2    0    4    0    6    0
>> c./c % create 0/0 values
Warning: Divide by zero.
Ans =
   NaN    1   NaN    1   NaN    1   NaN
>> 1./c % create 1/0 values
Warning: Divide by zero.
Ans =
    Inf    0.5    Inf    0.25    Inf    0.16667    Inf
```

7.3 SUMMARY

The following table identifies functions that pertain to the numeric data types supported in MATLAB 7:

Function	Description
double	Double-precision data type creation and conversion
single	Single-precision data type creation and conversion
int8, int16, int32, int64	Signed integer data type creation and conversion
uint8, uint16, uint32, uint64	Unsigned integer data type creation and conversion
isnumeric	True for integer or floating-point data types
isinteger	True for integer data types
isfloat	True for single or double data types
isa(x,'type')	True if x has class 'type', including 'numeric', 'integer', and 'float'

Function	Description
`cast(x,'type')`	Cast x to class `'type'`
`intmax('type')`	Maximum integer value for class `'type'`
`intmin('type')`	Minimum integer value for class `'type'`
`realmax('type')`	Maximum floating-point real value for class `'type'`
`realmin('type')`	Minimum floating-point real value for class `'type'`
`eps('type')`	eps value for floating-point value for class `'type'`
`eps(x)`	Distance between x and next larger representable value of same type as x
`zeros(...,'type')`	Create array containing zeros for class `'type'`
`ones(...,'type')`	Create array containing ones for class `'type'`
`eye(...,'type')`	Create identity array for class `'type'`

8

Cell Arrays and Structures

MATLAB Version 5 introduced two container data types called **cell arrays** and **structures**. These data types allow the grouping of dissimilar, but related, arrays into a single variable. Data management then becomes easier, since groups of related data can be organized and accessed through a cell array or structure. Because cell arrays and structures are *containers* for other data types, mathematical operations on them are not defined. The *contents* of a cell or structure must be addressed to perform mathematical operations.

One way to visualize cell arrays is to imagine a collection of post office boxes covering a wall at the post office. The collection of boxes is the cell array, with each box being one cell in the cell array. The contents of each post office box are different, just as the contents of each cell in a cell array are different types or sizes of MATLAB data, such as character strings or numerical arrays of varying dimensions. Just as each post office box is identified by a number, each cell in a cell array is indexed by a number. When you send mail to a post office box, you identify the box number you want them put in. When you put data in a particular cell, you identify the cell number you want them put in. The number is also used to identify which box or cell to take data out of.

Structures are almost identical to cell arrays, except that the individual post office boxes (or data-storage locations) are not identified by number. Instead, they are identified by name. Continuing the post office box analogy, the collection of post office boxes is the structure, and each box is identified by its owner's name, rather than by a number. To send mail to a post office box, you identify the name of the box you want it put in. To place data in a particular structure element, you identify the name (i.e., field) of the structure element to put the data in.

8.1 CELL ARRAY CREATION

Cell arrays are MATLAB arrays whose elements are cells. Each cell in a cell array can hold any MATLAB data type, including numerical arrays, character strings, symbolic objects, other cell arrays, and structures. For example, one cell of a cell array might contain a numerical array; another, an array of character strings; and another, a vector of complex values. Cell arrays can be created with any number of dimensions, just as numerical arrays can. However, in most cases, cell arrays are created as a simple vector of cells.

Cell arrays can be created by using assignment statements or by preallocating the array with the `cell` function and then assigning data to the cells. If you have trouble with these examples, it is very likely that you have another variable of the same name in the workspace. If any of the examples that follow give unexpected results, clear the array from the workspace and try again.

If you assign a cell to an existing variable that is not a cell, MATLAB will stop and report an error.

Like other kinds of arrays, cell arrays can be built by assigning data to individual cells, one at a time. There are two different ways to access cells. If you use standard array syntax to index the array, you must enclose the cell contents in curly braces, {}. These curly braces construct a cell array. They act like a call to the `cell` function:

```
>> clear A  % make sure A is not being used
>> A(1,1) = { [1 2 3; 4 5 6; 7 8 9] };
>> A(1,2) = { 2+3i };                  % semicolons suppress display
>> A(2,1) = { 'A character string' };
>> A(2,2) = { 12:-2:0 } % no semicolon, so display requested
A =

            [3x3 double]    [2.0000+ 3.0000i]
    'A character string'            [1x7 double]
```

Note that MATLAB shows that A is a 2-by-2 cell array, but it does not show the contents of all cells. Basically, MATLAB shows the cell contents if they do not take up much space, but only describes the cell contents if they do take up significant space. The curly braces on the right side of the equal sign indicate that the right-hand side expression returns a cell, rather than numerical values. This is called *cell indexing*. Alternatively, the following statements create the same cell array:

```
>> A{1,1} = [1 2 3; 4 5 6; 7 8 9];
>> A{1,2} = 2+3i;
```

```
>> A{2,1} = 'A character string';
>> A{2,2} = 12:-2:0
A =
            [3x3 double]      [2.0000+ 3.0000i]
    'A character string'               [1x7 double]
```

Here, the curly braces appear on the left-hand side of the equal sign. The term A{1,1} indicates that A is a cell array and that the contents of the first row and first column of the cell array are on the right-hand side of the equal sign. This is called *content addressing*. Both methods can be used interchangeably.

As the example demonstrates, curly braces { } are used to access or specify the contents of cells, whereas parentheses ( ) are used to identify cells, but not their contents. To apply the post office box analogy, curly braces are used to look at the contents of post office boxes, whereas parentheses are used to identify post office boxes without looking inside at their contents.

The two commands A(i,j) = {x}; and A{i,j} = x; both tell MATLAB to store the content of variable x in the (i,j) element of the cell array A. The notation A(i,j) is called **cell indexing**, and the notation A{i,j} is called **content addressing**. That is, curly braces { } access the contents of cells, whereas parentheses ( ) identify cells without looking at their contents.

The celldisp function forces MATLAB to display the contents of cells in the usual manner:

```
>> celldisp(A)
A{1,1} =
     1     2     3
     4     5     6
     7     8     9
A{2,1} =
    A character string
A{1,2} =
    2.0000+   3.0000i
A{2,2} =
    12    10     8     6     4     2     0
```

If a cell array has many elements, celldisp can produce a lot of output to the *Command* window. As an alternative, requesting the contents of the cell by using content addressing displays the contents of a single cell. This is different from cell indexing, which identifies the cell, but not its contents:

```
>> A{2,2} % content addressing
ans =
    12    10     8     6     4     2     0
>> A(2,2) % cell indexing
ans =
    [1x7 double]

>> A{1,:} % address contents of the first row
ans =
     1     2     3
     4     5     6
     7     8     9
ans =
    2.0000 + 3.0000i
>> A(1,:)
ans =
    [3x3 double]    [2.0000+ 3.0000i]
```

Note that the contents of all of the cells shown are generically named ans, because the cells store data that do not have associated variable names.

Square brackets were used in previous chapters to create numerical arrays. Curly braces work the same way for cells. In the following example, commas separate columns, and semicolons separate rows:

```
>> B = { [1 2], 'John Smith'; 2+3i, 5 }
B =
         [1x2 double]    'John Smith'
    [2.0000+ 3.0000i]    [         5]
```

When dealing with numerical arrays, it is common to preallocate an array with zeros and then fill the array as needed. The same can be done with cell arrays. The cell function creates a cell array and fills it with empty numerical matrices as shown in the following []:

```
>> C = cell(2,3)
C =
    []    []    []
    []    []    []
```

Once the cell array has been defined, both cell indexing and content addressing can be used to populate the cells. In the code

```
>> C(1,1) = 'This doesn''t work'
??? Conversion to cell from char is not possible.
```

the left-hand side uses cell indexing. As a result, the right-hand side must be a cell, but it is not, because it lacks curly braces surrounding its contents. This problem can be resolved by addressing the cell correctly:

```
>> C(1,1) = { 'This does work' }
C =
    'This does work'      []      []
                          []      []      []
>> C{2,3} = 'This works too'
C =
    'This does work'      []                    []
                          []      []      'This works too'
```

Because curly braces appear on the left-hand side of the last statement, MATLAB makes the character string the contents of the addressed cell. Once again, this is an example of content addressing, whereas the prior statement is an example of cell indexing.

8.2 CELL ARRAY MANIPULATION

The numerical array manipulation techniques presented in preceding chapters apply to cell arrays as well. In a sense, cell array manipulation is just a natural extension of these techniques to a different type of array. If you assign data to a cell outside of the dimensions of the current cell array, MATLAB automatically expands the array and fills the intervening cells with the empty numerical array [].

Square brackets are used to combine cell arrays into larger cell arrays, just as they are used to construct larger numerical arrays:

```
>> A  % recall prior cell arrays
A =
            [3x3 double]     [2.0000+ 3.0000i]
    'A character string'             [1x7 double]
>> B
B =
            [1x2 double]     'John Smith'
    [2.0000+ 3.0000i]     [            5]
```

```
>> C = [A;B]
C =
              [3x3 double]      [2.0000+ 3.0000i]
      'A character string'              [1x7 double]
              [1x2 double]      'John Smith'
      [    2.0000+ 3.0000i]      [                5]
```

A subset of cells can be extracted to create a new cell array by conventional array addressing techniques:

```
>> D = C([1 3],:) % first and third rows
D =
      [3x3 double]      [2.0000+ 3.0000i]
      [1x2 double]      'John Smith'
```

An entire row or column of a cell array can be deleted by using the empty array:

```
>> C(3,:) = []
C =
              [3x3 double]      [2.0000+ 3.0000i]
      'A character string'              [1x7 double]
      [    2.0000+ 3.0000i]      [                5]
```

Note that curly braces do not appear in either of the preceding expressions, because we are working with the cell array itself, not the contents of the cells. Once again, curly braces are used to address the contents of cells, whereas parentheses are used to identify the cells without regard for their content.

The reshape function can be used to change the configuration of a cell array, but cannot be used to add or remove cells. For example, in the code

```
>> X = cell(3,4);
>> size(X) % size of the cell array, not the contents
ans =
    3    4

>> Y = reshape(X,6,2);
>> size(Y)
ans =
    6    2
```

So the function reshape naturally reshapes any array, without regard to its type. Similarly, the size function returns the size of any array type.

The function `repmat` also works, even though its name implies that it replicates matrices. The function `repmat` was created when matrices were MATLAB's only data type. If MATLAB had been written after the introduction of cell arrays and other array types, perhaps this function would have been called `reparray`. The following example is illustrative:

```
>> Y  % recall data
Y =
        []        []
        []        []
        []        []
        []        []
        []        []
        []        []
>> Z = repmat(Y,1,3)
Z =
        []        []        []        []        []        []
        []        []        []        []        []        []
        []        []        []        []        []        []
        []        []        []        []        []        []
        []        []        []        []        []        []
        []        []        []        []        []        []
```

8.3 RETRIEVING CELL ARRAY CONTENT

To retrieve the contents of a cell in a cell array, you must apply content addressing, which involves the use of curly braces. For instance, in the code

```
>> B  % recall cell array
B =
        [1x2 double]      'John Smith'
    [2.0000+ 3.0000i]    [          5]
>> x = B{2,2} % content addressing uses { }
x =
     5
```

the variable x now contains the numerical value 5, making x a numerical array (a scalar, in this case). The `class` function can be used to confirm that this is true, as in the following example:

```
>> class(x) % return argument's data type
ans =
      double
```

Because numerical arrays are double precision, `class` returns a character string that identifies x as being `double`. If cell indexing had been used mistakenly, the result would be different:

```
>> y = B(2,2)    % cell indexing uses ( )
y =
    [5]
>> y = B(4)      % same as above using single index
y =
    [5]

>> class(y)      % y is not a double, but a cell!
ans =
cell
>> class(y{1})  % but the contents of y is a double!
ans =
double
```

By now, you are either bored with this distinction between cell indexing and content addressing, or your head is spinning from the confusion.

There are also other functions for testing variable types. The following examples return logical results, where True = 1 and False = 0:

```
>> iscell(y)        % yes, y is a cell
ans =
     1
>> iscell(y{1})     % contents of y is NOT a cell
ans =
     0
>> isa(y,'cell')    % yes, y is a cell
ans =
     1
>> isdouble(y{1})  % this function doesn't exist (yet?)
??? Undefined function or variable 'isdouble'
```

```
>> isnumeric(y{1}) % contents of y is numerical
ans =
    1
>> isfloat(y{1}) % contents of y is floating point
ans =
    1

>> isa(y{1},'double')  % contents of y is a double
ans =
    1
>> isa(y{1},'numeric') % contents of y is also numeric
ans =
    1
>> isa(y{1},'cell')    % contents of y is NOT a cell
ans =
    0
```

While you can display the contents of more than one cell array at a time, it is not possible to assign more than one at a time to a variable by using a standard assignment statement. Consider the following code:

```
>> B{:,2}
ans -
John Smith
ans =
    5
>> d = B{:,2}
??? Illegal right hand side in assignment. Too many elements.
```

If you think about it, this error message makes sense. How can two pieces of data be assigned to a single variable? If the method applied in the preceding example did work, d would be a 2-by-1 array with a character string in its first location and a scalar in its second. Extracting data from multiple cells at one time simply cannot be done so casually.

When addressing the contents of a single cell, it is possible to further address a subset of the contents by simply appending the desired subscript range:

```
>> A  % recall prior data
A =
          [3x3 double]    [2.0000+ 3.0000i]
```

```
          'A character string'          [1x7 double]
>> celldisp(A) % display contents
A{1,1} =
       1     2     3
       4     5     6
       7     8     9
A{2,1} =
A character string
A{1,2} =
   2.0000 + 3.0000i
A{2,2} =
      12    10     8     6     4     2     0
>> A{1,1}(3,:)  % third row of 3-by-3 array
ans =
       7     8     9
>> A{4}(2:5)    % second through fifth elements of A{2,2}
ans =
      10     8     6     4
>> A{1,2}(2)    % second element doesn't exist
??? Index exceeds matrix dimensions.
>> A{2,1}(3:11) % extract part of the character string
ans =
character
```

8.4 COMMA-SEPARATED LISTS

To extract the contents of more than one cell at a time, MATLAB provides comma-separated list syntax. This syntax applies in any place where variables or constants appear in a list separated by commas. For example, comma-separated lists appear in array construction:

```
>> a = ones(2,3);
>> b = zeros(2,1);
>> c = (3:4)';
>> d = [a,b,c] % same as [a b c]
```

```
d =

    1    1    1    0    3
    1    1    1    0    4
```

They also appear in function input and output argument lists:

```
>> d = cat(2,a,b,c)
d =

    1    1    1    0    3
    1    1    1    0    4
>> [m,n] = size(d)

m =

  2

n =

  5
```

Comma-separated list syntax is implemented as follows: Placing a content-addressed cell array in locations where comma-separated lists appear causes MATLAB to extract the contents of the addressed cell arrays and place them sequentially, separated by commas. In the code

```
>> F = {a b c}  % create a cell array
F =

    [2x3 double]    [2x1 double]    [2x1 double]
>> d = cat(2,F{:}) % same as cat(2,a,b,c)
d =

    1    1    1    0    3
    1    1    1    0    4
>> d = cat(2,F(:)) % not content addressing
d =

    [2x3 double]
    [2x1 double]
    [2x1 double]
```

cat(2,F{:}) is interpreted by MATLAB as cat(2,F{1},F{2},F{3}), which is equal to cat(2,a,b,c). That is, comma-separated list syntax dictates that F{:} is interpreted as a listing of all addressed parts of F{:}, separated by commas. Note that content addressing must be used. Cell indexing, as shown in the second case in the preceding example, acts on the cells, not on their contents. As a result, there is no comma-separated list in the second case, `cat(2,F(:))`. Consider this example:

```
>> d = [F{:}]
d =
        1       1       1       0       3
        1       1       1       0       4
>> d = [F{1},F{2},F{3}] % what is implied by the above
d =
        1       1       1       0       3
        1       1       1       0       4
>> e = [F{2:3}] % can also content address any subset
e =
        0       3
        0       4
>> e = [F{2},F{3}] % what is implied by the above
e =
        0       3
        0       4
```

At first, comma-separated list syntax may seem strange. However, once you become familiar with it, you will recognize its power. (For more information, see the on-line help for `lists`, or search for the phrase `comma separated list`.)

Given comma-separated list syntax, the function `deal` provides a way to extract the contents of numerous cells into separate variables. If you have ever used playing cards, you know that "to deal" means to pass out the cards in some organized way. The function `deal` does just that; it passes out cell array contents to individual variables. In the code

```
>> celldisp(F) % recall data
F{1} =
        1       1       1
        1       1       1
F{2} =
        0
        0
F{3} =
        3
        4
```

```
>> [r,s,t] = deal(F{:}) % deal out contents of F
r =

     1     1     1
     1     1     1
s =

     0
     0
t =

     3
     4
```

the variables r, s, and t are numerical variables, with r = F{1}, s = F{2}, t = F{3}. Hence, this example is equal to

```
>> [r,s,t] = deal(F{1},F{2},F{3})
r =

     1     1     1
     1     1     1
s =

     0
     0
t =

     3
     4
```

So, the function deal is, in reality, pretty simple. It just assigns the contents of the first input argument to the first output argument, the second to the second, and so on. Despite its simplicity, deal is a very powerful tool for extracting data from multiple cells with one statement.

Because the output of deal is also a comma-separated list, deal can be used to assign the contents of multiple cells in one statement:

```
>> [G{:}] = deal(a,b,c)
??? Error using ==> deal
The number of outputs should match the number of inputs.

>> [G{1:3}] = deal(a,b,c)
G =

     [2x3 double]    [2x1 double]    [2x1 double]
```

```
>> F = {a b c}
F =
    [2x3 double]     [2x1 double]     [2x1 double]
>> isequal(F,G) % True since F = G
ans =
    1
```

The first statement produced an error, as G{:} doesn't identify how many cell array elements to separate by commas. The second statement identifies the three cell array elements needed. Since the syntax {a b c} produces the same cell array, using deal to populate multiple cells with one statement is unnecessary (and slower).

8.5 CELL FUNCTIONS

Besides the functions celldisp, cell, iscell, isa, and deal, there are several other functions that are useful when dealing with cell arrays. The function cellfun provides a way to apply certain functions to all cells in a cell array, thereby eliminating the need to apply them to each cell individually:

```
>> A  % recall data
A =
              [3x3 double]     [2.0000+ 3.0000i]
    'A character string'              [1x7 double]

>> cellfun('isreal',A) % True=1 where not complex
ans =
    1     0
    1     1
>> cellfun('length',A) % length of contents
ans =
    3     1
   18     7
>> cellfun('prodofsize',A) % number of elements in each cell
ans =
    9     1
   18     7
>> cellfun('isclass',A,'char') % True for character strings
```

```
ans =
        0    0
        1    0
```

(The function cellfun offers other functions as well. For more information, see on-line help.)

Another function that is sometimes useful is num2cell. This function takes an array of any type (not just numbers, as the function name suggests) and fills a cell array with its components:

```
>> a = rand(3,6) % new numerical data
a =
    0.7266    0.2679    0.6833    0.6288    0.6072    0.5751
    0.4120    0.4399    0.2126    0.1338    0.6299    0.4514
    0.7446    0.9334    0.8392    0.2071    0.3705    0.0439
>> c = num2cell(a)    % c{i,j}=a(i,j)
c =
    [0.7266]    [0.2679]    [0.6833]    [0.6288]    [0.6072]    [0.5751]
    [0.4120]    [0.4399]    [0.2126]    [0.1338]    [0.6299]    [0.4514]
    [0.7446]    [0.9334]    [0.8392]    [0.2071]    [0.3705]    [0.0439]
>> d = num2cell(a,1) % d{i}=a(:,i)
d =
  Columns 1 through 4
    [3x1 double]    [3x1 double]    [3x1 double]    [3x1 double]
  Columns 5 through 6
    [3x1 double]    [3x1 double]
>> e = num2cell(a,2)    % e{i}=a(i,:)
e =
    [1x6 double]
    [1x6 double]
    [1x6 double]
```

With numerical data as input, num2cell(a) isn't useful in many applications, but packing larger pieces of an array into cells is often more useful, as illustrated in the last two cases.

8.6 CELL ARRAYS OF STRINGS

Although a rigorous discussion of character strings doesn't occur until the next chapter, the use of character strings and their storage in cell arrays are described in this section. If the concept of a character string is foreign to you, read the next chapter and then come back to this material.

In MATLAB, character strings are formed by enclosing characters within quotes. For example, in the code

```
>> s = 'Rarely is the question asked: is our children learning?'

s =

Rarely is the question asked: is our children learning?
```

the variable s contains a string of characters. In many applications, groups of character strings are associated with each other. When this occurs, it is convenient to group them into a cell array rather than store them in individual variables. In the code

```
>> cs = {'My answer is bring them on.'

s

'I understand small business growth. I was one.'}

cs =

    'My answer is bring them on.'

    'Rarely is the question asked: is our children learning?'

    'I understand small business growth. I was one.'

>> size(cs)  % a column cell array
ans =
     3     1
>> iscell(cs) % yes, it is a cell array
ans =
     1
```

the cell array cs has three cells, each containing a character string. In MATLAB, this is simply called a ***cell array of strings***. Because cell arrays are commonly used to store sets of character strings, MATLAB provides several functions to support them, like the one in the following example:

```
>> iscellstr(cs)

ans =

  1
```

The function iscellstr returns logical True (1) if all the cells in its cell array argument contain character strings; otherwise, it returns logical False (0).

Before cell arrays were introduced in MATLAB 5, groups of character strings were stored in character string arrays—2-D arrays just like the numerical arrays discussed earlier—with each string occupying a separate row in the array. Each character in the *string array* occupies its own location and is indexed just like a numerical array.

MATLAB provides functions to convert a cell array of strings to a string array and vice versa:

```
>> cs  % recall previous cell array of strings
cs =
    'My answer is bring them on.'
    'Rarely is the question asked: is our children learning?'
    'I understand small business growth. I was one.'
>> sa = char(cs) % convert to a string array
sa =
My answer is bring them on.
Rarely is the question asked: is our children learning?
I understand small business growth. I was one.
>> ischar(sa) % True for string array
ans =
    1
>> iscell(sa) % True for cell array
ans =
    0
>> size(sa)    % size of string array
ans =
    3    34
>> size(cs)    % size of cell array
ans =
    3    1
>> cst = cellstr(sa) % convert back to cell array
cst =
    'My answer is bring them on.'
    'Rarely is the question asked: is our children learning?'
    'I understand small business growth. I was one.'
>> iscell(cst)        % True for cell array
```

```
ans =
     1
>> isequal(cs,cst)    % True for equal variables
ans =
     1
>> isequal(cs,sa)     % cell array not equal to string array
ans =
     0
```

So, the MATLAB functions `char` and `cellstr` are inverses of each other. In addition, since a string array must have the same number of columns in each row, blank spaces are added to rows, as necessary, to make the string array rectangular.

8.7 STRUCTURE CREATION

Structures are like cell arrays in that they allow you to group collections of dissimilar data into a single variable. However, instead of addressing elements by number, structure elements are addressed by names called *fields*. Like cell arrays, structures can have any number of dimensions, but a simple scalar or vector array is the most common type.

Whereas cell arrays use curly braces to access data, structures use dot notation to access data in fields. Creating a structure can be as simple as assigning data to individual fields. For example, in the code

```
>> circle.radius = 2.5;     % semicolon, no display
>> circle.center = [0 1];
>> circle.linestyle = '--';
>> circle.color = 'red'     % no semicolon, so display
circle =
        radius: 2.5
        center: [0 1]
     linestyle: '--'
         color: 'red'
```

the data are stored in a structure variable called `circle`. The case-sensitive fields are entitled `radius`, `center`, `linestyle`, and `color`. Structure field names have the same restrictions as variable names: They can contain up to 63 characters, and they must begin with a letter. For example, in the statement

```
>> size(circle)
ans =
     1     1
```

```
>> whos
  Name          Size          Bytes  Class

   ans          1x2              16  double array
   circle       1x1             530  struct array

Grand total is 14 elements using 546 bytes
```

circle is a scalar structure, since size says that it is a 1-by-1 array. The whos command shows its size, the fact that it is a structure array, and the fact that it uses 530 bytes of memory. If there is more than one circle, it can be stored as a second element in the circle variable. For example, in the code

```
>> circle(2).radius = 3.4;
>> circle(2).color = 'green';
>> circle(2).linestyle = ':';
>> circle(2).center = [2.3 -1.2]
circle =
1x2 struct array with fields:
    radius
    center
    linestyle
    color
```

circle is a structure array having two elements. The (2) appears immediately after the variable name, because it is the variable that is having an element added to it. The .fieldname suffix identifies the field where data is to be placed. Note that the structure fields are filled in a different order this time and that the size of the data differs between the two elements. That is, the color fields are 'red' and 'green'. There are no restrictions on what can be placed in fields from one array element to the next. For instance, in the MATLAB statement

```
>> circle(2).radius = 'sqrt(2)'
circle =
1x2 struct array with fields:
    radius
    center
    linestyle
    color
```

```
>> circle.radius  % display radius contents
ans =

        2.5

ans =
sqrt(2)
```

Here, `circle(1).radius` holds numerical data and `circle(2).radius` holds a character string.

  If the value of structures is not apparent to you, consider how the data would be stored without structures:

```
>> Cradius = [2.5 3.4]; % ignore sqrt(2) change above
>> Ccenter = [0 1 ; 2.3 -1.2];
>> Clinestyle = {'--' ':'};  % cell array of strings
>> Ccolor = {'red' 'green'};
```

  Now, rather than having a single variable storing the information for two circles, there are four variables that must be indexed properly to extract data for each circle. It's easy to add another circle to the structure, but more cumbersome to add them to the four variables. Now, consider the following example:

```
>> circle(3).radius = 25.4;
>> circle(3).center = [-1 0];
>> circle(3).linestyle = '-.';
>> circle(3).color = 'blue'    % third circle added
circle =
1x3 struct array with fields:
    radius
    center
    linestyle
    color
>> Cradius(3) = 25.4
Cradius =

        2.5            3.4            25.4
>> Ccenter(3,:) = [-1 0]
Ccenter =

        0              1
      2.3            -1.2
```

```
             -1                      0
>> Cradius(3) = 25.4
Cradius =
            2.5            3.4            25.4


>> Clinestyle{3} = '-.'
Clinestyle =
     '--'      ':'      '-.'
>> Ccolor(3) = {'blue'}
Ccolor =
   'red'     'green'     'blue'
```

The clarity provided by structures should be readily apparent. In addition, consider passing the circle data to a function. For example, as a structure, the data are passed simply as myfunc(circle), whereas the other approach requires myfunc(Cradius,Ccenter,Clinestyle,Ccolor).

Suppose that, at some later date, you wanted to add another field to circle, as in

```
>> circle(1).filled = 'yes'
circle =
1x3 struct array with fields:
    radius
    center
    linestyle
    color
    filled
>> circle.filled  % display all .filled fields
ans =
yes
ans =
      []
ans =
      []
```

Now, all of the elements of circle have the field filled. Those not assigned by default contain the empty array []. The other filled fields are easily assigned:

```
>> circle(2).filled = 'no';
>> circle(3).filled = 'yes';
>> circle.filled
ans =
yes
ans =
no
ans =
yes
```

When structure creation by direct assignment isn't possible, MATLAB provides the function struct. For example, to re-create the previous structure, the following code is used:

```
>> values1 = {2.5 'sqrt(2)' 25.4}; % cell arrays with field data
>> values2 = {[0 1] [2.3 -1.2] [-1 0]};
>> values3 = {'--' ':' '-.'};
>> values4 = {'red' 'green' 'blue'};
>> values5 = {'yes' 'no' 'yes'};
>> CIRCLE = struct('radius',values1,'center',values2,...
            'linestyle',values3,'color',values4,'filled',values5)
CIRCLE =
1x3 struct array with fields:
    radius
    center
    linestyle
    color
    filled
>> isequal(circle,CIRCLE) % True since structures are equal
ans =
     1
```

8.8 STRUCTURE MANIPULATION

Structures are arrays and therefore can be combined and indexed like numerical arrays and cell arrays. When combining structure arrays, the only restriction is that the arrays combined must share the same fields. Thus, in the code

```
>> square.width = 5;  % a new structure
>> square.height = 14;
>> square.center = zeros(1,2);
>> square.rotation = pi/4
square =
        width: 5
       height: 14
       center: [0 0]
     rotation: 0.7854
>> A = [circle CIRCLE]
A =
1x6 struct array with fields:
    radius
    center
    linestyle
    color
    filled
>> B = [circle square]
??? Number of fields does not match in [] concatenation.
```

the structures `circle` and `CIRCLE` are both 1 by 3 and share the same fields; hence, concatenating them produces a 1-by-6 structure array. However, the structure `square` and `circle` do not have the same fields and therefore cannot be concatenated.

It is also possible to address a subarray of a structure. For example, in the code

```
>> C = [circle(1:2) CIRCLE(3)]
C =
1x3 struct array with fields:
    radius
    center
    linestyle
    color
    filled
>> isequal(C,circle) % True since equal
ans =
    1
```

C and `circle` are equal because `CIRCLE(3) = circle(3)`. Once again, basic array addressing and concatenation apply to structure arrays as well. For concatenation, field names must match exactly.

Although they are not as useful in this case, structures can also be manipulated with the functions `reshape` and `repmat`:

```
>> Aa = reshape(A,3,2)
Aa =
3x2 struct array with fields:
    radius
    center
    linestyle
    color
    filled
>> Aaa = reshape(A,1,2,3)
Aaa =
1x2x3 struct array with fields:
    radius
    center
    linestyle
    color
    filled
```

Since there is seldom a practical reason to have anything other than a structure array with vector orientation, `reshape` is seldom needed or used with structures. However, it is often convenient to create a structure array with default data in all fields of all array elements. The function `repmat` performs this task with ease:

```
>> S = repmat(square,3,1)
S =
3x1 struct array with fields:
    width
    height
    center
    rotation
>> S.width  % look at all width fields
ans =
     5
```

```
ans =

      5

ans =

      5
```

All three elements of the structure S contain the data originally assigned to the structure `square`. At this point, the structure fields can be modified, as needed, to describe different squares.

8.9 RETRIEVING STRUCTURE CONTENT

When you know the names of the fields associated with a structure array, retrieving the data in a particular structure element and field simply requires identifying them. For example, in the code

```
>> rad2 = circle(2).radius
rad2 =
sqrt(2)

>> circle(1).radius
ans =
          2.5

>> area1 = pi*circle(1).radius^2
area1 =
       19.635
```

`circle(1).radius` identifies the value 2.5, which is used to compute the area of the first circle.

When the contents of a field are an array, it is also possible to retrieve a subset of that field by appending an array index to the structure request:

```
>> circle(1).filled         % the entire field
ans =
yes
>> circle(1).filled(1)      % first element of field
ans =
y
>> circle(1).filled(2:end) % rest of field
ans =
es
```

As is true for cell arrays, retrieving the contents of more than one structure array element and field cannot be accomplished by direct addressing. For example, the command

```
>> col = circle.color
??? Illegal right hand side in assignment. Too many elements.
```

attempted to extract three pieces of data and store them in one variable, but failed. However, MATLAB solves this problem the same way it does for cell arrays by using comma-separated lists and `deal`.

When the structure field to be accessed is stored in a character string, MAT-LAB provides *dynamic addressing* to gain access to the data:

```
>> fldstr = 'color';  % store desired field in variable

>> circle.(fldstr)      % get all color fields
ans =
red
ans =
green
ans =
blue

>> fldstr = 'radius';
>> area=pi*circle(1).(fldstr)^2  % compute area of first circle
area =
        19.635
```

Dynamic addressing is similar to using subscripts to identify specific elements in an array. Both methods use parentheses to hold the chosen elements. For example, `A(r,c)` uses the information in variables `r` and `c` to identify the desired rows and columns of array `A`, whereas `circle.(fldstr)` uses the information in variable `fldstr` to identify the desired field in the structure `circle`.

8.10 COMMA-SEPARATED LISTS (AGAIN)

To extract the contents of more than one structure array element at a time, MATLAB provides comma-separated list syntax. This syntax applies in any place where variables or constants appear in a list, separated by commas. For example, it appears in array construction:

```
>> a = ones(2,3);
>> b = zeros(2,1);
>> c = (3:4)';
>> d = [a,b,c] % same as [a b c]
d =
   1    1    1    0    3
   1    1    1    0    4
```

It also appears in function input and output argument lists:

```
>> d = cat(2,a,b,c)
d =
     1    1    1    0    3
     1    1    1    0    4
>> [m,n] = size(d)
m =
  2
n =
  5
```

Building on this idea, comma-separated list syntax is implemented as follows: Placing a structure array with appended field name in locations where comma-separated lists appear causes MATLAB to extract the contents of the addressed fields and place them sequentially, separated by commas. In the code

```
>> cent = cat(1,circle.center) % comma-separated list syntax
cent =
          0              1
        2.3           -1.2
         -1              0
>> cent = cat(1,circle(1).center,circle(2).center,circle(3).center)
cent =
          0              1
        2.3           -1.2
         -1              0
>> some = cat(1,circle(2:end).center)
```

```
some =
         2.3           -1.2
         -1              0
```

the `cat` function concatenates the circle centers into rows of the numerical array `cent`. According to comma-separated list syntax, the first two statements in the preceding example are identical. The third statement shows that you can index and extract a subarray as well.

Since the `color` fields of the structure `circle` are character strings of different lengths, they cannot be extracted into a string array, but they can be extracted into a cell array, as in the following example:

```
>> circle.color
ans =
red
ans =
green
ans =
blue

>> col = cat(1,circle.color) % elements have different lengths
??? Error using ==> cat
CAT arguments dimensions are not consistent.
>> col = [circle.color] % no error but not much use!
col =
redgreenblue
>> col = {circle.color} % cell array of strings
col =
    'red'     'green'     'blue'

>> col = char(col) % if needed, convert to string array
col =
red
green
blue
```

MATLAB does not provide tools for extracting all the fields of a single structure array element. For instance, there is no way to retrieve the `radius`, `center`, linestyle, color, and filled fields of the first circle in one statement. In a sense, there is no need to have such a function, because each of these fields can be directly addressed and used in computations, such as area1 = pi*circle(1).radius^2.

Given comma-separated list syntax, the function deal provides a way to extract the contents of numerous structure elements into separate variables as in the following example:

```
>> [c1,c2,c3] = deal(circle.color) % get all colors
c1 =
red
c2 =
green
c3 =
blue
>> [rad1,rad3]=deal(circle([1 3]).radius) % 1st and 3rd radius
rad1 =
          2.5
rad3 =
        25.4
```

Because the output of deal is also a comma-separated list, deal can be used to assign the contents of multiple structure array elements, with a single field in one statement. For example, consider the following code:

```
>> [circle.radius] = deal(5,14,83)
circle =
1x3 struct array with fields:
    radius
    center
    linestyle
    color
    filled
>> circle.radius  % confirm assignments
ans =
      5
ans =
     14
```

```
ans =
    83
>> [triangle(:).type] = deal('right','isosceles','unknown')
??? Error using ==> deal
The number of outputs should match the number of inputs.
>> [triangle(1:3).type] = deal('right','isosceles','unknown')
triangle =
1x3 struct array with fields:
    type
>> triangle.type
ans =
right
ans =
isosceles
ans =
unknown
```

In the first statement, the structure `circle` already existed and has three elements. Therefore, the output argument was expanded into three elements. In the second statement containing `deal`, the structure `triangle` didn't exist. Because it wasn't possible to determine how many elements to create in a comma-separated list, an error was returned. However, in the last statement containing `deal`, the number of elements was given explicitly, and the `type` property of the newly created structure `triangle` was populated with the given data.

8.11 STRUCTURE FUNCTIONS

In the *Command* window, it is easy to identify the field names of a given structure by simply entering the structure name at the MATLAB prompt:

```
>> circle
circle =
1x3 struct array with fields:
    radius
    center
    linestyle
    color
    filled
```

```
>> square
square =
        width: 5
       height: 14
       center: [0 0]
     rotation: 0.7854
```

When a structure is passed to a function—for example, myfunc(circle)—the function internally must know or have some way to obtain the field names of the structure. (Writing functions in MATLAB is covered in Chapter 12.) In MATLAB, the function fieldnames provides this information:

```
>> fieldnames(circle)
ans =
    'radius'
    'center'
    'linestyle'
    'color'
    'filled'
```

The output of fieldnames is a cell array of strings identifying the fields that are associated with the input structure.

It is also possible to guess the field names and ask whether they exist by using the logical function isfield:

```
>> isfield(circle,'color') % True
ans =
    1
>> isfield(circle,'width') % False
ans =
    0
```

If you don't know whether a variable is a structure or not, the functions class and isstruct are helpful, as the following example shows:

```
>> class(square)     % ask for the class of variable square
ans =
struct
```

```
>> isstruct(circle) % True for structures
ans =
     1
>> d = pi;
>> isstruct(d)        % False for doubles
ans =
     0
```

When the field names of a structure are known, the function `rmfield` allows you to remove one or more fields from the structure. For example, the code

```
>> fnames = fieldnames(circle)
fnames =
     'radius'
     'center'
     'linestyle'
     'color'
     'filled'
```

stores the fnames field names of `circle` in a cell array. Arbitrarily choosing the last field name, we can remove the `filled` field from the structure by calling `rmfield`:

```
>> circle2 = rmfield(circle,fnames{5})
circle2 =
1x3 struct array with fields:
     radius
     center
     linestyle
     color
```

This code removes the field `filled` from `circle` and assigns the result to a new structure `circle2`.

Similarly,

```
>> circle3 = rmfield(circle,fnames{1:3})
circle3 =
1x3 struct array with fields:
     'color'
     'filled'
```

removes the fields radius, center, and linestyle from circle, and assigns the result to a new structure circle3.

In some cases, it is beneficial to rearrange the order in which the field names of a structure are presented. The function orderfields performs this task and places the fields in ASCII order:

```
>> circleA = orderfields(circle)
circleA =
1x3 struct array with fields:
    center
    color
    filled
    linestyle
    radius
```

Other orderings are possible as well, as in the following code:

```
>> circleB = orderfields(circleA,CIRCLE) % match fields of structure CIRCLE
circleB =
1x3 struct array with fields:
    radius
    center
    linestyle
    color
    filled
>> circleC = orderfields(circle,[2 5 1 4 3]) % provide permutation vector
circleC =
1x3 struct array with fields:
    center
    filled
    radius
    color
    linestyle
>> circleD = orderfields(circle,fnames(end:-1:1)) % reverse original order
circleD =
1x3 struct array with fields:
    filled
```

```
color
linestyle
center
radius
```

Finally, given the similarity between cell arrays and structures and applying the post office box analogy, it's not hard to believe that MATLAB provides the functions `cell2struct` and `struct2cell` to convert cell arrays to structures and back in an organized fashion. (More detailed explanations of these functions can be found in the on-line documentation.)

8.12 SUMMARY

MATLAB supports arrays with an unlimited number of dimensions. Cell arrays and structures can store any array type, including cell arrays and structures. So, it is possible to have a cell in a cell array that contains a structure with a field that contains another structure that has a field containing a cell array, of which one cell contains another cell array. Needless to say, there comes a point where the power of cell arrays and structures becomes indecipherable and of little practical use. Thus, to end this chapter, try to decipher the following legal MATLAB statements:

```
>> one(2).three(4).five = {circle}
one =
1x2 struct array with fields:
    three
>> test = {{{circle}}}
test =
    {1x1 cell}
```

How many structures are involved in the first statement? What is the total number of structure elements, including empty arrays, created by the statement? Can the structure `circle` be extracted from the second statement with a single MATLAB statement?

9

Character Strings

MATLAB's true power is in its ability to crunch numbers. However, there are times when it is desirable to manipulate text, such as when putting labels and titles on plots. In MATLAB, text is referred to as character strings, or simply strings. Character strings represent another variable class or data type in MATLAB.

9.1 STRING CONSTRUCTION

Character strings in MATLAB are special numerical arrays of ASCII values that are displayed as their character-string representation, as in the following code:

```
>> t = 'How about this character string?'
t =
How about this character string?
>> size(t)
ans =
     1    32
>> whos
  Name        Size          Bytes  Class

  ans         1x2              16  double array
  t           1x32             64  char array
Grand total is 34 elements using 80 bytes
```

A character string is simply text surrounded by single quotes. Each character in a string is one element in an array that requires 2 bytes per character for storage, which is different from the 8 bytes per element that are required for numerical or double arrays.

To see the underlying ASCII representation of a character string, you need only perform some arithmetic operation on the string or use the dedicated function double, as in the following code:

```
>> u = double(t)
u =
  Columns 1 through 12
    72   111   119    32    97    98   111   117   116    32   116   104
  Columns 13 through 24
   105   115    32    99   104    97   114    97    99   116   101   114
  Columns 25 through 32
    32   115   116   114   105   110   103    63
>> abs(t)
ans =
  Columns 1 through 12
    72   111   119    32    97    98   111   117   116    32   116   104
  Columns 13 through 24
   105   115    32    99   104    97   114    97    99   116   101   114
  Columns 25 through 32
    32   115   116   114   105   110   103    63
```

The function char performs the inverse transformation:

```
>> char(u)
ans =
How about this character string?
```

Numerical values less than 0 produce a warning message when converted to character; values greater than 255 simply address characters in the font beyond char(255):

```
>> a = double('a')
a =
    97
```

```
>> char(a)
ans =
a

>> char(a+256) % adding 256 does change the result
ans =
š

>> char(a-256) % negative value produces a blank character
Warning: Out of range or non-integer values truncated during conversion
to character.
ans =
```

Since strings are arrays, they can be manipulated with all the array-manipulation tools available in MATLAB, For example, in the code

```
>> u = t(16:24)
u =
character
```

strings are addressed just as arrays are. Here, elements 16 through 24 contain the word character.

```
>> u = t(24:-1:16)
u =
retcarahc
```

The result here is the word character spelled backwards.

Using the transpose operator changes the word character to a column:

```
>> u = t(16:24)'
u =
c
h
a
r
a
c
t
```

e

r

Single quotes within a character string are symbolized by two consecutive quotes:

```
>> v = 'I can''t find the manual!'
v =
I can't find the manual!
```

Also, string concatenation follows directly from array concatenation:

```
>> u = 'If a woodchuck could chuck wood,';
>> v = ' how much wood could a woodchuck chuck?';

>> w = [u v]
w =
If a woodchuck could chuck wood, how much wood could a woodchuck chuck?
```

The function disp allows you to display a string without printing its variable name:

```
>> disp(u)
If a woodchuck could chuck wood,
```

Note that the u = statement is suppressed. This feature is useful for displaying help text within a script file.

Like other arrays, character strings can have multiple rows, but each row must have an equal number of columns. Therefore, blanks are explicitly required to make all rows the same length, as in the following code:

```
>> v = ['Character strings having more than'
        'one row must have the same number '
        'of columns just like arrays!      ']
v =
Character strings having more than
one row must have the same number
of columns just like arrays!
```

The functions char and strvcat create multiple-row string arrays from individual strings of varying lengths:

```
>> legends = char('Wilt','Russel','Kareem','Bird','Magic','Jordan')
```

```
legends =
Wilt
Russel
Kareem
Bird
Magic
Jordan
>> legends = strvcat('Wilt','Russel','Kareem','Bird','Magic','Jordan')
legends =
Wilt
Russel
Kareem
Bird
Magic
Jordan
>> size(legends)
ans =
     6      6
```

The only difference between char and strvcat is that strvcat ignores empty string inputs, whereas char inserts blank rows for empty strings, as in the following example:

```
>> char('one','','two','three')
ans =
one

two
three

>> strvcat('one','','two','three')
ans =
one
two
three
```

Horizontal concatenation of string arrays that have the same number of rows is accomplished by the function `strcat`. Padded blanks are ignored. The following code is illustrative:

```
>> a = char('apples','bananas')
a =
apples
bananas
>> b = char('oranges','grapefruit')
b =
oranges
grapefruit
>> strcat(a,b)
ans =
applesoranges
bananasgrapefruit
```

Once a string array is created with padded blanks, the function `deblank` is useful for eliminating the extra blanks from individual rows extracted from the array:

```
>> c = legends(4,:)
c =
Bird

>> size(c)
ans =
     1     6

>> c = deblank(legends(4,:))
c =
Bird

>> size(c)
ans =
     1     4
```

9.2 NUMBERS TO STRINGS TO NUMBERS

There are numerous contexts in which it is desirable to convert numerical results to character strings and to extract numerical data from character strings. MATLAB provides the functions `int2str`, `num2str`, `mat2str`, `sprintf`, and `fprintf` for converting numerical results to character strings. Examples of the first three functions are in the following code:

```
>> int2str(eye(3)) % convert integer arrays
ans =
1  0  0
0  1  0
0  0  1

>> size(ans) % it's a character array, not a numerical matrix
ans =
     3     7

>> num2str(rand(2,4)) % convert noninteger arrays
ans =
0.95013    0.60684    0.8913    0.45647
0.23114    0.48598    0.7621    0.01850

>> size(ans) % again it is a character array
ans =
     2    40

>> mat2str(pi*eye(2)) % convert to MATLAB input syntax form!
ans =
[3.14159265358979 0; 0 3.14159265358979]

>> size(ans)
ans =
     1    40
```

The following code illustrates the two functions `sprintf` and `fprintf`:

```
>> fprintf('%.4g\n',sqrt(2)) % display in Command window
1.414
```

```
>> sprintf('%.4g',sqrt(2)) % create character string
ans =
1.414

>> size(ans)
ans =
      1      5
```

The functions `sprintf` and `fprintf` are general-purpose conversion functions that closely resemble their ANSI C language counterparts. As a result, these two functions offer the most flexibility. Normally `fprintf` is used to convert numerical results to ASCII format and append the converted results to a data file. However, if no file identifier is provided as the first argument to `fprintf` or if a file identifier of 1 is used, the resulting output is displayed in the *Command* window. The functions `sprintf` and `fprintf` are identical, except that `sprintf` simply creates a character array that can be displayed, passed to a function, or modified like any other character array. Because `sprintf` and `fprintf` are nearly identical, consider the usage of `sprintf` in the following example:

```
>> radius = sqrt(2);
>> area = pi * radius^2;

>>s = sprintf('A circle of radius %.5g has an area of %.5g.',radius,area)
s =
A circle of radius 1.4142 has an area of 6.2832.
```

Here, `%.5g`, the format specification for the variable `radius` indicates that five significant digits in general-conversion format are desired. The most common usage of `sprintf` is to create a character string for the purposes of annotating a graph, displaying numerical values in a graphical user interface, or creating a sequence of data file names. A rudimentary example of this last usage is as follows:

```
>> i = 3;
>> fname = sprintf('mydata%.0f.dat',i)
fname =
mydata3.dat
```

In the past, the functions `int2str` and `num2str` were nothing more than a simple call to `sprintf`, with `%.0f` and `%.4g` as format specifiers, respectively. In MATLAB 5, `int2str` and `num2str` were enhanced to work with numerical arrays. As a result of these functions' former simplicity, in prior versions of MATLAB it was common to use the following code:

```
s = ['A circle of radius ' num2str(radius) ' has an area of ' ...
num2str(area) '.']
s =
A circle of radius 1.4142 has an area of 6.2832.
```

Even though we obtain the same result when using int2str and num2str as we do when using sprintf, the latter form requires more computational effort, is more prone to typographical errors (such as missing spaces or single quotes), and requires more effort to read. As a result, it is suggested that usage of int2str and num2str be limited to the conversion of arrays, as illustrated earlier in this section. In almost all other cases, it is more productive to use sprintf directly.

The help text for sprintf concisely describes its use:

```
>> help sprintf
 SPRINTF Write formatted data to string.
    [S,ERRMSG] = SPRINTF(FORMAT,A,...) formats the data in the real
    part of matrix A (and in any additional matrix arguments), under
    control of the specified FORMAT string, and returns it in the
    MATLAB string variable S. ERRMSG is an optional output argument
    that returns an error message string if an error occurred or an
    empty matrix if an error did not occur. SPRINTF is the same as
    FPRINTF except that it returns the data in a MATLAB string
    variable rather than writing it to a file.

    FORMAT is a string containing C language conversion specifications.
    Conversion specifications involve the character %, optional flags,
    optional width and precision fields, optional subtype specifier, and
    conversion characters d, i, o, u, x, X, f, e, E, g, G, c, and s.
    See the Language Reference Guide or a C manual for complete details.

    The special formats \n,\r,\t,\b,\f can be used to produce linefeed,
    carriage return, tab, backspace, and formfeed characters respectively.
    Use \\ to produce a backslash character and %% to produce the percent
    character.

    SPRINTF behaves like ANSI C with certain exceptions and extensions.
    These include:
    1. The following non-standard subtype specifiers are supported for
       conversion characters o, u, x, and X.
```

t   - The underlying C datatype is a float rather than an
       unsigned integer.

b   - The underlying C datatype is a double rather than an
       unsigned integer.

For example, to print out in hex a double value use a format like
'%bx'.

2. SPRINTF is "vectorized" for the case when A is nonscalar. The
   format string is recycled through the elements of A (columnwise)
   until all the elements are used up. It is then recycled in a similar
   manner through any additional matrix arguments.

Examples

    sprintf('%0.5g',(1+sqrt(5))/2)   1.618
    sprintf('%0.5g',1/eps)   4.5036e+15
    sprintf('%15.5f',1/eps)   4503599627370496.00000
    sprintf('%d',round(pi))   3
    sprintf('%s','hello')   hello
    sprintf('The array is %dx%d.',2,3)   The array is 2x3.
    sprintf('\n') is the line termination character on all platforms.

See also FPRINTF, SSCANF, NUM2STR, INT2STR.

The following table shows how pi is displayed under a variety of conversion
specifications:

Command	Result
sprintf('%.0e',pi)	3e+000
sprintf('%.1e',pi)	3.1e+000
sprintf('%.3e',pi)	3.142e+000
sprintf('%.5e',pi)	3.14159e+000
sprintf('%.10e',pi)	3.1415926536e+000
sprintf('%.0f',pi)	3
sprintf('%.1f',pi)	3.1
sprintf('%.3f',pi)	3.142
sprintf('%.5f',pi)	3.14159
sprintf('%.10f',pi)	3.1415926536

sprintf('%.0g',pi)	3
sprintf('%.1g',pi)	3
sprintf('%.3g',pi)	3.14
sprintf('%.5g',pi)	3.1416
sprintf('%.10g',pi)	3.141592654
sprintf('%8.0g',pi)	3
sprintf('%8.1g',pi)	3
sprintf('%8.3g',pi)	3.14
sprintf('%8.5g',pi)	3.1416
sprintf('%8.10g',pi)	3.141592654

In the preceding table, the format specifier e signifies exponential notation, f signifies fixed-point notation, and g signifies the use of e or f, whichever is shorter. Note that, for the e and f formats, the number to the right of the decimal point indicates how many digits to the right of the decimal point are to be displayed. On the other hand, in the g format, the number to the right of the decimal specifies the total number of digits that are to be displayed. In addition, note that, in the last five entries, a width of eight characters is specified for the result, and the result is right justified. In the very last case, the 8 is ignored because more than eight digits were specified.

Although it is not as common, sometimes it is necessary to convert or extract a numerical value from a character string. The MATLAB functions str2num, sscanf, and str2double provide this capability:

```
>> s = num2str(pi*eye(2)) % create string data
s =
3.1416                0
     0        3.1416


>> ischar(s) % True for string
ans =
    1


>> m = str2num(s) % convert string to number
m =
        3.1416                0
            0        3.1416


>> isdouble(m)  % Oops, this function doesn't exist
??? Undefined function or variable 'isdouble'.
```

```
>> isnumeric(m) % True for numbers
ans =
     1
```

```
>> isfloat(m) % True for floating point numbers
ans =
     1
```

```
>> pi*eye(2) - m % accuracy is lost
ans =
  -7.3464e-006              0
            0 -7.3464e-006
```

The function str2num can contain expressions, but not variables, in the workspace:

```
>> x = pi; % create a variable
```

```
>> ss = '[sqrt(2) j; exp(1) 2*pi-x]' % string with variable x
ss =
[sqrt(2) j; exp(1) 2*pi-x]
```

```
>> str2num(ss) % conversion fails because of x
ans =
     []
```

```
>> ss = '[sqrt(2) j; exp(1) 2*pi-6]' % replace x with 6
ss =
[sqrt(2) j; exp(1) 2*pi-6]
```

```
>> str2num(ss) % now it works
ans =
  1.4142                           0 +            1i
  2.7183                    0.28319
```

```
>> class(ans)  % yes, its a double
ans =
double
```

The function sscanf is the counterpart to sprintf. The function sscanf reads data from a string under format control:

```
>> v = version      % get MATLAB version as a string
v =
7.0.0.51483 (R14)

>> sscanf(v,'%f')  % get floating point numbers
ans =
            7
            0
      0.51483

>> sscanf(v,'%f',1) % get just one floating point number
ans =
      7

>> sscanf(v,'%d')   % get an integer
ans =
      7

>> sscanf(v,'%s')   % get a string
ans =
7.0.0.51483(R14)
```

You can specify the format under which sscanf operates; this capability makes sscanf a very powerful and flexible function. (See the on-line help text under sscanf for more thorough information about the capabilities of this function.)

When the conversion to a single double-precision value is required, the function str2double is useful. While the function str2num performs this task as well, str2double is generally quicker because of its more limited scope:

```
>> str2double('Inf') % It does convert infinity
ans =
   Inf

>> class(ans)
ans =
double
```

```
>> str2double('34.6 - 23.2j') % complex numbers work
ans =
        34.6 -          23.2i

>> str2double('pi') % variables and expressions don't work
ans =
  NaN
```

9.3 STRING EVALUATION

There are some applications in which it is convenient to evaluate a character string as if it were a MATLAB expression. This is a generalization of what str2num and str2double do. These functions are limited to extracting numerical values from strings. The MATLAB functions eval and evalc bring in the entire MATLAB interpreter to evaluate any string that conforms to MATLAB syntax. For example, in the code

```
>> funs = char('ceil','fix','floor','round')
funs =
ceil
fix
floor
round
>> [deblank(funs(1,:)) '(pi)']  % display string to evaluate
ans =
ceil(pi)
>> f = eval([deblank(funs(1,:)) '(pi)'])
f =
     4
>> class(f) % data type of output is numeric double
ans =
double

>> fc = evalc([deblank(funs(1,:)) '(pi)']) % try evalc
fc =
ans =
     4
```

```
>> class(fc) % output of evalc is a character string
ans =
char
>> size(fc)
ans =
     1    13
```

the function `eval` uses the MATLAB command interpreter to evaluate a character-string input and return the results into its output argument. The function `evalc` also evaluates an input string, but its output is the character string representation of the output. In other words, it returns the results seen in the *Command* window.

    If at all possible, for the most part, the use of `eval` and `evalc` should be avoided. Since these functions bring in the entire MATLAB interpreter to evaluate a string expression, they incur significant overhead. In addition, the functions cannot be compiled by the MATLAB compiler, which is a MATLAB add-on product that converts MATLAB code into executable C code.

9.4 STRING FUNCTIONS

MATLAB provides a variety of string-related functions, some of which have been discussed already. The following table briefly describes many of the string functions in MATLAB:

Function	Description
`char(S1,S2, ...)`	Create character array from strings or cell arrays
`double(S)`	Convert string to ASCII representation
`cellstr(S)`	Create cell array of strings from character array
`blanks(n)`	Create string of n blanks
`deblank(S)`	Remove trailing blanks
`eval(S), evalc(S)`	Evaluate string expression with MATLAB interpreter
`ischar(S)`	True for string array
`iscellstr(C)`	True for cell array of strings
`isletter(S)`	True for letters of the alphabet
`isspace(S)`	True for white-space characters
`isstrprop(S, 'property')`	True for elements that have specified property
`strcat(S1,S2,...)`	Concatenate strings horizontally
`strvcat(S1,S2,...)`	Concatenate strings vertically, ignoring blanks
`strcmp(S1,S2)`	True if strings are equal

Function	Description
strncmp(S1,S2,n)	True if n characters of strings are equal
strcmpi(S1,S2)	True if strings are equal, ignoring case
strncmpi(S1,S2,n)	True if n characters of strings are equal, ignoring case
strtrim(S1)	Trim leading and trailing white space
findstr(S1,S2)	Find shorter string within longer string
strfind(S1,S2)	Find S2 string in S1 string
strjust(S1,type)	Justify string array left, right, or center
strmatch(S1,S2)	Find indices of matching strings
strrep(S1,S2,S3)	Replace occurrences of S2 in S1 with S3
strtok(S1,D)	Find tokens in string given delimiters
upper(S)	Convert to uppercase
lower(S)	Convert to lowercase
num2str(x)	Convert number to string
int2str(k)	Convert integer to string
mat2str(X)	Convert matrix to string for eval
str2double(S)	Convert string to double-precision value
str2num(S)	Convert string array to numerical array
sprintf(S)	Create string under format control
sscanf(S)	Read string under format control

Now consider some other examples of the usage of some of the functions that were defined in the preceding table. For instance, the function findstr returns the starting indices of one string within another:

```
>> b = 'Peter Piper picked a peck of pickled peppers';

>> findstr(b,' ')     % find indices of spaces
ans =
      6    12    19    21    26    29    37
>> findstr(b,'p')     % find the letter p
ans =
      9    13    22    30    38    40    41

>> find(b=='p')       % for single character searches find works too
ans =
      9    13    22    30    38    40    41
>> findstr(b,'cow')   % cow does not exist
```

```
ans =

    []
>> findstr(b,'pick') % find the string pick
ans =

    13    30
```

Note that findstr is case sensitive and returns the empty matrix when no match is found. findstr does not work on string arrays with multiple rows.

Tests on character strings include the following:

```
>> c = 'a2 : b_c'

c =

a2 : b_c
>> ischar(c)    % it is a character string
ans =

    1
>> isletter(c) % where are the letters?
ans =

    1    0    0    0    0    1    0    1
>> isspace(c)   % where are the spaces?
ans =

    0    0    1    0    1    0    0    0


>> isstrprop(c,'wspace') % same as above
ans =

    0    0    1    0    1    0    0    0


>> isstrprop(c,'digit') % True for digits
ans =

    0    1    0    0    0    0    0    0
```

To illustrate string comparison, consider the situation where a user types (perhaps into an editable text uicontrol) a string that must match, at least in part, one of a list of strings. The function strmatch provides this capability:

```
>> S = char('apple','banana','peach','mango','pineapple')
S =

apple
```

```
banana
peach
mango
pineapple
>> strmatch('pe',S)    % pe is in 3rd row
ans =
     3
>> strmatch('p',S)        % p is in 3rd and 5th rows
ans =
     3
     5
>> strmatch('banana',S) % banana is in 2nd row
ans =
     2
>> strmatch('Banana',S) % but Banana is nowhere
ans =
     []
>> strmatch(lower('Banana'),S) % changing B to b finds banana
ans =
     2
```

9.5 CELL ARRAYS OF STRINGS

The fact that all rows in string arrays must have the same number of columns is sometimes cumbersome, especially when the nonblank portions vary significantly from row to row. This issue of cumbersomeness is eliminated by using cell arrays. All data forms can be placed in cell arrays, but their most frequent use is with character strings. A cell array is simply a data type that allows you to name a group of data of various sizes and types, as in the following example:

```
>> C = {'How';'about';'this for a';'cell array of strings?'}
C =
    'How'
    'about'
    'this for a'
    'cell array of strings?'
```

```
>> size(C)
ans =
    4    1
```

Note that curly brackets { } are used to create cell arrays and that the quotes around each string are displayed. In this example, the cell array C has four rows and one column. However, each element of the cell array contains a character string of different length.

Cell arrays are addressed just as other arrays are:

```
>> C(2:3)
ans =
    'about'
    'this for a'
>> C([4 3 2 1])
ans =
    'cell array of strings?'
    'this for a'
    'about'
    'How'
>> C(1)
ans =
    'How'
```

Here, the results are still cell arrays. That is, C(indices) addresses given cells, but not the contents of those cells. To retrieve the contents of a particular cell, use curly brackets, as in the following example:

```
>> s = C{4}
s =
cell array of strings?
>> size(s)
ans =
    1    22
```

To extract more than one cell, the function deal is useful:

```
>> [a,b,c,d] = deal(C{:})
a =
How
```

```
b =
about
c =
this for a
d =
cell array of strings?
```

Here, C{:} denotes all the cells as a list. That is, it's the same as

```
>> [a,b,c,d] = deal(C{1},C{2},C{3},C{4})
a =
How
b =
about
c =
this for a
d =
cell array of strings?
```

Partial cell array contents can also be dealt, as in the following code:

```
>> [a,b] = deal(C{2:2:4}) % get 2nd and 4th cell contents
a =
about
b =
cell array of strings?
```

A subset of the contents of a particular cell array can be addressed as well:

```
>> C{4}(1:10) % 4th cell, elements 1 through 10
ans =
cell array
```

The multipurpose function char converts the contents of a cell array to a conventional string array, as in the following example:

```
>> s = char(C)
s =
How
```

about

this for a

cell array of strings?

```
>> size(s)  % result is a standard string array with blanks
ans =
     4    22
>> ss = char(C(1:2))  % naturally you can convert subsets
ss =
How
about
>> size(ss)  % result is a standard string array with blanks
ans =
     2     5
```

The inverse conversion is performed by the function `cellstr`, which deblanks the strings as well:

```
>> cellstr(s)
ans =
    'How'
    'about'
    'this for a'
    'cell array of strings?'
```

You can test whether a particular variable is a cell array of strings by using the function `iscellstr`:

```
>> iscellstr(C)     % True for cell arrays of strings
ans =
     1
>> ischar(C)        % True for string arrays not cell arrays of strings
ans =
     0
>> ischar(C{3})     % Contents of 3rd cell is a string array
ans =
     1
>> iscellstr(C(3)) % but 3rd cell itself is a cell
```

```
ans =
    1
>> ischar(C(3))      % and not a string array
ans =
    0

>> class(C)          % get data type or class string
ans =
cell
>> class(s)          % get data type or class string
ans =
char
```

Most of the string functions in MATLAB work with either string arrays or cell arrays of strings. In particular, the functions `deblank`, `strcat`, `strcmp`, `strncmp`, `strcmpi`, `strncmpi`, `strmatch`, and `strrep` all work with either string arrays or cell arrays of strings. (Further information regarding cell arrays in general can be found in Chapter 8).

9.6 SEARCHING WITH REGULAR EXPRESSIONS

The standard string functions, like `findstr`, are useful for searching for specific sequences of characters within strings and for replacing specific character sequences. More generally, you may want to find such *patterns* in strings as repeated characters, all uppercase letters, any capitalized words, or all dollar amounts (i.e., strings of digits preceded by a dollar sign and containing a decimal point). MATLAB supports ***regular expressions***, which is a powerful tool for searching for character strings within character strings. A regular expression is a formula for matching strings that follow some pattern. The Unix world has used regular expressions in many of its tools (such as `grep`, `awk`, `sed`, `vi`, and others) for years. Programmers who use Perl and other languages make extensive use of regular expressions. MATLAB's implementation of regular expressions is very comprehensive and includes many features, including some called *extended regular expressions*. And, if you are familiar with regular expressions, MATLAB's implementation will seem to you to follow naturally. If you are new to regular expressions, you can make use of the simpler features immediately and the more complex features later on.

Here are a few simple rules to get you started. A formula string, or *expression*, that describes the criteria for identifying matching portions of a character string is created. The expression consists of characters and optional modifiers that define the criteria to be used for the substring match. The simplest expression is a string of literal characters, as in the following code:

```
>> str = 'Peter Piper picked a peck of pickled peppers.'; % create a string

>> regexp(str,'pe')        % return the indices of substrings matching 'pe'
ans =
     9    22    38    41
```

Here, we found the letter combination 'pe' in the words Piper and peck and twice in the word peppers.

Character classes are used to match a specific type of character—such as a letter, a number, or a white-space character—or to match a specific set of characters. The most useful character class is a period (.), which represents *any single character*. Another useful character class is a list of characters, or range of characters, within square brackets ([]) that matches *any member of the list or range of characters in the set*. For example, consider searching for a sequence of three characters consisting of a p, followed by any single character, followed by the letter c or the letter r:

```
>> regexpi(str,'p.[cr]')
ans =
      9    13    22    30    41
>> regexp(str,'p.[cr]','match')            % list the substring matches
ans =
    'per'    'pic'    'pec'    'pic'    'per'
```

You could also use a range of characters to find all uppercase letters in the string:

```
>> regexp(str,'[A-Z]')                     % match any uppercase letter
ans =
     1    7
```

The character expressions in the following table match a single character with the indicated characteristics:

Character expression	Description and usage
.	Any single character (including white space)
[abcd35]	Any single character in the set enclosed by square brackets []
[a-zA-Z]	Any single character in the range a-z or A-Z; the dash - between characters defines an inclusive range of characters
[^aeiou]	Any single character NOT in the set of lowercase vowels; the ^ used as the first character in the set negates the sense of the set

Character expression	Description and usage
\s	Any white-space character (space, tab, form feed, new line, or carriage return); equivalent to the set [\t\f\n\r]
\S	Any nonwhite-space character: [^ \t\f\n\r]
\w	Any "word character" (an upper- or lowercase letter, a digit, or an underscore): [a-zA-Z_0-9]
\W	Any character that is not a "word character": [^a-zA-Z_0-9]
\d	Any numeric digit: [0-9]
\D	Any nondigit character: [^0-9]
\xN or \x{N}	The character with hexadecimal value N
\oN or \o{N}	The character with octal value N
\a	The alarm, bell, or beep character: \o007 or \x07
\b	The backspace character: \o010 or \x08
\t	The horizontal tab character: \o011 or \x09
\n	The line-feed or newline character: \o012 or \x0A
\v	The vertical tab character: \o013 or \x0B
\f	The form-feed character: \o014 or \x0C
\r	The carriage-return character: \o015 or \x0D
\e	The escape character: \o033 or \x1B
\	Use a backslash to force a literal match on the next character. This enables matching on characters that have special meaning in regular expressions; for example, the characters \.*\?\\ represent a literal period, asterisk, question mark, and backslash, respectively.

The character expressions in the preceding table can be modified by using regular expression *modifiers* or *quantifiers*. Here is an example of using the \w class, along with a modifier, to find all of the words in the string:

```
>> regexp(str,'\w+','match')              % find all individual words
ans =
      'Peter'    'Piper'    'picked'    'a'    'peck'    'of'
'pickled'    'peppers'
```

This example shows how the + modifier changes the sense of the expression. Without the modifier, the \w expression matches any single "word character." Adding the modifier changes the expression to match all words (groups of one or more "word characters") in the string.

Character modifiers are listed in the following table:

Modifier	Description and usage
?	Match the preceding element zero or one times
*	Match the preceding element zero or more times
+	Match the preceding element one or more times
{n}	Match the preceding element exactly n times
{n,}	Match the preceding element at least n times
{n,m}	Match the preceding element at least n times, but no more than m times

Character modifiers are considered *greedy* (i.e., they match the longest string they can). For example, in the code

```
>> regexp(str,'p.*p','match')
ans =
    'per picked a peck of pickled pepp'
```

the modifier matches a lowercase p, followed by zero or more of any character, followed by a lowercase p. This behavior can be changed by using the quantifier expressions listed in the following table:

Quantifier	Description and usage
q	Match as much of the string as possible (a *greedy quantifier*). This is the default. Here, q represents one of the modifiers in the previous table (e.g., ?, *, +, {n,m}).
q+	Match as many components of the string as possible, without rescanning any portions of the string should the initial match fail (a *possessive quantifier*). Again, q represents one of the modifiers in the previous table (e.g., ?, *, +, {n,m}).
q?	Match as little of the string as possible, while scanning the string (a *lazy quantifier*). Again, q represents one of the modifiers in the previous table (e.g., ?, *, +, {n,m}).

The following code matches shorter strings bracketed by lowercase p characters:

```
>> regexp(str,'p.*?p','match')
ans =
    'per p'     'peck of p'      'pep'
```

Notice that the string `'picked a p'` was not returned. Lazy quantifiers cause the string to be scanned from beginning to end. If a match is found, scanning continues, beginning with the next character in the string; previous portions are not rescanned.

Parentheses can be used to group patterns. For example, the expression in the following example matches a p, followed by one or more characters that are not the letter i:

```
>> regexp(str,'p[^i]+','match')
ans =

     'per p'     'peck of p'     'peppers'
```

The modifier applies to the preceding character. The sense of the expression changes slightly when parentheses are added. The expression in the following example matches one or more sequences of a p followed by a character that is not the letter i:

```
>> regexp(str,'(p[^i])+','match')
ans =

     'pe'     'pe'     'pepp'
```

The modifier now applies to the two-character sequence within the parentheses.

Matches can be made conditional to the context of a regular-expression match. MATLAB supports both *logical* operators and *lookaround* operators. Lookaround operators can be used to condition a match only if the string is preceded or followed by the presence or absence of another string match. The next example matches all words preceded by words ending in d:

```
>> regexp(str,'(?<=d\s)\w+','match')
ans =

     'a'     'peppers'
```

The expression matches groups of word characters only if preceded by the character d, followed by a white-space character.

The lookaround operators are listed in the following table:

Lookaround operator	Description and usage
p(?=q)	Lookahead. Match pattern p only if followed by a match for pattern q
p(?!q)	Negative lookahead. Match pattern p only if not followed by a match for pattern q
(?<=q)p	Lookbehind. Match pattern p only if preceded by a match for pattern q
(?<!q)p	Negative lookbehind. Match pattern p only if not preceded by a match for pattern q

Logical operators are also available. The following example matches words containing the string `'ip'` or the string `'ck'`:

```
>> regexp(str,'\w*(ip|ck)\w*','match')
ans =
    'Piper'    'picked'    'peck'    'pickled'
```

Logical operators are listed in the following table:

Logical operator	Description and usage
p|q	Match the pattern p or the pattern q
^p	Match the pattern p only if it occurs at the beginning of the string
p$	Match the pattern p only if it occurs at the end of the string
\<p	Match the pattern p only if it occurs at the beginning of a word
p\>	Match the pattern p only if it occurs at the end of a word
\<p\>	Match the pattern p only if it exactly matches a word

Repeated sequences can be matched by using tokens. When expressions contain parentheses, the string matching the expression within the parentheses (a *token*) is stored and can be reused in the expression. Up to 255 tokens can be used in an expression. Tokens can be referenced by using the syntax \d, where d is the index of the desired token. For example, \1 represents the first token, \2 represents the second token, and so on. The following example matches any doubled-word character:

```
>> regexp(str,'(\w)\1','match')
ans =
    'pp'
```

The next example matches any character followed by zero or more word characters, followed by a white-space character, followed by the same initial character, followed by zero or more word characters, followed by the same white space character:

```
>> regexp(str,'(.)\w*(\s)\1\w*\2','match')
ans =
    'Peter Piper '
```

Notice that the string `'pickled peppers'` did not match, since there is no space after the string `'peppers'`.

The following table lists the token expressions available in MATLAB:

Expression	Description and usage
(p)	Capture all characters matched by expression p in a token
(?:p)	Group all characters matched by expression p, but do not save in a token
(?>p)	Group atomically, but do not save in a token
(?#A Comment)	Insert a comment into an expression (The comment is ignored.)
\N	Match the Nth token in this expression (e.g., \1 is the first token, \2 is the second token, and so on)
$N	Insert a match for the Nth token in a replacement string (regexprep function only)
(?<name>p)	Capture all characters matched by the pattern p in a token and assigns a name to the token
\k<name>	Match the token referred to by name
(?(T)p)	If token T is generated (i.e., the match for token T was successful), then match pattern p. This is an IF/THEN construction. The token can be a named token or a positional token.
(?(T)p|q)	If token T is generated, then match pattern p; otherwise, match pattern q. This is an IF/THEN/ELSE construction. The token can be a named token or a positional token.

There are three different regular expression functions in MATLAB. The preceding examples used regexp. The regexpi function ignores cases when matching, whereas the regexprep function replaces strings by using regular expressions.

The following example searches for words beginning with the string 'pi', reverses the words, and returns the resulting string:

```
>> regexprep(str,'(pi\w*)(.*)(pi\w*)','$3$2$1')
ans =
    'Peter Piper pickled a peck of picked peppers'
```

This example generates three tokens: Token 1 is the string 'picked', token 2 is ' a peck of ', and token 3 is 'pickled'. The replacement string '$3$2$1' instructs regexprep to remove the part of the original string matched by the first argument, and replace it with a string consisting of the values of the tokens, in the reverse order. The rest of the original string is returned unchanged. Numbered tokens are referenced in the search pattern by using the \N notation. The $n notation is used to reference the tokens in the replacement string.

The regular expression functions are listed in the following table:

Function	Description and usage
regexp	Search for substrings by using regular expressions
regexpi	Search for substrings by using regular expressions, ignoring case
regexprep	Search and replace substrings by using regular expressions

There are many options available for applying and modifying the regular expression functions. One option is the `'match'` argument illustrated earlier. The `regexprep` function is normally case sensitive and replaces all matches it finds. Options are available to change that default behavior.

All three functions operate on cell arrays of strings, as well as on individual strings. One pattern can be used to search many strings, or different patterns can be used for different strings. Any or all of the input parameters can be a cell array of strings. (For more thorough information about these extremely powerful and complex capabilities, see the MATLAB help documentation.)

10

Relational and Logical Operations

In addition to traditional mathematical operations, MATLAB supports relational and logical operations. You may be familiar with these if you've had some experience with other programming languages. The purpose of these operators and functions is to provide answers to True/False questions. One important use of this capability is to control the flow or order of execution of a series of MATLAB commands (usually in a M-file) based on the results of True/False questions.

> As inputs to all relational and logical expressions, nonzero values are considered True, and zero values are considered False. The output of all relational and logical expressions produces logical arrays with True (1) and False (0).

Logical arrays are a special type of numerical array that can be used for logical array addressing, as shown in the chapter on arrays, as well as in any numerical expression. Internally, logical arrays are a separate variable class that uses one byte of storage per value.

10.1 RELATIONAL OPERATORS

MATLAB relational operators include all common comparisons, as shown in the following table:

Relational Operator	Description
<	Less than
<=	Less than or equal to
>	Greater than
>=	Greater than or equal to
==	Equal to (not to be confused with =)
~=	Not equal to

MATLAB relational operators can be used to compare two arrays of the same size or to compare an array with a scalar. In the latter case, scalar expansion is used to compare the scalar with each array element, and the result has the same size as the array. For example, the code

```
>> A = 1:9, B = 9-A
A =
     1    2    3    4    5    6    7    8    9
B =
     8    7    6    5    4    3    2    1    0
>> tf = A>4
tf =
     0    0    0    0    1    1    1    1    1
```

finds elements of A that are greater than 4. Zeros appear in the result where A<=4, and ones appear where A>4.

The code

```
>> tf = (A==B)
tf =
     0    0    0    0    0    0    0    0    0
```

finds elements of A that are equal to those in B.

> Note that = and == mean two different things: == compares two variables and returns ones (True) where they are equal and zeros (False) where they are not; =, on the other hand, is used to assign the output of an operation to a variable.

Note that testing for equality sometimes produces confusing results for floating point numbers:

```
>> tf = (-0.08 + 0.5 -0.42)==(0.5 - 0.42 - 0.08) % equal ?
tf =
      0
>> tf = (-0.08 + 0.5 -0.42)~=(0.5 - 0.42 - 0.08) % not equal ?
tf =
      1
>> (-0.08 + 0.5 -0.42)-(0.5 - 0.42 - 0.08) % not exactly equal!
ans =
  -1.3878e-017
```

This is the example we used in Chapter 2 to illustrate that arithmetic is not *exactly* commutative when using finite precision. We would expect the result to be True = 1; but it is not, because the two expressions differ by a number smaller than eps. On the MATLAB newsgroup, this fundamental fact of finite-precision arithmetic is regularly posed as a ***bug*** in MATLAB.

It is possible to combine relational expressions with mathematical expressions. For example, the code

```
>> tf = B - (A>2)
tf =
      8      7      5      4      3      2      1      0     -1
```

finds where A>2 and subtracts the resulting vector from B. The code

```
>> B = B + (B==0)*eps
B =
  Columns 1 through 7
    8.0000    7.0000    6.0000    5.0000    4.0000    3.0000    2.0000
  Columns 8 through 9
    1.0000    0.0000
```

is a demonstration of how to replace zero elements in an array with the special MATLAB number eps, which is approximately 2.2e-16 for double-precision values. This particular expression is sometimes useful to avoid dividing by zero, as in the code

```
>> x = (-3:3)/3
x =
  -1.0000   -0.6667   -0.3333        0    0.3333    0.6667    1.0000
```

```
>> sin(x)./x
Warning: Divide by zero.
ans =
    0.8415     0.9276     0.9816        NaN     0.9816     0.9276     0.8415
```

Computing the function sin(x)/x gives a warning because the fifth data point is zero. Since sin(0)/0 is undefined, MATLAB returns NaN (meaning Not-a-Number) at that location in the result. This can be avoided by replacing the zero with eps, as in the code

```
>> x = x + (x==0)*eps;
>> sin(x)./x
ans =
    0.8415     0.9276     0.9816     1.0000     0.9816     0.9276     0.8415
```

Now sin(x)/x for x=0 gives the correct limiting answer. Alternatively, you can avoid the computation at x=0, for example, by using the code

```
>> x = (-3:3)/3 % recreate x
x =
  Columns 1 through 6
        -1      -0.66667      -0.33333            0      0.33333      0.66667
  Column 7
         1
>> y = ones(size(x)) % create default output
y =
     1     1     1     1     1     1     1
>> tf = x~=0   % find nonzero locations
tf =
     1     1     1     0     1     1     1
>> y(tf) = sin(x(tf))./x(tf) % operate only on nonzeros
y =
  Columns 1 through 6
    0.84147       0.92755       0.98158            1      0.98158      0.92755
  Column 7
    0.84147
```

While this approach may seem cumbersome compared with adding eps to x, the concept of avoiding computations with selected components of an array is used often in efficient MATLAB programming.

10.2 LOGICAL OPERATORS

Logical operators provide a way to combine or negate relational expressions. MATLAB logical operators include those shown in the following table:

Logical Operator	Description
&	Element-by-element AND for arrays
|	Element-by-element OR for arrays
~	NOT
&&	Scalar AND with short-circuiting
||	Scalar OR with short-circuiting

The next four examples illustrate the use of logical operators. The code

```
>> A = 1:9; B = 9-A; % recall data
>> tf = A>4
tf =
     0    0    0    0    1    1    1    1    1
```

finds where A is greater than 4.
    The code

```
>> tf = ~(A>4)
tf =
     1    1    1    1    0    0    0    0    0
```

negates the preceding result; that is, it swaps the positions of the logical ones and zeros.
    The code

```
>> tf = (A>2) & (A<6)
tf =
     0    0    1    1    1    0    0    0    0
```

returns True where A is greater than 2 and less than 6.
    The code

```
>> tf= A<2 | A>7
tf =
     1    0    0    0    0    0    0    1    1
```

returns True where A is less than 2 or greater than 7.

The scalar short-circuiting logical operators permit an early exit from logical comparisons where the operator arguments are scalars. That is, when a logical result is known before performing all relational tests, unneeded comparisons are skipped. The following code is illustrative:

```
>> a=0; b=pi; % new data
>> a==0 || b~=1 % a=0 so this is True and b~=1 is not evaluated
ans =
    1
>> b==1 && a==0 % b~=1 so this is False and a==0 is not evaluated
ans =
    0
>> a==0 || (1/a)<1 % a==0 so 1/a is not computed
ans =
    1
```

This last example demonstrates the utility of short circuiting. Since a is equal to zero, the second expression is never evaluated because it does not change the True result. Therefore, 1/a is not computed. If it was, a divide-by-zero error would have been returned.

10.3 OPERATOR PRECEDENCE

When evaluating an expression, MATLAB follows a set of rules governing operator precedence. Operators having higher precedence are evaluated before operators of lower precedence. Operators of equal precedence are evaluated left to right. The following table illustrates the operator precedence rules used by MATLAB:

Operator	Precedence Level
Parentheses ( )	Highest
Transpose (.'), conjugate transpose ('), power (.^), matrix power (^)	
Unary plus (+), unary minus (−), negation (~)	
Multiplication (.*), matrix multiplication (*), right division (. /), left division (.\), matrix right division (/), matrix left division (\)	
Addition (+), subtraction (−), logical negation (~)	
Colon operator (:)	
Less than (<), less than or equal to (<=), greater than (>), greater than or equal to (>=), equal to (==), not equal to (~=)	

Operator	Precedence Level
Element-wise logical AND (&)	
Element-wise logical OR (|)	
Short-circuiting logical AND (&&)	
Short-circuiting logical OR (|)	Lowest

This order of precedence is similar, or exactly equal, to the order used by most computer programming languages. As a result, you are probably already comfortable with writing expressions that conform to these rules. As in other programming languages, parentheses can and should be used to control the order in which an expression is evaluated. Within each level of parentheses, the rules hold; for example, the code

```
>> 3|2&0
ans =
      1
>> 3|(2&0)
ans =
      1

>> (3|2)&0
ans =
      0
```

points out that the preceding order of precedence table is different from the tables included in MATLAB versions prior to Version 6. In Version 5 and prior versions of MATLAB, logical AND and logical OR shared the same order of precedence, whereas logical AND now has higher precedence than logical OR.

10.4 RELATIONAL AND LOGICAL FUNCTIONS

In addition to basic relational and logical operators, MATLAB provides a number of other relational and logical functions, including the following:

Function	Description
xor(x,y)	Exclusive OR operation. Returns True for each element where either x or y is nonzero. Returns False where both x and y are zero, or both are nonzero.
any(x)	Returns True if any element in a vector x is nonzero. Returns True for each column in an array x that has any nonzero elements.
all(x)	Returns True if all elements in a vector x are nonzero. Returns True for each column in an array x that has all nonzero elements.

In addition to these functions, MATLAB provides numerous functions that test for the existence of specific values or conditions and return logical results as shown in the following table:

Function	Description
ispc	True for the PC (Windows) version of MATLAB
isstudent	True for MATLAB student edition
isunix	True for the UNIX version of MATLAB
ismember	True for set member
isglobal	True for global variables
mislocked	True if an M-file cannot be cleared
isempty	True for an empty matrix
isequal	True if arrays are numerically equal
isequalwithequalnans	True if arrays are equal with NaNs considered equal
isfinite	True for finite elements
isfloatpt	True for floating-point numbers
isscalar	True for a scalar
isinf	True for infinite elements
islogical	True for a logical array
isnan	True for Not-a-Number
isnumeric	True for a numeric array
isreal	True for a real array
isprime	True for prime numbers
issorted	True if an array is sorted
automesh	True if the inputs should be automatically meshgridded
inpolygon	True for points inside a polygonal region
isvarname	True for a valid variable name
iskeyword	True for keywords or reserved words
issparse	True for a sparse matrix
isvector	True for a vector
isappdata	True if application-defined data exists
ishandle	True for graphics handles
ishold	True if the graphics hold state is On
figflag	True if a figure is currently displayed on screen
iscellstr	True for a cell array of strings
ischar	True for a character string array
isletter	True for letters of the alphabet

Function	Description
isspace	True for white-space characters
isa	True if an object is a given class
iscell	True for a cell array
isfield	True if a field is in structure array
isjava	True for Java object arrays
isobject	True for objects
isstruct	True for structures
isvalid	True for serial port objects that can be connected to hardware

10.5 NANS AND EMPTY ARRAYS

NaNs (Not-a-Numbers) and empty arrays ([]) require special treatment in MAT-LAB, especially when used in logical or relational expressions. According to IEEE mathematical standards, almost all operations on NaNs result in NaNs. Consider the following example:

```
>> a = [1 2 nan inf nan]  % note that NaN is not case-sensitive
a =
     1     2    NaN     Inf    NaN

>> b = 2*a
b =
     2     4    NaN     Inf    NaN
>> c = sqrt(a)
c =
    1.0000    1.4142        NaN        Inf        NaN
>> d = (a==nan)
d =
     0     0     0     0     0
>> f = (a~=nan)
f =
     1     1     1     1     1
```

The first two computations give NaN results for NaN inputs. However, the final two relational computations produce somewhat surprising results: (a==nan) produces all zeros or False results even when NaN is compared with NaN; at the same time, (a~=nan) produces all ones or True results. **Thus, individual NaNs are not equal to each other**. Because of this property of NaNs, MATLAB has a built-in logical function for finding NaNs, called isnan:

```
>> g = isnan(a)
g =
     0     0     1     0     1
```

Moreover, the isnan function makes it possible to find the indices of NaNs with the
find command:

```
>> i = find(isnan(a))  % find indices of NaNs
i =
     3     5
>> a(i) = zeros(size(i))  % changes NaNs in a to zeros
a =
     1     2     0   Inf     0
```

Whereas NaNs are well-defined mathematically by IEEE standards, empty
arrays are defined by the creators of MATLAB and have their own interesting
properties. Empty arrays are just that: They are MATLAB variables having zero
length in one or more dimensions, as in the following code:

```
>> size([])  % simplest empty array
ans =
     0     0
>> c = zeros(0,5)  % how about an empty array with multiple columns!
c =
   Empty matrix: 0-by-5

>> size(c)
ans =
     0     5
>> d = ones(4,0)  % an empty array with multiple rows!
d =
   Empty matrix: 4-by-0
>> size(d)
ans =
     4     0
>> length(d)  % it's length is zero even though it has 4 rows
ans =
     0
```

This may seem strange, but allowing an empty array to have zero length in any dimension is sometimes useful. [] is just the simplest empty array.

In MATLAB, many functions return empty arrays when no other result is appropriate. Perhaps the most common example is the find function;

```
>> x = -2:2  % new data
x =
    -2    -1     0     1     2
>> y = find(x>2)
y =
Empty matrix: 1-by-0
```

In this example, x contains no values greater than 2, and so, there are no indices to return. To test for empty results, MATLAB provides the logical function isempty:

```
>> isempty(y)
ans =
     1
```

When performing relational tests where empty arrays may appear, it is important to use isempty, as the following code shows:

```
>> c==[] % comparing 0-by-5 to 0-by-0 arrays produces an error
??? Error using ==> eq
Matrix dimensions must agree.

>> isempty(c) % isempty returns the desired result.
ans =
     1
>> a = []; % create an empty variable

>> a==[]  % comparing equal size empties gives empty results
ans =
     []

>> b=1; % create nonempty variable
>> b==[] % comparing nonempty to empty produces an empty result.
ans =
     []
```

```
>> b~=[] % even not equal comparison produces and empty result.
ans =

    []
```

The general rule is that relational operations on empty arrays produce either an error or an empty array result. Therefore, it is important to use isempty to consider the empty array case whenever it may appear.

11

Control Flow

Computer programming languages and programmable calculators offer features that allow you to control the flow of command execution using decision-making structures. If you have used these features before, this section will be very familiar to you. On the other hand, if control flow is new to you, the material may seem complicated the first time through.

Control flow is extremely powerful, since it lets past computations influence future operations. MATLAB offers five decision-making or control-flow structures: (1) For Loops, (2) While Loops, (3) If-Else-End constructions, (4) Switch-Case constructions, and (5) Try-Catch blocks. Because these constructions often encompass numerous MATLAB commands, they frequently appear in M-files, rather than having to be typed directly at the MATLAB prompt.

11.1 FOR LOOPS

For Loops allow a group of commands to be repeated for a fixed, predetermined number of times. The general form of a For Loop is

```
for x = array
    (commands)
end
```

The *(commands)* between the `for` and `end` statements are executed once for every **column** in `array`. At each iteration, `x` is assigned to the next column of `array`; that is, during the nth time through the loop, `x = array(:,n)`. For example, in the code

```
>> for n = 1:10
       x(n) = sin(n*pi/10);
   end
>> x
x =
  Columns 1 through 7
     0.3090     0.5878     0.8090     0.9511     1.0000     0.9511     0.8090
  Columns 8 through 10
     0.5878     0.3090     0.0000
```

the first statement says, "For n equals 1 to 10, evaluate all statements until the next end statement." The first time through the For Loop, n=1; the second time, n=2; and so on through the n=10 case. After the n=10 case, the For Loop ends, and any commands after the end statement are evaluated, which in this case results in the computed elements of x.

Since the loop variable is assigned to successive columns of the array on the right-hand side of the equal sign, arbitrary indexing, or inadvertent errors, can occur. For example, in the code

```
>> for n = 10:-1:1  % decrementing loop
       x(n) = sin(n*pi/10);
   end
>> x
x =
  Columns 1 through 6
     0.30902     0.58779     0.80902     0.95106          1     0.95106
  Columns 7 through 10
     0.80902     0.58779     0.30902  1.2246e-016
```

the loop variable n counts down from 10 to 1. The expression $10:-1:1$ is a standard array-creation statement that creates a row vector with multiple columns. Any numerical array can be used. In the code

```
>> i = 0; % count loop iterations
>> for n = (1:10)'
       i = i+1;
       x(n) = sin(n*pi/10);
   end
```

```
>> i   % Only one time through the loop!
i =
     1
>> x
x =
  Columns 1 through 6
     0.30902        0.58779        0.80902        0.95106            1       0.95106
  Columns 7 through 10
     0.80902        0.58779        0.30902   1.2246e-016
```

the For Loop executes only one pass! The expression $(1:10)'$ is a column vector, and so n is set equal to the entire array $(1:10)'$ on its first pass. Since there are no additional columns in the right-hand side array, the loop terminates. In the code

```
>> array = randperm(10)
array =
     8     2    10     7     4     3     6     9     5     1
>> for n = array
       x(n) = sin(n*pi/10);
    end
>> x
x =
  Columns 1 through 6
     0.30902        0.58779        0.80902        0.95106            1       0.95106
  Columns 7 through 10
     0.80902        0.58779        0.30902   1.2246e-016
```

the loop variable n takes on the numbers 1 to 10 in the random order given by array.

A For Loop cannot be terminated by reassigning the loop variable n within the loop:

```
>> for n = 1:10
       x(n) = sin(n*pi/10);
       n = 10;
    end
```

```
>> x

x =

   Columns 1 through 6

      0.30902       0.58779       0.80902       0.95106            1       0.95106

   Columns 7 through 10

      0.80902       0.58779       0.30902   1.2246e-016
```

To repeat, the right-hand-side array in the For Loop statement can be any valid array-creation statement:

```
>> i = 1;
>> for x = rand(4,5)
        y(i) = sum(x);
        i = i+1;
     end
>> y

y =

    1.7325        2.3954        1.6326        1.7201        2.0872
```

Here, the loop variable x is assigned to the successive 4-by-1 columns of a random array. Since the For Loop has no natural loop index, i was added.

Naturally, For Loops can be nested as desired:

```
>> for n = 1:5
        for m = 5:-1:1
            A(n,m) = n^2 + m^2;
        end
        disp(n)
     end
     1
     2
     3
     4
     5
>> A

A =

     2     5    10    17    26
     5     8    13    20    29
```

10	13	18	25	34
17	20	25	32	41
26	29	34	41	50

Just because the preceding examples were used to illustrate For Loop usage, it doesn't mean that they are examples of efficient MATLAB programming. Historically, For Loops represented poor programming practice whenever an equivalent array approach existed. The equivalent array approach, called a ***vectorized*** solution, often is orders of magnitude faster than the scalar approaches just shown. For example, in the code

```
>> n = 1:10;
>> x = sin(n*pi/10)
x =
  Columns 1 through 6
     0.30902       0.58779       0.80902       0.95106            1       0.95106
  Columns 7 through 10
     0.80902       0.58779       0.30902    1.2246e-016
```

the two statements duplicate the repeated example of computing the sine function at 10 angles. In addition to being orders of magnitude faster, the preceding vectorized solution is more intuitive, is easier to read, and requires less typing.

The earlier nested For Loop is equivalent to the following vectorized code:

```
>> n = 1:5;
>> m = 1:5;
>> [nn,mm] = meshgrid(n,m);

>> A = nn.^2 + mm.^2
A =
     2      5     10     17     26
     5      8     13     20     29
    10     13     18     25     34
    17     20     25     32     41
    26     29     34     41     50
```

As discussed in Chapter 5, arrays should be preallocated before a For Loop (or While Loop) is executed. Doing so minimizes the amount of memory allocation required. For example, in the first case considered in this section, every time the commands within the For Loop are executed, the size of the variable x is increased by 1, which forces MATLAB to take the time to allocate more memory for x every time it goes through the loop. To eliminate this step, the For Loop example should be rewritten, as follows:

```
>> x = zeros(1,10); % preallocated memory for x
>> for n = 1:10
      x(n) = sin(n*pi/10);
   end
```

In this case, only the values of x(n) need to be changed each time through the loop. Memory allocation occurs once outside the loop, so no memory allocation overhead bogs down the operations within the loop.

Starting with MATLAB 6.5, improvements were made to the MATLAB interpreter to minimize the processing overhead involved in executing loops. These improvements are collectively known as the ***JIT-Accelerator***. As stated in the documentation, such improvements will appear over a series of MATLAB releases. At this time, the JIT-Accelerator offers significant improvements for a subset of MATLAB syntax, data types, and array sizes. MATLAB code containing loops benefits from JIT-acceleration if it has the following features and properties:

1. The loop structure is a For Loop.
2. The loop contains only logical, character string, double-precision, and less than 64-bit integer data types.
3. The loop uses arrays that are three dimensional or less.
4. All variables within a loop are defined prior to loop execution.
5. Memory for all variables within the loop are preallocated, and maintain constant size and data type for all loop iterations.
6. Loop indices are scalar quantities, such as the index i in for i=1:N.
7. Only built-in MATLAB functions are called within the loop.
8. Conditional statements with *if-then-else* or *switch-case* constructions (introduced later in this chapter) involve scalar comparisons.
9. All lines within the block contain no more than one assignment statement.

JIT-acceleration provides the greatest benefit when the arrays that are addressed within the loop are relatively small. As array sizes increase, computational time increases and the percentage of time spent on processing overhead decreases, thereby leading to less dramatic improvements in overall execution time.

The following code demonstrates the capabilities of the JIT-Accelerator:

```
N = 1e5;
% generate sin(x) at 1e5 points by using array mathematics
% this is often called a 'vectorized' solution.
```

```
x = linspace(0,2*pi,N);
y = sin(x);        % vectorized solution requires two lines
% redo above using JIT-acceleration.
i = 0;
y = zeros(1,N);  % initialize all variables within loop
x = zeros(1,N);  % and allocate all memory
for i=1:N                   % scalar loop variable
   x(i) = 2*pi*(i-1)/N; % only built-in function calls
   y(i) = sin(x(i));
end
```

With JIT-acceleration, both approaches take approximately the same time to execute. However, prior to the existence of the JIT-Accelerator, the For Loop approach would have been orders of magnitude slower than the vectorized solution. For this particular case, the array mathematics approach is much shorter and much easier to read, so the JIT-acceleration approach has little value for solving this problem. JIT-acceleration proves its value in more substantial problems where it is difficult or impossible to compose a vectorized solution by using array mathematics.

11.2 WHILE LOOPS

As opposed to a For Loop that evaluates a group of commands a fixed number of times, a While Loop evaluates a group of statements an indefinite number of times.
   The general form of a While Loop is

```
while expression
   (commands)
end
```

The (commands) between the while and end statements are executed as long as *all* elements in *expression* are True. Usually, evaluation of *expression* gives a scalar result, but array results are also valid. In the array case, *all* elements of the resulting array must be True.
   One way of computing the double-precision value eps, which is the smallest number that can be added to 1 such that the result is greater than 1, using finite precision is

```
>> num = 0; EPS = 1;
>> while (1+EPS)>1
      EPS = EPS/2;
      num = num+1;
   end
```

```
>> num
num =
    53

>> EPS=2*EPS
EPS =
    2.2204e-16
```

Here, we used uppercase EPS so that the MATLAB value eps is not overwritten. In this example, EPS starts at 1. As long as (1+EPS)>1 is True (nonzero), the commands inside the While Loop are evaluated. Since EPS is continually divided in two, it eventually gets so small that adding EPS to 1 is no longer greater than 1. (Recall that this happens because a computer uses a fixed number of digits to represent numbers. Double precision specifies approximately 16 digits, so we would expect eps to be near 10^{-16}.) At this point, (1+EPS)>1 is False (zero) and the While Loop terminates. Finally, EPS is multiplied by 2, because the last division by 2 made it too small by a factor of 2.

For array expressions, the While Loop continues only when *all* elements in *expression* are True. If you want the While Loop to continue when *any* element is True, simply use the function any. For instance, while any(*expression*) returns a scalar logical True whenever any of its contents are True.

11.3 IF-ELSE-END CONSTRUCTIONS

Many times, sequences of commands must be conditionally evaluated on the basis of a relational test. In programming languages, this logic is provided by some variation of an If-Else-End construction. The simplest If-Else-End construction is

```
if expression
    (commands)
end
```

The *(commands)* between the if and end statements are evaluated if *all* elements in expression are True (nonzero).

In cases where *expression* involves several logical subexpressions, only the minimum number required to determine the final logical state are evaluated. For example, if *expression* is (*expression1* | *expression2*), then *expression2* is evaluated only if *expression1* is False. Similarly, if *expression* is (*expression1* & *expression2*), then expression2 is not evaluated if expression1 is False. Note that this short circuiting occurs in If-Else-End constructions even if the specific short-circuiting operators || or && are not used.

The following example is illustrative:

```
>> apples = 10;              % number of apples
>> cost = apples*25          % cost of apples
cost =
   250

>> if apples>5               % give 20% discount for larger purchases
      cost = (1-20/100)*cost;
  end
>> cost
cost =
   200
```

In cases where there are two alternatives, the If-Else-End construction is

```
if expression
    (commands evaluated if True)
else
    (commands evaluated if False)
end
```

Here, the first set of commands is evaluated if *expression* is True; the second set is evaluated if *expression* is False.

When there are three or more alternatives, the If-Else-End construction takes the form

```
if expression1
    (commands evaluated if expression1 is True)
elseif expression2
    (commands evaluated if expression2 is True)
elseif expression3
    (commands evaluated if expression3 is True)
elseif expression4
    (commands evaluated if expression4 is True)
elseif expression5
    .
    .
    .
```

```
else
```

    (commands evaluated if no other expression is True)

```
end
```

In this last form, only the commands associated with the first True expression encountered are evaluated; ensuing relational expressions are not tested; and the rest of the If-Else-End construction is skipped. Furthermore, the final `else` command may or may not appear.

    Now that we know how to make decisions with If-Else-End constructions, it is possible to show a legal way to break out of For Loops and While Loops:

```
>> EPS = 1;
>> for num = 1:1000
     EPS = EPS/2;
     if (1+EPS)<=1
         EPS = EPS*2
         break
     end
   end
EPS =
   2.2204e-16

>> num
num =
   53
```

This example demonstrates another way of estimating the double-precision value eps. In this case, the For Loop is instructed to run some sufficiently large number of times. The If-Else-End structure tests to see if EPS has gotten small enough. If it has, EPS is multiplied by 2, and the `break` command forces the For Loop to end prematurely, which, in this case is at num=53.

    Furthermore, when the `break` statement is executed, MATLAB jumps to the next statement outside of the loop in which the `break` statement appears. Therefore, it returns to the MATLAB prompt and displays EPS. If a `break` statement appears in a nested For Loop or While Loop structure, MATLAB only jumps out of the immediate loop in which the `break` statement appears. It does not jump all the way out of the entire nested structure.

    MATLAB Version 6 introduced the command `continue` for use in For Loops and While Loops. When MATLAB encounters a `continue` statement inside of a For Loop or While Loop, it immediately jumps to the end statement of the loop, bypassing all of the commands between the `continue` command and the `end` statement.

In doing so, the `continue` command moves immediately to the expression test for the next pass through the loop:

```
>> EPS = 1;
>> for num = 1:1000
      EPS = EPS/2;
      if (1+EPS)>1
        continue
      end
      EPS = EPS*2
      break
   end
EPS =
   2.2204e-016
```

Here, the previous example is rewritten to use the `continue` command. Note that the `continue` command has no affect on the If-End construction.

11.4 SWITCH-CASE CONSTRUCTIONS

When sequences of commands must be conditionally evaluated on the basis of repeated use of an equality test with one common argument, a Switch-Case construction is often easier. Switch-Case constructions have the form

```
switch expression
    case test_expression1
       (commands1)
    case {test_expression2, test_expression3, test_expression4}
       (commands2)
    otherwise
       (commands3)
end
```

where *expression* must be either a scalar or a character string. If *expression* is a scalar, *expression*==*test_expressionN* is tested by each case statement. If *expression* is a character string, `strcmp(`*expression*`,`*test_expression*`)` is tested. In this example, `expression` is compared with *test_expression1* at the first case statement. If they are equal, *(commands1)* are evaluated, and the rest of the statements before the end statement are skipped. If the first comparison is not true, the second is considered. In this example, *expression* is compared with *test_expression2*, *test_expression3*, and *test_expression4*, which are contained in a cell array. If any of these are equal to *expression*, *(commands2)* are evaluated, and the rest of the statements before end are skipped. If all case comparisons are False, *(commands3)* following the optional otherwise statement are executed. ***Note that this implementation of the Switch-Case construction allows at most one of the command groups to be executed.***

A simple example demonstrating the Switch-Case construction is

```
x = 2.7;
units = 'm';
switch units  % convert x to centimeters
   case {'inch','in'}
      y = x*2.54;
   case {'feet','ft'}
      y = x*2.54*12;
   case {'meter','m'}
      y = x/100;
   case {'millimeter','mm'}
     y = x*10;
   case {'centimeter','cm'}
      y = x;
   otherwise
      disp(['Unknown Units: ' units])
      y = nan;
end
```

Executing this code gives a final value of y=0.027.

11.5 TRY-CATCH BLOCKS

A Try-Catch block provides user-controlled error-trapping capabilities. That is, with a Try-Catch block, errors found by MATLAB are captured, giving the user the ability to control the way MATLAB responds to errors. Try-Catch blocks have the form

```
try
   (commands1)
catch
   (commands2)
end
```

Here, all MATLAB expressions in *(commands1)* are executed. If no MATLAB errors are generated, control is passed to the end statement. However, if a MAT-LAB error appears while executing *(commands1)*, control is immediately passed to the catch statement and subsequent expressions in *(commands2)*. The code within *(commands2)* can make use of the functions lasterr and lasterror to access information about the error and act accordingly.

Consider the following example, implemented in a script M-file for convenience:

```
x = ones(4,2);
y = 4*eye(2);
try
    z = x*y;
catch
    z = nan;
    disp('X and Y are not conformable.')
end
z
```

With the preceding data for x and y, this code segment produces the following output:

```
z =
        4       4
        4       4
        4       4
        4       4
```

In this case, only the code in the Try block was executed. Changing the variable y creates an error:

```
x = ones(4,2);
y = 4*eye(3);   % now wrong size
try
    z = x*y;
catch
    z = nan;
    disp('X and Y are not conformable.')
end
z
```

Executing the code this time generates the following output in the *Command* window:

```
X and Y are not conformable.
z =
   NaN
```

In addition, the function `lasterr` describes the error found:

```
>> lasterr
ans =
Error using ==> *
Inner matrix dimensions must agree.
```

More detailed information is returned by the structure output of the function `lasterror`:

```
>> errstr = lasterror
errstr =
        message: [1x53 char]
     identifier: 'MATLAB:innerdim'
>> errstr.message
ans =
Error using ==> *
Inner matrix dimensions must agree.
```

Here, the content of the `message` field is the same as the output from `lasterr`. The `indentifier` field describes the message identifier, which in this case classifies the error type as being generated by an inner-dimension error in MATLAB.

The Catch block also may perform tasks and then reissue the original error by using the `rethrow` function. For example, revising and rerunning the code from the previous example produces

```
x = ones(4,2);
y = 4*eye(3);   % now wrong size
try
   z = x*y;
catch
   z = nan;
   disp('X and Y are not conformable.')
```

```
      rethrow(lasterror) % process error as if Try-Catch did not happen.
end
z
```

In this case, the *Command* window displays

```
X and Y are not conformable.
??? Error using ==> *
Inner matrix dimensions must agree.
```

The rethrow function reissues the error, terminates execution, and therefore does not display the contents of z as requested by the last code line.

12

Functions

When you use MATLAB functions such as inv, abs, angle, and sqrt, MATLAB takes the variables that you pass to it, computes the required results using your input, and then passes these results back to you. The commands evaluated by the function, as well as any intermediate variables created by these commands, are hidden. All you see is what goes in and what comes out. In other words, a function is a black box.

This property makes functions very powerful tools for evaluating commands that encapsulate useful mathematical functions or sequences of commands that often appear when you are solving some larger problem. Because of the usefulness of this power, MATLAB provides several structures that enable you to create functions of your own. These structures include *M-file* functions, *anonymous* functions, and *inline* functions. Of these, M-file functions are the most common. M-file functions are text files that contain MATLAB code and a function header. The function mmempty is a good example of an M-file function:

```
function d=mmempty(a,b)
%MMEMPTY Substitute Value if Empty.
% MMEMPTY(A,B) returns A if A is not empty,
% otherwise B is returned.
%
% Example: The empty array problem in logical statements
% let a=[]; then use MMEMPTY to set default logical state
% (a==1) is [], but MMEMPTY(a,1)==1 is true
% (a==0) is [], but MMEMPTY(a,0)==0 is true
```

```
% Also:
% sum(a) is 0,  but sum(MMEMPTY(a,b))=sum(b)
% prod(a) is 1, but prod(MMEMPTY(a,b))=prod(b)
%
% See also ISEMPTY, SUM, PROD, FIND
if isempty(a)
     d=b;
else
     d=a;
end
```

A function M-file is similar to a script M-file in that it is a text file having a .m extension. Like script M-files, function M-files are not entered in the *Command* window, but rather are external text files created with a text editor (probably the Editor/Debugger that comes with MATLAB). A function M-file is different from a script file in that a function communicates with the MATLAB workspace only through the variables passed to it and through the output variables it creates. Intermediate variables within the function do not appear in, or interact with, the MATLAB workspace. As can be seen in the previous example, the first line of a function M-file defines the M-file as a function and specifies its name, which is the same as its filename without the .m extension. The first line also defines the M-file's input and output variables. The next continuous sequence of comment lines comprises the text displayed in response to the help command `help mmempty` or `helpwin mmempty`. The first help line, called the H1 line, is the line searched by the `lookfor` command. Finally, the remainder of the M-file contains MATLAB commands that create the output variables. Note that there is no `return` command in `mmempty`; the function simply terminates after it executes the last command. However, you can use the `return` command to terminate execution before reaching the end of the M-file.

12.1 M-FILE FUNCTION CONSTRUCTION RULES

Function M-files must satisfy a number of criteria and should have a number of desirable features:

1. The function M-file name and the function name (e.g., `mmempty`) that appear in the first line of the file should be identical. In reality, MATLAB ignores the function name in the first line and executes functions on the basis of the file name stored on disk.

2. Function M-file names can have up to 63 characters. This maximum may be limited by the operating system, in which case the lower limit applies. MATLAB ignores characters beyond the 63rd or the operating system limit, and so

longer names can be used, provided the legal characters point to a unique file name.

3. Function M-file names are case sensitive on UNIX platforms, and in MATLAB 7 they are now case sensitive on Windows platforms as well. To avoid platform dependencies across MATLAB versions, it is beneficial to use only lowercase letters in M-file names.

4. Function names must begin with a letter. Any combination of letters, numbers, and underscores can appear after the first character. Function names cannot contain spaces or punctuation characters. This naming rule is identical to that for variables.

5. The first line of a function M-file is called the *function-declaration line* and must contain the word `function` followed by the calling syntax for the function in its most general form. The input and output variables identified in the first line are variables local to the function. The input variables contain data passed to the function, and the output variables contain data passed back from the function. It is not possible to pass data back through the input variables.

6. The first set of contiguous comment lines after the function-declaration line are the help text for the function. The first comment line is called the H1 line and is the line searched by the `lookfor` command. The H1 line typically contains the function name in uppercase characters and a concise description of the function's purpose. Comment lines after the first describe possible calling syntaxes, algorithms used, and simple examples, if appropriate.

7. Function names appearing in the help text of a function are normally capitalized only to give them visual distinction. Functions are called by matching the exact case of the letters making up their filenames.

8. All statements following the first set of contiguous comment lines compose the body of the function. The body of a function contains MATLAB statements that operate on the input arguments and produce results in the output arguments.

9. A function M-file terminates after the last line in the file is executed or whenever a `return` statement is encountered. If an M-file contains nested functions, each function in the M-file requires a terminating `end` statement.

10. A function can abort operation and return control to the *Command* window by calling the function `error`. This function is useful for flagging improper function usage, as shown in the following code fragment:

```
if length(val) > 1
    error('VAL must be a scalar.')
end
```

When the function `error` in the preceding code is executed, the string `'VAL must be a scalar.'` is displayed in the *Command* window, after a line identifying the file that produced the error message. Passing an empty string to `error` (e.g., `error('')`) causes no action to be taken. After being issued, the error character string is passed to the functions `lasterror` and `lasterr` for later recall. It is also possible to pass numerical data to the displayed error string by using the function `error` as one would use the function `sprintf`. For example, the previous example could be revised as follows:

```
val = zeros(1,3);
if length(val) > 1
    error('VAL has %d elements but must be a scalar.',length(val))
end
??? VAL has 3 elements but must be a scalar.
```

The %d specification indicates that an integer format should be used to insert the value of `length(val)` into the given place in the error string. When MATLAB identifies an error, it also creates an error message identifier string that is returned by the function `lasterror`, as in the following code:

```
>> eig(eye(2,4))
??? Error using ==> eig
Matrix must be square.
>> lasterror
ans =
        message: [1x42 char]
     identifier: 'MATLAB:square'
```

Here, the `identifier` field identifies the error source as coming from MATLAB and having to do with square matrices. To include this information in an M-file function, simply add it as a first argument to the `error` function, as illustrated in the following code:

```
>> val = zeros(1,3);
>> msg = 'VAL has %d elements but must be a scalar.';
if length(val) > 1
    error('MyToolbox:scalar',msg,length(val))
end
??? VAL has 3 elements but must be a scalar.
>> lasterror
ans =
        message: 'VAL has 3 elements but must be a scalar.'
     identifier: 'MyToolbox:scalar'
```

The identifier field now shows that this error was flagged by a function in MyToolbox, rather than by MATLAB itself, and that the error was related to a scalar.

11. A function can report a warning and then continue operation by calling the function warning. This function has the same calling syntax as the function error. Warnings can contain simple character strings, strings containing formatted data, and an optional initial message-identifier string. The difference between the warning and error functions is that warnings can be turned Off globally, or have warnings associated with specific message identifiers turned Off. Warning states can also be queried. (See MATLAB documentation for more thorough information about the features of the function warning.)

12. Function M-files can contain calls to script files. When a script file is encountered, it is evaluated in the function's workspace, not in the MATLAB workspace.

13. Multiple functions can appear in a single function M-file. Additional functions, called subfunctions or local functions, are simply appended to the end of the primary function. Subfunctions begin with a standard function statement line and follow all function construction rules.

14. Subfunctions can be called by the primary function in the M-file, as well as by other subfunctions in the same M-file, but subfunctions cannot be called directly from outside the M-file. Like all functions, subfunctions have their own individual workspaces.

15. Subfunctions can appear in any order after the primary function in an M-file. Help text for subfunctions can be displayed by entering >> helpwin *func/subfunc*, where *func* is the main function name and *subfunc* is the subfunction name.

16. It is suggested that subfunction names begin with the word local, for example, local_myfun. This practice improves the readability of the primary function, because calls to local functions are clearly identifiable. All local function names can have up to 63 characters. The following function mmclass demonstrates the use of subfunctions:

```
function c=mmclass(arg)
%MMCLASS MATLAB Object Class Existence.
% MMCLASS returns a cell array of strings containing the
% names of MATLAB object classes available with this license.
%
% MMCLASS('ClassName') returns logical True (1) if the class
% having the name 'ClassName' exists with this license.
% Otherwise logical False (0) is returned.
%
% MMCLASS searches the MATLABPATH for class directories.
```

```matlab
% Classes not on the MATLABPATH are ignored.
%
% See also CLASS, ISA, METHODS, ISOBJECT
persistent clist  % save data for future calls

if isempty(clist) % clist contains no data, so create it
   clist=local_getclasslist;
end
if nargin==0
   c=clist;
elseif ischar(arg)
   c=~isempty(strmatch(arg,clist));
else
   error('Character String Argument Expected.')
end

function clist=local_getclasslist
%LOCAL_GETCLASSLIST Get list of MATLAB classes
%
% LOCAL-GETCLASSLIST returns a list of all MATLAB classes
% in a cell array of strings.
clist=cell(0);
cstar=[filesep '@*'];
dlist=[pathsep matlabpath];
sidx=findstr(pathsep,dlist)+1;      % path segment starting indices
eidx=[sidx(2:end)-2 length(dlist)]; % path segment ending indices

for i=1:length(sidx)-1   % look at each path segment
   cdir=dir([dlist(sidx(i):eidx(i)) cstar]); % dir @* on segment
   clist=[clist {cdir.name}];               % add results to list
end
cstr=char(clist);         % convert to string array
cstr(:,1)=[];             % eliminate initial '@'
cstr=unique(cstr,'rows');% alphabetize and make unique
clist=cellstr(cstr);      % back to a cell array
% end of subfunction
```

17. In addition to subfunctions, M-files can call private M-files, which are standard function M-files that reside in a subdirectory of the calling function entitled `private`. Only functions in the immediate parent directory of private M-files have access to private M-files. Private subdirectories are meant to contain utility functions useful to several functions in the parent directory. Private function M-file names need not be unique, because of their higher precedence and limited scope.

18. It is suggested that private M-file names begin with the word `private`, for example, `private_myfun`. This practice improves the readability of the primary function, because calls to private functions are clearly identifiable. Like other function names, the names of all private M-files can have up to 63 characters.

12.2 INPUT AND OUTPUT ARGUMENTS

MATLAB functions can have any number of input and output arguments. The features of, and criteria for, these arguments are as follows:

1. Function M-files can have zero input arguments and zero output arguments.

2. Functions can be called with fewer input and output arguments than are specified in the function-definition line in the M-file. Functions cannot be called with more input or output arguments than the M-file specifies.

3. The number of input and output arguments used in a function call can be determined by calls to the functions `nargin` and `nargout`, respectively. Since `nargin` and `nargout` are functions, not variables, one cannot reassign them with statements such as `nargin = nargin - 1`. The function `mmdigit` illustrates the use of `nargin`:

```
function y=mmdigit(x,n,b,t)
%MMDIGIT Round Values to Given Significant Digits.
% MMDIGIT(X,N,B) rounds array X to N significant places in base B.
% If B is not given, B=10 is assumed.
% If X is complex the real and imaginary parts are rounded separately.
% MMDIGIT(X,N,B,'fix') uses FIX instead of ROUND.
% MMDIGIT(X,N,B,'ceil') uses CEIL instead of ROUND.
% MMDIGIT(X,N,B,'floor') uses FLOOR instead of ROUND.

if nargin<2
    error('Not enough input arguments.')
elseif nargin==2
    b=10;
```

```
    t='round';
elseif nargin==3
    t='round';
end
n=round(abs(n(1)));
if isempty(b), b=10;
else            b=round(abs(b(1)));
end
if isreal(x)
    y=abs(x)+(x==0);
    e=floor(log(y)./log(b)+1);
    p=repmat(b,size(x)).^(n-e);
    if strncmpi(t,'round',1)
        y=round(p.*x)./p;
    elseif strncmpi(t,'fix',2)
        y=fix(p.*x)./p;
    elseif strncmpi(t,'ceil',1)
        y=ceil(p.*x)./p;
    elseif strncmpi(t,'floor',2)
        y=floor(p.*x)./p;
    else
        error('Unknown rounding requested')
    end
else % complex input
    y=complex(mmdigit(real(x),n,b,t),mmdigit(imag(x),n,b,t));
end
```

In mmdigit, nargin is used to assign default values to input arguments not supplied by the user.

4. When a function is called, the input variables are not copied into the function's workspace, but their values are made *readable* within the function. However, if any values in the input variables are changed, the array is then copied into the function's workspace. Thus, to conserve memory and increase speed, it is better to extract elements from large arrays and then modify them, rather than

to force the entire array to be copied into the function's workspace. Note that using the same variable name for both an input and an output argument causes an immediate copying of the contents of the variable into the function's workspace. For example, `function y=myfunction(x,y,z)` causes the variable y to be immediately copied into the workspace of `myfunction`.

5. If a function declares one or more output arguments, but no output is desired, simply do not assign the output variable (or variables) any values. Alternatively, the function `clear` can be used to delete the output variables before terminating the function.

6. Functions can accept a variable and an unlimited number of input arguments by specifying `varargin` as the last input argument in the function-declaration line. The argument `varargin` is a predefined cell array whose *i*th cell is the *i*th argument, starting from where `varargin` appears. For example, consider a function having the following function-declaration line:

`function a=myfunction(varargin)`

If this function is called as `myfunction(x,y,z)`, then `varargin{1}` contains the array x, `varargin{2}` contains the array y, and `varargin{3}` contains the array z. Likewise, if the function is called as `myfunction(x)`, then `varargin` has length 1 and `varargin{1}=x`. Every time `myfunction` is called, it can be called with a different number of arguments.

In cases where one or more input arguments are fixed, `varargin` must appear as the last argument:

`function a=myfunction(x,y,varargin)`

If this function is called as `myfunction(x,y,z)`, then, inside of the function, x and y are available, and `varargin{1}` contains z. In any case, the function `nargin` returns the actual number of input arguments used. (For further information on cell arrays, see Chapter 8.)

7. Functions can accept a variable, unlimited number of output arguments by specifying `varargout` as the last output argument in the function-declaration line. The argument `varargout` is a predefined cell array whose *i*th cell is the *i*th argument, starting from where `varargout` appears. For example, consider a function having the following function-declaration line:

`function varargout=myfunction(x)`

If this function is called as `[a,b]=myfunction(x)`, then, inside of the function, the contents of `varargout{1}` must be assigned to the data that become the variable a, and the contents of `varargout{2}` must be assigned to the data that become the variable b. As with `varargin`, discussed previously, the length of `varargout` is equal to the number of output arguments used and `nargout` returns this length. In cases where one or more output arguments are fixed, `varargout` must appear as the last argument in the function-declaration line, that is, `function [a,b,varargout]=myfunction(x)`. (For further information on cell arrays, see Chapter 8.)

8. The functions `nargchk` and `nargoutchk` provide simple error checking for the number of valid input and output arguments, respectively. Since functions automatically return an error if called with more input or output arguments than appear in their function definitions, these functions have limited value. They may be useful, however, when a function definition declares an arbitrary number of input or output arguments.

12.3 FUNCTION WORKSPACES

As stated earlier, functions are black boxes. They accept inputs, act on these inputs, and create outputs. Any and all variables created within the function are hidden from the MATLAB or base workspace. Each function has its own temporary workspace that is created with each function call and deleted when the function completes execution. MATLAB functions can be called recursively, and each call has a separate workspace. In addition to furnishing input and output arguments, MATLAB provides several techniques for communicating among function workspaces and the MATLAB or base workspace:

1. Functions can share variables with other functions, the MATLAB workspace, and recursive calls to themselves if the variables are declared `global`. To gain access to a global variable within a function or the MATLAB workspace, the variable must be declared `global` within each desired workspace—for example, `global myvariable`.

> As a matter of programming practice, the use of global variables is discouraged whenever possible. However, if they are used, it is suggested that global variable names be long, contain all capital letters, and optionally start with the name of the M-file where they appear; for example, `MYFUN_ALPHA`. If followed, these suggestions will minimize unintended conflicts among global variables.

2. In addition to sharing data through global variables, function M-files can have restricted access to variables for repeated or recursive calls to themselves by declaring a variable `persistent`, such as `persistent myvariable`. Persistent variables act like global variables whose scope is limited to the function where they are declared. Persistent variables exist as long as an M-file remains in memory in MATLAB. The function `mmclass` illustrates the use of persistent variables:

```
function c=mmclass(arg)
%MMCLASS MATLAB Object Class Existence.
% MMCLASS returns a cell array of strings containing the
```

```
% names of MATLAB object classes available with this license.
%
% MMCLASS('ClassName') returns logical True (1) if the class
% having the name 'ClassName' exists with this license.
% Otherwise logical False (0) is returned.
%
% MMCLASS searches the MATLABPATH for class directories.
% Classes not on the MATLABPATH are ignored.
%
% See also CLASS, ISA, METHODS, ISOBJECT
persistent clist  % save data for future calls

if isempty(clist) % clist contains no data, so create it
   clist=cell(0);
   cstar=[filesep '@*'];
   dlist=[pathsep matlabpath];
   sidx=findstr(pathsep,dlist)+1;      % path segment starting indices
   eidx=[sidx(2:end)-2 length(dlist)]; % path segment ending indices
   for i=1:length(sidx)-1     % look at each path segment
      cdir=dir([dlist(sidx(i):eidx(i)) cstar]); % dir @* on segment
      clist=[clist {cdir.name}];                % add results to list
   end
   cstr=char(clist);          % convert to string array
   cstr(:,1)=[];              % eliminate initial '@'
   cstr=unique(cstr,'rows'); % alphabetize and make unique
   clist=cellstr(cstr);      % back to a cell array
end
if nargin==0
   c=clist;
elseif ischar(arg)
   c=~isempty(strmatch(arg,clist));
else
   error('Character String Argument Expected.')
end
```

In `mmclass`, the variable `clist` is declared persistent. The first time `mmclass` is called during a MATLAB session, `clist` is created as an empty array. When the function finds it empty, it fills it with data in the first If-End construction in the function. In future calls to `mmclass`, `clist` exists because of its persistence, and re-creating `clist` in the future is unnecessary. For these subsequent function calls, the data previously stored in `clist` is simply reused.

3. MATLAB provides the function `evalin` to allow you to reach into another workspace, evaluate an expression, and return the result to the current workspace. The function `evalin` is similar to `eval`, except that the string is evaluated in either the ***caller*** or the ***base*** workspace. The caller workspace is the workspace where the current function was called from. The base workspace is the MATLAB workspace in the *Command* window. For example, `A=evalin('caller', 'expression')` evaluates `'expression'` in the caller workspace and returns the results to the variable A in the current workspace. Alternatively, `A=evalin('base', 'expression')` evaluates `'expression'` in the MATLAB workspace and returns the results to the variable A in the current workspace. The function `evalin` also provides error trapping with the syntax `evalin('workspace', 'try', 'catch')`, where `'workspace'` is either `'caller'` or `'base'`; `'try'` is the first expression evaluated; and `'catch'` is an expression that is evaluated in the ***current*** workspace if the evaluation of `'try'` produces an error.

4. Since you can evaluate an expression in another workspace, it makes sense that you can also assign the results of some expression in the current workspace to a variable in another workspace. The function `assignin` provides this capability. For example, `assignin('workspace', 'vname', X)`, where `'workspace'` is either `'caller'` or `'base'`, assigns the contents of the variable *X* in the current workspace to a variable named `'vname'` in the `'caller'` or `'base'` workspace.

5. The function `inputname` provides a way to determine the variable names used when a function is called. For example, suppose a function is called as

```
>> y = myfunction(xdot,time,sqrt(2))
```

Issuing `inputname(1)` inside of `myfunction` returns the character string `'xdot'`, `inputname(2)` returns `'time'`, and `inputname(3)` returns an empty array, because `sqrt(2)` is not a variable, but rather an expression that produces an unnamed temporary result.

The function `mmswap` illustrates the use of `evalin`, `assignin`, and `inputname`:

```
function mmswap(x,y)
%MMSWAP Swap Two Variables.
% MMSWAP(X,Y) or MMSWAP X Y  swaps the contents of the
% variable X and Y in the workspace where it is called.
```

```
% X and Y must be variables not literals or expressions.
%
% For example: Rat=ones(3); Tar=pi; MMSWAP(Rat,Tar) or MMSWAP Rat
Tar
% swaps the contents of the variables named Rat and Tar in the
% workspace where MMSWAP is called giving Rat=pi and Tar=ones(3).
if nargin~=2
   error('Two Input Arguments Required.')
end
if ischar(x) & ischar(y)   % MMSWAP X Y 'string input arguments'

   % check existence of arguments in caller
   estr=sprintf('[exist(''%s'',''var'')
exist(''%s'',''var'')]',x,y);
   t=evalin('caller',estr);
   if all(t)                   % both x and y are valid
      xx=evalin('caller',x);   % get contents of x
      yy=evalin('caller',y);   % get contents of y
      assignin('caller',y,xx)  % assign contents of x to y
      assignin('caller',x,yy)  % assign contents of y to x

   elseif isequal(t,[0 1])     % x is not valid
      error(['Undefined Variable: ''' x ''''])

   elseif isequal(t,[1 0])     % y is not valid
      error(['Undefined Variable: ''' y ''''])

   else                        % neither is valid
      error(['Undefined Variables: ''' x ''' and ''' y ''''])
   end
else                  % MMSWAP(X,Y) 'numerical input arguments'
   xname=inputname(1);   % get x argument name if it exists
   yname=inputname(2);   % get x argument name if it exists
```

```
    if ~isempty(xname) & ~isempty(yname)    % both x and y are valid
        assignin('caller',xname,y)          % assign contents of y to x
        assignin('caller',yname,x)          % assign contents of x yo y

    else
        error('Arguments Must be Valid Variables.')
    end
end
```

6. The name of the M-file being executed is available within a function in the variable mfilename. For example, when the M-file myfunction.m is being executed, the workspace of the function contains the variable mfilename, which contains the character string 'myfunction'. This variable also exists within script files, in which case it contains the name of the script file being executed.

12.4 FUNCTIONS AND THE MATLAB SEARCH PATH

Function M-files and their powerful features are among the fundamental strengths of MATLAB. They allow you to encapsulate sequences of useful commands and apply them over and over. Since M-files exist as text files on disk, it is important that MATLAB maximize the speed at which the files are found, opened, and executed. The following are the techniques that MATLAB uses to maximize speed:

1. The first time MATLAB executes a function M-file, it opens the corresponding text file and *compiles* the commands into an internal pseudocode representation in memory that speeds execution for all later calls to the function. If the function contains references to other M-file functions and script M-files, they too are compiled into memory.

2. The function inmem returns a cell array of strings containing a list of functions and script files that are currently compiled into memory.

3. Issuing the command mlock within an M-file locks the compiled function so that it cannot be cleared from memory. For example, clear functions does not clear a locked function from memory. By locking an M-file, persistent variables declared in a function are guaranteed to exist from one call to the next. The function call munlock('FUN') unlocks the function FUN, so that it can be cleared from memory. The function mislocked('FUN') returns True if the function FUN is currently locked in memory. By default, function M-files are unlocked.

4. It is possible to store the compiled pseudocode, or P-code, version of a function M-file to disk by using the pcode command. When this is done, MATLAB loads the compiled function, rather than the M-file, into memory. For most functions, this step does not significantly shorten the amount of time required to execute a function for the first time. However, it can speed up large M-files associated with complex graphical user-interface functions. P-code files are created by issuing the command

```
>> pcode myfunction
```

where *myfunction* is the M-file name to be compiled. P-code files are encrypted, platform-independent binary files that have the same name as the original M-file, but end in .p rather than in .m. P-code files provide a level of security, since they are visually indecipherable and can be executed without the corresponding M-file. Furthermore, it is not possible to convert a P-code file into an M-file. Because of their binary nature, P-code files are not backward compatible across MATLAB versions. That is, a P-code file created by using MATLAB 7.0 will not run on MATLAB 6.X. However, a P-code file created using MATLAB 6.X will generally run on MATLAB 7.0.

5. When MATLAB encounters a name it doesn't recognize, it follows a set of precedence rules to determine what to do. For example, when you enter cow at the MATLAB prompt, or if MATLAB encounters a reference to cow in a script or function M-file,

  a. MATLAB checks to see if cow is a ***variable*** in the current workspace; if not,
  b. it checks to see if cow is a ***subfunction*** in the file in which cow appears; if not,
  c. it checks to see if cow is a ***private*** function to the file in which cow appears; if not,
  d. it checks to see if cow exists in the ***current directory***; if not,
  e. it checks to see if cow exists in each directory specified on the ***MATLAB search path***, by searching in the order in which the search path is specified.

  MATLAB uses the first match it finds. In addition, in steps d and e, it prioritizes by type by considering MEX-files first, followed by P-code files, followed by M-files. So if cow.*mex*, cow.p, and cow.m exist, MATLAB uses cow.*mex*, where *mex* is replaced by the platform-dependent MEX-file extension. If cow.p and cow.m exist, MATLAB uses cow.p. If cow.m exists and cow is a *built-in* function, the built-in function is executed. If cow.m exists and is not a built-in function, the M-file is executed.

6. When MATLAB is started, it ***caches*** the name and location of all M-files stored within the toolbox subdirectory and within all subdirectories of the toolbox directory. This allows MATLAB to find and execute function M-files much faster.

M-file functions that are cached are considered read-only. If they are executed and then later altered, MATLAB simply executes the function that was
previously compiled into memory, ignoring the changed M-files. Moreover, if
new M-files are added within the `toolbox` directory after MATLAB is already running, their presence will not be noted in the cache, and thus they will
be unavailable for use.

As a result, as you develop M-file functions, it is best to store them outside of the `toolbox` directory, perhaps in the `MATLAB` directory, until you
consider them to be complete. When they are complete, move them to a subdirectory inside of the read-only `toolbox` directory. Finally, make sure the
MATLAB search path cache is changed to recognize their existence.

7. When new M-files are added to a ***cached*** location, MATLAB finds them only
if the cache is refreshed by the command `rehash  toolbox`. On the other
hand, when cached M-files are modified, MATLAB recognizes the changes
only if a previously compiled version is dumped from memory as a result of
issuing the `clear` command; for example, >> `clear  myfun` clears the M-file
function *myfun* from memory. Alternatively, >> `clear functions` clears all
unlocked, compiled functions from memory.

8. MATLAB keeps track of the modification date of M-files outside of the
`toolbox` directory. As a result, when an M-file function that was previously
compiled into memory is encountered, MATLAB compares the modification
date of the compiled M-file with that of the M-file on disk. If the two dates are
the same, MATLAB executes the compiled M-file. On the other hand, if the
M-file on disk is newer, MATLAB dumps the previously compiled M-file and
compiles the newer, revised M-file for execution.

9. It is possible to check all of the file dependencies for a function M-file by using
the function `depfun`. This function rigorously parses an M-file for all calls to
other M-file functions, built-in functions, and function calls in `eval` strings and
callbacks, and identifies variable and Java classes used. This function is helpful
for identifying function dependencies in M-files that are being shared with
others who may not have the same *Toolboxes* installed. The function `depdir`
uses `depfun` to return a listing of the dependent directories of an M-file.

12.5 CREATING YOUR OWN TOOLBOX

It is common to organize a group of M-files into a subdirectory on the MATLAB
search path. If the M-files are considered complete, the subdirectory should be
placed in the `toolbox` directory so that the M-file names are cached, as described
earlier. When a `toolbox` subdirectory is created, it is beneficial to include two additional script M-files containing only MATLAB comments (i.e., lines that begin with
a percent sign %). These M-files, named `Readme.m` and `Contents.m`, have the following properties:

1. The script file `Readme.m` typically contains comment lines that describe late-
breaking changes or descriptions of undocumented features. Issuing the

command >> `whatsnew` *MyToolbox* (where *MyToolbox* is the name of the directory containing the group of M-files) displays this file in the *Help* window. If the Toolbox is posted to *MATLAB Central* at *The Mathworks* website, the `Readme.m` file should include a disclaimer, such as the following, to avoid legal problems:

```
% These M-files are User Contributed Routines that are being redistributed
% by The Mathworks, upon request, on an "as is" basis. A User Contributed
% Routine is not a product of The Mathworks, Inc. and The Mathworks assumes
% no responsibility for any errors that may exist in these routines.
```

2. The script file `Contents.m` contains comment lines that list all the M-files in the Toolbox. Issuing the command >> `helpwin` *MyToolbox* (where *MyToolbox* is the name of the directory containing the group of M-files) displays the file listing in the *Help* window. The first line in the `Contents.m` file should specify the name of the Toolbox, and the second line should state the Toolbox version and date, as follows:

```
% Toolbox Description
% Version xxx dd-mmm-yyyy
```

This information is used by the `ver` command, which lists installed toolbox information.

3. When writing a collection of M-files to form a Toolbox, it is sometimes convenient to allow the user to maintain a set of preferences for Toolbox use or to set preferences for one or more functions in the Toolbox. While it is always possible to store this information in a MAT-file and retrieve it in later MATLAB sessions, doing so requires choosing a directory location for the preferences file and guaranteeing that the file isn't moved or deleted between sessions. To eliminate these weaknesses of the data file approach, MATLAB provides the functions `getpref`, `setpref`, `addpref`, and `rmpref`. These functions allow you to get, set, add, and remove preferences, respectively. Preferences are organized in groups so that preferences for multiple activities are supported. Within each group, individual preferences are named with character strings, and the values stored can be any MATLAB variable. This is similar to the way structures work. For example, `group.prefname=values` is a way to store `values` for `prefname` in a structure named `group`. When these functions are used, the handling of preference files is hidden from the user. Where they are stored is system dependent, and they remain persistent from one MATLAB session to the next.

12.6 COMMAND-FUNCTION DUALITY

In addition to creating function M-files, it is also possible to create MATLAB ***commands***. Examples of MATLAB commands include `clear`, `who`, `dir`, `ver`, `help`, and `whatsnew`. MATLAB commands are very similar to functions. In fact, there are only two differences between commands and functions:

1. Commands do not have output arguments.
2. Input arguments to commands are not enclosed in parentheses.

For example, `clear  functions` is a command that accepts the input argument `functions` without parentheses, performs the action of clearing all compiled functions from memory, and produces no output. A function, on the other hand, usually places data in one or more output arguments, and must have its input arguments separated by commas and enclosed in parentheses (e.g., `a=atan2(x,y)`).

In reality, MATLAB commands are function calls that obey the two differences. For example, the command `whatsnew` is a function M-file. When called from the MATLAB prompt as

```
>> whatsnew MyToolbox
```

MATLAB interprets the command as a call to the function `whatsnew` with the following syntax:

```
>> whatsnew('MyToolbox')
```

In other words, as long as there are no output arguments requested, MATLAB interprets command arguments as character strings, places them in parentheses, and then calls the requested function. This interpretation applies to all MATLAB commands.

Both command and function forms can be entered at the MATLAB prompt, although the command form generally requires less typing. A function M-file can also be interpreted as a function, if it obeys the rules for calling functions. For example, the command

```
>> which fname
```

displays the directory path string to the M-file *fname*, and the function call

```
>> s = which('fname')
```

returns the directory path string in the variable `s`. At the same time, the code

```
>> s = which fname
??? s = which fname
            |
Error: Missing MATLAB operator.
```

causes an error because it mixes function and command syntaxes. *Whenever MATLAB encounters an equal sign, it interprets the rest of the statement as a function, which requires a comma-separated list of arguments enclosed in parentheses.*

To summarize, both commands and functions call functions. Commands are translated into function calls by interpreting command arguments as character strings, placing them in parentheses, and then calling the requested function. Any function call can be made in the form of a command if it produces no output and if it requires only character-string input.

12.7 FUNCTION HANDLES AND ANONYMOUS FUNCTIONS

There are a number of occasions when the identity of a function must be passed to another function for evaluation. For example, many of the numerical analysis functions in MATLAB evaluate a function provided by the user as part of the function's input arguments. For example, Chapter 24 discusses the function quad, which, when called as quad(Fun, low, high), computes the area under the function Fun over the range from low to high. Historically, the function argument Fun was specified by the character-string name of the function to be evaluated (e.g., $sin(x)$ was denoted as 'sin'). This method works for both built-in functions and M-file functions. Alternatively, MATLAB 5 introduced *inline functions*, which create functions from character-string expressions. For example, in the code

```
>> il_humps = inline('1./((x-.3).^2 +.01) + 1./((x-.9).^2 +.04) - 6','x')
il_humps =

    Inline function:
    il_humps(x) = 1./((x-.3).^2 +.01) + 1./((x-.9).^2 +.04) - 6
```

the function inline creates an inline function object from the character string, with x as the variable. Using an inline function definition, Fun in quad(Fun,low,high) is simply replaced by the name of the inline function (e.g., quad(il_humps,low,high)). To evaluate functions specified by their character-string name or inline function object, MATLAB provides the function feval. For example, in the code

```
>> y = feval('sin',pi*(0:4)/4)
y =
             0      0.70711          1      0.70711   1.2246e-016
>> z = feval(il_humps,[-1 0 1])
z =
       -5.1378        5.1765             16
```

feval simply evaluates the function described by its first argument, using whatever data are supplied as additional arguments. The preceding example demonstrates the use of feval with one vector input argument and one vector output argument. In general, any number of input or output arguments can be used.

In MATLAB 7, the use of strings identifying function names (e.g., 'sin') and the use of in-line functions (e.g., il_humps) remain supported, but are discouraged in favor of using **anonymous functions** and their corresponding **function handles**.

Anonymous functions were introduced in MATLAB 7 as a replacement for inline functions and are created as shown in the following example:

```
>> af_humps =@(x) 1./((x-.3).^2 +.01) +1./((x-.9).^2 +.04) -6;
```

Here, the @ symbol identifies that the left-hand side is to be a *function handle*. The (x) defines the list of function arguments, and the remainder of the line describes the function expression. Evaluation of this function can use feval, as in the code

```
>> z = feval(af_humps,[-1 0 1])
z =
        -5.1378        5.1765               16
```

However, the function feval is no longer needed or recommended. It is preferable to use the function handle name itself to perform function evaluation:

```
>> z = af_humps([-1 0 1])
z =
        -5.1378        5.1765               16
```

The definition of an anonymous function can access any MATLAB function, as well as the present content of variables that exist in the workspace where the anonymous function is created. For example, in the code

```
>> a = -.3; b = -.9;
>> af_humpsab = @(x) 1./((x+a).^2 + .01) + 1./((x+b).^2 + .04) - 6;

>> af_humpsab([-1 0 1])
ans =
         -5.1378         5.1765              16
```

the values of the previously defined values of a and b become part of the anonymous function definition. If the values of a or b change, the anonymous function does not change. The function handle af_humpsab captures and holds a snapshot of the function at the time it is created:

```
>> a = 0; % changing the value of a does not change the function
>> af_humpsab([-1 0 1]) % evaluate again, get the same results
ans =
        -5.1378        5.1765               16
```

The concept of creating function handles applies to built-in and M-file functions as well. For example, in the code

```
>> fh_Mfile = @humps % function handle for M-file function
fh_Mfile =
    @humps

>> fh_Mfile(1) % evaluate humps(1)
ans =
    16
>> fh_builtin = @cos % function handle for built-in function
fh_builtin =
    @cos

>> fh_builtin(pi) % evaluate cos(pi)
ans =
    -1
```

a function handle for an M-file function or built-in function is created by using the @ symbol, followed immediately by the name of the function to be converted. Function handles can also be placed in cell arrays and evaluated by using the same approach:

```
>> fhan = {@humps @cos}
fhan =
    [@humps]      [@cos]

>> fhan{1}(1) % evaluate humps(1)
ans =
    16
>> fhan{2}(pi) % evaluate cos(pi)
ans =
    -1
```

Here, the cell array fhan contains a function handle to the M-file function humps.m and a function handle to the built-in function cos. Evaluation of the two functions follows the approach outlined earlier, with fhan{1} and fhan{2} addressing the first and second function handles, respectively. In MATLAB 6, function handles were standard arrays, so the preceding handles could be created using standard brackets, as in fhan = [@humps @cos]. This standard-array format for function handles is no longer supported.

To support function handles, MATLAB offers a number of useful functions:

```
>> functions(fh_Mfile)
ans =
    function: 'humps'
        type: 'simple'
        file: 'C:\matlab7\toolbox\matlab\demos\humps.m'

>> functions(fh_builtin)
ans =
    function: 'cos'
        type: 'simple'
        file: 'C:\matlab7\toolbox\matlab\elfun\cos.m'

>> functions(af_humps)
ans =
     function: '@(x) 1./((x-.3).^2 +.01) +1./((x-.9).^2 +.04) -6'
         type: 'anonymous'
         file: ''
    workspace: [1x1 struct]
```

The function `functions` returns information about the function handle. (For the most part, this information is used for debugging purposes only. MATLAB warns that the content of the structure returned by `functions` is subject to change.)

When the name of a function is stored in a character-string variable, the function `str2func` provides a means for creating a function handle:

```
>> myfunc = 'humps' % place name of humps.m in a string variable
myfunc =
humps
>> fh2 = @myfunc % this doesn't work!
??? Error: "myfunc" was previously used as a variable,
conflicting with its use here as the name of a function.
>> fh2 = str2func(myfunc) % this works
```

```
fh2 =

    @humps

>> isequal(fh2,fh_Mfile) % these are equal function handles
ans =

     1
```

The inverse operation of str2func is func2str. As shown next, it simply extracts the name of the function or the string identifying the anonymous function:

```
>> func2str(fh2) % M-file function
ans =
humps
>> func2str(af_humps) % anonymous function
ans =
@(x) 1./((x-.3).^2 +.01) +1./((x-.9).^2 +.04) -6
>> class(ans) % output is a character string
ans =
char

>> isa(fh2,'function_handle') % True for function handles
ans =

     1
```

Function handles are an extremely powerful and beneficial feature in MATLAB. First, they capture all of the information needed to evaluate the function at the time of function handle creation. As a result, when a function defined by a function handle is evaluated, MATLAB does not need to search for it on the MATLAB path. It immediately evaluates it. For example, as shown before, the file location for humps.m is stored as part of the function handle fh_Mfile. The overhead time required to find the function is eliminated, which improves performance, especially when a function is evaluated repeatedly.

Another powerful and beneficial feature of function handles is that they can point to subfunctions, private functions, and nested functions (to be discussed next) that normally are not visible from outside of the functions where they appear or are referenced. That is, if the output of a function contains a function handle to a subfunction, private function, or nested function that is visible or within the scope of the function returning the function handle, the returned function handle can be evaluated. For example, the following M-file function skeleton demonstrates returning a function handle to a subfunction:

```
function out=myfunction(select)
%MYFUNCTION Return function handle to a subfunction.
% Example function demonstrating function handles to subfunctions.

switch select
case 'case1'
    out=@local_subfun1;
case 'case2'
    out=@local_subfun2;
otherwise
    out=[];
    error('Unknown Input.')
end
function a=local_subfun1(b,c)
%LOCAL_SUBFUN Some function operation.
% code that operates on the input arguments b and c
% and returns content in the variable a

% end of local_subfun1
function d=local_subfun2(e,f)
%LOCAL_SUBFUN Some function operation.
% code that operates on the input arguments e and f
% and returns content in the variable d

% end of local_subfun2
```

On the basis of this function skeleton, the following sample code creates a function handle to a subfunction in myfunction and then evaluates the subfunction outside of the context of the original function myfunction:

```
>> h_subfun = myfunction('case2'); % handle to local_subfun2
>> dout = h_subfun(x,y); % execute local_subfun2(e,f)
```

This works because all of the information required to execute either subfunction is captured by the function handle when it is created. Therefore, the subfunction `local_subfun2` can be evaluated, because `h_subfun` has complete knowledge of the content and location of `local_subfun2`.

12.8 NESTED FUNCTIONS

Nested functions are new in MATLAB 7. If you are unfamiliar with the concept of nested functions, they can appear to be strange and confusing, and they can promote poor programming practice. Indeed, they can be all of these things if used improperly. However, they can also be immensely helpful in some situations.

Fundamentally, nested functions provide a way to pass information to and from a function without using global variables, and without passing information through the input and output arguments. The following is the basic form of a simple nested function:

```
function out=primary_function (...)
%PRIMARYFUNCTION primary function.

% code in primary function
% this code can call nested functions

    function nout1=nested_function1(...)
    % Code in nested_function1.
    % In addition to variables passed through the input arguments
    % this nested function has access to all variables in existence
    % in the primary function at the point where this function
    % definition appears. This access permits reading and writing
    % to all variables available.

    end % required to mark the end of nested_function1
% other code in primary_function, including other
% nested functions terminated with end statements

end % end statement required here for primary_function
```

As shown in the preceding code, nested functions are functions that are fully contained within the definition of another function, with end statements marking the end of each nested function and the primary function as well. As opposed to subfunctions, which are functions appended to the primary function, nested functions have access not only to data passed through the nested function input arguments, but also to variables in the parent function. A nested function has its own workspace that includes read and write access to the primary function workspace. In addition, the primary function has read and write access to variables defined in functions nested within it.

In general, a primary function can have any number of nested functions. In addition, nested functions can have nested functions within them. Arbitrary nesting is permitted, but usually is not that useful. In addition, a nested function has access to variables passed to it, and to the workspaces of all functions in which it is nested. If this sounds confusing to you, you are not alone. Nested functions add a level of complexity that can make M-file debugging difficult. Moreover, it may not be clear what beneficial purpose this apparent complexity provides.

The example that follows illustrates a prime use for nested functions. In this rational polynomial example, the primary function returns a function handle to a nested function, which obtains data directly from the primary function workspace:

```
function fhandle=nestexample(num,den)
%NESTEXAMPLE Example Nested Function.
%   NESTEXAMPLE(Num,Den) returns a function handle to a function that can
%   be used to evaluate a rational polynomial. Num and Den are vectors
%   containing the numerator and denominator polynomial coefficients.
%
%   For example, ratpoly=nestexample([1 2],[1 2 3]) returns a function
%   handle that facilitates evaluation of the rational polynomial
%
%        x + 2
%   ------------
%   x^2 + 2x + 3
if ~isnumeric(num) || ~isnumeric(den)
    error('Num and Den Must be Numeric Vectors.')
end
```

```
num=reshape(num,1,[]); % make num into a row vector
den=reshape(den,1,[]); % make den into a row vector
fhandle=@nested_ratpoly; % create function handle for return
   function out=nested_ratpoly(x)
   % Nested function that evaluates a rational polynomial, where the
   % numerator and denominator coefficients are obtained from the primary
   % function workspace. Only the evaluation points x appear as an input
   % argument.

   out = polyval(num,x)./polyval(den,x);

end % nested function terminated with an end statement
end % primary function terminated with an end statement too!
```

Using this function, function handles to specific rational polynomials can be created as demonstrated by the following code:

```
>> ratpoly1 = nestexample([1 2],[1 2 3]) % (x + 2)/(x^2 + 2x + 3);
>> ratpoly2 = nestexample([2 1],[3 2 1]) % (2x +1)/(3x^2 + 2x +1);
>> x = linspace(-10,10); % independent variable data
>> y1 = ratpoly1(x);       % evaluate first rational polynomial
>> y2 = ratpoly2(x);       % evaluate second rational polynomial
>> plot(x,y1,x,y2) % plot created data
>> xlabel('X')
>> ylabel('Y')
>> title('Figure 12.1: Rational Polynomial Evaluation')
```

The two calls to nestexample in the preceding code return function handles to two different rational polynomials. Because each rational polynomial obtained its coefficients from the input arguments to nestexample, and these coefficients were available to the nested function through a shared workspace, the coefficients are not needed for evaluating the rational polynomials. The coefficients are contained in their respective function handles.

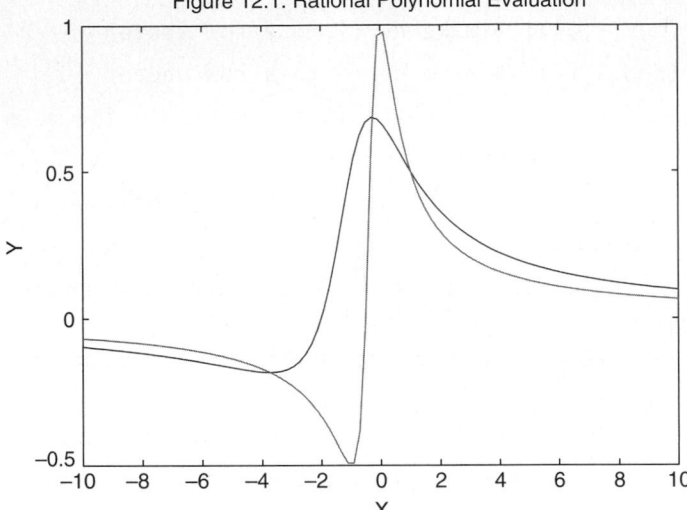

Figure 12.1: Rational Polynomial Evaluation

Prior to the existence of nested functions, rational polynomial evaluation using function handles could have been implemented with the following function:

```
function out=ratpolyval(num,den,x)
%RATPOLYVAL Evaluate Rational Polynomial.
% RATPOLYVAL(Num,Den,X) evaluates the rational polynomial, whose
% numerator and denominator coefficients are given by the vectors
% Num and Den respectively, at the points given in X.
if ~isnumeric(num) || ~isnumeric(den)
    error('Num and Den Must be Numeric Vectors.')
end
num=reshape(num,1,[]); % make num into a row vector
den=reshape(den,1,[]); % make den into a row vector
out=polyval(num,x)./polyval(den,x);
```

Evaluation of the two rational polynomials used earlier proceeds as follows:

```
>> yy1 = ratpolyval([1 2],[1 2 3],x); % same as y1 above

>> yy2 = ratpolyval([2 1],[3 2 1],x); % same as y2 above
```

In this case, the numerator and denominator polynomials are required every time a rational polynomial is evaluated. While this is straightforward and common in MATLAB, the nested function approach allows you to encapsulate all information required for a particular rational polynomial into a function handle that is easily evaluated.

Because rational polynomial evaluation can be written as a single statement, this example can also be implemented by using anonymous functions, while retaining the features of the nested function approach. That is, the two function handles ratpoly1 and ratpoly2 could be created by the following statements:

```
>> num = [1 2]; den = [1 2 3]; % specify numerator and denominator
>> ratpoly1 = @(x) polyval(num,x)./polyval(den,x); create anonymous fun

>> num = [2 1]; den = [3 2 1]; % redefine numerator and denominator
>> ratpoly2 = @(x) polyval(num,x)./polyval(den,x); create anonymous fun

>> y1 = ratpoly1(x);      % evaluate first rational polynomial
>> y2 = ratpoly2(x);      % evaluate second rational polynomial
```

This process of creating handles to nested functions and using function handles to evaluate functions is extremely powerful. It allows you to create multiple instances of a function, each with a shared purpose (such as rational polynomial evaluation), while permitting distinct properties as well (such as different numerator and denominator coefficients). Equally important is the fact that each instance has its own workspace, and therefore, each operates independently of all others.

13

M-File Debugging and Profiling

In the process of developing function M-files, it is inevitable that errors (i.e., **bugs**) appear. MATLAB provides a number of approaches and functions to assist in **debugging** M-files. MATLAB also provides a tool to help you improve the execution speed of M-files. In **profiling** the execution of an M-file, MATLAB identifies which lines of code take the most time to complete.

13.1 DEBUGGING TOOLS

Two types of errors can appear in MATLAB expressions: syntax errors and run-time errors. Syntax errors (such as misspelled variables or function names or missing quotes or parentheses) are found when MATLAB evaluates an expression or when a function is compiled into memory. MATLAB flags these errors immediately and provides feedback on the type of error encountered and the line number in the M-file where the error occurred. Given this feedback, these errors are usually easy to spot. An exception to this situation occurs in syntax errors within GUI callback strings. These errors are not detected until the strings themselves are evaluated during the operation of the GUI.

Run-time errors, on the other hand, are generally more difficult to find, even though MATLAB flags them also. When a run-time error is found, MATLAB returns control to the *Command* window and the MATLAB workspace. Access to the function workspace where the error occurred is lost, and so you cannot interrogate the contents of the function workspace in an effort to isolate the problem.

According to the authors' experience, the most common run-time errors occur when the result of some operation leads to empty arrays or to NaNs. All operations on NaNs return NaNs, so if NaNs are a possible result, it is good to use the logical function isnan to perform some default action when NaNs occur. Addressing arrays that are empty always leads to an error, since empty arrays have a zero dimension. The find function represents a common situation where an empty array may result. If the empty array output of the find function is used to index some other array, the result returned will also be empty. That is, empty matrices tend to propagate empty matrices, as, for example, in the following code:

```
>> x = pi*(1:4)  % example data
x =
    3.1416    6.2832    9.4248    12.5664
>> i = find(x>20)  % use find function
i =
   Empty matrix: 1-by-0
>> y = 2*x(i)  % propagate the empty matrix
y =
   Empty matrix: 1-by-0
```

Clearly, when y is expected to have a finite dimension and values, a run-time error is likely to occur. When performing operations or using functions that can return empty results, it helps to use the logical function isempty to define a default result for the empty matrix case, thereby avoiding a run-time error.

There are several approaches to debugging function M-files. For simple problems, it is straightforward to use a combination of the following:

1. Remove semicolons from selected lines within the function so that intermediate results are displayed in the *Command* window.
2. Add statements that display variables of interest within the function.
3. Place the keyboard command at selected places in the M-file to give temporary control to the keyboard. By doing so, the function workspace can be interrogated and values changed as necessary. Resume function execution by issuing a return command, K>>return, at the keyboard prompt.
4. Change the function M-file into a script M-file by placing a % before the function definition statement at the beginning of the M-file. When executed as a script file, the workspace is the MATLAB workspace, and thus it can be interrogated after the error occurs.

When the M-file is large, recursive, or highly nested (that is, it calls other M-file functions that call still other functions, and so on), it is more convenient to use the MATLAB graphical debugging functions found on the **Debug** and **Breakpoints** menus of the Editor/Debugger. *Command* window equivalents of these functions exist, but are more cumbersome to use. If you insist on using these functions rather than the graphical debugger, see the on-line help text for debug—that is,

```
>> doc debug
```

The graphical debugging tools in MATLAB allow you to stop at user-set break-points, at MATLAB warnings and errors, and at expressions that create NaNs and Infs. You can also set conditional breakpoints on the basis of some condition tested in the breakpoint itself. When a breakpoint is reached, MATLAB stops execution before the affected line completes execution and returns results. While the program is stopped, the *Command* window keyboard prompt K>> appears, allowing you to interrogate the function workspace, change the values of variables in the workspace, and so on, as required to track down the bug. In addition, the *Editor/Debugger* window shows the line where execution stopped and provides the means for stepping into the workspace of other functions that were called by the M-file being debugged. When you are ready to move beyond a breakpoint, the Editor/Debugger provides menu items for single-stepping, continuing until the next breakpoint, continuing until the cursor position is reached, or terminating the debugging activity.

Describing the use of the graphical debugging tools is difficult because of their graphical nature, and because the process of debugging is unique to each debugging session. However, the authors have found the debugging tools to be intuitive and easy to use. Once the basic steps are mastered, the graphical debugging features of MATLAB are extremely powerful and productive in assisting the process of creating good MATLAB code.

13.2 SYNTAX CHECKING AND FILE DEPENDENCIES

During the process of creating and debugging an M-file, it is advantageous to check the code for syntax errors and runtime errors. In the past, this was accomplished by executing the M-file and watching for warnings and errors reported in the *Command* window. Alternatively, it was done by using the pcode command to create a P-code file from the M-file. In MATLAB 7, the function mlint parses M-files for syntax errors and other possible problems and inefficiencies. For example, mlint will point out variables that are defined but never used, input arguments that are not used, output arguments that are not assigned, obsolete usages, statements that are unreachable, etc. In addition, mlint will make suggestions to improve execution time. This function is commonly called as a command—for example,

```
>> mlint myfunction
```

where *myfunction* is the name of an M-file anywhere on the MATLAB search path. This function can also be called using function syntax, in which case the output can be captured in a number of different ways. For example, the function

```
>> out = mlint('myfunction','-struct');
```

returns the output of `mlint` in the structure variable `out`. In addition to checking a single function by using `mlint`, the function `mlintrpt` accepts a directory name as an input argument, applies `mlint` to all M-files in the directory, and provides a report for each function in a separate HTML-based window. This feature is also available as a choice from the *Current Directory* window in the MATLAB desktop.

Given the ease with which M-files can be transferred electronically, it is not uncommon to run an M-file only to find out that it fails to run because it calls one or more M-file functions not found on your MATLAB path. MATLAB provides the function `depfun`, which parses an M-file for file dependencies. This function recursively searches for all function dependencies, including the dependencies within functions called by the function in question as well as any dependencies found in the callbacks of Handle Graphics objects. Because of the large amount of output produced by this function, it is difficult to demonstrate here and is best explored firsthand.

13.3 PROFILING M-FILES

Even when a function M-file works correctly, there may be ways to fine-tune the code to avoid unnecessary calculations or function calls. Performance improvements can be obtained by simply storing the result of a calculation to avoid a complex recalculation, by using vectorization techniques to avoid an iterative procedure such as a For Loop, or by using MATLAB JIT-acceleration features. When writing a function, it is difficult to guess where most execution time is spent. In today's high-speed processors, with integrated floating-point units, it may be faster to calculate a result more than once than to store it in a variable and recall it again later. There is an inherent tradeoff between memory usage and the number of computations performed. Depending on the data being manipulated, it may be faster to use more memory to store intermediate results, or it may be faster to perform more computations. Furthermore, the tradeoff between memory and computations is almost always dependent on the size of the data set being considered. If a function operates on large data sets, the optimum implementation may be much different from that of a small data set. Complicating things even further is the fact that the best computer implementation of a given algorithm is often much different from how one writes the algorithm on paper.

MATLAB provides ***profiling*** tools to optimize the execution of M-file functions. These tools monitor the execution of M-files and identify which lines consume the greatest amount of time relative to the rest of the code. For example, if one line (or function call) consumes 50 percent of the time in a given M-file, the attention paid to this line (or function call) will have the greatest impact on overall execution speed. Sometimes, you can rewrite the code to eliminate the offending line or lines. Other times, you can minimize the amount of data manipulated in the line, thereby speeding it up. And still other times, there may be nothing you can do to increase speed. In any case, a great deal of insight is gained by profiling the operation of M-file functions.

In addition to using the profiler through the *Profiler* window on the MATLAB desktop, MATLAB uses the `profile` command to determine which lines of code in an M-file take the most time to execute. Using `profile` is straightforward. For example, the execution profile of `myfunction` is found by executing the following commands:

```
>> profile on
```

```
>> for i=1:100
     out = myfunction(in);
   end
```

```
>> profile viewer
```

First, the profiler is turned On; then, `myfunction` is executed some sufficient number of times to gather sufficient data; finally, a profile report is generated. The profile report is an HTML file, displayed in the *Profiler* window in the MATLAB Desktop, which in turn is displayed by choosing it from the **Desktop** menu or by issuing the last command in the preceding example. The profile report generates a variety of data. Clicking on various links in the window gives different information.

14

File and Directory Management

MATLAB opens and saves data files in a variety of file formats. Some are formats custom to MATLAB, others are industry standards, and still others are file formats custom to other applications. The techniques used to open and save data files include GUIs, as well as *Command* window functions.

    Like most modern applications, MATLAB uses the current directory as the default location for data files and M-files. Implementing directory management tools and changing the current directory are accomplished through GUIs, as well as through the *Command* window functions.

    This chapter covers file and directory management features in MATLAB.

14.1 NATIVE DATA FILES

Variables in the MATLAB workspace can be saved in a format native to MATLAB by using the `save` command. For example,

```
>> save
```

stores all variables in the the MATLAB workspace, in MATLAB binary format, and in the file `matlab.mat` in the current directory. These native binary MAT-files maintain full double precision, as well as the names of the variables saved. MAT-files are not platform-independent, but are completely cross-platform compatible. Variables saved on one platform can be opened on other MATLAB platforms without any special treatment.

The `save` command can be used to store specific variables as well. For example, the code

```
>> save var1 var2 var3
```

saves just the variables `var1`, `var2`, and `var3` to `matlab.mat`. The file name can be specified as a first argument to `save`, as in the code

```
>> save filename var1 var2 var3
```

which saves `var1`, `var2`, and `var3` to the file named `filename.mat`.

Using command-function duality, the preceding command form can also be written in function form as

```
>> save('filename','var1','var2,','var3')
```

This particular format is useful if the file name is stored in a MATLAB character string:

```
>> fname = 'myfile';
>> save(fname,'var1','var2,','var3')
```

Here, the named variables are stored in a file named `myfile.mat`.

In addition to these above simple forms, the `save` command supports options for saving in compressed format and in ASCII text format, and the command can be used to append data to a file that already exists. (For help with these features refer to the on-line help.)

The complement to `save` is the `load` command. This command opens data files that were created by the `save` command or that are compatible with the `save` command. For example,

```
>> load
```

loads all variables found in `matlab.mat`, wherever it is first found in the current directory or on the MATLAB search path. The variable names originally stored in `matlab.mat` are restored in the workspace, and they overwrite any like-named variables that may exist there.

To load specific variables from a MAT-file, you must include the file name and a variable list:

```
>> load filename var1 var2 var3
>> load('filename','var1','var2','var3')
```

Here, `filename.mat` is opened, and variables `var1`, `var2`, and `var3` are loaded into the workspace. The second statement demonstrates the functional form of the `load` command, which allows the data file to be specified as a character string. Although not shown, the `filename` string can include a complete or a partial directory path, thereby restricting `load` to a search in a specific directory for the data file.

The latter example provides a way to open a sequence of enumerated data files, such as `mydata1.mat, mydata2.mat`, and so on, as in the following example:

```
for i=1:N
    fname=sprintf('mydata%d',i);
    load(fname)
end
```

This code segment uses `sprintf` to create file-name strings inside of a For Loop, so that a sequence of data files is loaded into the workspace.

When you do not wish to overwrite workspace variables, you can write the `load` command in function form and give it an output argument. For example,

```
>> vnew = load('filename','var1','var2');
```

opens the file `filename.mat` and loads the variables `var1` and `var2` into a ***structure*** variable named `vnew` that has fields `var1` and `var2`—that is, `vnew.var1 = var1` and `vnew.var2 = var2`.

The `load` command can also open ASCII text files. In particular, if the data file consists of MATLAB comment lines and rows of space-separated values, the syntax

```
>> load filename.ext
```

opens the file `filename.ext` and loads the data into a single double-precision data array named `filename`. (For further information regarding the load command, see the on-line help.)

To find out whether a data file exists and what variables it holds, the MATLAB commands `exist` and `whos` are valuable. For example, the command

```
>> exist('matlab.mat','file')
```

returns 0 if the file doesn't exist and 2 if it does, and the command

```
>> whos -file matlab.mat
```

returns the standard `whos` *Command* window display for the variables contained in the file `matlab.mat`. Alternatively, the code

```
>> w = whos('-file','matlab.mat')
w =
3x1 struct array with fields:
    name
```

```
size
bytes
class
```

returns a structure array with fields named for the columns of the whos display. Used in this way, the variable names, sizes, memory, and class are stored in variables.

Last, but not least, data files can be deleted by using the *Command* window command delete. For example,

```
>> delete filename.ext
```

deletes the file named filename.ext.

In MATLAB Version 7, data file management functions can be accessed from the *Workspace* browser, as well as from the *Import* wizard. The *Workspace* browser can be viewed by choosing **Workspace** from the **View** menu on the *MATLAB* desktop. The *Import* wizard, which appears by selecting **Import Data...** from the **File** menu or by typing uiimport in the *Command* window, is a general-purpose GUI that facilitates loading data in a variety of formats, not just MATLAB's native MAT-file format.

14.2 DATA IMPORT AND EXPORT

In addition to supporting MATLAB's native MAT-file format and conventional ASCII text format, MATLAB supports a variety of industry standard formats and other custom file formats. Some formats are restricted to reading, others to writing. Some formats are restricted to images, others to multimedia or spreadsheets. These data import and export functions and capabilities make it possible for MATLAB to exchange data with other programs.

Figure window images can be saved in a native MATLAB FIG-file format using the **Save** item on the *Figure* window **File** menu. In addition, *Figure* window images can be exported to a variety of formats by selecting **Export...** on the **File** menu of a *Figure* window. The *Command* window function saveas provides an alternative to this GUI-based approach. (For assistance in using saveas, see the on-line documentation.)

Data-specific import and export functions available in MATLAB include those listed in the following table:

Function	Description
dlmread	Reads delimited text file.
dlmwrite	Writes delimited text file.
textread	Reads formatted text from file.
textscan	Reads formatted text from file after opening with fopen.

wk1read	Reads spreadsheet file.
wk1write	Writes spreadsheet file.
xlsread	Reads spreadsheet file.
aviread	Reads movie file.
imread	Reads image file.
imwrite	Writes image file.
auread	Reads Sun sound file.
auwrite	Writes Sun sound file.
wavread	Reads Microsoft sound file.
wavwrite	Writes Microsoft sound file.
hdf	Gateway to HDF file capabilities.
cdfepoch	Constructs object for Common Data File (CDF) format export.
cdfinfo	Gets information about a CDF file.
cdfread	Reads from a CDF file.
cdfwrite	Writes to a CDF file.

(The help text for each of these functions provides information on their use.)

The functions imread and imwrite in particular support multiple formats, including JPEG, TIFF, BMP, PNG, HDF, PCX, and XWD. The help text for fileformats provides a more complete listing of the file formats supported by MATLAB, such as the following:

```
>> help fileformats
  Readable file formats.

  Data formats                      Command     Returns
   MAT  - MATLAB workspace          load        Variables in file.
   CSV  - Comma separated numbers   csvread     ·Double array.
   DAT  - Formatted text            importdata  Double array.
   DLM  - Delimited text            dlmread     Double array.
   TAB  - Tab separated text        dlmread     Double array.


  Spreadsheet formats
   XLS  - Excel worksheet           xlsread     Double array and cell array.
   WK1  - Lotus 123 worksheet       wk1read     Double array and cell array.
  Scientific data formats
   CDF  - Common Data Format        cdfread     Cell array of CDF records
```

```
 FITS - Flexible Image          fitsread  Primary table data
 HDF  - Hierarchical Data Format hdfread  HDF or HDF-EOS data set
Movie formats
 AVI  - Movie                   aviread   MATLAB movie.
Image formats
 TIFF - TIFF image              imread    image.
 PNG  - PNG image               imread    image.
 HDF  - HDF image               imread    Truecolor or indexed image(s).
 BMP  - BMP image               imread    Truecolor or indexed image.
 JPEG - JPEG image              imread    Truecolor or grayscale image.
 GIF  - GIF image               imread    Indexed image.
 PCX  - PCX image               imread    Indexed image.
 XWD  - XWD image               imread    Indexed image.
 CUR  - Cursor image            imread    Indexed image.
 ICO  - Icon image              imread    Indexed image.
 RAS  - Sun raster image        imread    Truecolor or indexed.
 PBM  - PBM image               imread    Grayscale image.
 PGM  - PGM image               imread    Grayscale image.
 PPM  - PPM image               imread    Truecolor image.


Audio formats
 AU   - NeXT/Sun sound          auread    Sound data and sample rate.
 SND  - NeXT/Sun sound          auread    Sound data and sample rate.
 WAV  - Microsoft Wave sound    wavread   Sound data and sample rate.
See also IOFUN
```

14.3 LOW-LEVEL FILE I/O

Because an infinite variety of file types exists, MATLAB provides low-level file I/O functions for reading or writing any binary or formatted ASCII file imaginable. These functions closely resemble their ANSI C programming language counterparts, but do not necessarily exactly match their characteristics. In fact, many of the special-purpose file I/O commands described previously use these commands internally. The low-level file I/O functions in MATLAB are given in the following table:

Category	Function	Description/Syntax Example
File opening and closing	fopen	Opens file *fid* = fopen*('filename','permission')*
	fclose	Closes file *status* = fclose*(fid)*
Binary I/O	fread	Reads part or all of a binary file *A* = fread*(fid, num, precision)*
	fwrite	Writes array to a binary file *count* = fwrite*(fid, array, precision)*
Formatted I/O	fscanf	Reads formatted data from file *A* = fscanf*(fid, format, num)*
	fprintf	Writes formatted data to file *count* = fprintf*(fid, format, A)*
	fgetl	Reads line from file; discard newline character *line* = fgetl*(fid)*
	fgets	Reads line from file, keep newline character *line* = fgets*(fid)*
String conversion	sprintf	Writes formatted data to string *S* = sprintf*(format, A)*
	sscanf	Reads string under format control *A* = sscanf*(string, format, num)*
File positioning	ferror	Inquires about file I/O status *message* = ferror *(fid)*
	feof	Tests for end of file *TF* = feof*(fid)*
	fseek	Sets file position indicator *status* = fseek*(fid, offset, origin)*
	ftell	Gets file position indicator *position* = ftell*(fid)*
	frewind	Rewinds file frewind*(fid)*

In this table, *fid* is a file identifier number and *permission* is a character string identifying the permissions requested. Possible strings include 'r' for reading only, 'w' for writing only, 'a' for appending only, and 'r+' for both reading and writing. Since the PC distinguishes between text and binary files, a 'b' must often be appended when you are working with binary files—for example, 'rb'. In the preceding table, format is a character string defining the desired formatting. Note that format follows ANSI standard C very closely. (More information regarding the use of these functions can be found in the on-line documentation for each function.)

14.4 DIRECTORY MANAGEMENT

With all the windows in MATLAB, it makes sense to have management of the current directory and its files available in a GUI. The *Current Directory* window, displayed by choosing **Current Directory** from the **View** menu in the *MATLAB* desktop window, performs these tasks. In addition to traversing the directory tree, this GUI allows you to preview the files in the current directory, see their modification dates, search for text in M-files, create new directories and new M-files, and so on. Because of the fundamental utility of knowing the current directory, the *Current Directory* window is also displayed in a pop-up menu in the *MATLAB* desktop window. Therefore, the current directory always can be known simply by looking at the toolbar on the desktop.

Prior to MATLAB Version 6, directory management was conducted through the use of *Command* window functions. Although these functions are not as important now, they still serve a valuable purpose. In particular, most functions have the capability of returning directory and file information in MATLAB variables, thereby allowing complex manipulation of files and directories to be completed within function M-files. The *Command* window directory management functions available in MATLAB are summarized in the following table:

Function	Description
cd, pwd	Show present working directory
S = cd;	Return present working directory as a string in S
cd *dirname*	Change present working directory to *dirname*
copyfile (*oldname*, *dirname*)	Copy file *oldname* to directory *dirname*
copyfile (*oldname*, *newname*)	Copy file *oldname* to *newname*
delete *filename.ext*	Delete file *filename.ext*
dir, ls	Display files in current directory
S = dir;	Return directory information in structure S
fileattrib	Get or set file attributes
mkdir *dirname*	Make directory *dirname* in current directory
movefile(*source*, *destination*)	Move source file or directory to a new destination
rmdir *dirname*	Remove directory *dirname*

what	Display an organized listing of all MATLAB files in the current directory
S = what;	Return listing information in structure S
which *filename*	Display directory path to *filename*
S = which ('*filename*');	Return directory path to *filename* as a string in S
who	Display variables in workspace
who -file *filename*	Display variables in MAT-file *filename.mat*
S = who('-file','*filename*');	Return variables names in *filename.mat* in a cell array S
whos	Display variables, size, and class in workspace
whos -file *filename*	Display variables, size, and class in MAT-file *filename.mat*
S = whos ('-file','*filename*');	Return variables, size, and class in *filename.mat* in structure S
help *filename*	Display help text for *filename* in *Command* window
S = help('*filename*');	Return help text for *filename* in a character string S
type *filename*	Display M-file *filename* in *Command* window

Most of these functions require only partial path information to locate a particular file. That is, `filename` may or may not include part of its directory path. If no directory path is included in `filename`, then the MATLAB search path is used to find the requested file. If some part of the directory path is provided, MATLAB traverses the MATLAB search path to find the subdirectory and file specified. For example, if `filename`='mystuff/myfile', MATLAB restricts its search to a subdirectory named `mystuff` on the MATLAB path.

To illustrate the usefulness of these functions, consider the function `mmbytes`:

```
function y=mmbytes(arg)
%MMBYTES Variable Memory Usage.
% MMBYTES and MMBYTES('base') returns the total memory in bytes
% currently used in the base workspace.
% MMBYTES('caller') returns the total memory in bytes currently
% used in the workspace where MMBYTES is called from.
% MMBYTES('global') returns the total memory in bytes currently
% used in the global workspace.
```

```
if nargin==0
    arg='base';
end
if strcmp(arg,'global')
    x=evalin('base','whos("global")');
else
    x=evalin(arg,'whos');
end
y=sum(cat(1,x.bytes));
```

This function uses the whos function to gather information about the variables that exist within MATLAB. The output of the whos function generates a structure array x. The bytes field of this structure contains the memory allocated to all variables. The final statement in the function concatentates all of the memory-allocated numbers into a vector, which is summed using the sum function.

Because of its varied uses, the function exist was not listed in the preceding table. This function tests for the existence of variables, files, directories, and so on. The help text for this function describes its many uses:

```
>> help exist
 EXIST  Check if variables or functions are defined.
    EXIST('A') returns:
      0 if A does not exist
      1 if A is a variable in the workspace
      2 if A is an M-file on MATLAB's search path.  It also returns 2 when
          A is the full pathname to a file or when A is the name of an
          ordinary file on MATLAB's search path
      3 if A is a MEX-file on MATLAB's search path
      4 if A is a MDL-file on MATLAB's search path
      5 if A is a built-in MATLAB function
      6 if A is a P-file on MATLAB's search path
      7 if A is a directory
      8 if A is a Java class
    EXIST('A') or EXIST('A.EXT') returns 2 if a file named 'A' or 'A.EXT'
```

exists and the extension isn't a P or MEX function extension.

EXIST('A','var') checks only for variables.

EXIST('A','builtin') checks only for built-in functions.

EXIST('A','file') checks for files or directories.

EXIST('A','dir') checks only for directories.

EXIST('A','class') checks only for Java classes.

EXIST returns 0 if the specified instance isn't found.

To facilitate the manipulation of character strings containing directory paths and filenames, MATLAB provides several useful functions, summarized in the following table:

Function	Description
addpath('*dirname*')	Prepend directory *dirname* to the MATLAB search path
[path,name,ext] = ... fileparts(*filename*)	Return path, name, and extension for file *filename*
filesep	Return file separator character for this computer platform. The file separator is the character used to separate directories and the file name. For example, on the PC the file separator is '\'.
fullfile (*d1, d2, ...,* *filename*)	Return full path and file specification for *filename* by using directory tree strings, *d1, d2, ...*
matlabroot	Return a string containing the path to the root of directory MATLAB
mexext	Return MEX-file extension for this computer platform
pathsep	Return path separator for this platform. The path separator is the character used to separate entries on the MATLAB search path string
prefdir	Return MATLAB preferences directory for this platform
rmpath('*dirname*')	Remove directory *dirname* from the MATLAB search path
tempdir	Return name of a temporary directory for this platform
tempname	Return name of a temporary file for this platform

These functions, along with those in the preceding table, facilitate the creation of file and directory management functions as M-files. (More assistance for each function can be found by consulting the on-line help.)

14.5 FTP FILE OPERATIONS

MATLAB 7 now includes built-in FTP capabilities. From within MATLAB, the function ftp creates an *FTP server object* that acts like the file identifier used in low-level file I/O. For example,

```
>> ftp_object = ftp('ftp.somewhere.org')
```

establishes an FTP connection to the site ftp.somewhere.org and returns an identifier to this site in the variable ftp_object. Once this FTP session is established, common FTP commands can be used to change directories and perform file operations. Many of these operations use the same functions used for local file and directory manipulation. For instance,

```
>> dir(ftp_object)
```

returns a directory listing of the default directory. Once FTP operations are completed, the function disconnect closes the FTP session. For example,

```
>> disconnect(ftp_object)
```

terminates the ftp session at *ftp.somewhere.org*.

The following table identifies a list of FTP functions available in MATLAB:

Function	Description
ascii	Set FTP transfer type to ASCII
binary	Set FTP transfer type to binary
cd	Change current directory
connect	Open connection that timed out
delete	Delete file
dir	List directory contents
disconnect	Close connection
disp	Set display format
ftp	Connect to FTP server
isconnected	Detect if connection is valid
mget	Download file
mkdir	Create new directory
mput	Upload file or directory
rename	Rename file
rmdir	Remove directory

15

Set, Bit, and Base Functions

15.1 SET FUNCTIONS

Since arrays are ordered collections of values, they can be thought of as sets. With that understanding, MATLAB provides several functions for testing and comparing sets. The simplest test is for equality, as in the following code:

```
>> a = rand(2,5);  % random array
>> b = randn(2,5); % a different random array
>> isequal(a,b)    % a and b are not equal
ans =
     0
>> isequal(a,a)    % but a is certainly equal to a
ans =
     1

>> isequal(a,a(:)) % a with a as a column
ans =
     0
```

For two arrays to be equal, they must have the same dimensions and the same contents. This function applies to all MATLAB data types, not just numerical arrays:

```
>> a = 'a string';
>> b = 'a String';
>> isequal(a,b)  % character string equality
ans =
     0
>> a = {'one' 'two' 'three'};
>> b = {'one' 'two' 'four'};
>> isequal(a,b)  % cell array equality
ans =
     0
>> isequal(a,a)
ans =
     1
>> a.one = 'one';
>> a.two = 2;
>> a.three = pi;
>> b.two = 2;
>> b.one = 'one';
>> b.three = pi;
>> isequal(a,b) % structure equality
ans =
     1
>> isequal(a,a)
ans =
     1
```

MATLAB variables are equal if they have the same size and exactly the same content.

The function unique removes duplicate items from a given set:

```
>> a = [2:2:8;4:2:10] % new data
a =
     2     4     6     8
     4     6     8    10
```

```
>> unique(a) % unique elements sorted into a column
ans =
    2
    4
    6
    8
   10
```

The unique function returns a sorted column vector because removal of duplicate values makes it impossible to maintain the array dimensions. The function unique also applies to cell arrays of strings, as in the following example:

```
>> c = {'Shania' 'Britney' 'Dixie' 'Shania' 'Faith'};
>> unique(c)
ans =
    'Britney'    'Dixie'    'Faith'    'Shania'
```

Set membership is determined with the function ismember, as in the following code:

```
>> a = 1:9
a =
    1    2    3    4    5    6    7    8    9
>> b = 2:2:9
b =
    2    4    6    8
>> ismember(a,b) % which elements in a are in b
ans =
    0    1    0    1    0    1    0    1    0
>> ismember(b,a) % which elements in b are in a
ans =
    1    1    1    1
```

For vector arguments, ismember returns a logical array the same size as its first argument, with ones appearing at the indices where the two vectors share common values:

```
>> A = eye(3);  % new data
>> B = ones(3);
```

```
>> ismember(A,B) % those elements in A that are also in B
ans =
     1
     0
     0
     0
     1
     0
     0
     0
     1
>> ismember(B,A)
ans =
     1
     1
     1
     1
     1
     1
     1
     1
     1
```

For 2-D arrays, ismember returns a column vector, with ones appearing at the *single* indices of the first argument where the two arguments share common values. The function ismember also applies to cell arrays of strings:

```
>> ismember(c,'Dixie')
ans =
     0     0     1     0     0
```

Set arithmetic is accomplished with the functions union, intersect, setdiff, and setxor. Examples of the use of these functions include the following:

```
>> a,b                 % recall prior data
a =
     1     2     3     4     5     6     7     8     9
```

```
b =

    2    4    6    8
>> union(a,b) % union of a and b
ans =

    1    2    3    4    5    6    7    8    9
>> intersect(a,b) % intersection of a and b
ans =

    2    4    6    8
>> setxor(a,b)    % set exclusive or of a and b
ans =

    1    3    5    7    9
>> setdiff(a,b)   % values in a that are not in b
ans =

    1    3    5    7    9
>> setdiff(b,a)   % values in b that are not in a
ans =

    []
>> union(A,B,'rows') % matrix inputs produce combined rows with no repetitions
ans =

    0    0    1
    0    1    0
    1    0    0
    1    1    1
```

Like prior functions discussed in this chapter, these set functions also apply to cell arrays of strings.

15.2 BIT FUNCTIONS

In addition to the logical operators discussed in the last chapter, MATLAB provides functions that allow logical operations on individual bits of any unsigned integer data. For backward compatibility, these bit functions also work on floating-point integers, namely integers stored in double-precision floating-point variables. The MATLAB bitwise functions `bitand`, `bitcmp`, `bitor`, `bitxor`, `bitset`, `bitget`, and `bitshift` work on integers. Examples of bit operations include:

```
>> format hex
>> intmax ('uint16') % largest unsigned 16-bit number
ans =
      ffff
>> a = uint16 (2^10 -1) % first data value
a =
        03ff
>> b = uint16 (567) % second data value
b =
        0237
>> bitand (a, b) % (a & b)
ans =
        0237
>> bitor (a, b) % (a | b)
ans =
         03ff
>> bitcmp (a) % complement a
ans =
        fc00
>> bitxor (a,b) % xor (a, b)
ans =
        01c8
>> bitget (b, 7) % get 7th bit of b
ans =
        0000
>> bitset (b, 7) % set 7th bit to 1
ans =
      0277
>> format short g % reset display format
```

15.3 BASE CONVERSIONS

MATLAB provides a number of utility functions for converting decimal numbers to other bases *in the form of character strings*. Conversions between decimals and binary numbers are performed by the functions dec2bin and bin2dec, as in the following example:

```
>> a = dec2bin(17) % find binary representation of 17
a =
10001
>> class(a) % result is a character string
ans =
char
>> bin2dec(a) % convert a back to decimal
ans -
    17
>> class(ans) % result is a double precision decimal
ans =
double
```

Conversions between decimals and hexadecimals are performed by dec2hex and hex2dec, as in the following example:

```
>> a = dec2hex(2047) % hex representation of 2047
a =
7FF
>> class(a) % result is a character string
ans =
char
>> hex2dec(a) % convert a back to decimal
ans =
        2047
>> class(ans) % result is a double precision decimal
ans =
double
```

Conversions between decimals and any base between 2 and 36 are performed by dec2base and base2dec:

```
>> a = dec2base(26,3)
a =
222
>> class(a)
ans =
char
>> base2dec(a,3)
ans =
     26
```

Base 36 is the maximum usable base, because it uses the numbers 0 through 9 and the letters A through Z to represent the 36 distinct digits of a base 36 number.

16

Time Computations

MATLAB offers a number of functions to manipulate time. You can do arithmetic with dates and times, print calendars, and find specific days. MATLAB does this by storing the date and time as a double-precision number representing the number of days since the beginning of year zero. For example, midnight, January 1, 2000, is represented as 730486, and the same day at noon is 730486.5. This format may make calculations easier for a computer, but it is difficult to interpret visually. That's why MATLAB supplies a number of functions to convert between date numbers and character strings, and to manipulate dates and times.

16.1 CURRENT DATE AND TIME

The function clock returns the current date and time in an array:

```
>> T = clock
T =
      2004           2           2          18           4      27.893
```

This was the time when this part of the text was written. The preceding data are organized as T = [year month day hour minute seconds], so the time shown is the year 2004, the 2nd month, 2nd day, 18th hour, 4th minute, and 27.893 seconds.

The function now returns the current date and time as a double-precision date number, or simply a date number:

```
>> format long

>> t = now
```

```
t =
    7.319797537603473e+005
```

```
>> format short g
```

Both T and t represent essentially the same information.

The function date returns the current date as a character string in the dd-mmm-yyyy format:

```
>> date
ans =
02-Feb-2004
>> class(ans)
ans =
char
```

16.2 DATE FORMAT CONVERSIONS

In general, mathematics with time involves converting times to date number format, performing standard mathematical operations on the date numbers, and then converting the result back to a format that makes human sense. As a result, converting time among different formats is very important. MATLAB supports three formats for dates: (1) double-precision date number, (2) date (character) strings in a variety of styles, and (3) numerical date vector, where each element contains a different date component—that is, [year,month,day,hour,minute,seconds].

The function datestr converts the date number to a date string. The syntax for using datestr is datestr(date,dateform), where dateform is described by the help text for datestr:

```
>> help datestr
 DATESTR String representation of date.
    DATESTR(D,DATEFORM) converts D, a date vector (as returned by DATEVEC)
    or serial date number (as returned by DATENUM), or a free format date
    string into a date string with a format specified by format number or
    string DATEFORM (see table below). By default, DATEFORM is 1, 16, or 0
    depending on whether D contains dates, times or both. Date strings
    with 2 character years are interpreted to be within the 100 years
    centered around the current year.

    DATESTR(D,DATEFORM,PIVOTYEAR) uses the specified pivot year as the
    starting year of the 100-year range in which a two-character year
```

resides. The default pivot year is the current year minus 50 years.
DATEFORM = -1 uses the default format.

DATEFORM number	DATEFORM string	Example
0	'dd-mmm-yyyy HH:MM:SS'	01-Mar-2000 15:45:17
1	'dd-mmm-yyyy'	01-Mar-2000
2	'mm/dd/yy'	03/01/00
3	'mmm'	Mar
4	'm'	M
5	'mm'	03
6	'mm/dd'	03/01
7	'dd'	01
8	'ddd'	Wed
9	'd'	W
10	'yyyy'	2000
11	'yy'	00
12	'mmmyy'	Mar00
13	'HH:MM:SS'	15:45:17
14	'HH:MM:SS PM'	3:45:17 PM
15	'HH:MM'	15:45
16	'HH:MM PM'	3:45 PM
17	'QQ-YY'	Q1-96
18	'QQ'	Q1
19	'dd/mm'	01/03
20	'dd/mm/yy'	01/03/00
21	'mmm.dd,yyyy HH:MM:SS'	Mar.01,2000 15:45:17
22	'mmm.dd,yyyy'	Mar.01,2000
23	'mm/dd/yyyy'	03/01/2000
24	'dd/mm/yyyy'	01/03/2000
25	'yy/mm/dd'	00/03/01
26	'yyyy/mm/dd'	2000/03/01
27	'QQ-YYYY'	Q1-1996
28	'mmmyyyy'	Mar2000
29 (ISO 8601)	'yyyy-mm-dd'	2000-03-01
30 (ISO 8601)	'yyyymmddTHHMMSS'	20000301T154517
31	'yyyy-mm-dd HH:MM:SS'	2000-03-01 15:45:17

See also DATE, DATENUM, DATEVEC, DATETICK.

Some examples of datestr usage include the following:

```
>> t = now
t =
    7.319797588851968e+005
>> datestr(t)
ans =
02-Feb-2004 18:12:48

>> class(ans)
ans =
char
>> datestr(t,12)
ans =
Feb04
>> datestr(t,23)
ans =
02/02/2004
>> datestr(t,25)
ans =
04/02/02
>> datestr(t,13)
ans =
18:12:48
>> datestr(t,29)
ans =
2004-02-02
```

The function datenum is the inverse of datestr. That is, datenum converts a date string to a date number using the form datenum(str). Alternatively, it converts individual date specifications using the form datenum(year,month,day) or datenum(year,month,day,hour,minute,second), as in the following example:

```
>> t = now
t =
    7.319797588851968e+005
```

```
>> ts = datestr(t)
ts =
02-Feb-2004 18:12:48
>> datenum(ts)
ans =
    7.319797588888889e+005
>> datenum(2004,5,14,16,48,07)
ans =
    7.320817000810185e+005
>> datenum(2004,5,14)
ans =
      732081
```

The datevec function converts a date string using datestr formats 0, 1, 2, 6, 13, 14, 15, or 16 to a numerical vector containing the date components. Alternatively, it converts a date number to a numerical vector of date components, as in the following code:

```
>> c = datevec('12/24/1984')
c =
    1984    12    24    0    0    0

>> [yr,mo,day,hr,min,sec] = datevec('24-Dec-1984 08:22')
yr =
    1984

mo =
    12

day =
    24

hr =
    8

min =
    22

sec =
    0
```

16.3 DATE FUNCTIONS

The numerical day of the week can be found from a date string or a date number by using the function weekday (***MATLAB uses the convention that Sunday is day 1 and Saturday is day 7***):

```
>> [d,w] = weekday(728647)
d =
    2
w =
    Mon


>> [d,w] = weekday('21-Dec-1994')
d =
    4
w =
    Wed
```

The last day of any month can be found by using the function eomday. Because of leap year, both the year and the month are required, as in the following example:

```
>> eomday(1996,2) % divisible by 4 is a leap year
ans =
    29


>> eomday(1900,2) % divisible by 100 not a leap year
ans =
    28


>> eomday(2000,2) % divisible by 400 is a leap year
ans =
    29
```

MATLAB can generate a calendar for any month you request and display it in the *Command* window or place it in a 6-by-7 matrix by using the function calendar:

```
>> calendar(date)
                      Feb 2004
      S     M    Tu     W    Th     F     S
      1     2     3     4     5     6     7
      8     9    10    11    12    13    14
     15    16    17    18    19    20    21
     22    23    24    25    26    27    28
     29     0     0     0     0     0     0
      0     0     0     0     0     0     0
>> calendar(1954,9)
                      Sep 1954
      S     M    Tu     W    Th     F     S
      0     0     0     1     2     3     4
      5     6     7     8     9    10    11
     12    13    14    15    16    17    18
     19    20    21    22    23    24    25
     26    27    28    29    30     0     0
      0     0     0     0     0     0     0
>> x = calendar(2004,2)
x =
      1     2     3     4     5     6     7
      8     9    10    11    12    13    14
     15    16    17    18    19    20    21
     22    23    24    25    26    27    28
     29     0     0     0     0     0     0
      0     0     0     0     0     0     0

>> class(x)
ans =
double
```

16.4 TIMING FUNCTIONS

The functions tic and toc are used to time a sequence of MATLAB operations. The function tic starts a stopwatch, while toc stops the stopwatch and displays the elapsed time:

```
>> tic; plot(rand(50,5)); toc
elapsed_time =
    0.33

>> tic; plot(rand(50,5)); toc
elapsed_time =
    0.11
```

Note the difference in elapsed times for identical `plot` commands. The second `plot` was significantly faster, because MATLAB had already created the *Figure* window and compiled the functions it needed into memory.

The function `cputime` returns the amount of central processing unit (CPU) time, in seconds, that MATLAB has used since the current session was started. The function `etime` calculates the elapsed time between two time vectors, in six-element row vector form such as that returned by the functions `clock` and `datevec`. Both `cputime` and `etime` can be used to compute the time it takes for an operation to be completed. In fact, the functions `tic` and `toc` simply automate the use of `clock` and `etime` to compute elapsed time. Usage of `cputime` and `etime` are demonstrated by the following examples, in which `myoperation` is a script file containing a number of MATLAB commands:

```
>> t0 = cputime; myoperation; cputime-t0
ans =
    0.149999999999991

>> t1 = clock; myoperation; etime(clock,t1)
ans =
    11.284853
```

16.5 PLOT LABELS

Sometimes it is useful to plot data and use dates or time strings for one or more of the axis labels. The `datetick` function automates this task. *Use of this function requires that the axis to be marked be plotted with a vector of date numbers, such as the output of the `datenum` function*. The following code is illustrative:

```
>> t = (1900:10:1990)';
>> p = [ 75.995;  91.972; 105.711; 123.203; 131.669;
        150.697; 179.323; 203.212; 226.505; 249.633];
>> plot(datenum(t,1,1),p)
>> datetick('x','yyyy')    % use 4-digit year on the x-axis
>> title('Figure 16.1: Population by Year')
```

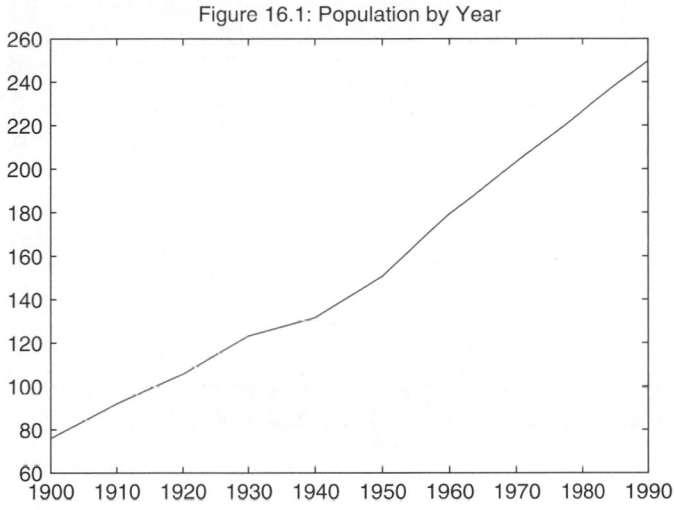

Next, we create a bar chart of company sales from November 1998 to December 1999:

```
>> y = [1998 1998 1999*ones(1,12)]';
>> m = [11 12 (1:12)]';
>> s = [1.1 1.3 1.2 1.4 1.6 1.5 1.7 1.6 1.8 1.3 1.9 1.7 1.6 1.95]';
>> bar(datenum(y,m,1),s)

>> datetick('x','mmmyy')
>> ylabel('$ Million')
>> title('Figure 16.2: Monthly Sales')
```

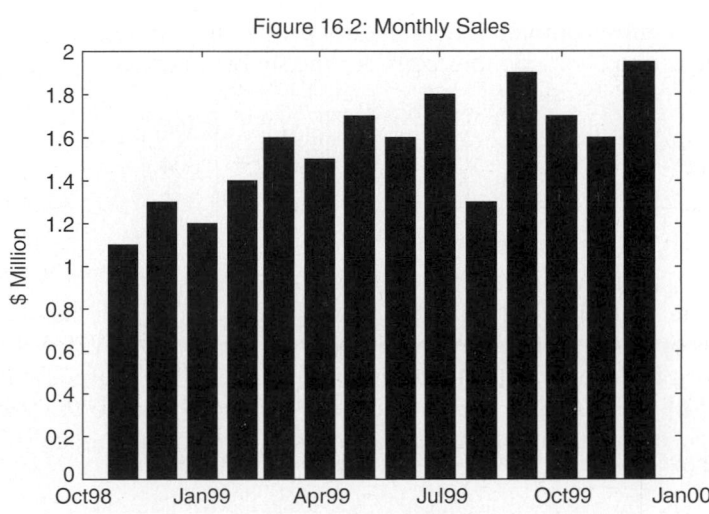

17

Matrix Algebra

MATLAB was originally written to provide an easy-to-use interface to professionally developed numerical linear algebra subroutines. As it has evolved over the years, other features, such as graphics and graphical user interfaces, have made the numerical linear algebra routines less prominent. Nevertheless, MATLAB offers a wide range of valuable matrix algebra functions.

> It is important to note that, while MATLAB supports n-dimensional arrays, matrix algebra is defined only for 2-D arrays—that is, vectors and matrices.

17.1 SETS OF LINEAR EQUATIONS

One of the most common linear algebra problems is finding the solution of a linear set of equations. For example, consider the set of equations

$$\begin{bmatrix} 1 & 2 & 3 \\ 4 & 5 & 6 \\ 7 & 8 & 0 \end{bmatrix} \begin{bmatrix} x_1 \\ x_2 \\ x_3 \end{bmatrix} = \begin{bmatrix} 366 \\ 804 \\ 351 \end{bmatrix}$$

$$A \cdot x = y$$

where the mathematical multiplication symbol ($\cdot$) is now defined in the matrix sense, as opposed to the array sense discussed earlier. In MATLAB, this matrix multiplication is denoted with the asterisk notation *. The above equations define the product of the matrix A and the vector x as being equal to the vector y. The very existence of solutions of this equation is a fundamental issue in linear algebra.

Moreover, when a solution does exist, there are numerous approaches to finding
that solution, such as Gaussian elimination, LU factorization, or the direct use of
A^{-1}. Clearly, it is beyond the scope of this text to discuss the many analytical and
numerical issues of matrix algebra. We wish only to demonstrate how MATLAB can
be used to solve problems like the one in the preceding example.

To solve the problem at hand, it is necessary to enter A and y:

```
>> A = [1 2 3;4 5 6
7 8 0]
A =
     1     2     3
     4     5     6
     7     8     0
>> y = [366;804;351]
y =
   366
   804
   351
```

As discussed earlier, the entry of the matrix A shows the two ways that MATLAB
distinguishes between rows. The semicolon between the 3 and 4 signifies the start of
a new row, as does the new line between the 6 and 7. The vector y is a column,
because each semicolon signifies the start of a new row.

If you have a background in linear algebra, you will find it easy to see that this
problem has a unique solution if the rank of A and the rank of the augmented matrix
[A y] are both equal to 3. Alternatively, you can check the condition number of A.
If the condition number is not excessively large, then A has an inverse with good
numerical properties. Testing this problem produces the following result:

```
>> rank(A)
ans =
     3
>> rank([A y])
ans =
     3

>> cond(A) % close to one is best
ans =
     35.106
```

Since the rank and condition number tests hold, MATLAB can find the solution of $A \cdot x = y$ in two ways, one of which is preferred. The less favorable method is to take $x = A^{-1} \cdot y$ literally, as in the following code:

```
>> x = inv(A)*y % Avoid this approach if possible
x =
        25
        22
        99
```

Here, `inv(A)` is a MATLAB function that computes A^{-1}, and the multiplication operator * is matrix multiplication. The preferable solution is found by using the matrix left-division operator, or backward slash, as in this example:

```
>> x = A\y % Recommended approach to solving sets of equations
x =
        25
        22
        99
```

This equation uses an LU factorization approach and expresses the answer as the left division of A into y. The left-division operator \ has no preceding dot, as this is a matrix operation, not an element-by-element array operation. There are many reasons to prefer this second solution. Of these, the simplest is that the latter method requires fewer floating-point operations and, as a result, is significantly faster. In addition, this solution is generally more accurate, especially for larger problems. In either case, if MATLAB cannot find a solution or cannot find one accurately, it displays a warning message.

   If the transpose of the above set of linear equations is taken (i.e., $(A \cdot x)' = y'$) then the previous set of linear equations can be written as $x' \cdot A' = y'$, where x' and y' are now row vectors. As a result, it is equally valid to express a set of linear equations in terms of the product of a row vector and a matrix being equal to another row vector—for example, $u \cdot B = v$. In MATLAB, this case is solved by the same internal algorithms, by using the matrix right-division operator, or forward slash, as in `u = v/B`.

   It is important to note that when MATLAB encounters a forward slash (/) or backward slash (\), it checks the structure of the coefficient matrix to determine what internal algorithm to use to find the solution. In particular, if the matrix is upper or lower triangular or a permutation of an upper or lower triangular matrix, MATLAB does not refactor the matrix, but rather just performs the forward or backward substitution steps required to find the solution. As a result, MATLAB makes use of the properties of the coefficient matrix to compute the solution as quickly as possible.

When the structure of the coefficient matrix is known ahead of time, the function linsolve solves the set of equations by using matrix structure information provided by the user. By using this function, MATLAB does not spend time analyzing the structure of the coefficient matrix and thereby minimizes the time required to solve the set of equations. The calling syntax for this function is linsolve(A,y,options), where options is a structure containing fields that can be set to True or False to identify the structure of the A matrix.

If you've studied linear algebra rigorously, you know that when the number of equations and number of unknowns differ, a single unique solution usually does not exist. However, with further constraints, a practical solution can usually be found. In MATLAB, when $\text{rank}(A) = \min(r,c)$, where r and c are the number of rows and columns in A, respectively, and there are more equations than unknowns $(r > c)$ (i.e., the **overdetermined** case), a division operator / or \ automatically finds the solution that minimizes the norm of the squared residual error $e = A \cdot x - y$. This solution is of great practical value and is called the **least-squares solution**. The following example is illustrative:

```
>> A = [1 2 3;4 5 6;7 8 0;2 5 8]   % 4 equations in 3 unknowns
A =
     1     2     3
     4     5     6
     7     8     0
     2     5     8
>> y = [366 804 351 514]'   % a new r.h.s. vector
y =
   366
   804
   351
   514
>> x = A\y       % least squares solution
x =
      247.98
     -173.11
      114.93
>> e = A*x-y     % this residual has the smallest norm.
e =
  -119.4545
    11.9455
     0.0000
    35.8364
```

```
>> norm(e)
ans =
   125.2850
```

In addition to the least squares solution that is computed with the left- and right-division operators, MATLAB offers the functions lscov and lsqnonneg. The function lscov solves the weighted least-squares problem when the covariance matrix (or weighting matrix) of the data is known, and lsqnonneg finds the nonnegative least squares solution where all solution components are constrained to be positive.

When there are fewer equations than unknowns ($r < c$) (i.e., the **_underdetermined_** case) an infinite number of solutions exist. Of these solutions, MATLAB computes two in a straightforward way. Use of the division operator gives a solution that has a maximum number of zeros in the elements of **_x_**. Alternatively, computing x=pinv(A)*y gives a solution where the length or norm of x is smaller than all other possible solutions. This solution, based on the _pseudoinverse_, also has great practical value and is called the **_minimum norm solution_**:

```
>> A = A'  % create 3 equations in 4 unknowns
A =
     1      4      7      2
     2      5      8      5
     3      6      0      8
>> y = y(1:3) % new r.h.s. vector
y =
   366
   804
   351
>> x = A\b   % solution with maximum zero elements
x =
          0
  -165.9000
    99.0000
   168.3000
>> xn = pinv(A)*b % minimum norm solution
xn =
    30.8182
  -168.9818
    99.0000
   159.0545
```

```
>> norm(x)  % norm of solution with zero elements
ans =
   256.2200
>> norm(xn) % minimum norm solution has smaller norm!
ans =
   254.1731
```

17.2 MATRIX FUNCTIONS

In addition to the solution of linear sets of equations, MATLAB offers numerous matrix functions that are useful for solving numerical linear algebra problems. A thorough discussion of these functions is beyond the scope of this text. In general, MATLAB provides functions for all common, and some uncommon, numerical linear algebra problems. A brief description of many of the matrix functions is given in the following table:

Function	Description
/ and \	Solve Ax=y (much better than using inv(A)*y)
accumarray(ind,val)	Construct array with accumulation
A^n	Exponentiation, e.g., A^3 = A*A*A
balance(A)	Scale to improve eigenvalue accuracy
[V,D]=cdf2rdf(V,D)	Complex diagonal form to real block diagonal form
chol(A)	Cholesky factorization
cholinc(A,DropTol) cholinc(A,Options)	Incomplete Cholesky factorization
cholupdate(R,X)	Rank 1 update to Cholesky factorization
cond(A)	Matrix condition number using singular value decomposition
condest(A)	1-norm condition number estimate
[V,D,s]=condeig(A)	Condition number with respect to repeated eigenvalues
det(A)	Determinant
dmperm(A)	Dulmage–Mendelsohn permutation
eig(A)	Vector of eigenvalues
[V,D]=eig(A)	Matrix of eignenvectors, and diagonal matrix containing eigenvalues
expm(A)	Matrix exponential
funm(A)	General matrix function
gsvd(A,B)	Generalized singular values

Function	Description
`[U,V,X,C,S]=gsvd(A)`	Generalized singular value decomposition
`hess(A)`	Hessenburg form of a matrix
`inv(A)`	Matrix inverse (use only when / or \ won't do)
`linsolve(A,y,options)`	Solve Ax=y quickly when structure of A is given by `options`
`logm(A)`	Matrix logarithm
`lscov(A,y,V)`	Weighted least squares with covariance matrix
`lsqnonneg(A,y)`	Nonnegative least squares solution
`[L,U,P]=lu(A)`	LU decomposition
`minres(A,y)`	Minimum residual method
`norm(A,type)`	Matrix and vector norms
`null(A)`	Null space
`orth(A)`	Orthogonal range space using singular value decomposition
`pinv(A)`	Pseudoinverse using singular value decomposition
`planerot(X)`	Plane rotation
`poly(A)`	Characteristic polynomial
`polyeig(A0,A1,. . .)`	Polynomial eigenvalue solution
`polyvalm(A)`	Evaluate matrix polynomial
`qr(A)`	Orthogonal-triangular decomposition
`qrdelete(Q,R,J)`	Delete column or row from QR factorization
`qrinsert(Q,R,J,X)`	Insert column or row into QR factorization
`qrupdate(Q,R,U,V)`	Rank 1 update to QR factorization
`qz(A,B)`	Generalized eigenvalues
`rank(A)`	Matrix rank using singular value decomposition
`rcond(A)`	LAPACK reciprocal condition estimator
`rref(A)`	Reduced row echelon form
`rsf2csf(A)`	Real Schur form to complex Schur form
`schur(A)`	Schur decomposition
`sqrtm(A)`	Matrix square root
`subspace(A,B)`	Angle between two subspaces
`svd(A)`	Singular values
`[U,S,V]=svd(A)`	Singular value decomposition
`trace(A)`	Sum of matrix diagonal elements

17.3 SPECIAL MATRICES

MATLAB offers a number of special matrices; some of them are general utilities, while others are matrices of interest to specialized disciplines. These and other special matrices include those given in the following table (use the on-line help to learn more about these matrices):

Matrix	Description
`[ ]`	Empty matrix
`blkdiag(a0,a1,...)`	Block diagonal concatenation of input arguments
`compan(P)`	Companion matrix to polynomial
`eye(r,c)`	Identity matrix
`gallery`	More than 50 test matrices
`hadamard(n)`	Hadamard matrix of order n
`hankel(C)`	Hankel matrix
`hilb(n)`	Hilbert matrix of order n
`invhilb(n)`	Inverse Hilbert matrix of order n
`magic(n)`	Magic matrix of order n
`ones(r,c)`	Matrix containing all ones
`pascal(n)`	Pascal matrix of order n
`rand(r,c)`	Uniformly distributed random matrix with elements between 0 and 1
`randn(r,c)`	Normally distributed random matrix with elements having zero mean and unit variance
`rosser`	Classic symmetric eigenvalue test problem
`toeplitz(C,R)`	Toeplitz matrix
`vander(C)`	Vandermonde matrix
`wilkinson(n)`	Wilkinson's eigenvalue test matrix of order n
`zeros(r,c)`	Matrix containing all zeros

17.4 SPARSE MATRICES

In many practical applications, matrices are generated that contain only a few nonzero elements. As a result, these matrices are said to be *sparse*. For example, circuit simulation and finite element analysis programs routinely deal with matrices containing less than 1 percent nonzero elements. If a matrix is large—for example, `max(size(A)) > 100`—and has a high percentage of zero elements, it is wasteful, both of computer storage to store the zero elements, and of computational power to perform arithmetic operations by using the zero elements. To eliminate the storage of zero elements, it is common to store only the nonzero elements of a sparse matrix and two sets of indices identifying the row and column positions of these elements.

Similarly, to eliminate arithmetic operations on the zero elements, special algorithms have been developed to solve typical matrix problems (such as solving a set of linear equations in which operations involving zeros are minimized, and intermediate matrices have minimum nonzero elements).

The techniques used to optimize sparse matrix computations are complex in implementation, as well as in theory. Fortunately, MATLAB hides this complexity. In MATLAB, sparse matrices are stored in variables, just as regular, full matrices are. Moreover, most computations with sparse matrices use the same syntax as that used for full matrices. In particular, all of the array manipulation capabilities of MATLAB work equally well on sparse matrices. For example, s(i,j) = *value* assigns value to the ith row and jth column of the sparse matrix s.

In this text, only the creation of sparse matrices and the conversion to and from sparse matrices are illustrated. In general, operations on full matrices produce full matrices and operations on sparse matrices produce sparse matrices. In addition, operations on a mixture of full and sparse matrices generally produce sparse matrices, unless the operation makes the result too densely populated with nonzeros to make sparse storage efficient.

Sparse matrices are created using the MATLAB function sparse. For example, the code

```
>> As = sparse(1:10,1:10,ones(1,10))
As =
    (1,1)        1
    (2,2)        1
    (3,3)        1
    (4,4)        1
    (5,5)        1
    (6,6)        1
    (7,7)        1
    (8,8)        1
    (9,9)        1
   (10,10)       1
```

creates a 10-by-10 identity matrix. In this usage, sparse(i,j,s) creates a sparse matrix whose kth nonzero element is s(k), which appears in the row i(k) and column j(k). Note the difference in how sparse matrices are displayed. Nonzero elements and their row and column positions are displayed. The preceding sparse matrix can also be created by conversion. For example, the code

```
>> As = sparse(eye(10))
As =
    (1,1)        1
```

(2,2)	1
(3,3)	1
(4,4)	1
(5,5)	1
(6,6)	1
(7,7)	1
(8,8)	1
(9,9)	1
(10,10)	1

creates the 10-by-10 identity matrix again, this time by converting the full matrix eye(10) to sparse format. While this method of creating a sparse matrix works, it is seldom used in practice, because the initial full matrix wastes a great deal of memory.

Given a sparse matrix, the function full generates the conventional full matrix equivalent. The following example converts the sparse matrix As back to its full form:

```
>> A = full(As)
A =
```

1	0	0	0	0	0	0	0	0	0
0	1	0	0	0	0	0	0	0	0
0	0	1	0	0	0	0	0	0	0
0	0	0	1	0	0	0	0	0	0
0	0	0	0	1	0	0	0	0	0
0	0	0	0	0	1	0	0	0	0
0	0	0	0	0	0	1	0	0	0
0	0	0	0	0	0	0	1	0	0
0	0	0	0	0	0	0	0	1	0
0	0	0	0	0	0	0	0	0	1

To compare sparse matrix storage to full matrix storage, consider the following example:

```
>> B = eye(200);
>> Bs = sparse(B);
>> whos
  Name      Size            Bytes  Class
```

```
B          200x200          320000  double array
Bs         200x200            3204  sparse array
```

```
Grand total is 40200 elements using 323204 bytes
```

Here, the sparse matrix Bs contains only 0.5 percent nonzero elements and requires 3204 bytes of storage. On the other hand, B, the same matrix in full matrix form, requires two orders of magnitude more bytes of storage!

17.5 SPARSE MATRIX FUNCTIONS

MATLAB provides numerous sparse matrix functions. Many involve different aspects of, and techniques for, the solution of sparse simultaneous equations. A discussion of these functions is beyond the scope of this text. The functions available are listed in the following table:

Sparse Matrix Function	Description
bicg	Biconjugate gradient iterative linear equation solution
bicgstab	Biconjugate gradient stabilized iterative linear equation solution
cgs	Conjugate gradients squared iterative linear equation solution
cholinc	Incomplete Cholesky factorization
colamd	Column approximate minimum degree reordering method
colamdtree	colamd followed by column elimination tree postordering
colmmd	Column minimum degree reordering method
colperm	Column permutation
condest	1-norm condition number estimate
dmperm	Dulmage–Mendelsohn reordering method
eigs	A few eigenvalues using ARPACK
etree	Elimination tree
etreeplot	Plot elimination tree
find	Find indices of nonzero elements
full	Convert sparse matrix to full matrix
gmres	Generalized minimum residual iterative linear equation solution
gplot	Construct graph theory plot
issparse	True for sparse matrix

`lsqr`	LSQR implementation of conjugate gradients on normal equations
`luinc`	Incomplete LU factorization
`minres`	Minimum residual iterative linear equation solution
`nnz`	Number of nonzero matrix elements
`nonzeros`	Nonzero matrix elements
`normest`	Estimate of matrix 2-norm
`nzmax`	Storage allocated for nonzero elements
`pcg`	Preconditioned conjugate gradients iterative linear equation solution
`qmr`	Quasi-minimal residual iterative linear equation solution
`randperm`	Random permutation
`spalloc`	Allocate space for sparse matrix
`sparse`	Create sparse matrix
`spaugment`	Form least squares augmented system
`spconvert`	Import from sparse matrix external format
`spdiags`	Sparse matrix formed from diagonals
`speye`	Sparse identity matrix
`spfun`	Apply function to nonzero elements
`spones`	Replace nonzeros with ones
`spparms`	Set parameters for sparse matrix routines
`sprand`	Sparse uniformly distributed matrix
`sprandn`	Sparse normally distributed matrix
`sprandsym`	Sparse random symmetric matrix
`sprank`	Structural rank
`spy`	Visualize sparsity pattern
`svds`	A few singular values
`symbfact`	Symbolic factorization analysis
`symamd`	Symmetric approximate minimum degree reordering method
`symamdtree`	`symamd` followed by symmetric elimination tree postordering
`symmd`	Symmetric minimum degree reordering method
`symmlq`	Symmetric LQ iterative linear equation solution
`symrcm`	Symmetric reverse Cuthill–Mckee reordering method
`treelayout`	Lay out tree or forest
`treeplot`	Plot tree

18

Data Analysis

Because of its array orientation, MATLAB readily performs statistical analyses on data sets. While MATLAB, by default, considers data sets stored in column-oriented arrays, data analysis can be conducted along any specified dimension.

> Unless specified otherwise, each column of an array represents a different measured variable and each row represents individual samples or observations.

18.1 BASIC STATISTICAL ANALYSIS

For example, let's assume that the daily high temperature (in Celsius) of three cities over a 31-day month was recorded and assigned to the variable `temps` in a script M-file. Running the M-file puts the variable `temps` in the MATLAB workspace. When this work is done, the variable `temps` contains the following data:

```
>> temps

temps =
      12      8      18
      15      9      22
      12      5      19
      14      8      23
      12      6      22
```

11	9	19
15	9	15
8	10	20
19	7	18
12	7	18
14	10	19
11	8	17
9	7	23
8	8	19
15	8	18
8	9	20
10	7	17
12	7	22
9	8	19
12	8	21
12	8	20
10	9	17
13	12	18
9	10	20
10	6	22
14	7	21
12	5	22
13	7	18
15	10	23
13	11	24
12	12	22

Each row contains the high temperatures for a given day. Each column contains the high temperatures for a different city. To visualize the data, plot it with the following code:

```
>> d = 1:31;    % number the days of the month
>> plot(d,temps)

>> xlabel('Day of Month'), ylabel('Celsius')
>> title('Figure 18.1: Daily High Temperatures in Three Cities')
```

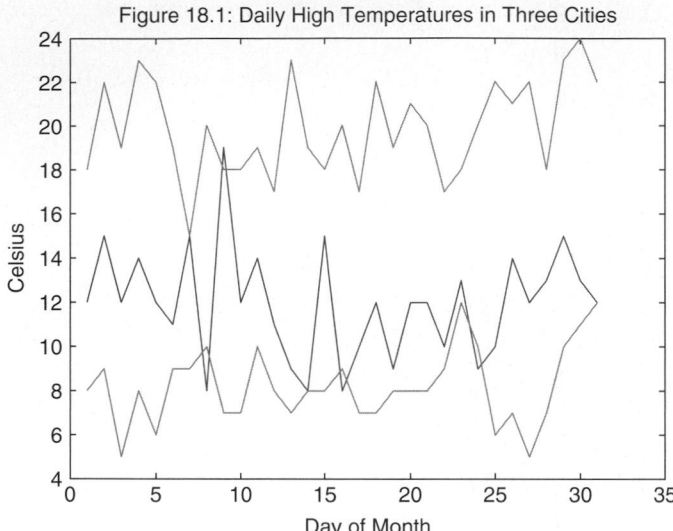

Figure 18.1: Daily High Temperatures in Three Cities

The `plot` command in this example illustrates yet another form of `plot` command usage. The variable d is a vector of length 31, whereas `temps` is a 31-by-3 matrix. Given this data, the `plot` command plots each column of `temps` versus d.

To illustrate some of the data-analysis capabilities of MATLAB, consider the following commands based on the temperature data stored in `temps`:

```
>> avg_temp = mean(temps)
avg_temp =
   11.968    8.2258   19.871
```

This result shows that the third city has the highest average temperature. Here, MATLAB found the average of each column individually. Taking the average again gives

```
>> avg_avg = mean(avg_temp)
avg_avg =
   13.3548
```

which returns the overall average temperature of the three cities.

> When the input to a data analysis function is a row or column vector, MATLAB simply performs the operation on the vector, returning a scalar result.

Alternatively, you can specify the dimension to work on:

```
>> avg_temp = mean(temps,1)  % same as above, work down the rows
avg_temp =
       11.968         8.2258         19.871
>> avg_tempr = mean(temps,2)  % compute means across columns
avg_tempr =
       12.667
       15.333
          12
          15
       13.333
          13
          13
       12.667
       14.667
       12.333
       14.333
          12
          13
       11.667
       13.667
       12.333
       11.333
       13.667
          12
       13.667
       13.333
          12
       14.333
          13
       12.667
          14
          13
       12.667
```

                                16
                                16
                        15.333

This output lists the three-city average temperature on each day. The scalar second argument to mean dictates the dimension to be analyzed; for example, 1 is the row dimension number, so MATLAB performs the analysis accumulating information down the rows; 2 is the column dimension number, so MATLAB performs the analysis accumulating information across the columns.

    If temps were n-dimensional, the mean and other statistical functions could be used along any dimension, as in the following example:

```
>> temps2 = temps+round(2*rand(size(temps))-1);
>> temps3 = cat(3,temps,temps2);
>> size(temps3)
ans =
    31     3     2
>> mean(temps3)  % same as mean(temps3,1)
ans(:,:,1) =
       11.968        8.2258        19.871
ans(:,:,2) =
       11.677        8.0968        19.903

>> squeeze(mean(temps3))' % squeeze to two dimensions
ans =
       11.968        8.2258        19.871
       11.677        8.0968        19.903

>> reshape(mean(temps3),3,2)' % alternate squeeze
ans =
       11.968        8.2258        19.871
       11.677        8.0968        19.903
>> mean(temps3,3)
ans =
           12            8            18
         14.5            9          22.5
           12          5.5            19
           14          7.5          23.5
```

11.5	6.5	21.5
10.5	9.5	18.5
15	8.5	15.5
8	9.5	20
18.5	6.5	18.5
11.5	6.5	18
13.5	10	19
11	7.5	17
8.5	7.5	22.5
8	7.5	18.5
15	7.5	17.5
8	9.5	20
10	7	17.5
11.5	7	22
9	8	19.5
12	8	21
12	8	20
9.5	9	17
13	11.5	18
9	9.5	20.5
10	6	21.5
14	7.5	21
12	5.5	21.5
12.5	7	18
15	9.5	23
13	11.5	24.5
12.5	11.5	22

Here, `temps2` contains randomly generated data for the temperatures in the three cities for a second month. The variable `temps3` contains the first month's temperatures on page 1 and the second month's temperatures on page 2. The function call `mean(temps3)` computes the means down each column on each page, giving a result that has one row, three columns, and two pages. This data can be squeezed into two rows and three columns by using `squeeze` or `reshape`. The function call `mean(temps3,3)` computes the mean along the page dimension, which is the month-to-month mean of the temperatures on a given day, in a given city. The result is an array having 31 rows, three columns, and one page (i.e., a 2-D array).

Going back to the 2-D case, consider the problem of finding the daily deviation from the mean of each city. That is, avg_temp(i) must be subtracted from column i of temps. You cannot simply issue the statement

```
>> avg_temp
avg_temp =
      11.968        8.2258        19.871
```

```
>> temps - avg_temp
??? Error using ==> -
Matrix dimensions must agree.
```

because the operation is not a defined array operation (temps is 31 by 3, and avg_temp is 1 by 3). Perhaps the most straightforward approach is to use a For Loop:

```
>> for c=1:3
      tdev(:,c) = temps(:,c) - avg_temp(c);
   end
```

While this approach works, it was much slower before the introduction of JIT-Acceleration in MATLAB. An alternative is to use the array manipulation features of MATLAB. In this case, you duplicate avg_temp to make it the size of temps and then do the subtraction:

```
>> tdev = temps - repmat(avg_temp,31,1)
tdev =
    0.0323    -0.2258    -1.8710
    3.0323     0.7742     2.1290
    0.0323    -3.2258    -0.8710
    2.0323    -0.2258     3.1290
    0.0323    -2.2258     2.1290
   -0.9677     0.7742    -0.8710
    3.0323     0.7742    -4.8710
   -3.9677     1.7742     0.1290
    7.0323    -1.2258    -1.8710
    0.0323    -1.2258    -1.8710
    2.0323     1.7742    -0.8710
   -0.9677    -0.2258    -2.8710
```

```
   -2.9677     -1.2258      3.1290
   -3.9677     -0.2258     -0.8710
    3.0323     -0.2258     -1.8710
   -3.9677      0.7742      0.1290
   -1.9677     -1.2258     -2.8710
    0.0323     -1.2258      2.1290
   -2.9677     -0.2258     -0.8710
    0.0323     -0.2258      1.1290
    0.0323     -0.2258      0.1290
   -1.9677      0.7742     -2.8710
    1.0323      3.7742     -1.8710
   -2.9677      1.7742      0.1290
    0.0323     -3.2258      2.1290
    1.0323     -1.2258     -1.8710
    3.0323      1.7742      3.1290
    1.0323      2.7742      4.1290
    0.0323      3.7742      2.1290
```

Here, repmat(avg_temp,31,1) replicates the row vector avg_temp, 31 times in the row dimension and once in the column dimension, creating a 31-by-3 matrix the same size as temps. The first column of repmat(avg_temp,31,1) contains avg_temp(1), the second column contains avg_temp(2), and the third column contains avg_temp(3).

MATLAB can also find minima and maxima. For example, the code

```
>> max_temp = max(temps)
max_temp =
    19    12    24
```

finds the maximum high temperature of each city over the month, while

```
>> [max_temp,maxday] = max(temps)
max_temp =
    19    12    24
maxday =
     9    23    30
```

finds the maximum high temperature of each city and the row index maxday where the maximum appears. In this example, maxday identifies the day of the month when the highest temperature occurred.

The code

```
>> min_temp = min(temps)
min_temp =
    8     5    15
```

finds the minimum high temperature of each city, and the code

```
>> [min_temp,minday] = min(temps)
min_temp =
    8     5    15
minday =
    8     3     7
```

finds the minimum high temperature of each city and the row index minday where the minimum appears. For this example, minday identifies the day of the month when the lowest high temperature occurred.

Other standard statistical measures are also provided in MATLAB, including the following:

```
>> s_dev = std(temps) % standard deviation in each city
s_dev =
    2.5098    1.7646    2.2322
>> median(temps) % median temperature in each city
ans =
    12     8    20
>> cov(temps) % covariance
ans =
       6.2989       0.04086      -0.13763
       0.04086       3.114        0.063441
      -0.13763       0.063441      4.9828
>> corrcoef(temps) % correlation coefficients
ans =
           1       0.0092259     -0.024567
       0.0092259        1         0.016106
      -0.024567      0.016106          1
```

You can also compute differences from day to day by using the function `diff`:

```
>> daily_change = diff(temps)
daily_change =
      3     1     4
     -3    -4    -3
      2     3     4
     -2    -2    -1
     -1     3    -3
      4     0    -4
     -7     1     5
     11    -3    -2
     -7     0     0
      2     3     1
     -3    -2    -2
     -2    -1     6
     -1     1    -4
      7     0    -1
     -7     1     2
      2    -2    -3
      2     0     5
     -3     1    -3
      3     0     2
      0     0    -1
     -2     1    -3
      3     3     1
     -4    -2     2
      1    -4     2
      4     1    -1
     -2    -2     1
      1     2    -4
      2     3     5
     -2     1     1
     -1     1    -2
```

This function computes the difference between daily high temperatures and describes how much the daily high temperature varied from day to day. For example, between the first and second days of the month, the first row of daily_change is the amount that the daily high changed. As with other functions, the difference can be computed along other dimensions as well:

```
>> city_change = diff(temps,1,2)
city_change =
     -4      10
     -6      13
     -7      14
     -6      15
     -6      16
     -2      10
     -6       6
      2      10
    -12      11
     -5      11
     -4       9
     -3       9
     -2      16
      0      11
     -7      10
      1      11
     -3      10
     -5      15
     -1      11
     -4      13
     -4      12
     -1       8
     -1       6
      1      10
     -4      16
     -7      14
     -7      17
     -6      11
     -5      13
```

−2	13
0	10

Here, `diff(temps,1,2)` says to compute a first-order difference along dimension 2. Therefore, the result is the city difference. The first column is the difference between city 2 and city 1; the second column is the difference between city 3 and city 2.

It is common to use the value NaN to signify missing data. When this is done, most statistical functions require special treatment, because operations on NaNs generally produce NaNs, as in the following example:

```
>> temps4 = temps;     % copy data
>> temps4(5,1) = nan; % insert some NaNs
>> temps4(29,2) = nan;
>> temps4(13:14,3) = nan
temps4 =
```

12	8	18
15	9	22
12	5	19
14	8	23
NaN	6	22
11	9	19
15	9	15
8	10	20
19	7	18
12	7	18
14	10	19
11	8	17
9	7	NaN
8	8	NaN
15	8	18
8	9	20
10	7	17
12	7	22
9	8	19
12	8	21
12	8	20
10	9	17
13	12	18

```
   9     10    20
  10      6    22
  14      7    21
  12      5    22
  13      7    18
  15    NaN    23
  13     11    24
  12     12    22
>> max(temps4) % max and min ignore NaNs
ans =
  19     12    24
>> mean(temps4) % other statistical functions propagate NNs
ans =
  NaN    NaN   NaN
>> std(temps4)
ans =
  NaN    NaN   NaN
```

One solution to this NaN problem is to write your own functions that exclude NaN elements:

```
>> m = zeros(1,3); % preallocate memory for faster results

>> for j=1:3 % find mean column by column
       idx = ~isnan(temps4(:,j));
       m(j)=mean(temps4(idx,j));
   end
>> m
m =
           13        7.3333      19.667
```

Here, the function isnan is used to locate elements containing NaN. Then, the mean down each column is found by indexing only non-NaN elements. It is not possible to exclude NaN elements over all columns simultaneously, because not all columns have the same number of NaN elements.

An alternative to the preceding approach is to replace the NaNS with zeros, and compute the mean accordingly, as in the following example:

```
>> lnan = isnan(temps4); % logical array identifying NaNs
>> temps4(lnan) = 0;      % change all NaNs to zero
```

```
>> n = sum(~1nan);        % number of nonNaN elements per column


>> m = sum(temps4)./n    % find mean for all columns
m =
            13          7.3333         19.667
```

In this example, a logical array 1nan containing True where the elements of temps4 are NaN was found. The NaN values were then set to zero. Then the mean was computed by dividing the column sums by the number of nonNaN elements per column.

18.2 BASIC DATA ANALYSIS

In addition to statistical data analysis, MATLAB offers a variety of general-purpose data analysis functions. For example, the temperature data temps can be filtered by using the function filter(b,a,data):

```
>> filter(ones(1,4),4,temps)
ans =
            3             2           4.5
         6.75          4.25           10
         9.75           5.5         14.75
        13.25           7.5          20.5
        13.25           7            21.5
        12.25           7           20.75
           13           8           19.75
         11.5           8.5           19
        13.25          8.75           18
         13.5          8.25         17.75
        13.25           8.5         18.75
           14           8             18
         11.5           8           19.25
         10.5          8.25          19.5
        10.75          7.75         19.25
           10           8             20
        10.25           8           18.5
        11.25          7.75         19.25
         9.75          7.75          19.5
        10.75           7.5         19.75
```

11.25	7.75	20.5
10.75	8.25	19.25
11.75	9.25	19
11	9.75	18.75
10.5	9.25	19.25
11.5	8.75	20.25
11.25	7	21.25
12.25	6.25	20.75
13.5	7.25	21
13.25	8.25	21.75
13.25	10	21.75

Here, the filter implemented is $4y_n = x_n + x_{n-1} + x_{n-2} + x_{n-3}$, or, equivalently, $y_n = (x_n + x_{n-1} + x_{n-2} + x_{n-3})/4$. In other words, each column of temps is passed through a ***moving-average filter*** of length 4. Any realizable filter structure can be applied by specifying different coefficients for the input and output coefficient vectors.

The function y=filter(b,a,x) implements the following general tapped delay-line algorithm:

$$\sum_{k=0}^{N} a_{k+1} y_{n-k} = \sum_{k=0}^{M} b_{k+1} x_{n-k}$$

Here, the vector a is the tap weight vector a_{k+1} on the output, and the vector b is the tap weight vector b_{k+1} on the input. For $N = 2$ and $M = 3$, the preceding equation is equivalent to

$$a_1 y_n + a_2 y_{n-1} + a_3 y_{n-2} = b_1 x_n + b_2 x_{n-1} + b_3 x_{n-2} + b_4 x_{n-3}$$

The filter function uses a difference equation description of a filter. When a state space description of a filter is known, the built-in function ltitr is useful. The help text of this function contains the following text:

```
LTITR    Linear time-invariant time response kernel.

X = LTITR(A,B,U) calculates the time response of the system:
        x[n+1] = Ax[n] + Bu[n]
to input sequence U.  The matrix U must have as many columns as
there are inputs u.  Each row of U corresponds to a new time
point.  LTITR returns a matrix X with as many columns as the
number of states x, and with LENGTH(U) rows.
```

LTITR(A,B,U,Xo) can be used if initial conditions Xo exist.

Here is what it implements, in high speed:
```
for i=1:n
    x(:,i) = xo;
    xo = a * xo + b * u(i,:).';
end
x = x.';
```

In this case, if the state space output equation is $y[n] = Cx[n] + Du[n]$, where C and D are matrices of appropriate dimensions, x=ltitr(A,B,u) can be used to compute the state response $x[n]$, followed by

```
y = x*C.' + u*D.';
```

to obtain the filter output. Here, the array y will have as many columns as there are outputs and as many rows as there are time points.

In MATLAB, data can also be sorted:

```
>> data = rand(10,1) % create some data
data =
        0.61543
        0.79194
        0.92181
        0.73821
        0.17627
        0.40571
        0.93547
         0.9169
        0.41027
        0.89365
>> [sdata,sidx] = sort(data) % sort in ascending order
sdata =
        0.17627
        0.40571
        0.41027
```

```
        0.61543
        0.73821
        0.79194
        0.89365
         0.9169
        0.92181
        0.93547
sidx =
     5
     6
     9
     1
     4
     2
    10
     8
     3
     7
```

The second output of the sort function is the sorted index order. That is, the fifth element in data has the lowest value, and the seventh element in data has the highest value.

Sometimes, it is important to know the **rank** of the data. For example, what is the rank or position of the *i*th data point in the unsorted array with respect to the sorted array? With MATLAB array indexing, the rank is found by the single statement

```
>> ridx(sidx) = 1:10 % ridx is rank
ridx =
     4     6     9     5     1     2    10     8     3     7
```

That is, the first element of the unsorted data appears fourth in the sorted data, and the last element is seventh. If a descending sort is required, it is simply a matter of calling the sort function as [sdata, sidx] = sort (data, 'descend').

When the array to be sorted is a matrix, as temps in the preceding example, each column is sorted, and each column produces a column in the optional index matrix. As with the other data analysis functions, the dimension to analyze can be specified as a second input argument before the optional final 'ascend' or 'descend' argument.

Very often, it is desirable to use the results of sorting one column of an array by applying that sort order to all remaining columns, as in the following example:

```
>> newdata = randn(10,4) % new data for sorting
newdata =
     -0.43256       -0.18671        0.29441       -0.39989
     -1.6656         0.72579       -1.3362         0.69
      0.12533       -0.58832        0.71432        0.81562
      0.28768        2.1832         1.6236         0.71191
     -1.1465        -0.1364        -0.69178        1.2902
      1.1909         0.11393        0.858          0.6686
      1.1892         1.0668         1.254          1.1908
     -0.037633       0.059281      -1.5937        -1.2025
      0.32729       -0.095648      -1.441         -0.01979
      0.17464       -0.83235        0.57115       -0.15672
>> [tmp,idx] = sort(newdata(:,2)); % sort second column
>> newdatas = newdata(idx,:) % shuffle rows using idx from 2nd column
newdatas =
      0.17464       -0.83235        0.57115       -0.15672
      0.12533       -0.58832        0.71432        0.81562
     -0.43256       -0.18671        0.29441       -0.39989
     -1.1465        -0.1364        -0.69178        1.2902
      0.32729       -0.095648      -1.441         -0.01979
     -0.037633       0.059281      -1.5937        -1.2025
      1.1909         0.11393        0.858          0.6686
     -1.6656         0.72579       -1.3362         0.69
      1.1892         1.0668         1.254          1.1908
      0.28768        2.1832         1.6236         0.71191
```

Here, the second column of the random array is sorted in increasing order. Then, the sort index is used to shuffle the rows in all columns. For example, the last row of newdata is now the first row in newdatas, because the last element in the second column is the smallest element in the second column.

A vector is strictly monotonic if its elements either always increase or always decrease as one proceeds down the array. The function diff is useful for determining monotonicity:

```
>> A = diff(data) % check random data
A =
      0.1765
```

```
        0.12988
       -0.18361
       -0.56194
        0.22944
        0.52976
      -0.018565
       -0.50663
        0.48338
>> mono = all(A>0) | all(A<0) % as expected, not monotonic
mono =
        0
>> B = diff(sdata) % check random data after sorting
B =
        0.22944
       0.004564
        0.20516
        0.12277
        0.05373
        0.10171
       0.023255
      0.0049085
       0.013657
>> mono = all(B>0) | all(B<0) % as expected, monotonic
mono =
        1
```

Furthermore, a monotonic vector is equally spaced if

```
>> all( diff( diff(sdata) )==0 ) % random data is not equally spaced
ans =
     0
>> all( diff( diff(1:25) )==0 ) % but numbers from 1 to 25 are equally
spaced
ans =
     1
```

18.3 DATA ANALYSIS AND STATISTICAL FUNCTIONS

By default, data analysis in MATLAB is performed on column-oriented matrices. Different variables are stored in individual columns, and each row represents a different observation of each variable. Many data analysis functions work along any dimension, provided that the dimension is specified as an input argument. The data analysis and statistical functions in MATLAB are listed in the following table:

Function	Description
corrcoef(A)	Correlation coefficients
cov(A)	Covariance matrix
cplxpair(v)	Sorts vector into complex conjugate pairs
cumprod(A)	Cumulative product of elements
cumsum(A)	Cumulative sum of elements
cumtrapz(A)	Cumulative trapezoidal integration
del2(A)	Discrete Laplacian (surface curvature)
diff(A)	Differences between elements
gradient(Z,dx,dy)	Approximate surface gradient
histc(X,edges)	Histogram count and bin locations using bins marked by edges
issorted(A)	True if A is sorted
max(A)	Maximum values
median(A)	Median values
mean(A)	Mean values
min(A)	Minimum values
prod(A)	Product of elements
sort(A)	Sorts in ascending or descending order
sortrows(A)	Sorts rows in ascending order, i.e., dictionary sort
std(A)	Standard deviation
sum(A)	Sum of elements
trapz(A)	Trapezoidal integration
var(A)	Variance, i.e., square of standard deviation

19

Data Interpolation

Interpolation is a way of estimating values of a function between those given by some set of data points. In particular, interpolation serves as a valuable tool when you cannot quickly evaluate the function at the desired intermediate points—for example, when the data points are the result of some experimental measurements or lengthy computational procedure. MATLAB provides tools for interpolating in any number of dimensions by using multidimensional arrays. To illustrate interpolation, only 1- and 2-D interpolations are considered in depth here. However, the functions used for higher dimensions are briefly discussed.

19.1 ONE-DIMENSIONAL INTERPOLATION

Perhaps the simplest example of interpolation is MATLAB plots. By default, MATLAB draws straight lines connecting the data points used to make a plot. This linear interpolation guesses that intermediate values fall on a straight line between the entered points. Certainly, as the number of data points increases and the distance between them decreases, linear interpolation becomes more accurate:

```
>> x1 = linspace(0,2*pi,60);
>> x2 = linspace(0,2*pi,6);
>> plot(x1,sin(x1),x2,sin(x2),'--')
>> xlabel('x'),ylabel('sin(x)')
>> title('Figure 19.1: Linear Interpolation')
```

Of the two plots of the sine function shown, the one using 60 points is much more accurate between the data points than the one using only 6 points.

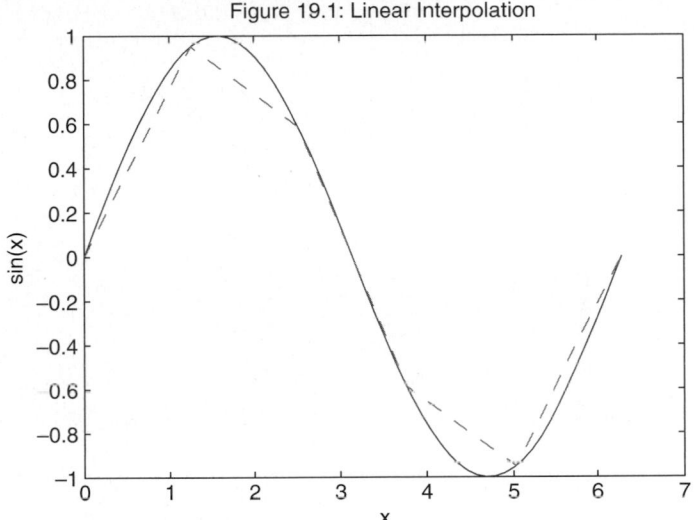

Figure 19.1: Linear Interpolation

To understand 1-D interpolation, consider the following illustration: The threshold of audibility (i.e., the lowest perceptible sound level) of the human ear varies with frequency. Typical data are as follows:

```
>> Hz=[20:10:100 200:100:1000 1500 2000:1000:10000];   % frequencies in Hertz
>> spl =[76 66 59 54    49  46  43 40 38 22 ...   % sound pressure level in dB
        14   9   6 3.5 2.5 1.4 0.7   0 -1 -3 ...
        -8 -7 -2    2    7    9   11 12];
```

The sound pressure levels are normalized, so that 0 dB appears at 1000 Hz. Since the frequencies span such a large range, plot the data using a logarithmic x-axis:

```
>> semilogx(Hz,spl,'-o')
>> xlabel('Frequency, Hz')
>> ylabel('Relative Sound Pressure Level, dB')
>> title('Figure 19.2: Threshold of Human Hearing')
>> grid on
```

According to this plot, the human ear is most sensitive to tones around 3 kHz. Given these data, let's use the function `interp1` to estimate the sound pressure level in several different ways at a frequency of 2.5 kHz:

```
>> s = interp1(Hz,spl,2.5e3)            % linear interpolation
s =

        -5.5
```

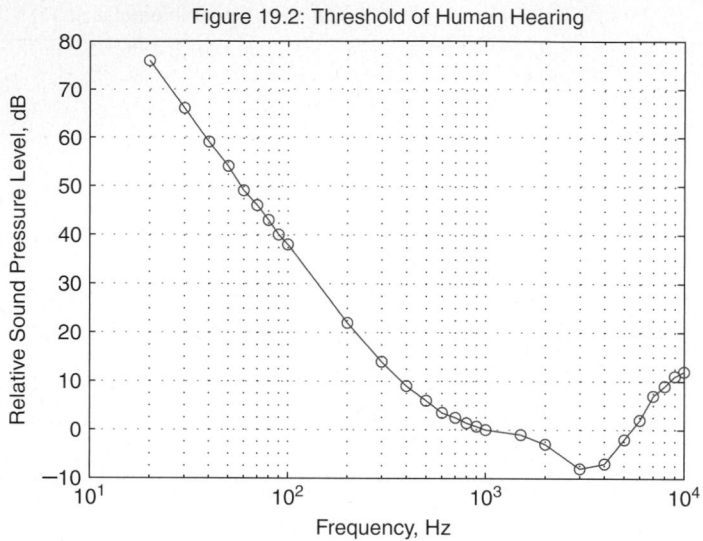

Figure 19.2: Threshold of Human Hearing

```
>> s = interp1(Hz,spl,2.5e3,'linear')  % linear interpolation again
s =
        -5.5
>> s = interp1(Hz,spl,2.5e3,'cubic')    % cubic interpolation
s =
      -6.0488
>> s = interp1(Hz,spl,2.5e3,'spline')  % cubic spline interpolation
s =
      -5.869
>> s = interp1(Hz,spl,2.5e3,'nearest') % nearest neighbor interpolation
    s =
        -8
```

Note the differences in these results. The first two results return exactly what is
shown in the figure at 2.5 kHz, since MATLAB linearly interpolates between data
points on plots. Cubic and spline interpolation fit cubic, that is, third-order, polynomials to each data interval by using different constraints. Cubic interpolation
maintains data monotonicity, whereas spline interpolation exhibits the greatest
smoothness. The crudest interpolation in this case is the nearest-neighbor method,
which returns the input data point nearest the given value.

So how do you choose an interpolation method for a given problem? In many cases, linear interpolation is sufficient. In fact, that's why it is the default method. While the nearest-neighbor method produced poor results here, it is often used when speed is important or when the data set is large. The most time-consuming method is `'spline'`, but it frequently produces the most desirable results.

While the preceding case considered only a single interpolation point, `interp1` can handle any arbitrary number of points. In fact, one of the most common uses of cubic or spline interpolation is to smooth data. That is, given a set of data, use interpolation to evaluate the data at a finer interval:

```
>> Hzi = linspace(2e3,5e3);  % look closely near minimum
>> spli = interp1(Hz,spl,Hzi,'spline');  % interpolate near minimum
>> i = find(Hz>=2e3 & Hz<=5e3);  % find original data indices near minimum

>> semilogx(Hz(i),spl(i),'--o',Hzi,spli)  % plot old and new data
>> xlabel('Frequency, Hz')
>> ylabel('Relative Sound Pressure Level, dB')
>> title('Figure 19.3: Threshold of Human Hearing')
>> grid on
```

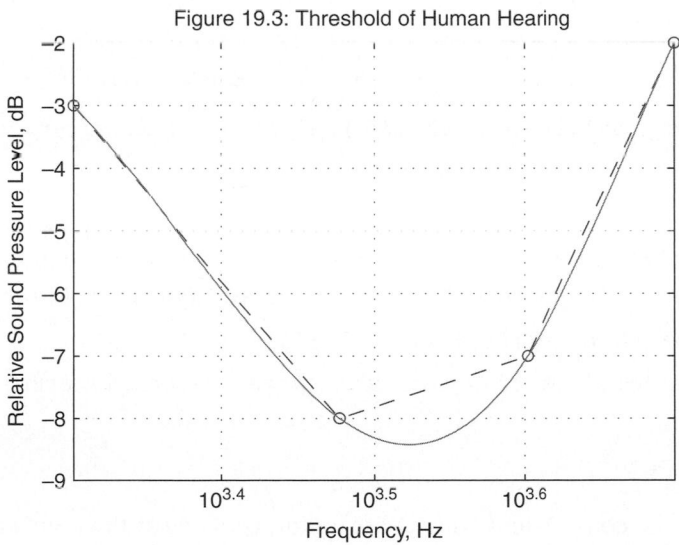

In the plot, the dashed line is the linear interpolation, the solid line is the cubic interpolation, and the original data are marked with `'o'`. By asking for a finer resolution on the frequency axis, and by using spline interpolation, we have a smoother estimate of the sound pressure level. In particular, note how the slope of the spline solution does not change abruptly at the data points.

With these data, we can make a better estimate of the frequency of greatest sensitivity, using for example, the following code:

```
>> [spl_min,i] = min(spli) % minimum and index of minimum
spl_min =
     -8.4245
i =
    45
>> Hz_min = Hzi(i)  % frequency at minimum
Hz_min =
       3333.3
```

According to this analysis, the human ear is most sensitive to tones near 3.33 kHz.

It is important to recognize the major restriction enforced by `interp1`, namely, that the independent variable must be monotonic. That is, the first variable must always increase or must always decrease. In our example, `Hz` is monotonic.

Finally, it is possible to interpolate more than one data set at a time, because the function `interp1` supports multidimensional input (for example, if y is a column-oriented data array). That is, if x is a vector, either y can be a vector, as shown previously, or it can be an array having `length(x)` rows and any number of columns. For example, in the code

```
>> x = linspace(0,2*pi,11)'; % example data
>> y = [sin(x) cos(x) tan(x)];
>> size(y) % three columns
ans =
    11     3
>> xi = linspace(0,2*pi); % interpolate on a finer scale
>> yi = interp1(x,y,xi,'cubic');
>> size(yi) % result is all three columns interpolated
ans =
   100    31
```

`sin(x)`, `cos(x)`, and `tan(x)` are all interpolated at the points in `xi`.

19.2 TWO-DIMENSIONAL INTERPOLATION

Two-dimensional interpolation is based on the same underlying ideas as 1-D interpolation. However, as the name implies, 2-D interpolation interpolates functions of two variables: $z = f(x, y)$. To illustrate this added dimension, consider the following illustration: An exploration company is using sonar to map the ocean floor. At points every 0.5 km on a rectangular grid, the ocean depth in meters is recorded for later analysis. A portion of the data collected is entered into MATLAB in the script M-file ocean.m, as shown in the following:

```
% ocean.m, example test data

% ocean depth data

x = 0:.5:4;  % x-axis (varies across the rows of z)
y = 0:.5:6;  % y-axis (varies down the columns of z)

z=[100     99   100     99   100     99     99     99   100
   100     99     99     99   100     99   100     99     99
    99     99     98     98   100     99   100    100    100
   100     98     97     97     99    100    100    100     99
   101    100     98     98    100    102    103    100    100
   102    103    101    100    102    106    104    101    100
    99    102    100    100    103    108    106    101     99
    97     99    100    100    102    105    103    101    100
   100    102    103    101    102    103    102    100     99
   100    102    103    102    101    101    100     99     99
   100    100    101    101    100    100    100     99     99
   100    100    100    100    100     99     99     99     99
   100    100    100     99     99    100     99    100     99];
```

A plot of this data can be displayed by entering

```
>> mesh(x,y,z)
>> xlabel('X-axis, km')
>> ylabel('Y-axis, km')
>> zlabel('Ocean Depth, m')
>> title('Figure 19.4: Ocean Depth Measurements')
```

With these data, the depth at arbitrary points within the rectangle can be found by using the function interp2, as in the following code:

```
>> zi = interp2(x,y,z,2.2,3.3)
```

Figure 19.4: Ocean Depth Measurements

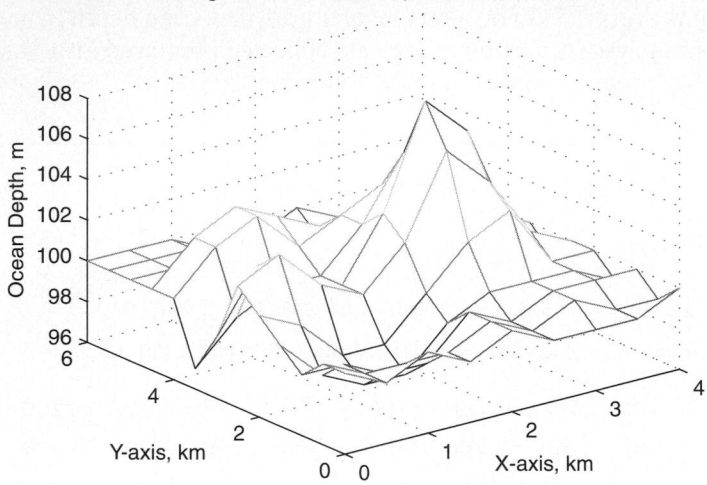

```
zi =

        103.92
>> zi = interp2(x,y,z,2.2,3.3,'linear')
zi =

        103.92
>> zi = interp2(x,y,z,2.2,3.3,'cubic')
zi =

        104.19
>> zi = interp2(x,y,z,2.2,3.3,'nearest')
zi =

    102
```

As was the case with 1-D interpolation, several interpolation methods are available, with the default method being linear.

Once again, we can interpolate on a finer scale, or mesh, to smooth the plot:

```
>> xi = linspace(0,4,30);  % finer x-axis
>> yi = linspace(0,6,40);  % finer y-axis
```

For each value in xi, we wish to interpolate at all values in yi. That is, we wish to create a grid of all combinations of the values of xi and yi, and then interpolate at all of these points. The function meshgrid accepts two vectors and produces two arrays, each containing duplicates of its inputs, so that all combinations of the inputs are considered:

```
>> xtest = 1:5
xtest =
     1    2    3    4    5

>> ytest = 6:9
ytest =
     6    7    8    9

>> [xx,yy] = meshgrid(xtest,ytest)
xx =
     1    2    3    4    5
     1    2    3    4    5
     1    2    3    4    5
     1    2    3    4    5

yy =
     6    6    6    6    6
     7    7    7    7    7
     8    8    8    8    8
     9    9    9    9    9
```

As shown in this code, xx has length(ytest) rows, each containing xtest, and yy has length(xtest) columns, each containing ytest. With this structure, xx(i,j) and yy(i,j) for all i and j cover all combinations of the original vectors, xtest and ytest.

Applying meshgrid to our ocean depth example produces

```
>> [xxi,yyi] = meshgrid(xi,yi);  % grid of all combinations of xi and yi
>> size(xxi) % xxi has 40 rows each containing xi
ans =
    40    30
>> size(yyi) % yyi has 30 columns each containing yi
ans =
    40    30
```

Given xxi and yyi, the ocean depth can now be interpolated on the finer scale by entering

```
>> zzi = interp2(x,y,z,xxi,yyi,'cubic');  % interpolate
>> size(zzi) % zzi is the same size as xxi and yyi
```

```
ans =
    40    30
>> mesh(xxi,yyi,zzi)  % plot smoothed data
>> hold on
>> [xx,yy] = meshgrid(x,y); % grid original data
>> plot3(xx,yy,z+0.1,'ok')  % plot original data up a bit to show nodes
>> hold off

>> xlabel('X-axis, km')
>> ylabel('Y-axis, km')
>> zlabel('Ocean Depth, m')
>> title('Figure 19.5: 2-D Smoothing')
```

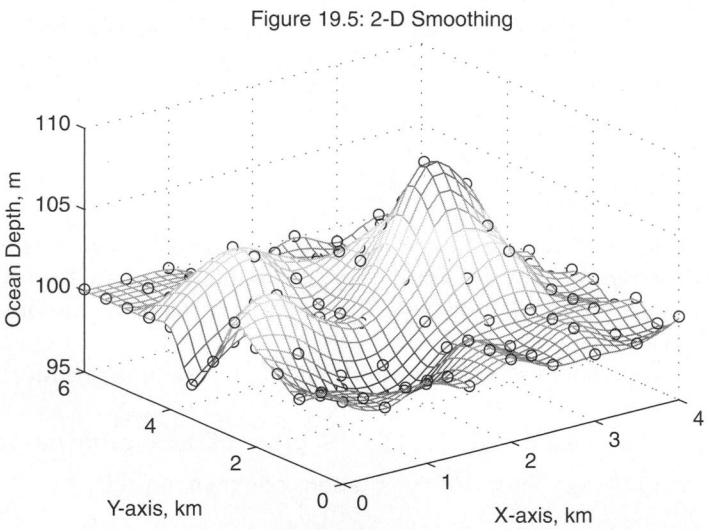

Figure 19.5: 2-D Smoothing

Using these data, we can now estimate the peak and its location:

```
>> zmax = max(max(zzi))
zmax =
        108.05
>> [i,j] = find(zmax==zzi);
>> xmax = xi(j)
xmax =
        2.6207
```

```
>> ymax = yi(i)
ymax =
        2.9231
```

The concepts discussed in these first two sections extend naturally to higher dimensions, where `ndgrid`, `interp3`, and `interpn` apply. The function `ndgrid` is the multidimensional equivalent of `meshgrid`. Multidimensional interpolation uses multidimensional arrays in a straightforward way to organize the data and perform the interpolation. The function `interp3` performs interpolation in 3-D space, and `interpn` performs interpolation in higher order dimensions. Both `interp3` and `interpn` offer method choices of `'linear'`, `'cubic'`, and `'nearest'`. (For more information regarding these functions, see the MATLAB documentation and the on-line help.)

19.3 TRIANGULATION AND SCATTERED DATA

In a number of applications, such as those involving geometric analysis, data points are often scattered, rather than appearing on a rectangular grid, like the ocean data in the example discussed in the last section. For example, consider the following 2-D random data:

```
>> x = randn(1,12);
>> y = randn(1,12);
>> z = zeros(1,12); % no z component for now
>> plot(x,y,'o')
>> title('Figure 19.6: Random Data')
```

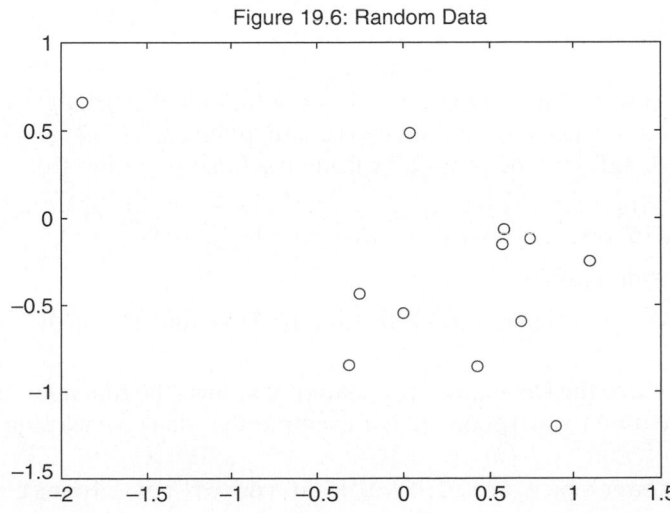

Figure 19.6: Random Data

Given these scattered data, it is common to apply ***Delaunay triangulation***, which returns a set of triangles connecting the data points such that no data points are contained within any triangle. The MATLAB function delaunay accepts data points and returns a list of indices that identify the triangle vertices. For the random data, delaunay returns

```
>> tri = delaunay(x,y)
tri =
    12     9     4
    12     2     4
     7     9     3
     8    12     9
     8     7     9
     8     7    11
     5     2     6
     1     8    12
     1     8    11
     1    12     2
     1     5     2
    10     7    11
    10     1    11
    10     1     5
    10     7     3
    10     6     3
    10     5     6
```

Each row contains indices into x and y that identify triangle vertices. For example, the first triangle is described by the data points in x([12 9 4]) and y([12 9 4]). The triangles can be plotted by using the function trimesh:

```
>> hold on, trimesh(tri,x,y,z), hold off
>> hidden off
>> title('Figure 19.7: Delaunay Triangulation')
```

Once the Delaunay triangulation is known, the functions tsearch and dsearch can be used to interpolate it. For example, the triangle enclosing the origin is

```
>> tsearch(x,y,tri,0,0) % find row of tri closest to (0,0)
```

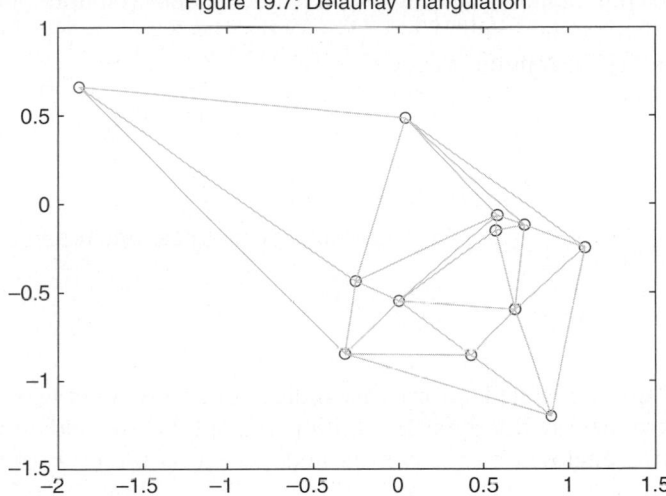

Figure 19.7: Delaunay Triangulation

```
ans =

    4
>> tri(ans,:) % vertices of triangle closest to (0,0)
ans =

    8    12    9
```

Naturally, tsearch accepts multiple values—for example,

```
>> tsearch(x,y,tri,[-.5 1],[.1 .5])
ans =

    1    NaN
```

Here, the first triangle encloses the point $(-0.5, 0.1)$, and no triangle encloses the point $(1, 0.5)$.

Rather than returning the triangle enclosing one or more data points, the function dsearch returns the indices into x and y that are closest to the desired points:

```
>> dsearch(x,y,tri,[-.5 1],[.1 .5])
ans =

    12    7
```

Here, the point (x(12),y(12)) is closest to the point $(-0.5, 0.1)$, and the point (x(7),y(7)) is closest to $(1, 0.5)$.

In addition to interpolating the data, it is often useful to know which points form the outer boundary or ***convex hull*** for the set. The function convhull returns indices into x and y that describe the convex hull, as in the following example:

```
>> [k,a] = convhull(x,y);
>> k'
k  =
      2    6    3    9    4    2
>> a  % optional 2nd argument give area enclosed
a =
       2.7551
```

Note that convhull returns the indices of a closed curve, since the first and last index values are the same. In addition, the optional second output argument returns the area enclosed by the convex hull. Based on the basis of the data returned by convhull, the boundary can be drawn with the following code:

```
>> plot(x,y,'o',x(k),y(k))
>> title('Figure 19.8: Convex Hull')
```

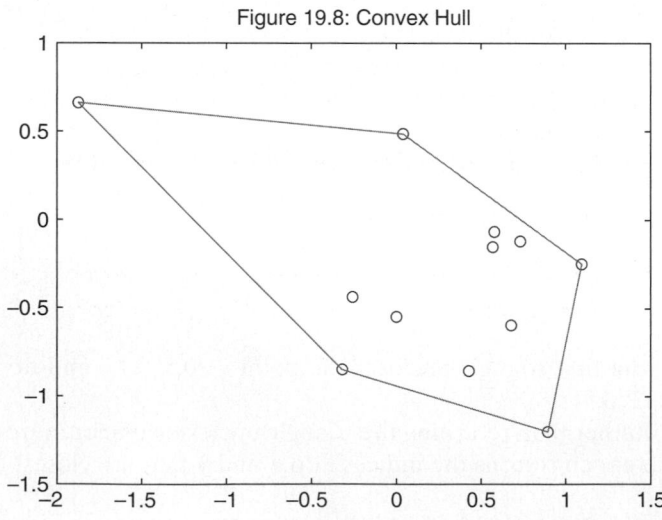

Figure 19.8: Convex Hull

It is also possible to define and draw the lines that separate regions in the plane that are closest to a particular data point. These lines form what is called a **_Voronoi polygon_**. In MATLAB, these lines are drawn by the function voronoi:

```
>> voronoi(x,y)
>> title('Figure 19.9: Voronoi Diagram')
```

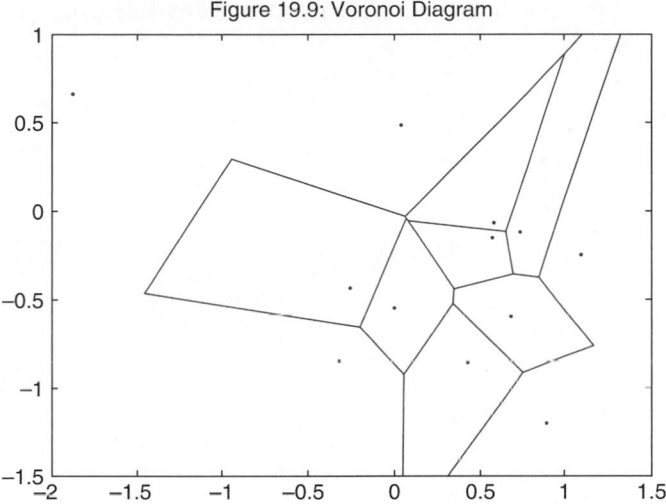

Figure 19.9: Voronoi Diagram

Finally, it is possible to interpolate a Delaunay triangulation to produce interpolated points on a rectangular grid by using the function griddata. In particular, this step is required to use functions such as surf and other standard plotting routines. These plotting routines require data that contain information at a complete sequence of points along two coordinate axes, rather than the scattered data that trimesh uses. Think about a map analogy. Delaunay triangulation allows you to identify specific scattered points on a map. The function griddata uses this information to construct an approximation to the rest of the map, filling in data in a user-specified rectangular region in two coordinate directions:

```
>> z = rand(1,12); % now use some random z axis data
>> xi = linspace(min(x),max(x),30);  % x interpolation points
>> yi = linspace(min(y),max(y),30);  % y interpolation points

>> [Xi,Yi] = meshgrid(xi,yi);        % create grid of x and y

>> Zi = griddata(x,y,z,Xi,Yi);       % grid data at Xi,Yi points

>> mesh(Xi,Yi,Zi)
>> hold on
>> plot3(x,y,z,'ko')  % show original data as well
>> hold off
>> title('Figure 19.10: Griddata Example')
```

Figure 19.10: Griddata Example

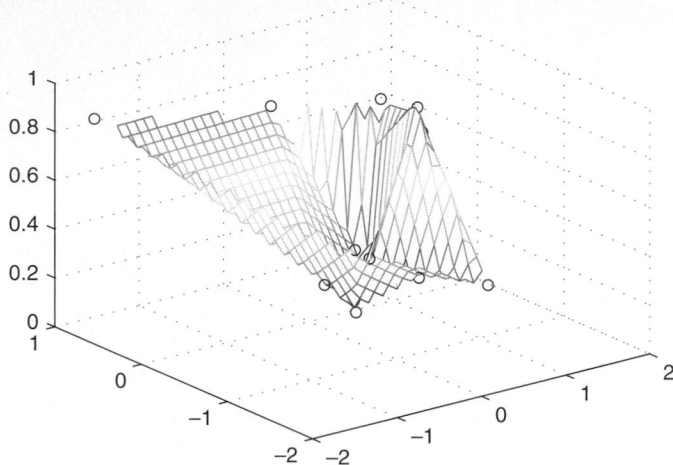

In this example, the information at the 12 scattered data points was interpolated to produce an array of data on a 30-by-30 grid in the x-y plane. The variable Zi contains a 30-by-30 array of points linearly interpolated from the triangulation of the data in x, y and z. Just as interp1 and interp2 support other interpolations, griddata also supports others:

```
>> Zi = griddata(x,y,z,Xi,Yi,'linear')  % same as above (default)
>> Zi = griddata(x,y,z,Xi,Yi,'cubic')   % triangle based cubic interpolation
>> Zi = griddata(x,y,z,Xi,Yi,'nearest') % triangle based nearest neighbor
```

19.4 SUMMARY

In MATLAB, the triangulation functions discussed in this chapter have n-dimensional forms as well. Many of the data interpolation functions have other features and options that cannot be covered here. The data interpolation functions in MATLAB are summarized in the following table:

Function	Description
convhull	Convex hull
convhulln	n-D convex hull
delaunay	Delaunay triangulation
delaunay3	3-D Delaunay tesselation
delaunayn	n-D Delaunay tesselation

dsearch	Nearest point search in Delaunay triangulation
griddata	2-D rectangular gridding
griddata3	3-D rectangular gridding
griddatan	n-D rectangular gridding
interp1	1-D interpolation
interp1q	1-D quick interpolation (no error checking)
interp2	2-D interpolation
interp3	3-D interpolation
interpft	1-D interpolation using FFT method
interpn	n-D interpolation
meshgrid	Generate X and Y matrices for 3-D functions
ndgrid	Generate arrays for multidimensional functions
tetramesh	Tetrahedron mesh plot
trimesh	Triangular mesh plot
triplot	2-D triangular plot
trisurf	Triangular surface plot
tsearch	Closest triangle search
tsearchn	n-D closest triangle search
voronoi	Voronoi diagram
voronoin	n-D Voronoi diagram

20

Polynomials

MATLAB provides a number of functions for manipulating polynomials. Polynomials are easily differentiated and integrated, and it is straightforward to find polynomial roots. However, higher-order polynomials pose numerical difficulties in a number of situations and therefore should be used with caution.

20.1 ROOTS

Finding the roots of a polynomial—that is, the values for which the polynomial is zero—is a problem common to many disciplines. MATLAB solves this problem and provides other polynomial manipulation tools as well. In MATLAB, a polynomial is represented by a row vector of its coefficients in descending order. For example, the polynomial $x^4 - 12x^3 + 0x^2 + 25x + 116$ is entered as

```
>> p = [1 -12 0 25 116]
p =
     1    -12     0    25   116
```

Note that terms with zero coefficients must be included. MATLAB has no way of knowing which terms are zero unless you specifically identify them. Given this form, the roots of a polynomial are found by using the function roots:

```
>> r = roots(p)
r =
  11.7473
   2.7028
```

```
-1.2251 + 1.4672i

-1.2251 - 1.4672i
```

Since in MATLAB both a polynomial and its roots are vectors, MATLAB adopts the convention that *polynomials are row vectors and roots are column vectors*.

Given the roots of a polynomial, it is also possible to construct the associated polynomial. In MATLAB, the command `poly` performs this task:

```
>> pp = poly(r)

pp =

            1          -12 -1.7764e-014           25          116

>> pp(abs(pp)<1e-12) = 0   % change small element to zero!

pp =

            1          -12            0           25          116
```

Because of truncation errors, it is not uncommon for the results of `poly` to have near-zero components or to have components with small imaginary parts. As already shown, near-zero components can be corrected by array manipulation. Similarly, eliminating spurious imaginary parts is simply a matter of using the function `real` to extract the real part of the result.

20.2 MULTIPLICATION

Polynomial multiplication is supported by the function `conv` (which performs the convolution of two arrays). Consider the product of the two polynomials $a(x) = x^3 + 2x^2 + 3x + 4$ and $b(x) = x^3 + 4x^2 + 9x + 16$:

```
>> a = [1 2 3 4];   b = [1 4 9 16];

>> c = conv(a,b)

c =

     1     6    20    50    75    84    64
```

This result is $c(x) = x^6 + 6x^5 + 20x^4 + 50x^3 + 75x^2 + 84x + 64$. Multiplication of more than two polynomials requires repeated use of `conv`.

20.3 ADDITION

MATLAB does not provide a direct function for adding polynomials. Standard array addition works if both polynomial vectors are the same order, as in the code

```
>> d = a + b

d =

     2     6    12    20
```

which represents $d(x) = 2x^3 + 6x^2 + 12x + 20$. When two polynomials are of different orders, the polynomial of lower order must be padded with leading zeros so that it has the same effective order as the higher-order polynomial. Consider the addition of the preceding polynomials c and d:

```
>> e = c + [0 0 0 d]

e =

     1     6    20    52    81    96    84
```

The resulting polynomial is $e(x) = x^6 + 6x^5 + 20x^4 + 52x^3 + 81x^2 + 96x + 84$. Leading zeros, rather than trailing zeros, are required, because coefficients associated with like powers of x must line up. The following M-file function automates polynomial addition:

```
function p=mmpadd(a,b)
%MMPADD Polynomial Addition.
% MMPADD(A,B) adds the polynomials A and B.

if nargin<2
    error('Not Enough Input Arguments.')
end

a=reshape(a,1,[]); % make sure inputs are polynomial row vectors
b=b(:).';          % this makes a row as well

na=length(a);  % find lengths of a and b
nb=length(b);

p=[zeros(1,nb-na) a]+[zeros(1,na-nb) b];  % pad with zeros as necessary
```

To illustrate the use of mmpadd, consider again the preceding example

```
>> f = mmpadd(c,d)

f =

     1     6    20    52    81    96    84
```

which is the same as our earlier e. Of course, mmpadd can also be used for subtraction, as in

```
>> g = mmpadd(c,-d)
g =
      1     6     20     48     69     72     44
```

The resulting polynomial is $g(x) = x^6 + 6x^5 + 20x^4 + 48x^3 + 69x^2 + 72x + 44$.

20.4 DIVISION

In some special cases, it is necessary to divide one polynomial into another. In MAT-LAB, this is accomplished with the function deconv:

```
>> [q,r] = deconv(c,b)
q =
      1     2     3     4
r =
      0     0     0     0     0     0     0
```

This result says that b divided into c gives the quotient polynomial q and the remainder r, which is zero in this case, since the product of b and q is exactly c. Another example gives a remainder:

```
>> [q,r] = deconv(f,b)
q =
      1     2     3     6
r =
      0     0     0     0    -2    -6    -12
```

Here, b divided into f gives the quotient polynomial q and the remainder r. The leading zeros in r simply make r the same length as f. In this case, the quotient is $q(x) = x^3 + 2x^2 + 3x + 6$ and the remainder term is $r(x) = -2x^2 - 6x - 12$.

20.5 DERIVATIVES AND INTEGRALS

Because differentiation of a polynomial is simple to express, MATLAB offers the function polyder for polynomial differentiation:

```
>> g    % recall polynomial
g =
      1     6     20     48     69     72     44
```

```
>> h = polyder(g)
h =
     6    30    80   144   138    72
```

Similarly, the integral of a polynomial is easy to express. Given an integration constant, the function `polyint` returns the integral:

```
>> polyint(h,44) % get g back from h=polyder(g)
ans =
     1     6    20    48    69    72    44
```

20.6 EVALUATION

Given that you can add, subtract, multiply, divide, and differentiate polynomials on the basis of row vectors of their coefficients, you should be able to evaluate them also. In MATLAB, this is accomplished with the function `polyval`, as in the following example:

```
>> p = [1 4 -7 -10];    % the polynomial
>> x = linspace(-1,3);  % evaluation points
>> v = polyval(p,x);    % evaluate p at points in x
>> plot(x,v)            % plot results
>> title('Figure 20.1: x{^3} + 4x{^2} - 7x -10')
>> xlabel('x')
```

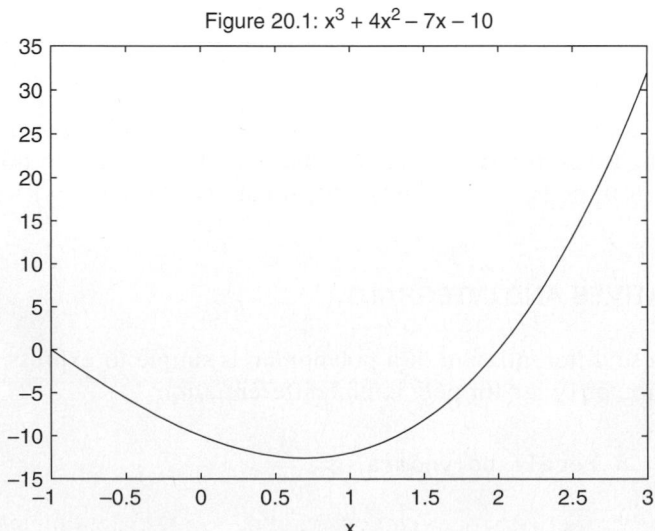

Figure 20.1: $x^3 + 4x^2 - 7x - 10$

20.7 RATIONAL POLYNOMIALS

Sometimes one encounters ratios of polynomials—for example, transfer functions and Pade approximations to functions. In MATLAB, these are manipulated by considering the numerator and denominator polynomials separately, as in the following example:

```
>> n = [1 -10 100]  % a numerator
n =
      1    -10    100
>> d = [1 10 100 0]  % a denominator
d =
      1     10    100     0
>> z - roots(n)  % the zeros of n(x)/d(x)
z =
         5 +        8.6603i
         5 -        8.6603i

>> p = roots(d)  % the poles of n(x)/d(x)
p =

         0
        -5 +        8.6603i
        -5 -        8.6603i
```

The derivative of this rational polynomial with respect to x is found by using `polyder`:

```
>> [nd,dd] = polyder(n,d)
nd =
         -1          20         -100        -2000       -10000
dd =
  Columns 1 through 6
          1          20          300         2000        10000           0
  Column 7
          0
```

Here, nd and dd are the respective numerator and denominator polynomials of the derivative.

Another common operation is to find the partial-fraction expansion of a rational polynomial—for example,

```
>> [r,p,k] = residue(n,d)
r =
    9.7954e-17 +        1.1547i
    9.7954e-17 -        1.1547i
           1

p =
           -5 +        8.6603i
           -5 -        8.6603i
            0

k =
      []
```

In this case, the `residue` function returns the residues or partial-fraction expansion coefficients r, their associated poles p, and the direct term polynomial k. Since the order of the numerator is less than that of the denominator, there are no direct terms. For this example, the partial fraction expansion of the rational polynomial is

$$\frac{n(x)}{d(x)} = \frac{1.1547i}{x + 5 - 8.6603i} + \frac{-1.1547i}{x + 5 + 8.6603i} + \frac{1}{x}$$

Given this information, the original rational polynomial is found by using `residue` yet again:

```
>> [nn,dd] = residue(r,p,k)
nn =
            1          -10          100

dd =
            1           10          100            0
```

So, in this case, the function `residue` performs two operations that are inverses of one another depending on how many input and output arguments are used.

20.8 CURVE FITTING

In numerous application areas, you are faced with the task of fitting a curve to measured data. Sometimes the chosen curve passes through the data points, but at other times the curve comes close to, but does not necessarily pass through, the data points. In the most common situation, the curve is chosen so that the sum of the squared errors at the data points is minimized. This choice results in a *least squares* curve fit. While least squares curve fitting can be done by using any set of basis functions, it is straightforward and common to use a truncated power series—that is, a polynomial.

In MATLAB, the function `polyfit` solves the least squares polynomial curve-fitting problem. To illustrate the use of this function, let's start with the following data:

```
>> x = [0 .1 .2 .3 .4 .5 .6 .7 .8 .9 1];
>> y = [-.447 1.978 3.28 6.16 7.08 7.34 7.66 9.56 9.48 9.30 11.2];
```

To use `polyfit`, we must supply this data and the order, or degree, of the polynomial we wish to best fit to the data. If we choose $n = 1$ as the order, the best straight line approximation will be found. This is often called *linear regression*. On the other hand, if we choose $n = 2$ as the order, a quadratic polynomial will be found. For now, let's choose a quadratic polynomial:

```
>> n = 2;
>> p = polyfit(x,y,n)

p =

    -9.8108    20.1293    -0.0317
```

The output of `polyfit` is a row vector of the polynomial coefficients. Here the solution is $y(x) = -9.8108x^2 + 20.1293x - 0.0317$. To compare the curve-fit solution to the data points, let's plot both:

```
>> xi = linspace(0,1,100);
>> yi = polyval(p,xi);
>> plot(x,y,'-o',xi,yi,'-')
>> xlabel('x'), ylabel('y=f(x)')
>> title('Figure 20.2:  Second Order Curve Fitting')
```

This plot contains the original data x and y, marking the data points with `'o'` and connecting them with straight lines. In addition, the evaluated polynomial data xi and yi are plotted with a dashed line (`'--'`).

The choice of polynomial order is somewhat arbitrary. It takes two points to define a straight line or first-order polynomial. (If this isn't clear to you, mark two points and draw a straight line between them.) It takes three points to define a quadratic, or second-order, polynomial. Following this progression, it takes $n + 1$ data points to uniquely specify an nth-order polynomial. Thus, in the preceding case, where there are 11 data points, we could choose up to a 10th-order polynomial. However, given the poor numerical properties of higher order polynomials, you

Figure 20.2: Second Order Curve Fitting

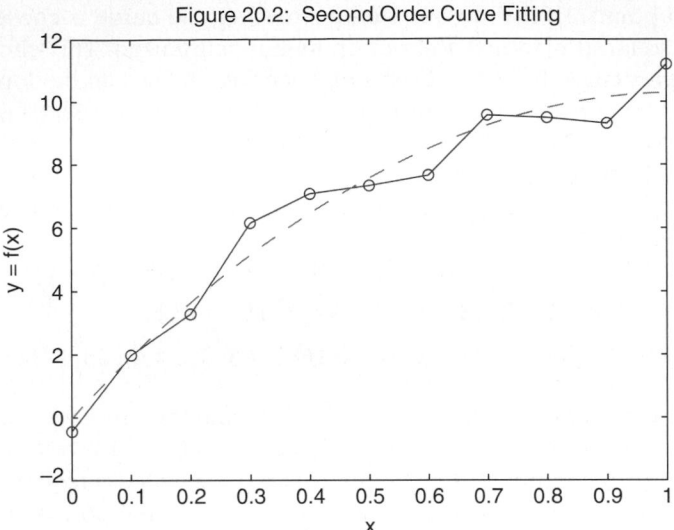

should not choose a polynomial order that is any higher than necessary. In addition, as the polynomial order increases, the approximation becomes less smooth, since higher order polynomials can be differentiated more times before they become zero. For example, consider choosing a 10th-order polynomial:

```
>> pp = polyfit(x,y,10);
>> pp.'  % display polynomial coefficients as a column
ans =
 -4.6436e+005
  2.2965e+006
 -4.8773e+006
  5.8233e+006
 -4.2948e+006
  2.0211e+006
 -6.0322e+005
  1.0896e+005
       -10626
       435.99
       -0.447
```

Note the size of the polynomial coefficients in this case, compared with those of the earlier quadratic fit. Note also the seven orders of magnitude difference between the smallest (−0.447) and largest (−4.6436e+005) coefficients and the alternating signs on the coefficients. To see how this polynomial differs from the quadratic fit shown earlier, consider a plot of both:

```
>> y10 = polyval(pp,xi);  % evaluate 10th order polynomial
>> plot(x,y,'o',xi,yi,'--',xi,y10) % plot data
>> xlabel('x'), ylabel('y=f(x)')
>> title('Figure 20.3: 2nd and 10th Order Curve Fitting')
```

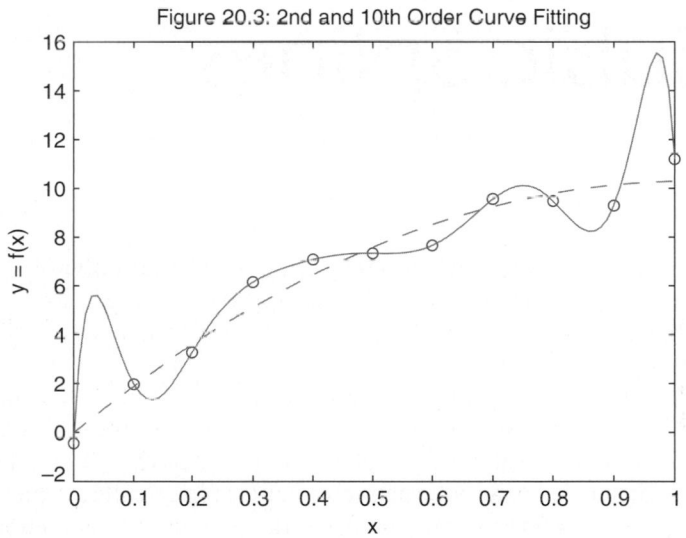

Figure 20.3: 2nd and 10th Order Curve Fitting

In this plot, the original data are marked with `'o'`, the quadratic curve fit is dashed, and the 10th-order fit is solid. Note the wavelike ripples that appear between the data points at the left and right extremes in the 10th-order fit. This example clearly demonstrates the difficulties with higher order polynomials.

21

Cubic Splines

It is well known that interpolation using high-order polynomials often produces ill-behaved results. There are numerous approaches to eliminating this poor behavior. Of these approaches, cubic splines are very popular. In MATLAB, basic cubic splines interpolation is accomplished by the functions `spline`, `ppval`, `mkpp`, and `unmkpp`. Of these, only `spline` appears in the MATLAB documentation. However, help text is available with all of these functions. In the following sections, the basic features of cubic splines as implemented in these M-file functions is demonstrated. Also considered is an alternative to cubic splines called a piecewise cubic Hermite interpolating polynomial. This piecewise polynomial is computed by the function `pchip` and returns a piecewise polynomial, just as `spline` does.

21.1 BASIC FEATURES

In cubic splines, cubic polynomials are found to approximate the curve between each pair of data points. In the language of splines, these data points are called breakpoints. Since a straight line is uniquely defined by two points, an infinite number of cubic polynomials can be used to approximate a curve between two points. Therefore, in cubic splines, additional constraints are placed on the cubic polynomials to make the result unique. By constraining the first and second derivatives of each cubic polynomial to match at the breakpoints, all internal cubic polynomials are well defined. Moreover, both the slope and curvature of the approximating polynomials are continuous across the breakpoints. However, the first and last cubic polynomials do not have adjoining cubic polynomials beyond the first and last breakpoints. As a result, the remaining constraints must be determined by some other means. The most common approach, which is adopted by the function `spline`, is to adopt a ***not-a-knot*** condition. This condition forces the third derivative of the first and second cubic polynomials to be identical, and likewise for the last and second-to-last cubic polynomials.

Based on this description, you could guess that finding cubic spline polynomials requires solving a large set of linear equations. In fact, given n breakpoints, there are $n - 1$ cubic polynomials to be found, each having 4 unknown coefficients. Thus, the set of equations to be solved involves $4(n - 1)$ unknowns. By writing each cubic polynomial in a special form and by applying the constraints, the cubic polynomials can be found by solving a reduced set of n equations in n unknowns. Thus, if there are 50 breakpoints, there are 50 equations in 50 unknowns. Luckily, these equations can be concisely written and solved using sparse matrices, which describes what the function spline uses to compute the unknown coefficients.

21.2 PIECEWISE POLYNOMIALS

In its most simple use, spline takes data x and y and desired values xi, finds the cubic spline interpolation polynomials that fit x and y, and then evaluates the polynomials to find the corresponding yi values for each xi value. This approach matches the use of yi = interp1(x,y,xi,'spline'):

```
>> x = 0:12;
>> y = tan(pi*x/25);
>> xi = linspace(0,12);

>> yi - spline(x,y,xi);

>> plot(x,y,'o',xi,yi)
>> title('Figure 21.1: Spline Fit')
```

This approach is appropriate if only one set of interpolated values is required. However, if more values are needed from the same set of data, it doesn't make sense to recompute the same set of cubic spline coefficients a second time. In this situation, one can call spline with only the first two arguments:

```
>> pp = spline(x,y)
pp =
        form: 'pp'
      breaks: [0 1 2 3 4 5 6 7 8 9 10 11 12]
       coefs: [12x4 double]
      pieces: 12
       order: 4
         dim: 1
```

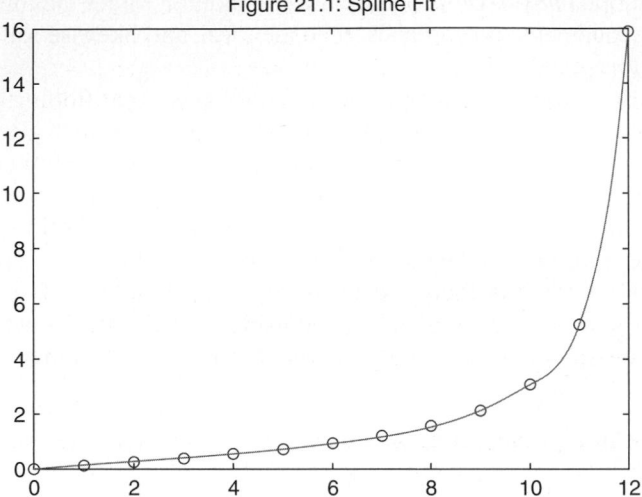

Figure 21.1: Spline Fit

When called in this way, `spline` returns a structure containing the ***pp-form***, or piecewise polynomial form, of the cubic splines. This structure contains all of the information necessary to evaluate the cubic splines for any set of desired interpolation values. The pp-form structure is also compatible with the optional *Spline Toolbox* available with MATLAB. Given the pp-form, the function `ppval` evaluates the cubic splines. For example,

```
>> yi = ppval(pp,xi);
```

computes the same `yi` values computed earlier. Similarly,

```
>> xi2 = linspace(10,12);
>> yi2 = ppval(pp,xi2);
```

uses the pp-form again to evaluate the cubic splines over a finer spacing, restricted to the region between 10 and 12. The code

```
>> xi3 = 10:15;
```

```
>> yi3 = ppval(pp,xi3)
yi3 =
     3.0777     5.2422    15.8945    44.0038    98.5389   188.4689
```

```
>> yi4 = ppval(xi3,pp) % can be called with arguments reversed
yi4 =
     3.0777     5.2422    15.8945    44.0038    98.5389   188.4689
```

shows that cubic splines can be evaluated outside of the region over which the cubic polynomials were computed. When data appears beyond the last or before the first breakpoint, the last and first cubic polynomials are used, respectively, to find interpolated values. In addition, this example shows that ppval can be called with its input arguments reversed. This permits creating a function handle for ppval and passing it as an argument to functions that evaluate a user-supplied function as part of their work. For example, the function

```
>> quad(@ppval,0,10,[],[],pp)
ans =
        9.3775
```

computes the area under the spline given by pp over the range 0 to 10. (More information about the function quad can be found in Chapter 24.)

The cubic splines pp-form just given stores the breakpoints and polynomial coefficients, as well as other information regarding the cubic splines representation. This form is a convenient data structure in MATLAB, since all information is stored in a single structure. When a cubic spline representation is evaluated, the various fields in the pp-form must be extracted. In MATLAB, this process is conveniently performed by the function unmkpp. Using this function on the preceding pp-form gives

```
>> [breaks,coefs,npolys,ncoefs,dim] = unmkpp(pp)
breaks -
  Columns 1 through 12
      0     1     2     3     4     5     6     7     8     9    10    11
  Column 13
     12
coefs =
     0.0007    -0.0001     0.1257          0
     0.0007     0.0020     0.1276     0.1263
     0.0010     0.0042     0.1339     0.2568
     0.0012     0.0072     0.1454     0.3959
     0.0024     0.0109     0.1635     0.5498
     0.0019     0.0181     0.1925     0.7265
     0.0116     0.0237     0.2344     0.9391
    -0.0083     0.0586     0.3167     1.2088
     0.1068     0.0336     0.4089     1.5757
    -0.1982     0.3542     0.7967     2.1251
```

```
     1.4948     -0.2406      0.9102       3.0777
     1.4948      4.2439      4.9136       5.2422
npolys =
     12
ncoefs =
      4
dim =
      1
```

Here, breaks contains the breakpoints, coefs is a matrix whose *i*th row is the *i*th cubic polynomial, npolys is the number of polynomials, ncoefs is the number of coefficients per polynomial, and dim is the spline dimension. Note that this pp-form is sufficiently general that the spline polynomials need not be cubic. This fact is useful when the spline is integrated or differentiated.

In MATLAB versions prior to version 6, the pp-form was stored in a single numerical array rather than in a structure. As a result, unmkpp was valuable in separating the parts of the pp-form from the numerical array. Given the simplicity of the structure form, you can easily address the fields directly and avoid using unmkpp entirely. However, unmkpp continues to support the prior numerical array pp-form, thereby making the process of extracting the parts of a pp-form transparent to the user.

Given the broken-apart form, the function mkpp restores the pp-form:

```
>> pp = mkpp(breaks,coefs)
pp =
        form: 'pp'
      breaks: [0 1 2 3 4 5 6 7 8 9 10 11 12]
       coefs: [12x4 double]
      pieces: 12
       order: 4
         dim: 1
```

Since the size of the matrix coefs determines npolys and ncoefs, they are not needed by mkpp to reconstruct the pp-form.

21.3 CUBIC HERMITE POLYNOMIALS

When the underlying data to be interpolated represents a smooth function, cubic splines return appropriate values. However, when the underlying data are not so smooth, cubic splines can predict minima and maxima that do not exist and can destroy monotonicity. Therefore, for nonsmooth data, a different piecewise polynomial interpolation is called for. In MATLAB, the function pchip returns a piecewise cubic polynomial that has the properties described in the following help text:

```
>> help pchip


 PCHIP  Piecewise Cubic Hermite Interpolating Polynomial.
  PP = PCHIP(X,Y) provides the piecewise polynomial form of a certain
  shape-preserving piecewise cubic Hermite interpolant, to the values
  Y at the sites X, for later use with PPVAL and the spline utility UNMKPP.
  X must be a vector.
  If Y is a vector, then Y(j) is taken as the value to be matched at X(j),
  hence Y must be of the same length as X.
  If Y is a matrix or ND array, then Y(:,...,:,j) is taken as the value to
  be matched at X(j),  hence the last dimension of Y must equal length(X).
  YY = PCHIP(X,Y,XX) is the same as YY = PPVAL(PCHIP(X,Y),XX), thus
  providing, in YY, the values of the interpolant at XX.
  The PCHIP interpolating function, p(x), satisfies:
  On each subinterval,  X(k) <= x <= X(k+1),  p(x) is the cubic Hermite
      interpolant to the given values and certain slopes at the two endpoints.
  Therefore, p(x) interpolates Y, i.e., p(X(j)) = Y(:,j), and
        the first derivative, Dp(x), is continuous, but
        D^2p(x) is probably not continuous; there may be jumps at the X(j).
  The slopes at the X(j) are chosen in such a way that
      p(x) is "shape preserving" and "respects monotonicity". This means that,
  on intervals where the data is monotonic, so is p(x);
  at points where the data have a local extremum, so does p(x).
 Comparing PCHIP with SPLINE:
   The function s(x) supplied by SPLINE is constructed in exactly the same way,
   except that the slopes at the X(j) are chosen differently, namely to make
   even D^2s(x) continuous. This has the following effects.
   SPLINE is smoother, i.e., D^2s(x) is continuous.
   SPLINE is more accurate if the data are values of a smooth function.
   PCHIP has no overshoots and less oscillation if the data are not smooth.
```

PCHIP is less expensive to set up.

The two are equally expensive to evaluate.

The following example demonstrates the similarities and differences between spline and pchip:

```
>> x = [0 2 4 5 7.5 10];% sample data
>> y = exp(-x/6).*cos(x);

>> cs = spline(x,y);    % cubic spline
>> ch = pchip(x,y);     % cubic Hermite

>> xi = linspace(0,10);
>> ysi = ppval(cs,xi);  % interpolate spline
>> yhi = ppval(ch,xi);  % interpolate Hermite

>> plot(x,y,'o',xi,ysi,':',xi,yhi)
>> legend('data','spline','hermite')
>> title('Figure 21.2: Spline and Hermite Interpolation')
```

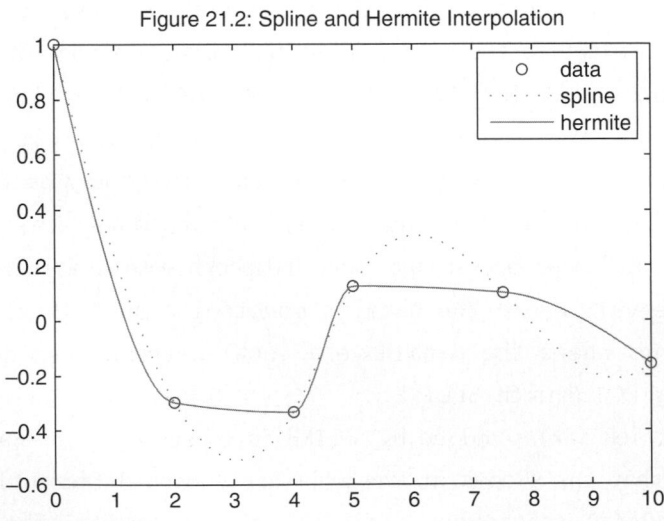

Figure 21.2: Spline and Hermite Interpolation

21.4 INTEGRATION

In many situations, it is desirable to know the area under a function described by piecewise polynomials as a function of the independent variable x. That is, if the piecewise polynomials are denoted $y = s(x)$, we are interested in computing

$$S(x) = \int_{x_1}^{x} s(x)\, dx + C$$

where x_1 is the first breakpoint and C is the integration constant. Since $s(x)$ is composed of connected cubic polynomials, with the kth cubic polynomial being

$$s_k(x) = a_k(x - x_k)^3 + b_k(x - x_k)^2 + c_k(x - x_k) + d_k, \quad x_k \leq x \leq x_{k+1}$$

and whose integral or area over the range $[x_k, x]$, where $x_k \leq x \leq x_{k+1}$, is

$$S_k(x) = \int_{x_k}^{x} s_k(x)\, dx = \frac{a_k}{4}(x - x_k)^4 + \frac{b_k}{3}(x - x_k)^3 + \frac{c_k}{2}(x - x_k)^2 + d_k(x - x_k)$$

the area under a piecewise polynomial is easily computed as

$$S(x) = \sum_{i=1}^{k-1} S_i(x_{i+1}) + S_k(x)$$

where $x_k \leq x \leq x_{k+1}$. The summation term is the cumulative sum of the areas under all preceding cubic polynomials. As such, it is readily computed and forms the constant term in the polynomial describing $S(x)$, since $S_k(x)$ is a polynomial. With this understanding, the integral itself can be written as a piecewise polynomial. In this case, it is a quartic piecewise polynomial, because the individual polynomials are of order four.

Because the pp-form used in MATLAB can support piecewise polynomials of any order, the preceding piecewise polynomial integration is embodied in the function mmppint. The body of this function is as follows:

```
function ppi=mmppint(pp,c)
%MMPPINT Cubic Spline Integral Interpolation.
% PPI=MMPPINT(PP,C) returns the piecewise polynomial vector PPI
% describing the integral of the cubic spline described by
% the piecewise polynomial in PP and having integration constant C.

if prod(size(c))~=1
   error('C Must be a Scalar.')
end
[br,co,npy,nco]=unmkpp(pp);          % take apart pp
```

```
sf=nco:-1:1;                          % scale factors for integration
ico=[co./sf(ones(npy,1),:) zeros(npy,1)];    % integral coefficients
nco=nco+1;                            % integral spline has higher order
ico(1,nco)=c;                         % integration constant
for k=2:npy                           % find constant terms in polynomials
        ico(k,nco)=polyval(ico(k-1,:),br(k)-br(k-1));
end
ppi=mkpp(br,ico);                     % build pp form for integral
```

Consider the following example using mmppint:

```
>> x = (0:.1:1)*2*pi;
>> y = sin(x); % create rough data
>> pp = spline(x,y);        % pp-form fitting rough data
>> ppi = mmppint(pp,0);     % pp-form of integral
>> xi = linspace(0,2*pi); % finer points for interpolation
>> yi = ppval(pp,xi);       % evaluate curve
>> yyi = ppval(ppi,xi);     % evaluate integral
>> plot(x,y,'o',xi,yi,xi,yyi,'--') % plot results
>> title('Figure 21.3: Spline Integration')
```

Note that this plot qualitatively shows the identity

$$\int_0^x \sin(x)\, dx = 1 - \cos(x)$$

21.5 DIFFERENTIATION

Just as you may be interested in piecewise polynomial integration, you may find that the derivative or slope of a function described by piecewise polynomials is also useful. Given that the kth cubic polynomial is

$$s_k(x) = a_k(x - x_k)^3 + b_k(x - x_k)^2 + c_k(x - x_k) + d_k, \quad x_k \le x \le x_{k+1}$$

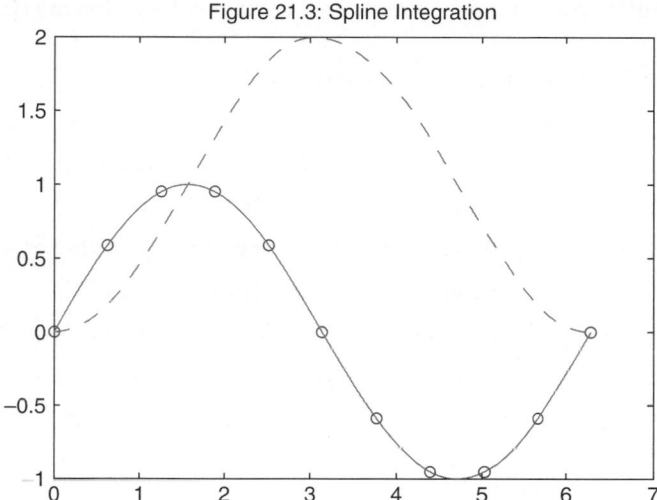

Figure 21.3: Spline Integration

the derivative of $s_k(x)$ is easily written as

$$\frac{ds_k(x)}{dx} = 3a_k(x - x_k)^2 + 2b_k(x - x_k) + c_k$$

where $x_k \leq x \leq x_{k+1}$. As with integration, the derivative is also a piecewise polynomial. However, in this case it is a quadratic piecewise polynomial, since the order of the polynomial is two.

Based on this expression, the function mmppder performs piecewise polynomial differentiation. The body of this function is as follows:

```
function ppd=mmppder(pp)
%MMPPDER Cubic Spline Derivative Interpolation.
% PPD=MMPPDER(PP) returns the piecewise polynomial vector PPD
% describing the cubic spline derivative of the curve described
% by the piecewise polynomial in PP.
[br,co,npy,nco]=unmkpp(pp);              % take apart pp
sf=nco-1:-1:1;                           % scale factors for differentiation
dco=sf(ones(npy,1),:).*co(:,1:nco-1);    % derivative coefficients
ppd=mkpp(br,dco);                        % build pp form for derivative
```

To demonstrate the use of mmppder, consider the following example:

```
>> x = (0:.1:1)*2*pi; % same data as earlier
>> y = sin(x);
>> pp = spline(x,y);      % pp-form fitting rough data
>> ppd = mmppder(pp);     % pp-form of derivative
>> xi = linspace(0,2*pi); % finer points for interpolation
>> yi = ppval(pp,xi);     % evaluate curve
>> yyd = ppval(ppd,xi);   % evaluate derivative
>> plot(x,y,'o',xi,yi,xi,yyd,'--') % plot results
>> title('Figure 21.4: Spline Differentiation')
```

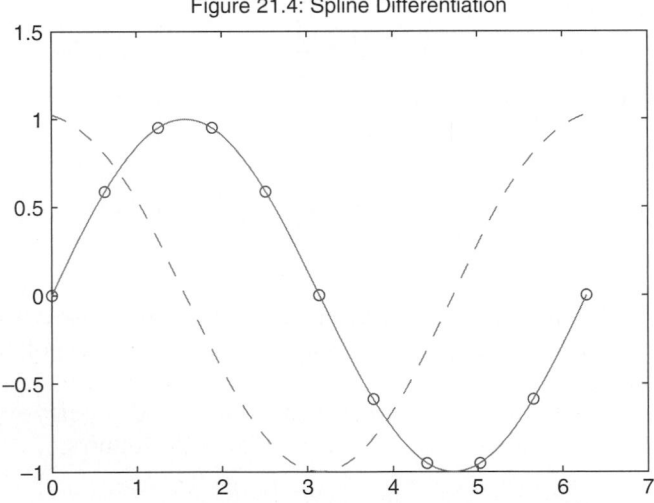

Figure 21.4: Spline Differentiation

Note that this plot qualitatively shows the identity

$$\frac{d}{dx}\sin(x) = \cos(x)$$

21.6 SPLINE INTERPOLATION ON A PLANE

Spline interpolation, as implemented by the function spline, assumes that the independent variable is monotonic. That is, the spline $y = s(x)$ describes a continuous function. When it is not monotonic, there is no one-to-one relationship between x and y, and the function ppval has no way of knowing what y value to return for a given x. A common situation where this occurs is a curve defined on a plane—for example,

```
>> t = linspace(0,3*pi,15);
>> x = sqrt(t).*cos(t);
>> y = sqrt(t).*sin(t);


>> plot(x,y)
>> xlabel('X')
>> ylabel('Y')
>> title('Figure 21.5: Spiral Y=f(X)')
```

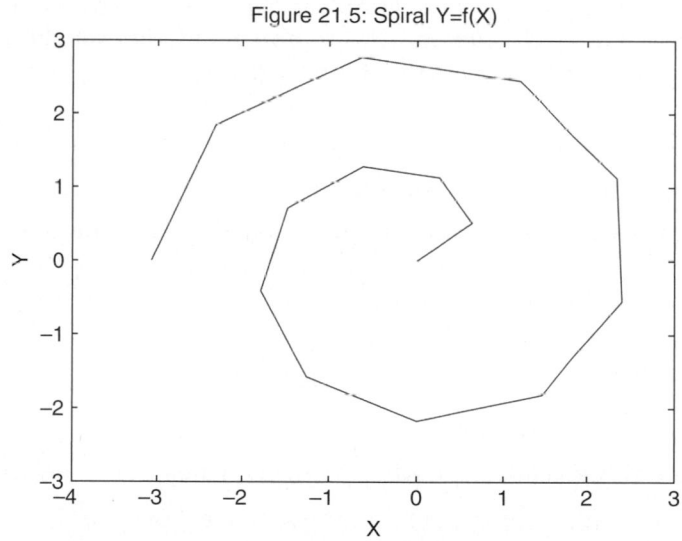

Figure 21.5: Spiral Y=f(X)

It is not possible to compute a cubic spline for the spiral as a function of x, since there are multiple y values for each x near the origin. However, it is possible to compute a spline for each axis with respect to the variable or parameter t. This can be accomplished in two ways in MATLAB. First, you could make two calls to spline to fit a spline to $x(t)$ and then make another call to fit a spline to $y(t)$. Alternatively, the spline function can fit both splines simultaneously and return a single pp-form structure containing both fits. The following example demonstrates the latter approach:

```
>> ppxy = spline(t,[x;y])
ppxy =
      form: 'pp'
    breaks: [1x15 double]
     coefs: [28x4 double]
    pieces: 14
     order: 4
       dim: 2
```

Here, the second argument to `spline` is an array containing two ***rows***, each of which is fit with a spline by using the independent variable t, which is monotonic. Elsewhere in MATLAB, data arrays are column-oriented, with different columns representing different variables. However, the function `spline` adopts a row-oriented approach in which different rows represent different variables. Not recognizing this subtle fact can lead to errors—for example,

```
>> ppz = spline(t,[x;y]') % try "normal" column oriented data
??? Error using ==> spline
Abscissa and ordinate vector should be of the same length.
```

In addition, the pp-form structure returned now identifies `ppxy.dim` = 2, meaning that ppxy describes a 2-D spline.

Given this spline fit, the original data can be interpolated as desired—for example,

```
>> ti = linspace(0,3*pi); % total range, 100 points
>> xy = ppval(ppxy,ti);   % evaluate both splines
>> size(xy)               % results are row-oriented too!
ans =
      2    100
>> plot(x,y,'d',xy(1,:),xy(2,:))
>> xlabel('X')
>> ylabel('Y')
>> title('Figure 21.6: Interpolated Spiral Y=f(X)')
```

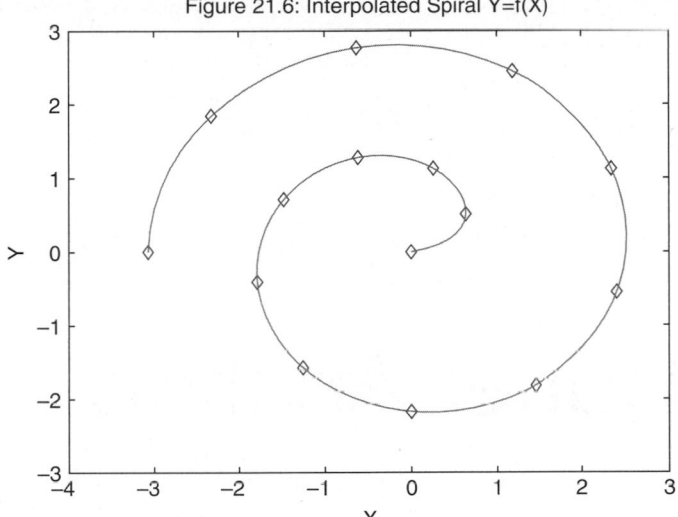

Figure 21.6: Interpolated Spiral Y=f(X)

In the example, the results of ppval are also row-oriented, the first row being associated with the first spline, and so on. Therefore, to plot y versus x, the second row xy(2,:) is plotted versus the first row xy(1,:).

Finally, the preceding approach is not limited to two dimensions. The pp-form structure and MATLAB piecewise polynomial functions all handle n-dimensional splines.

22

Fourier Analysis

Frequency-domain tools, such as Fourier series, Fourier transforms, and their discrete time counterparts, form a cornerstone in signal processing. These transforms decompose a signal into a sequence or continuum of sinusoidal components that identify the frequency-domain content of the signal. MATLAB provides the functions `fft`, `ifft`, `fft2`, `ifft2`, `fftn`, `ifftn`, `fftshift`, and `ifftshift` for Fourier analysis. This collection of functions performs the discrete Fourier transform and its inverse in one or more dimensions. (More extensive signal-processing tools are available in the optional *Signal Processing Toolbox*.)

Because signal processing encompasses such a diverse area, it is beyond the scope of this text to illustrate even a small sample of the type of problems that can be solved using the discrete Fourier transform functions in MATLAB. Therefore, only one example of using the function `fft` to approximate the Fourier transform of a continuous time signal is illustrated. In addition, the use of the function `fft` to approximate the Fourier series of a periodic continuous time signal is demonstrated.

22.1 DISCRETE FOURIER TRANSFORM

In MATLAB, the function `fft` computes the discrete Fourier transform of a signal. In cases where the length of the data is a power of 2, or a product of prime factors, fast Fourier transform (FFT) algorithms are employed to compute the discrete Fourier transform.

> Because of the substantial increase in computational speed that occurs when data length is a power of 2, whenever possible it is important to choose data lengths equal to a power of 2, or to pad data with zeros to give it a length equal to a power of 2.

The fast Fourier transform implemented in MATLAB follows that commonly used in engineering texts:

$$F(k) = FFT\{f(n)\} = \sum_{n=0}^{N-1} f(n)e^{-j2\pi nk/N} \quad k = 0, 1, \ldots, N-1$$

Since MATLAB does not support zero indices, the values are shifted by one index value to

$$F(k) = FFT\{f(n)\} = \sum_{n=1}^{N} f(n)e^{-j2\pi(n-1)(k-1)/N} \quad k = 1, 2, \ldots, N$$

The inverse transform follows accordingly as

$$f(n) = FFT^{-1}\{F(k)\} = \frac{1}{N}\sum_{k=1}^{N} F(k)e^{j2\pi(n-1)(k-1)/N} \quad n = 1, 2, \ldots, N$$

Specific details on the use of the `fft` function are described in its help text:

```
>> help fft
 FFT Discrete Fourier transform.
    FFT(X) is the discrete Fourier transform (DFT) of vector X.  For
    matrices, the FFT operation is applied to each column. For N-D
    arrays, the FFT operation operates on the first non-singleton
    dimension.

    FFT(X,N) is the N-point FFT, padded with zeros if X has less
    than N points and truncated if it has more.
    FFT(X,[],DIM) or FFT(X,N,DIM) applies the FFT operation across the
    dimension DIM.
    For length N input vector x, the DFT is a length N vector X,
    with elements
                 N
    X(k) =      sum   x(n)*exp(-j*2*pi*(k-1)*(n-1)/N), 1 <= k <= N.
                n=1

    The inverse DFT (computed by IFFT) is given by
                 N
    x(n) = (1/N) sum   X(k)*exp( j*2*pi*(k-1)*(n-1)/N), 1 <= n <= N.
                k=1

    See also IFFT, FFT2, IFFT2, FFTSHIFT.
```

To illustrate the use of the FFT, consider the problem of estimating the continuous Fourier transform of the signal

$$f(t) = 2e^{-3t} \quad t \geq 0$$

Analytically, the Fourier transform of $f(t)$ is given by

$$F(\omega) = \frac{2}{3 + j\omega}$$

Although using the FFT has little real value in this case (since the analytical solution is known), this example illustrates an approach to estimating the Fourier transform of less common signals, especially those whose Fourier transform is not readily found analytically. The following MATLAB statements estimate $|F(\omega)|$ by using the FFT and graphically compare it to the preceding analytical expression:

```
N = 128;                % choose a power of 2 for speed
t = linspace(0,3,N);    % time points for function evaluation
f = 2*exp(-3*t);        % evaluate function, minimize aliasing: f(3) ~ 0
Ts = t(2) - t(1);       % the sampling period
Ws = 2*pi/Ts;           % the sampling frequency in rad/sec
F = fft(f);             % compute the fft
Fc = fftshift(F)*Ts;    % shift and scale

W = Ws*(-N/2:(N/2)-1)/N;% frequency axis
Fa = 2./(3+j*W);        % analytical Fourier transform
plot(W,abs(Fa),W,abs(Fc),'.') % generate plot, 'o' marks fft
xlabel('Frequency, Rad/s')
ylabel('|F(\omega)|')
title('Figure 22.1: Fourier Transform Approximation')
```

The function fftshift flips the halves of F so that the $(N/2) + 1$ element of Fc is the DC component of the result. Elements less than this are negative frequency components, while those greater are positive frequency components. Using this fact, W creates the appropriate analog frequency axis with W(N/2 + 1) = 0. Graphically, the FFT approximation is good at low frequencies, but demonstrates some aliasing at higher frequencies near the Nyquist frequency.

The FFT-related functions in MATLAB include those listed in the following table:

Function	Description
conv	Convolution
conv2	2-D convolution
convn	n-D convolution
deconv	Deconvolution
filter	Discrete time filter
filter2	2-D discrete time filter
fft	Discrete Fourier transform
fft2	2-D discrete Fourier transform
fftn	n-D discrete Fourier transform
ifft	Inverse discrete Fourier transform
ifft2	2-D inverse discrete Fourier transform
ifftn	n-D inverse discrete Fourier transform
fftshift	Shift FFT results so that negative frequencies appear first
ifftshift	Undo actions performed by fftshift
abs	Magnitude of complex array
angle	Radian angle of complex array
unwrap	Remove phase angle jumps
cplxpair	Sort vector into complex conjugate pairs
nextpow2	Next higher power of 2

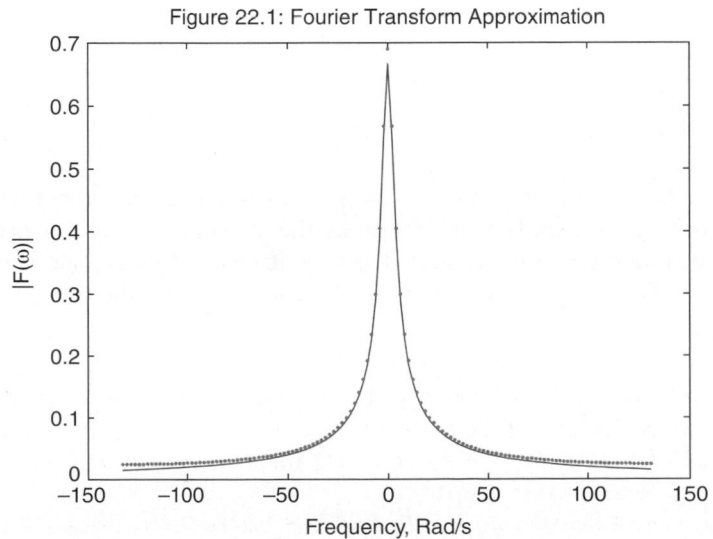

Figure 22.1: Fourier Transform Approximation

22.2 FOURIER SERIES

MATLAB itself offers no functions specifically tailored to Fourier series analysis and manipulation. However, these functions can be easily added when you understand the relationship between the discrete Fourier transform of samples of a periodic signal and its Fourier series.

The Fourier series representation of a real-valued periodic signal $f(t)$ can be written in complex exponential form as

$$f(t) = \sum_{n=-\infty}^{\infty} F_n e^{jn\omega_o t}$$

where the Fourier series coefficients are given by

$$F_n = \frac{1}{T_o} \int_t^{t+T_o} f(t) e^{-jn\omega_o t} \, dt$$

and the fundamental frequency is $\omega_o = 2\pi/T_o$, where T_o is the period. The complex exponential form of the Fourier series can be rewritten in trigonometric form as

$$f(t) = A_o + \sum_{n=1}^{\infty} \{A_n \cos(n\omega_o t) + B_n \sin(n\omega_o t)\}$$

where the coefficients are given by

$$A_o = \frac{1}{T_o} \int_t^{t+T_o} f(t) \, dt$$

$$A_n = \frac{2}{T_o} \int_t^{t+T_o} f(t) \cos(n\omega_o t) \, dt$$

$$B_n = \frac{2}{T_o} \int_t^{t+T_o} f(t) \sin(n\omega_o t) \, dt$$

Of these two forms, the complex exponential Fourier series is generally easier to manipulate analytically, whereas the trigonometric form provides a more intuitive understanding because it makes it easier to visualize sine and cosine waveforms. The relationships between the coefficients of the two forms are

$$A_o = F_o$$

$$A_n = 2 \, \mathrm{Re}\{F_n\}$$

$$B_n = -2 \, \mathrm{Im}\{F_n\}$$

$$F_n = F^*_{-n} = (A_n - jB_n)/2$$

Using these relationships, you can use the complex exponential form analytically and then convert results to the trigonometric form for display.

The discrete Fourier transform can be used to compute the Fourier series coefficients, provided that the time samples are appropriately chosen and the transform output is scaled. For example, consider computing the Fourier series coefficients of the sawtooth waveform shown next.

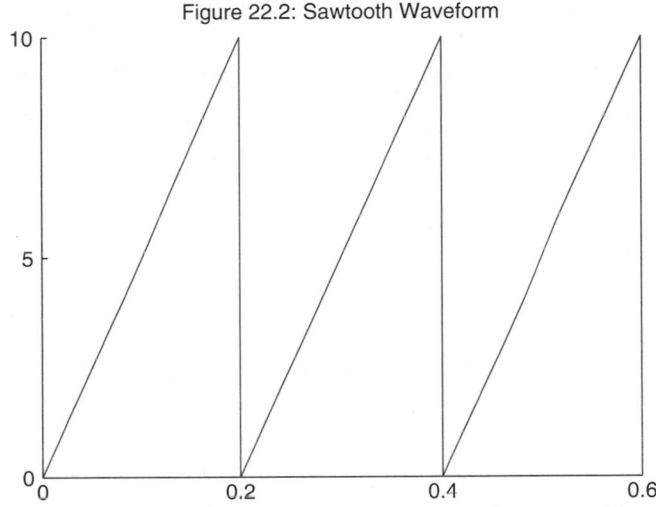

Figure 22.2: Sawtooth Waveform

First, you must create a function to evaluate the sawtooth at arbitrary points:

```
function f=sawtooth(t,To)
%SAWTOOTH Sawtooth Waveform Generation.
% SAWTOOTH(t,To) computes values of a sawtooth having
% a period To at the points defined in the vector t.
f = 10*rem(t,To)/To;
f(f==0 | f==10) = 5; % must average value at discontinuity!
```

To minimize aliasing, it is necessary to compute enough harmonics so that the highest harmonic amplitude is negligible. In this case, choose

```
>> N = 25;    % number of harmonics
>> To = 0.2; % choose period
```

The number of terms to consider in the discrete Fourier transform is twice the number of harmonics, since the discrete Fourier transform computes both positive and negative harmonics:

```
>> n = 2*N;
```

The function must be evaluated at n points over one period in such a manner that the $(n + 1)$th point is one period away from the first point:

```
>> t = linspace(0,To,n+1); % (n+1)th point is one period away
>> t(end) = [];            % throw away undesired last point
>> f = sawtooth(t,To);     % compute sawtooth
```

We are now ready to compute the transform, rearrange the components, and scale the results:

```
>> Fn = fft(f);                              % compute FFT
>> Fn = [conj(Fn(N+1)) Fn(N+2:end) Fn(1:N+1)]; % rearrange values
>> Fn = Fn/n;                                % scale results
```

The vector Fn now contains the complex exponential Fourier series coefficients in ascending order. That is, Fn(1) is F_{-25}; Fn(N+1) is F_o, the DC component; and Fn(2*N+1) is F_{25}, the 25th harmonic component.

From these data, the trigonometric Fourier series coefficients are as follows:

```
>> A0 = Fn(N+1) % DC component
A0 =

    5
>> An = 2*real(Fn(N+2:end)) % Cosine terms
An =

  1.0e-015 *
  Columns 1 through 7
  -0.1176   -0.0439   -0.2555    0.3814    0.0507   -0.2006    0.1592
  Columns 8 through 14
  -0.1817    0.0034         0    0.0034   -0.1141   -0.1430   -0.0894
  Columns 15 through 21
  -0.0685   -0.0216    0.0537   -0.0496   -0.0165         0   -0.0165
  Columns 22 through 25
  -0.0079    0.2405    0.3274    0.2132
```

```
>> Bn = -2*imag(Fn(N+2:end)) % Sine terms
Bn =

  Columns 1 through 7
    -3.1789    -1.5832    -1.0484    -0.7789    -0.6155    -0.5051    -0.4250
  Columns 8 through 14
    -0.3638    -0.3151    -0.2753    -0.2418    -0.2130    -0.1878    -0.1655
  Columns 15 through 21
    -0.1453    -0.1269    -0.1100    -0.0941    -0.0792    -0.0650    -0.0514
  Columns 22 through 25
    -0.0382    -0.0253    -0.0126         0
```

Since the sawtooth waveform has odd symmetry (except for its DC component), it makes sense that the cosine coefficients An are negligible. (Note that they are scaled by 10^{-15}.) Comparing the actual Fourier series coefficients for this sawtooth waveform with the preceding Bn terms gives a relative error of

```
>> idx = -N:N;          % harmonic indices
>> Fna = 5j./(idx*pi); % complex exponential terms
>> Fna(N+1) = 5;
>> Bna = -2*imag(Fna(N+2:end)); % sine terms

>> Bn_error = (Bn-Bna)./Bna     % relative error
Bn_error =

  Columns 1 through 7
    -0.0013    -0.0053    -0.0119    -0.0211    -0.0331    -0.0478    -0.0653
  Columns 8 through 14
    -0.0857    -0.1089    -0.1352    -0.1645    -0.1971    -0.2330    -0.2723
  Columns 15 through 21
    -0.3152    -0.3620    -0.4128    -0.4678    -0.5273    -0.5917    -0.6612
  Columns 22 through 25
    -0.7363    -0.8174    -0.9051    -1.0000
```

As with the earlier Fourier transform example, aliasing causes errors that increase with increasing frequency. Since all practical signals are not band-limited, aliasing is inevitable, and a decision must be made about the degree of aliasing that can be tolerated in a given application. As the number of requested harmonics increases, the degree of aliasing decreases. Therefore, to minimize aliasing, you can

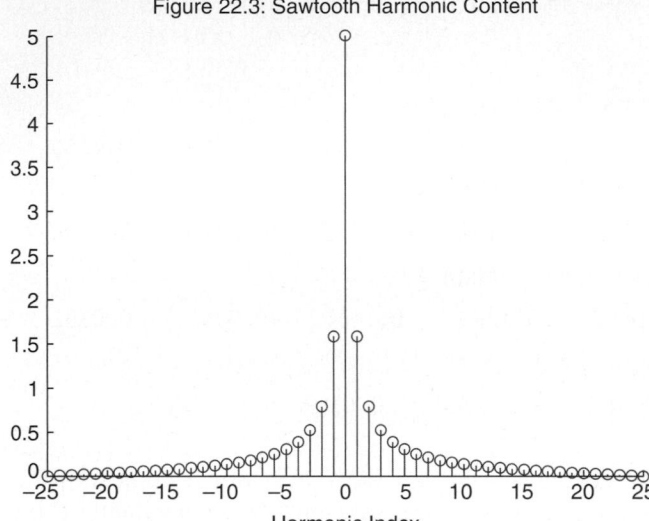

Figure 22.3: Sawtooth Harmonic Content

request a larger number of harmonics and then choose a subset of them to view and further manipulate.

Finally, the line spectra of the complex exponential Fourier series can be plotted using the `stem` function:

```
>> stem(idx,abs(Fn))
>> xlabel('Harmonic Index')
>> title('Figure 22.3: Sawtooth Harmonic Content')
>> axis tight
```

23

Optimization

Optimization in the context of this chapter refers to the process of determining where a function $y = g(x)$ takes on either specific or extreme values. When a function is defined simply, the corresponding inverse function $x = g^{-1}(y)$ can often be found, in which case you can determine what x values produce a given y by evaluating the inverse function. On the other hand, many functions, including many simple ones, have no inverse. When this is the case, you must estimate the x that produces a known y by some iterative procedure. In practice, this iterative procedure is called *zero finding*, because finding x such that $y = g(x)$ for some y is equivalent to finding x such that $f(x) = 0$, where $f(x) = y - g(x)$.

In addition to knowing where a function takes on specific values, it is also common to know its extreme values—that is, where it achieves maximum or minimum values. As before, there are numerous times when these extreme values must be estimated by some iterative procedure. Since a function maximum is the minimum of its negative (i.e., $\max f(x) = \min\{-f(x)\}$), iterative procedures for finding extreme values typically find only minimum values, and the procedures are called *minimization* algorithms.

In this chapter, the optimization functions available in basic MATLAB are covered. (Many more functions are available in the optional *Optimization Toolbox*.)

23.1 ZERO FINDING

Finding a zero of a function can be interpreted in a number of ways, depending on the function. When the function is 1-D, the MATLAB function `fzero` can be used to find a zero. The algorithm used by this function is a combination of bisection and inverse quadratic interpolation. When the function is multidimensional—that is, the function definition consists of multiple scalar functions of a vector variable—you

359

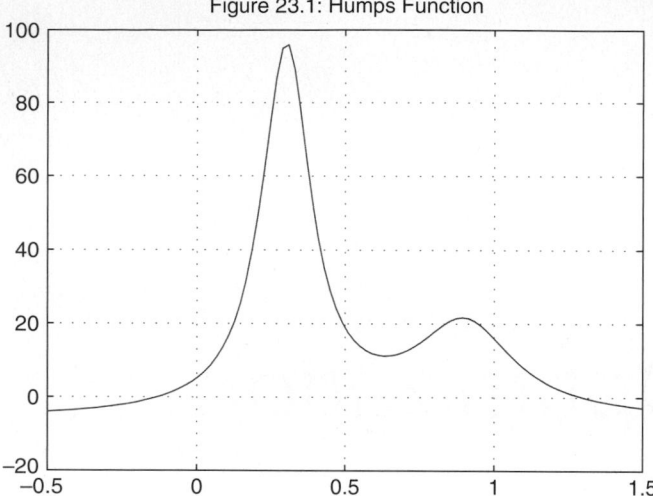

must look beyond basic MATLAB for a solution. The *Optimization Toolbox* or other third-party toolbox is required to solve the multidimensional case.

To illustrate the use of the function `fzero`, consider the function `humps`:

```
>> x = linspace(-.5,1.5);
>> y = humps(x);
>> plot(x,y)
>> grid on
>> title('Figure 23.1: Humps Function')
```

The `humps` M-file evaluates the function

$$humps(x) = \frac{1}{(x - 0.3)^2 + 0.01} + \frac{1}{(x - 0.9)^2 + 0.04} - 6$$

which crosses zero near $x = -0.2$ and $x = 1.3$. The function `fzero` provides a way to find a better approximation to these zero crossings:

```
>> format long  % display more precision

>> H_humps = @humps;  % create function handle to humps.m function.

>> x = fzero(H_humps,1.3)
x =

    1.29954968258482
>> humps(x) % how close is it to 0?
```

```
ans =

    0

>> H_humps(x) % evaluate humps through its handle as well
 ans =

    0

>> [x,value] = fzero(H_humps,-0.2)
x =

  -0.13161801809961
value =

    8.881784197001252e-016
```

Here, the two zeros of the function were found. The first zero is very close to 1.3, and evaluation of function at the zero produced zero. The second zero was found to be close to −0.13. In this call to `fzero`, a second output argument was supplied, which returned the function evaluated at the zero. Therefore, it wasn't necessary to call `humps(x)` to check the accuracy of the solution found by `fzero`. It is important to note that `fzero` returns just one zero—the zero found closest to the initial guess. So, if a function has more than one zero, it is up to the user to call `fzero` multiple times with different initial guesses.

When initially called, the function `fzero` searches on either side of the initial guess for a sign change in the function. When a sign change is found, the two end-points that produced the sign change form a ***bracket*** on the number line. If a function is continuous, it must cross through zero somewhere in the bracket. Knowing this, the function `fzero` then searches for the zero crossing and returns the value of *x* that comes closest to making this happen.

In many cases, supplying an initial guess or estimate of the zero is unnecessary, because a bracket is already known from the properties of the problem to be solved. When this occurs, you can simply supply `fzero` with the bracket rather than with an initial guess of the zero location:

```
>> [x,value] = fzero(H_humps,[-2 0])
Zero found in the interval: [-2, 0].
x =

  -0.13161801809961
value =

    0


>> [x,value] = fzero(H_humps,[0 1.2])
??? Error using ==> fzero
The function values at the interval endpoints must differ in sign.
```

In the first example, [−2 0] is a bracket around the zero near −0.13. As a result, fzero finds the zero. In the second example, [0 1.2] is not a bracket around a zero, forcing fzero to report an error and abort its search. So, if a two-element array is supplied to fzero, it must bracket a zero, or else the function terminates without doing a zero search.

In the preceding examples, the function to be searched was provided to fzero as a function handle, which uniquely identifies the function M-file. As discussed in Chapter 12, the function to be searched can also be supplied as an **anonymous function**, an **in-line function** object, or a **string expression**. While all three of these work, the use of function handles is encouraged. In-line function objects and string expression definitions are obsolete, but remain operable in MATLAB for the time being. (To learn more about these alternatives, see the MATLAB documentation.)

All functions in this chapter have various settable parameters. These functions, as well as those in the *Optimization Toolbox*, share the same format for managing parameters. The functions optimset and optimget are used, respectively, to set and get parameters for all functions. For fzero, there are two settable parameters, 'Display' and 'TolX'. The first parameter controls the amount of detail returned while the function is working; the second sets a tolerance range for accepting the final answer. The following code is illustrative:

```
>> options = optimset('Display','iter'); % show iteration history
>> [x,value] = fzero(H_humps,[-2 0],options)
  Func-count          x              f(x)              Procedure
      1                   -2        -5.69298           initial
      2                    0         5.17647           initial
      3            -0.952481        -5.07853           interpolation
      4            -0.480789        -3.87242           interpolation
      5            -0.240394        -1.94304           bisection
      6            -0.120197         0.28528           bisection
      7            -0.135585        -0.0944316         interpolation
      8            -0.131759        -0.00338409        interpolation
      9            -0.131618         1.63632e-006      interpolation
     10            -0.131618        -7.14819e-010      interpolation
     11            -0.131618                0          interpolation
Zero found in the interval: [-2, 0].
x =
   -0.13161801809961
value =

    0
```

```
>> options = optimset('Display','final'); % display successful interval

>> [x,value] = fzero(H_humps,[-2 0],options)
Zero found in the interval: [-2, 0].
x =

  -0.13161801809961
value =

     0

>> options = optimset('TolX',0.1);

>> [x,value] = fzero(H_humps,[-2 0],options)
x =
  -0.24039447250762
value =
  -1.94303825972565

>> options = optimset('Display','iter','TolX',0.1); % set both
>> [x,value] = fzero(H_humps,[-2 0],options)
 Func-count        x              f(x)           Procedure
     1              -2          -5.69298         initial
     2               0           5.17647         initial
     3        -0.952481         -5.07853         interpolation
     4        -0.480789         -3.87242         interpolation
     5        -0.240394         -1.94304         bisection
Zero found in the interval: [-2, 0].
x =

  -0.24039447250762
value =
  -1.94303825972565
```

In the preceding code, an options structure was created with the desired parameters and then passed as a third argument to fzero. The 'Display' option has four settings: 'final', 'iter', 'notify', and 'off'. The setting 'notify' is the default, and it means "Display information only if no solution is found." The 'TolX' option sets the tolerance for the final answer, which is equal to eps by default. (See the online help for optimset and optimget for more information regarding parameters for MATLAB optimization functions.)

23.2 MINIMIZATION IN ONE DIMENSION

In addition to the visual information provided by plotting, it is often necessary to determine other, more specific attributes of a function. Of particular interest in many applications are function extremes—that is, a function's maxima (peaks) and minima (valleys). Mathematically, these extremes are found analytically by determining where the derivative (slope) of a function is zero. This idea can be readily understood by inspecting the slope of the humps plot at its peaks and valleys. Clearly, when a function is simply defined, this process often works. However, even for many simple functions that can be differentiated readily, it is often impossible to find where the derivative is zero. In these cases and in cases where it is difficult or impossible to find the derivative analytically, it is necessary to search for function extremes numerically. MATLAB provides two functions that perform this task, fminbnd and fminsearch. These two functions find minima of 1-D and *n*-D functions, respectively. The function fminbnd employs a combination of golden-section search and parabolic interpolation. Since a maximum of *f(x)* is equal to a minimum of $-f(x)$, fminbnd and fminsearch can be used to find both minima and maxima. If this notion is not clear, visualize the preceding humps(*x*) plot flipped upside down. In the upside-down state, peaks become valleys and valleys become peaks.

To illustrate 1-D minimization and maximization, consider the preceding humps(*x*) example once again. From the figure, there is a maximum near $x = 0.3$ and a minimum near $x = 0.6$. With fminbnd, these extremes can be found with more accuracy:

```
>> H_humps=@humps;   % create handle to humps.m function.

>> [xmin,value] = fminbnd(H_humps,0.5,0.8)
xmin =
   0.63700821196362
value =
   11.25275412587769
>> options=optimset('Display','iter');
>> [xmin,value] = fminbnd(H_humps,0.5,0.8,options)
```

Func-count	x	f(x)	Procedure
1	0.61459	11.4103	initial
2	0.68541	11.9288	golden
3	0.57082	12.7389	golden
4	0.638866	11.2538	parabolic
5	0.637626	11.2529	parabolic
6	0.637046	11.2528	parabolic
7	0.637008	11.2528	parabolic
8	0.636975	11.2528	parabolic

```
Optimization terminated successfully:
the current x satisfies the termination criteria using
OPTIONS.TolX of 1.000000e-004
xmin =
   0.63700821196362
value =
   11.25275412587769
```

In this example, in the two calls to fminbnd, 0.5 and 0.8 denote the range over which to search for minimum. In the second case, options were set to display the iterations performed by fminbnd.

To find the maximum near $x = 0.3$, we can either modify the humps.m file to negate the expression, or we can create an anonymous function. The following example demonstrates the latter approach:

```
>> AH_humps = @(x) -1./((x-.3).^2 +.01) - 1./((x-.9).^2 +.04) + 6;
>> [xmax,value] = fminbnd(AH_humps,0.2,0.4,options)
 Func-count        x            f(x)            Procedure
     1         0.276393        -91.053          initial
     2         0.323607        -91.4079         golden
     3         0.352786        -75.1541         golden
     4         0.300509        -96.5012         parabolic
     5         0.300397        -96.5014         parabolic
     6         0.300364        -96.5014         parabolic
     7         0.300331        -96.5014         parabolic
Optimization terminated successfully:
the current x satisfies the termination criteria using
OPTIONS.TolX of 1.000000e-004
xmax =
   0.30036413790024
value =
  -96.50140724387050
```

On termination, the maximum is found to be very close to 0.3, and the peak has an amplitude of +96.5. The value returned by fminbnd is the negative of the actual value, because fminbnd computes the minimum of $-\text{humps}(x)$.

23.3 MINIMIZATION IN HIGHER DIMENSIONS

As described previously, the function `fminsearch` provides a simple algorithm for minimizing a function of several variables. That is, `fminsearch` attempts to find the minimum of $f(x)$, where $f(x)$ is a scalar function of a vector argument x. The function `fminsearch` implements the Nelder-Mead simplex search algorithm, which modifies the components of x to find the minimum of $f(x)$. This algorithm is not as efficient on smooth functions as some other algorithms are, but, on the other hand, it does not require gradient information that is often expensive to compute. It also tends to be more robust on functions that are not smooth, where gradient information is less valuable. If the function to be minimized is inexpensive to compute, the Nelder-Mead algorithm usually works very well.

To illustrate usage of `fminsearch`, consider the banana function, also called Rosenbrock's function:

$$f(x) = 100(x_2 - x_1{}^2)^2 + (1 - x_1)^2$$

This function can be visualized by creating a 3-D mesh plot with x_1 as the x-dimension, and x_2 as the y-dimension:

```
x = [-1.5:0.125:1.5];  % range for x1 variable
y = [-.6:0.125:2.8];   % range for x2 variable

[X,Y] = meshgrid(x,y); % grid of all x and y
Z = 100.*(Y-X.*X).^2 + (1-X).^2; % evaluate banana

mesh(X,Y,Z)
hidden off
xlabel('x(1)')
ylabel('x(2)')
title('Figure 23.2: Banana Function')

hold on
plot3(1,1,1,'k.','markersize',30)
hold off
```

As shown in the plot, the banana function has a unique minimum of zero at $x = [1; 1]$. To find the minimum of this function, it must be rewritten in terms of $x_1 = $ `x(1)` and $x_2 = $ `x(2)`, as shown before mathematically. It can be entered as the M-file

Figure 23.2: Banana Function

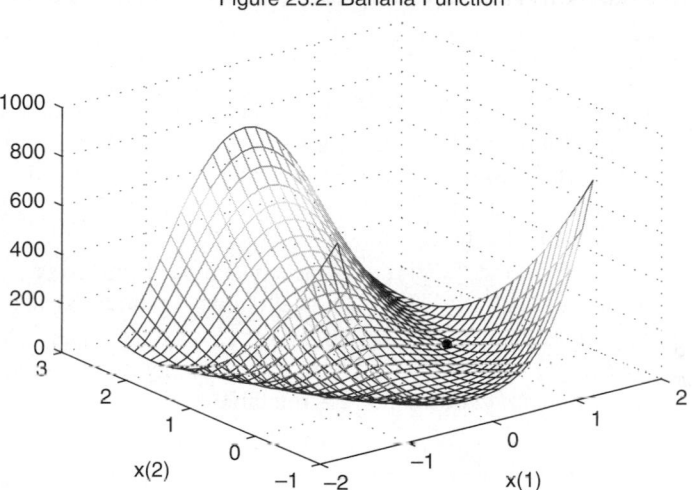

```
function f=banana(x)
% Rosenbrock's banana function
    f=100*(x(2)-x(1)^2)^2 + (1-x(1))^2;
```

or as the anonymous function

```
>> AH_banana = @(x) 100*(x(2)-x(1)^2)^2 + (1-x(1))^2;
```

Using either of these representations, fminsearch produces

```
>> [xmin,value,flag,output] = fminsearch(AH_banana,[-1.9,2])
xmin =
    1.00001666889480    1.00003447386277
value =
    4.068551535063419e-010
flag =
     1
output =
    iterations: 114
     funcCount: 210
```

```
algorithm: 'Nelder-Mead simplex direct search'
message: [1x196 char]
```

Here, four output parameters are shown: the minimum found, the function evaluated at the minimum, a flag signifying success, and finally an algorithm statistics structure. Finding the minimum with a tolerance of 1e-4 required 114 iterations and 210 banana function evaluations. If less output is desired, it is simply a matter of providing fewer output variables.

As with fminbnd, fminsearch accepts an options structure. The options that can be set for fminsearch are listed in the next table. As shown earlier, preferences are set by calling optimset as options = optimset('Name',value, 'Name',value, ...). Setting 'Display' to 'iter' in fminsearch can lead to a tremendous amount of output to the *Command* window. Here is the table:

Option Name	Description	Default Value
'Display'	Displays frequency, 'iter','final','notify', or 'off'	'notify' (displays information only if no solution is found)
'MaxFunEvals'	Maximum function evaluations	200*length(x)
'MaxIter'	Maximum algorithm iterations	200*length(x)
'TolFun'	Function solution tolerance	1.00E-004
'TolX'	Variable solution tolerance	1.00E-004

To demonstrate how to use the options shown in the table, consider finding the solution to the previous problem with tighter function and variable tolerances:

```
>> options = optimset('TolFun',1e-8,'TolX',1e-8);
>> [xmin,value,flag,output] = fminsearch(AH_banana,[-1.9,2],options)
xmin =
    1.00000000126077    1.00000000230790
value =
    6.153858843361103e-018
flag =
    1
output =
    iterations: 144
    funcCount: 266
    algorithm: 'Nelder-Mead simplex direct search'
    message: [1x196 char]
```

With the tighter tolerances, xmin is now within 1e-8 of the actual minimum, the function evaluates to well within eps of the minimum of zero, and the number of algorithm iterations and function evaluations increases by approximately 26 percent. From this information, it is clear that fminsearch requires a lot of function evaluations and therefore can be slow for functions that are computationally expensive.

23.4 PRACTICAL ISSUES

Iterative solutions, such as those found by fzero, fminbnd, and fminsearch, all make some assumptions about the function to be iterated. Since there are essentially no limits to the function provided, it makes sense that these ***function functions*** may not converge or may take many iterations to converge. At worst, these functions can produce a MATLAB error that terminates the iteration without producing a result. And even if they do terminate promptly, there is no guarantee that they have stopped at the desired result. To make the most efficient use of these function functions, consider applying the following points:

1. Start with a good initial guess. This is the most important consideration. A good guess keeps the problem in the neighborhood of the solution, where its numerical properties are hopefully stable.

2. If components of the solution (e.g., in fminsearch) are separated by several orders of magnitude or more, consider scaling them to improve iteration efficiency and accuracy. For example, if x(1) is known to be near 1, and x(2) is known to be near 1e6, scale x(2) by 1e-6 in the function definition, and then scale the returned result by 1e6. When you do so, the search algorithm uses numbers that are all around the same order of magnitude.

3. If the problem is complicated, look for ways to simplify it into a sequence of simpler problems that have fewer variables.

4. Make sure your function cannot return complex numbers, Inf, or NaN, which usually results in convergence failure. The functions isreal, isfinite, and isnan can be used to test results before returning them.

5. Avoid functions that are discontinuous. Functions such as abs, min, and max all produce discontinuities that can lead to divergence.

6. Constraints on the allowable range of x can be included by adding a penalty term to the function to be iterated, such that the algorithm is persuaded to avoid out-of-range values.

24

Integration and Differentiation

Integration and differentiation are fundamental tools in calculus. Integration computes the area under a function, and differentiation describes the slope or gradient of a function. MATLAB provides functions for numerically approximating the integral and slope of a function. Functions are provided for making approximations when functions exist as M-files and as anonymous functions, as well as when functions are tabulated at uniformly spaced points over the region of interest.

24.1 INTEGRATION

MATLAB provides four functions for computing integrals of functions: quad, quadl, dblquad, and triplequad.

To illustrate integration, consider the function humps(x), as shown in Figure 24.1. As is apparent in the figure, the sum of the trapezoidal areas approximates the integral of the function. Clearly, as the number of trapezoids increases, the fit between the function and the trapezoids gets better, leading to a better integral or area approximation.

Using tabulated values from the humps function, the MATLAB function trapz approximates the area using the trapezoidal approximation. Duplicating the trapezoids shown in the preceding figure produces

```
>> x = -1:.17:2;
>> y = humps(x);
>> area = trapz(x,y)
```

370

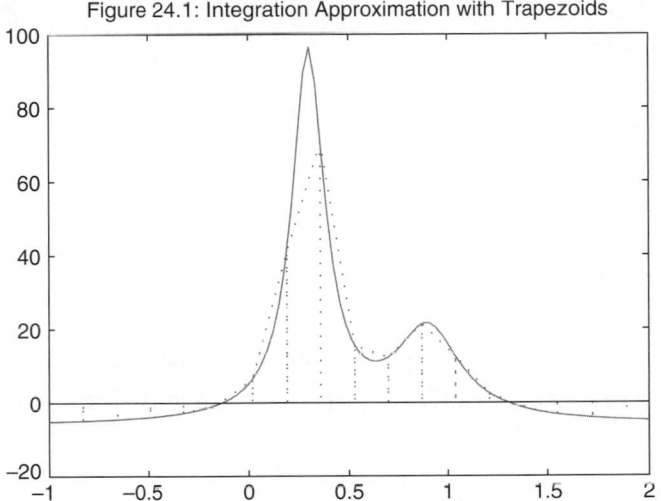

Figure 24.1: Integration Approximation with Trapezoids

area =

        25.917

In comparison to the figure, this is probably not a very accurate estimate of the area. However, when a finer discretization is used, more accuracy is achieved:

```
>> x = linspace(-1,2,100);
>> y = humps(x);
>> format long
>> area = trapz(x,y)
area =

        26.34473119524596
```

This area agrees with the analytical integral through five significant digits.

Sometimes, we are interested in the integral as a function of x—that is,

$$\int_{x_1}^{x} f(x)\,dx$$

where x_1 is a known lower limit of integration. The definite integral from x_1 to any point x is then found by evaluating the function at x. Using the trapezoidal rule, tabulated values of the cumulative integral are computed using the function cumtrapz, as in the following example:

```
>> x = linspace(-1,2,100);
>> y = humps(x);
```

```
>> z = cumtrapz(x,y);
>> size(z)
ans =
     1    100


>> plotyy(x,y,x,z)
>> grid on
>> xlabel('x')
>> ylabel('humps(x) and integral of humps(x)')
>> title('Figure 24.2: Cumulative Integral of humps(x)')
```

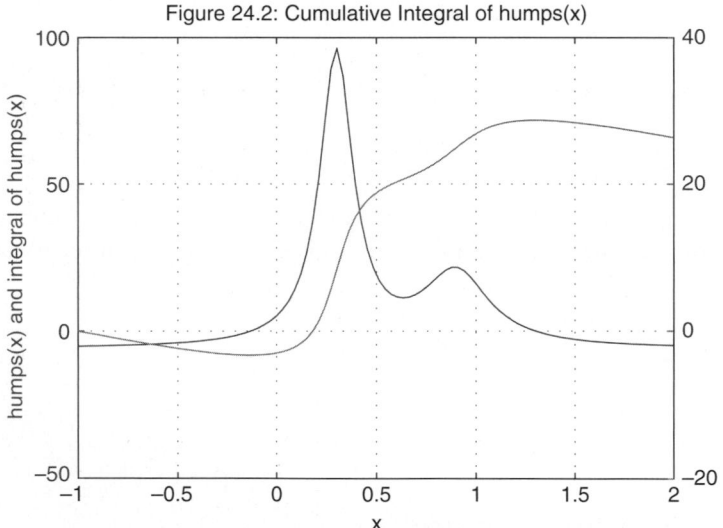

Depending on the properties of the function at hand, it may be difficult to determine an optimum trapezoidal width. Clearly, if you could somehow vary the individual trapezoid widths to match the characteristics of the function, much greater accuracy could be achieved.

The MATLAB functions quad and quadl, which are based on a mathematical concept called quadrature, take this approach. These integration functions operate in the same way. Both evaluate the function to be integrated at whatever intervals are necessary to achieve accurate results. Moreover, both functions make higher order approximations than a simple trapezoid, with quadl being more rigorous than quad. As an example, consider computing the integral of the humps function again:

```
>> z(end)                % cumtrapz result
ans =
   26.34473119524596
```

```
>> H_humps = @humps;  % create function handle
```

```
>> quad(H_humps,-1,2)
ans =
   26.34496050120123
```

```
>> quadl(H_humps,-1,2)
ans =
   26.34496047137897
```

For this example, quad and quadl both return the same result to eight significant digits, with these digits exhibiting eight-digit accuracy with respect to the true solution. On the other hand, cumtrapz achieves only five significant digit accuracy.

The function to be integrated (i.e., the integrand) must support a vector input argument. That is, the integrand must return a vector of outputs for a vector of inputs. Doing so means using dot-arithmetic operators, $.\char94$, $./$, $.\backslash$, and $.\char94$. For example, the function in humps.m is given by the statement

```
y = 1./((x-.3).^2 + .01) + 1./((x-.9).^2 + .04) - 6;
```

The functions quad and quadl also allow you to specify an absolute error tolerance as a fourth input argument, with the default absolute tolerance being 10^{-6}.

In addition to 1-D integration, MATLAB supports 2-D integration with the function dblquad. That is, dblquad approximates the integral

$$\int_{y_{min}}^{y_{max}} \int_{x_{min}}^{x_{max}} f(x, y) \, dx \, dy$$

To use dblquad, you must first create a function that evaluates $f(x,y)$. For example, consider the function myfun.m:

```
function z=myfun(x,y)
%MYFUN(X,Y) an example function of two variables
z = sin(x).*cos(y) + 1; % must handle vector x input
```

This function can be plotted by issuing the commands

```
>> x = linspace(0,pi,20);    % xmin to xmax
>> y = linspace(-pi,pi,20); % ymin to ymax

>> [xx,yy] = meshgrid(x,y); % create grid of point to evaluate at

>> zz = myfun(xx,yy);        % evaluate at all points

>> mesh(xx,yy,zz)
>> xlabel('x'), ylabel('y')
>> title('Figure 24.3: myfun.m plot')
```

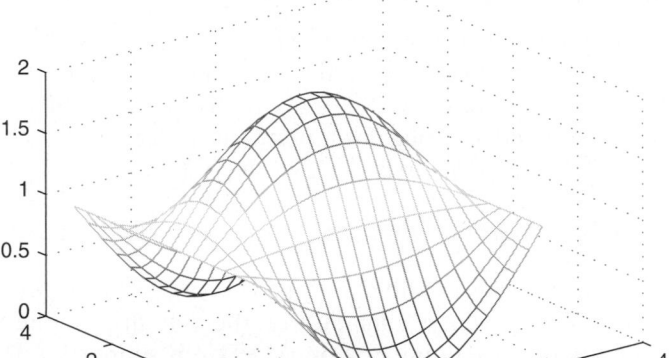

Figure 24.3: myfun.m plot

The volume under this function is computed by calling dblquad as

```
>> area = dblquad(@myfun,0,pi,-pi,pi)
area =
   19.73920880609102
>> relerr = (area-2*pi^2)/(2*pi^2)
relerr =
   1.981996941074027e-010
```

Here, dblquad is called as dblquad(Fname,xmin,xmax,ymin,ymax). Based on the relative error computed above, we can say that the results produced by dblquad are highly accurate, even though the function quad is called by dblquad to do the actual integration.

The function triplequad extends the above quadrature integration schemes to triple integration, or volume integration. In this case, triplequad(Fname, xmin,xmax,ymin,ymax,zmin,zmax) integrates the function *Fname(x,y,z)* over the limits specified by the remainder of the triplequad arguments. (For further information regarding these integration function, see the MATLAB documentation.)

24.2 DIFFERENTIATION

As opposed to integration, numerical differentiation is much more difficult. Integration describes an overall or macroscopic property of a function, whereas differentiation describes the slope of a function at a point, which is a microscopic property of a function. As a result, integration is not sensitive to minor changes in the shape of a function, whereas differentiation is. Any small change in a function can easily create large changes in its slope in the neighborhood of the change.

Because of this inherent sensitivity in differentiation, numerical differentiation is avoided whenever possible, especially if the data to be differentiated is obtained experimentally. In this case, it is best to perform a least-squares curve fit to the data and then differentiate the resulting polynomial. Alternatively, you could fit cubic splines to the data and then find the spline representation of the derivative, as discussed in Chapter 21. For example, consider again the example from Chapter 20:

```
>> x = [0 .1 .2 .3 .4 .5 .6 .7 .8 .9 1];
>> y = [-.447 1.978 3.28 6.16 7.08 7.34 7.66 9.56 9.48 9.30 11.2];
>> n = 2; % order of fit
>> p = polyfit(x,y,n)  % find polynomial coefficients
p =
    -9.8108    20.1293    -0.0317


>> xi = linspace(0,1,100);
>> yi = polyval(p,xi); % evaluate polynomial


>> plot(x,y,'-o',xi,yi,'-')
>> xlabel('x'), ylabel('y=f(x)')
>> title('Figure 24.4: Second Order Curve Fitting')
```

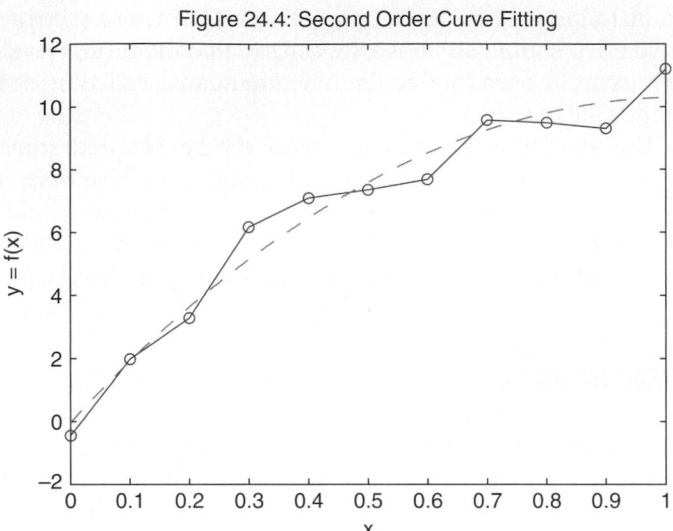

Figure 24.4: Second Order Curve Fitting

The derivative in this case is found by using the polynomial derivative function `polyder`:

```
>> pd = polyder(p)
pd =
  -19.6217    20.1293
```

The derivative of $y(x) = -9.8108x^2 + 20.1293x - 0.0317$ is $dy/dx = -19.6217x + 20.1293$. Since the derivative of a polynomial is yet another polynomial of the next-lowest order, the derivative can also be evaluated at any point. In this case, the polynomial fit is second order, making the resulting derivative first order. As a result, the derivative is a straight line, meaning that it changes linearly with x.

MATLAB provides a function for computing an approximate derivative, given tabulated data describing some function. This function, named `diff`, computes the difference between elements in an array. Since differentiation is defined as

$$\frac{dy}{dx} = \lim_{\Delta x \to 0} \frac{f(x + \Delta x) - f(x)}{\Delta x}$$

the derivative of $y = f(x)$ can be approximated by

$$\frac{dy}{dx} \approx \frac{\Delta y}{\Delta x} = \frac{f(x + \Delta x) - f(x)}{\Delta x}$$

which is the *forward* finite difference in *y* divided by the finite difference in *x*. Since diff computes differences between array elements, differentiation can be approximated in MATLAB. Continuing with the previous example, we have

```
>> dyp = polyval(dp,x);    % poly derivative for comparison
```

```
>> dy = diff(y)./diff(x);  % compute differences and use array division
>> xd = x(1:end-1);        % new x axis array since dy is shorter than y
```

```
>> plot(xd,dy,x,dyp,':')
>> ylabel('dy/dx'), xlabel('x')
>> title('Figure 24.5: Forward Difference Derivative Approximation')
```

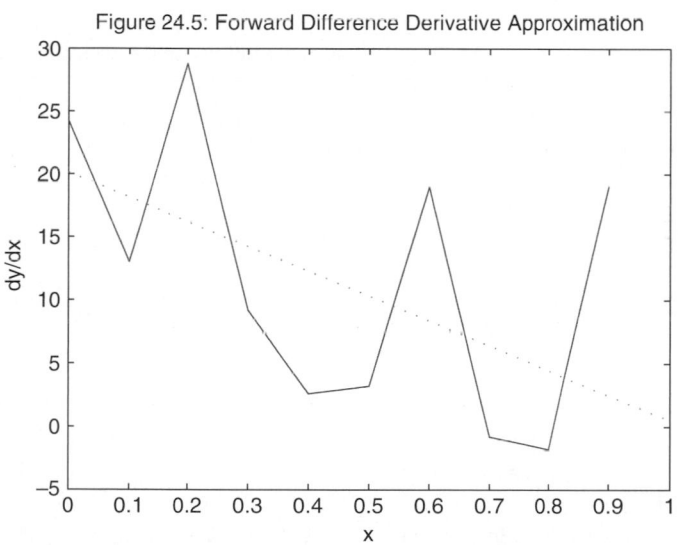

Since diff computes the difference between elements of an array, the resulting output contains one less element than the original array. Thus, to plot the derivative, one element of the x array must be thrown out. When the first element of x is thrown out, the this procedure gives a *backward* difference approximation, which uses information at x(n−1) and x(n) to approximate the derivative at x(n). On the other hand, throwing out the last element gives a *forward* difference approximation, which uses x(n+1) and x(n) to compute results at x(n). Comparing the derivative found using diff with that found from polynomial approximation, it is overwhelmingly apparent that approximating the derivative by finite differences can lead to poor results, especially if the data originates from experimental or noisy measurements.

When the data used do not have uncertainty, the results of using diff can be acceptable, especially for visualization purposes, as in this example:

```
>> x = linspace(0,2*pi);
>> y = sin(x);

>> dy = diff(y)/(x(2)-x(1));
>> xd = x(2:end);

>> plot(x,y,xd,dy)
>> axis tight
>> xlabel('x'), ylabel('sin(x) and cos(x)')
>> title('Figure 24.6: Backward Difference Derivative Approximation')
```

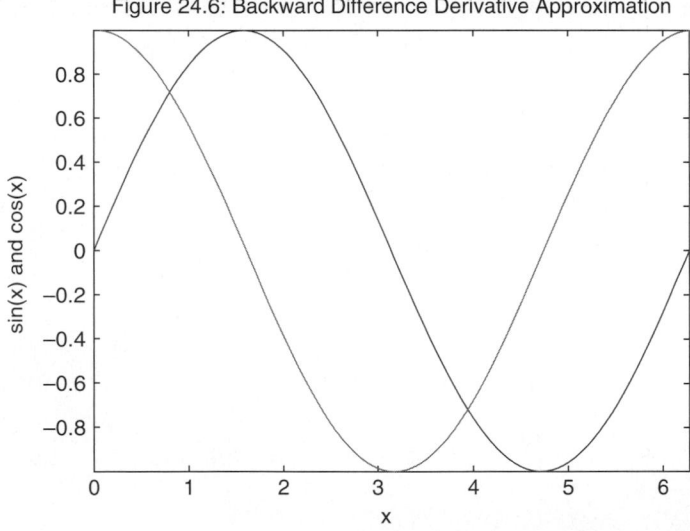

Figure 24.6: Backward Difference Derivative Approximation

Here, x was linearly spaced, so dividing by $x(2)-x(1)$ gives the same answer as diff(x), which is required if x is not linearly spaced. In addition, the first element in x is thrown out, making the result a backward-difference derivative approximation. Visually, the derivative in this example is quite accurate. In fact, the maximum error is

```
>> max(abs(cos(xd)-dy))
ans =
    0.0317
```

When forward- or backward-difference approximations are not sufficient, *central* differences can be computed by performing the required array operations directly. The first central difference for equally spaced data is given by

$$\frac{dy(x_n)}{dx} \approx \frac{f(x_{n+1}) - f(x_{n-1})}{x_{n+1} - x_{n-1}}$$

Therefore, the slope at x_n is a function of its neighboring data points. Repeating the previous example gives

```
>> dy = (y(3:end)-y(1:end-2)) / (x(3)-x(1));
>> xd - x(2:end-1);
>> max(abs(cos(xd)-dy))

ans =

    0.00067086
```

In this case, the first and last data points do not have a central-difference approximation, because there are no data at n = 0 and n = 101, respectively. However, at all intermediate points, the central-difference approximation is nearly two orders of magnitude more accurate than the forward- or backward-difference approximations.

When dealing with 2-D data, the function gradient uses central differences to compute the slope in each direction at each tabulated point. Forward differences are used at the initial points, and backward differences are used at the final points, so that the output has the same number of data points as the input. The function gradient is used primarily for graphical data visualization:

```
>> [x,y,z] = peaks(20);    % simple 2-D function
>> dx = x(1,2) - x(1,1);   % spacing in x direction
>> dy = y(2,1) - y(1,1);   % spacing in y direction


>> [dzdx,dzdy] = gradient(z,dx,dy);


>> contour(x,y,z)
>> hold on
>> quiver(x,y,dzdx,dzdy)
>> hold off
>> title('Figure 24.7: Gradient Arrow Plot')
```

In this example, gradient computes *dz/dx* and *dz/dy* from the tabulated data output of peaks. These data are supplied to the function quiver, which draws arrows normal to the underlying surface, with lengths scaled by the slope at each point.

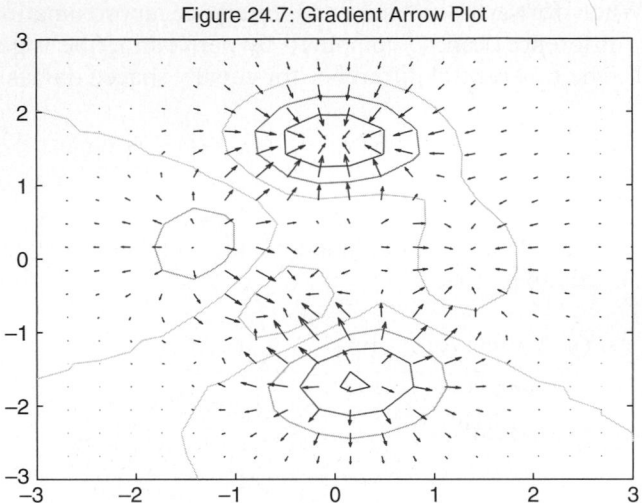

Figure 24.7: Gradient Arrow Plot

In addition to the gradient, it is sometimes useful to know the curvature of a surface. The curvature, or change in slope, at each point is calculated by the function de12, which computes the discrete approximation to the Laplacian:

$$\nabla^2 z(x, y) = \frac{d^2 z}{dx^2} + \frac{d^2 z}{dy^2}$$

In its simplest form, this value is computed by taking each surface element and subtracting from it the average of its four neighbors. If the surface is flat in each direction at a given point, the element does not change. Since MATLAB version 5, central second differences have been used at interior points to produce more accurate results. In the following example, the absolute surface curvature influences the color of the surface:

```
>> [x,y,z] = peaks; % default output of peaks
>> dx = x(1,2) - x(1,1); % spacing in x direction
>> dy = y(2,1) - y(1,1); % spacing in y direction
```

```
>> L = del2(z,dx,dy);
>> surf(x,y,z,abs(L))
>> shading interp
>> title('Figure 24.8: Discrete Laplacian Color')
```

Figure 24.8: Discrete Laplacian Color

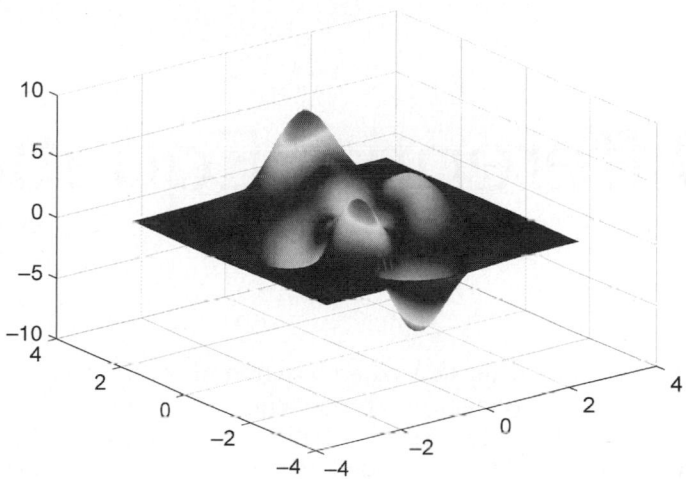

25

Differential Equations

In 1995, MATLAB introduced a collection of M-files called the MATLAB ODE suite for solving ordinary differential equations (ODEs). With the introduction of MATLAB 5, the MATLAB ODE suite became a standard part of MATLAB. In MATLAB 6, the ODE suite added two initial value problem (IVP) solvers. In addition, functions have been added in MATLAB 6 to solve boundary value problems (BVPs) and partial differential equations (PDEs). Since MATLAB 6, the ODE suite has added the ability to solve delay differential equations (DDEs) and to solve implicit differential equations.

Taken as a whole, MATLAB now has the capability to solve a wide variety of problems involving differential equations. However, discussing each of these is beyond the scope of this text. Because initial value problems appear most often in applications, they are discussed here.

25.1 IVP FORMAT

The initial value problem solvers in MATLAB compute the time history of a set of coupled first-order differential equations with known initial conditions. In mathematical terms, these problems have the form

$$\dot{y} = f(t, y) \qquad y(t_o) = y_o$$

which is vector notation for the set of differential equations

$$\dot{y}_1 = f_1(t, y_1, y_2, \ldots, y_n) \quad y_1(t_o) = y_{1o}$$
$$\dot{y}_2 = f_2(t, y_1, y_2, \ldots, y_n) \quad y_2(t_o) = y_{2o}$$
$$\vdots \qquad\qquad\qquad \vdots$$
$$\dot{y}_n = f_n(t, y_1, y_2, \ldots, y_n) \quad y_n(t_o) = y_{no}$$

where $\dot{y}_i = dy_i/dt$, n is the number of first-order differential equations, and y_{io} is the initial condition associated with the ith equation. When an initial value problem is not specified as a set of first-order differential equations, it must be rewritten as one. For example, consider the classic van der Pol equation

$$\ddot{x} - \mu(1 - x^2)\dot{x} + x = 0$$

where μ is a parameter greater than zero. If we choose $y_1 = x$ and $y_2 = dx/dt$, the van der Pol equation becomes

$$\dot{y}_1 = y_2$$
$$\dot{y}_2 = \mu(1 - y_1^2)y_2 - y_1$$

This initial value problem is used throughout this chapter to demonstrate aspects of the IVP solvers in MATLAB.

25.2 ODE SUITE SOLVERS

The MATLAB ODE suite offers eight initial value problem solvers. Each has characteristics appropriate for different initial value problems. The calling syntax for each solver is identical, making it relatively easy to change solvers for a given problem. A description of each solver is given in the following table:

Solver	Description
ode23	An explicit one-step Runge-Kutta low-order (2nd- to 3rd-order) solver. Suitable for problems that exhibit mild stiffness, problems where lower accuracy is acceptable, or problems where $f(t,y)$ is not smooth (e.g., discontinuous).
ode23s	An implicit one-step modified Rosenbrock solver of order two. Suitable for stiff problems where lower accuracy is acceptable or where $f(t,y)$ is discontinuous. ***Stiff problems are generally described as problems in which the underlying time constants vary by several orders of magnitude or more.***
ode23t	An implicit, one-step trapezoidal rule using a *free* interpolant. Suitable for moderately stiff problems. Can be used to solve differential-algebraic equations (DAEs).
ode23tb	An implicit trapezoidal rule followed by a backward differentiation of order two. Similar to ode23s. Can be more efficient than ode15s for crude tolerances.
ode45	An explicit one-step Runge Kutta medium-order (4th- to 5th-order) solver. Suitable for nonstiff problems that require moderate accuracy. ***This is typically the first solver to try on a new problem.***
ode113	A multistep Adams-Bashforth-Moulton PECE solver of varying order (1st- to 13th-order). Suitable for nonstiff problems that require moderate to high accuracy involving problems where $f(t,y)$ is expensive to compute. Not suitable for problems where $f(t,y)$ is not smooth.

Solver	Description
ode15s	An implicit, multistep numerical differentiation solver of varying order (1st- to 5th-order). Suitable for stiff problems that require moderate accuracy. ***This is typically the solver to try if*** ode45 ***fails or is too inefficient.***
ode15i	A solver of varying order (1st- to 5th-order) for solving fully implicit differential equations.

This table uses terminology—for example, *explicit, implicit,* and *stiff*—that requires a substantial theoretical background to understand. If you understand the terminology, the table describes the basic properties of each solver. If you don't understand the terminology, just follow the guidelines presented in the table, and apply ode45 and ode15s, respectively, as the first and second solvers to be tried on a given problem.

It is important to note that the MATLAB ODE suite is provided as a set of M-files that can be viewed. In addition, these same solvers are included internally in SIMULINK for the simulation of dynamic systems.

25.3 BASIC USE

Before a set of differential equations can be solved, they must be coded in a function M-file as ydot=odefile(t,y). That is, the file must accept a time t and a solution y and return values for the derivatives. For the van der Pol equation, this ODE file can be written as follows:

```
function ydot=vdpol(t,y)
%VDPOL van der Pol equation.
% Ydot=VDPOL(t,Y)
% Ydot(1) = Y(2)
% Ydot(2) = mu*(1-Y(1)^2)*Y(2)-Y(1)
% mu = 2

mu = 2;
ydot = [y(2); mu*(1-y(1)^2)*y(2)-y(1)];
```

Note that the input arguments are t and y, but that this particular function does not use t. Note also that the output ydot must be a column vector.

Given the preceding ODE file, this set of ODEs is solved by using the following commands:

```
>> tspan = [0 20];  % time span to integrate over

>> yo = [2; 0];     % initial conditions (must be a column)
>> [t,y] = ode45(@vdpol,tspan,yo);
>> size(t)            % number of time points
ans =
   333     1
>> size(y)            % (i) th column is y(i) at t(i)
ans =

   333     2
>> plot(t,y(:,1),t,y(:,2),'--')
>> xlabel('time')
>> title('Figure 25.1: van der Pol Solution')
```

Note that the use of function handles is appropriate in this case, since differential equations are written as M-files, and they are evaluated many times in the process of generating a solution over a reasonable time span.

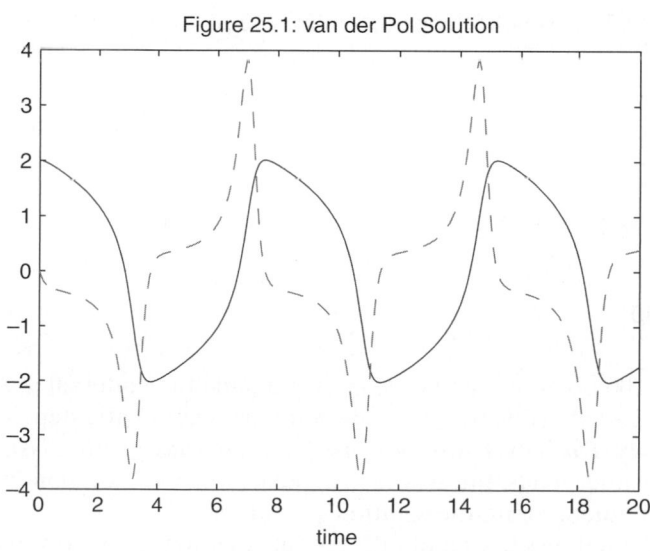

Figure 25.1: van der Pol Solution

By default, if a solver is called with no output arguments (e.g., ode45(@vdpol, tspan,yo)) the solver generates no output variables, but generates a time plot similar to that shown in the previous figure. Alternatively, if a solver is called with one output argument, that argument is a structure containing all of the information needed to evaluate the solution at arbitrary time points by using the function deval:

```
>> sol = ode45(@vdpol, tspan,yo)
sol =
          x: [1x84 double]
          y: [2x84 double]
     solver: 'ode45'
      idata: [1x1 struct]
```

The help text for deval provides information about the contents of the output variable sol. It is also possible to use this solution structure to extend or extrapolate the solution by using the function odextend.

In addition to specifying the initial and final time points in tspan, you can identify the desired specific solution time points by simply adding them to tspan, as in the following example:

```
>> yo = [2; 0];
>> tspan = linspace(0,20,100);

>> [t,y] = ode45(@vdpol,tspan,yo);

>> size (t)
ans =
   100       1
>> size(y)
ans =
   100       2
```

Here, 100 points are gathered over the same time interval as in the earlier example. When called in this way, the solver still uses automatic step-size control to maintain accuracy. ***The solver does not use fixed step integration.*** To gather the solution at the desired points, the solver interpolates its own solution in an efficient way that does not deteriorate the solution's accuracy.

Sometimes the set of differential equations to be solved contains a set of parameters that the user wishes to vary. Rather than open the ODE file and change the parameters before each call to a solver, the parameters can be added to the solver and ODE file input arguments:

```
function ydot=vdpol(t,y,mu)
%VDPOL van der Pol equation.
% Ydot=VDPOL(t,Y,mu)
% Ydot(1) = Y(2)
% Ydot(2) = mu*(1-Y(1)^2)*Y(2)-Y(1)

% mu = ?; now passed as an input argument

if nargin<3 % supply default if not given
    mu=2;
end
ydot = [y(2); mu*(1-y(1)^2)*y(2)-y(1)];
```

```
>> mu = 10 % set mu in Command window
mu =
     10
>> ode45(@vdpol, tspan,yo,[],mu)
>>title('Figure 25.2 van der Pol Solution, \mu=10')
```

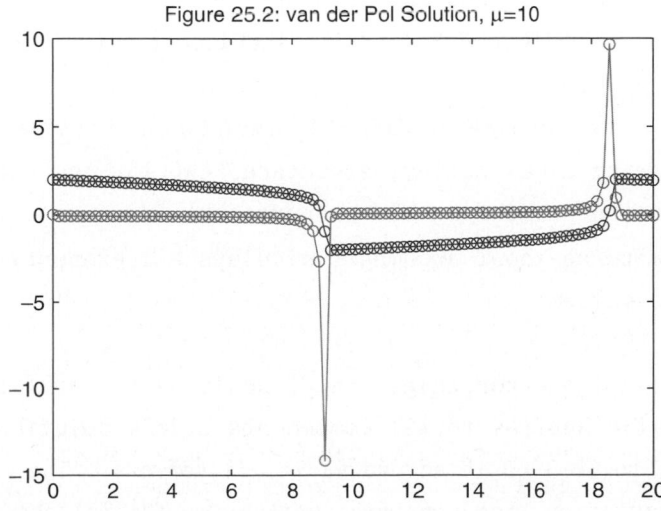

Figure 25.2: van der Pol Solution, μ=10

The preceding solves the van der Pol equation with $\mu = 10$. Here, mu is added as a third input argument to vdpol.m. Then the differential equations are solved by adding two input arguments to the ode45 function call. The first added argument is an empty array; this tells the function to use default solver options. The second added argument contains the value of the parameter. Since no output arguments were specified, the plotted solution is automatically created. While the above example describes the use of one parameter, any number of parameters can be added after the first one shown. In addition, each parameter can be an array of any data type.

25.4 SETTING OPTIONS

Up to this point, we have just accepted all default tolerances and options. When these are not sufficient, an options structure can be passed to a solver as a fourth input argument. The MATLAB ODE suite contains the functions odeset and odeget to manage this options structure. The function odeset works similarly to the Handle Graphics set function, in that parameters are specified in name/value pairs—for example, options = odeset('Name1',Value1,'Name2',Value2,...);. The available parameter names and values are described in the on-line help for odeset:

```
>> help odeset
 ODESET Create/alter ODE OPTIONS structure.
    OPTIONS = ODESET('NAME1',VALUE1,'NAME2',VALUE2,...) creates an integrator
    options structure OPTIONS in which the named properties have the
    specified values. Any unspecified properties have default values. It is
    sufficient to type only the leading characters that uniquely identify the
    property. Case is ignored for property names.
    OPTIONS = ODESET(OLDOPTS,'NAME1',VALUE1,...) alters an existing options
    structure OLDOPTS.
    OPTIONS = ODESET(OLDOPTS,NEWOPTS) combines an existing options structure
    OLDOPTS with a new options structure NEWOPTS. Any new properties
    overwrite corresponding old properties.
    ODESET with no input arguments displays all property names and their
    possible values.
 ODESET PROPERTIES
 RelTol - Relative error tolerance   [ positive scalar {1e-3} ]
    This scalar applies to all components of the solution vector, and
    defaults to 1e-3 (0.1% accuracy) in all solvers.  The estimated error in
    each integration step satisfies e(i) <= max(RelTol*abs(y(i)),AbsTol(i)).
```

AbsTol - Absolute error tolerance [positive scalar or vector {1e-6}]
  A scalar tolerance applies to all components of the solution vector.
  Elements of a vector of tolerances apply to corresponding components of
  the solution vector. AbsTol defaults to 1e-6 in all solvers. See RelTol.
NormControl - Control error relative to norm of solution [on | {off}]
  Set this property 'on' to request that the solvers control the error in
  each integration step with norm(e) <= max(RelTol*norm(y),AbsTol). By
  default the solvers use a more stringent component-wise error control.
Refine - Output refinement factor [positive integer]
  This property increases the number of output points by the specified
  factor producing smoother output. Refine defaults to 1 in all solvers
  except ODE45, where it is 4. Refine does not apply if length(TSPAN) > 2
  or the ODE solver returns the solution as a structure.
OutputFcn - Installable output function [function]
  This output function is called by the solver after each time step. When
  a solver is called with no output arguments, OutputFcn defaults to the
  function odeplot. Otherwise, OutputFcn defaults to [].
OutputSel - Output selection indices [vector of integers]
  This vector of indices specifies which components of the solution vector
  are passed to the OutputFcn. OutputSel defaults to all components.
Stats - Display computational cost statistics [on | {off}]
Jacobian - Jacobian function [function | constant matrix]
  Set this property to a function FJac (if FJac(t,y) returns dF/dy) or to
  the constant value of dF/dy.
JPattern - Jacobian sparsity pattern [sparse matrix]
  Set this property to a sparse matrix S with S(i,j) = 1 if component i of
  F(t,y) depends on component j of y, and 0 otherwise.
Vectorized - Vectorized ODE function [on | {off}]
  Set this property 'on' if the ODE function F is coded so that
  F(t,[y1 y2 ...]) returns [F(t,y1) F(t,y2) ...].
Events - Locate events [function]
  To detect events, set this property to the event function.
Mass - Mass matrix [constant matrix | function]
  For problems M*y' = f(t,y) set this property to the value of the constant

mass matrix. For problems with time- or state-dependent mass matrices, set this property to a function that evaluates the mass matrix.

MStateDependence - Dependence of the mass matrix on y [none | {weak} | strong]
  Set this property to 'none' for problems M(t)*y' = F(t,y). Both 'weak' and 'strong' indicate M(t,y), but 'weak' will result in implicit solvers using approximations when solving algebraic equations.

MassSingular - Mass matrix is singular [yes | no | {maybe}]
  Set this property to 'no' if the mass matrix is not singular.

MvPattern - dMv/dy sparsity pattern [sparse matrix]
  Set this property to a sparse matrix S with S(i,j) = 1 if for any k, the (i,k) component of M(t,y) depends on component j of y, and 0 otherwise.

InitialSlope - Consistent initial slope yp0 [vector]
  yp0 satisfies M(t0,y0)*yp0 = F(t0,y0).

InitialStep - Suggested initial step size [positive scalar]
  The solver will try this first. By default the solvers determine an initial step size automatically.

MaxStep - Upper bound on step size [positive scalar]
  MaxStep defaults to one-tenth of the tspan interval in all solvers.

BDF - Use Backward Differentiation Formulas in ODE15S [on | {off}]
  This property specifies whether the Backward Differentiation Formulas (Gear's methods) are to be used in ODE15S instead of the default Numerical Differentiation Formulas.

MaxOrder - Maximum order of ODE15S [1 | 2 | 3 | 4 | {5}]

The following examples use the preceding options:

```
>> odeset
           AbsTol: [ positive scalar or vector {1e-6} ]
           RelTol: [ positive scalar {1e-3} ]
       NormControl: [ on | {off} ]
        OutputFcn: [ function ]
        OutputSel: [ vector of integers ]
           Refine: [ positive integer ]
            Stats: [ on | {off} ]
      InitialStep: [ positive scalar ]
```

```
         MaxStep: [ positive scalar ]
             BDF: [ on | {off} ]
        MaxOrder: [ 1 | 2 | 3 | 4 | {5} ]
        Jacobian: [ matrix | function ]
        JPattern: [ sparse matrix ]
      Vectorized: [ on | {off} ]
            Mass: [ matrix | function ]
 MStateDependence: [ none | weak | strong ]
       MvPattern: [ sparse matrix ]
    MassSingular: [ yes | no | {maybe} ]
    InitialSlope: [ vector ]
          Events: [ function ]
```

Invoking odeset without input or output arguments returns an option listing, their possible values, and default values in braces:

```
>> tspan = [0 20]; % set time span to solve
>> yo = [2; 0];    % intial conditions
>> mu = 10;        % parameter mu
>> options = odeset('AbsTol',1e-12,'RelTol',1e-6);

>> [t,y] = ode45(@vdpol,tspan,yo,[],mu);  % default tolerances
>> length(t)
ans =
  593
>> [t,y] = ode(@vdpol,tspan,yo,options,mu); % tight tolerances
>> length(t)
ans =
       1689


>> [t,y] = ode15s(@vdpol,tspan,yo,[],mu); % new solver, default tols
>> length(t)
ans =
  232
```

```
>> [t,Y] = ode15s(@vdpol,tspan,yo,options,mu);  %  new solver, tight tols
>> length(t)
ans =
   651
```

Here, the steps needed to integrate the first 20 seconds are shown. Default tolerances, (i.e., AbsTol=1e−6 and RelTol=1e−3) forces ode45 to take 593 time steps.
Decreasing the tolerances to AbsTol=1e−12 and RelTol=1e−6 takes 1689 time
steps. However, changing to the stiff solver ode15s requires only 651 steps at the
tighter tolerances.

    The stiff solvers ode15s, ode23s, ode23t, and ode23tb allow you to specify an
analytical Jacobian, rather than a numerically computed approximation, which is the
default. The Jacobian is a matrix of partial derivatives that has the form

$$
\begin{bmatrix}
\dfrac{\partial f_1}{\partial y_1} & \dfrac{\partial f_1}{\partial y_2} & \cdots & \dfrac{\partial f_1}{\partial y_n} \\
\dfrac{\partial f_2}{\partial y_1} & \dfrac{\partial f_2}{\partial y_2} & \cdots & \dfrac{\partial f_2}{\partial y_n} \\
\vdots & \vdots & \ddots & \vdots \\
\dfrac{\partial f_n}{\partial y_1} & \dfrac{\partial f_n}{\partial y_2} & \cdots & \dfrac{\partial f_n}{\partial y_n}
\end{bmatrix}
$$

This matrix or a numerical approximation of it is used by the stiff solvers to compute
the solution of a set of nonlinear equations at each time step. If at all possible, an
analytical Jacobian should be supplied, as shown in the following code:

```
function jac=vdpoljac(t,y,mu)
%VDPOLJAC van der Pol equation Jacobian.

% mu = ?; passed as an input argument
if nargin<3 % supply default if not given
    mu=2;
end
jac = [         0                      1
        (-2*mu*y(1)*y(2)-1) (mu*(1-y(1)^2))];
```

```
>> options = odeset(options,'Jacobian',@vdpoljac);
>> [t,y] = ode15s(@vdpol,tspan,yo,options,mu);
>> length (t)
ans =
    670
```

In this example, the function handle of the M-file that computes the Jacobian is added to the options structure set discussed earlier. Running the solver now shows that it takes 670 steps. While this is more than the 651 steps shown earlier, providing the analytical Jacobian still significantly increases execution speed.

When it is not possible to supply an analytical Jacobian, it is beneficial to supply a *vectorized* ODE file. Vectorizing the ODE file usually means replacing y(i) with y(i,:) and using array operators. Doing so allows computation of the numerical Jacobian to proceed as fast as possible, as in this example:

```
function ydot=vdpol(t,y,mu)
%VDPOL van der Pol equation.
% Ydot=VDPOL(t,Y)
% Ydot(1) = Y(2)
% Ydot(2) = mu*(1-Y(1)^2)*Y(2)-Y(1)

% mu = ?; now passed as an input argument
if nargin<3 % supply default if not given
   mu=2;
end
ydot = [y(2,:); mu*(1-y(1,:)^2)*y(2,:)-y(1,:)];
```

The 'Refine' property determines how much output data to generate. It does not affect the step sizes chosen by the solvers, or the solution accuracy. It merely dictates how many intermediate points to interpolate the solution at within each integration step, as in the following example:

```
>> options = odeset('Refine',1);
>> [t,y] = ode45(@vdpol,tspan,yo,options,mu);
>> length (t) % # of time points
ans =
    149
```

```
>> options = odset('Refine',4);
>> [t,y] = ode45(@vdpol,tspan,yo,options,mu);
>> length(t)  % # of time points
ans =
    593
```

With 'Refine' set to 1, 149 time points are returned over the 20-second time span.
Setting 'Refine' to 4 increases the number of time points returned to 593.

The 'Events' property allows you to flag one or more events that occur as an
ODE solution evolves in time. For example, a simple event could be when some
solution component reaches a maximum, a minimum, or crosses through zero.
Optionally, the occurrence of an event can force a solver to stop integrating. To
make use of this feature, you must simply supply a function handle that computes
the values of the events to be tracked. At each time step, the solver computes the
events and marks in time those that cross through zero—for example,

```
function [value,isterminal,direction]=vdpolevents(t,y,mu)
%VDPOLEVENTS  van der Pol equation events.
value(1)=abs(y(2))-2; % find where |y(2)|=2
isterminal(1)=0;        % don't stop integration
direction(1)=0;         % don't care about crossing direction
```

```
>> mu = 2;
>> options = odeset('Events',@vdpolevents);
>> [t,y,te,ye] = ode45(@vdpol,tspan,yo,options,mu);
>> plot (t,y,te,ye(:,2),'o')
>> title('Figure 25.3: van der Pol Solution, |y(2)|=2')
```

The function vdpolevents receives the same three arguments as the other functions and returns three numerical vectors. The first is the value of the events, the second is a logical array dictating whether the solver should terminate execution on a
zero crossing in one or more of the computed events, and the third gives the user the
ability to specify whether the direction of event crossing should be considered. In
the preceding example, the points where $abs(y_2(t)) = 2$ are selected as the only
event. The solver is told to not halt on sensing events and to not care about the
direction of event crossing. The generated plot shows the system solution and the
points where the chosen events occur.

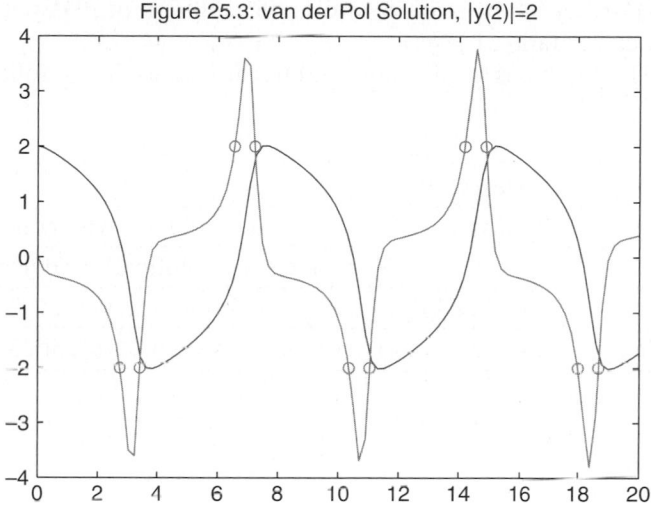

Figure 25.3: van der Pol Solution, |y(2)|=2

25.5 BVPs, PDEs, AND DDEs

In addition to the seven MATLAB solvers for solving initial value ordinary differential equations, MATLAB includes functions for the solution of boundary value problems (BVPs), partial differential equations (PDEs) and delay differential equations (DDEs). The functions used to solve BVPs are shown in the following table:

Function	Description
bvp4c	BVP solver
bvpget	Gets BVP options structure
bvpinit	Forms the initial solution guess, which is refined by bvp4c
bvpset	Sets the BVP options structure
deval	Evaluates/interpolates the solution found by using bvp4c

(Further information regarding the solution of BVPs can be found in the MATLAB documentation.)

The functions used to solve PDEs are shown in this table:

Function	Description
pdepe	Solves IVPs for parabolic-elliptic PDEs in one dimension
pdeval	Evaluates/interpolates the solution found using pdepe

(Further information regarding the solution of PDEs can be found in the on-line documentation.)

The functions used to solve DDEs are shown in the following table:

Function	Description
dde23	Solves DDE initial value problems with constant delays
deval	Evaluates/interpolates the solution found using dde23
ddeget	Gets DDE options from options structure
ddeset	Creates or alters options structure for dde23

26

Two-Dimensional Graphics

Throughout this text, several of MATLAB's graphics features have been introduced. In this and the next several chapters, the graphics features in MATLAB are more rigorously illustrated. Many of the features and capabilities illustrated here are available as menu items from the top of a *Figure* window. They are also available as buttons on the *Figure* or *Camera* toolbar, which appears by default when plots are generated or can be chosen via the **View** menu in a *Figure* window.

    The general rule is to use the toolbar and menu features of *Figure* windows if you want to customize a single *Figure*. Otherwise, use *Command* window functions to automate the process of customizing plots. This text concentrates on the *Command* window functions, since they perform the actions taken when menu and toolbar items are used.

26.1 THE plot FUNCTION

As you have seen in earlier examples, the most common function for plotting 2-D data is the plot function. This versatile function plots sets of data arrays on appropriate axes and connects the points with straight lines:

```
>> x = linspace(0,2*pi,30);
>> y = sin(x);
>> plot(x,y), title('Figure 26.1: Sine Wave')
```

This example creates 30 data points over $0 \le x \le 2\pi$ to form the horizontal axis of the plot and creates another vector y containing the sine of the data points in x. The plot function opens a graphics window, called a *Figure* window, scales the

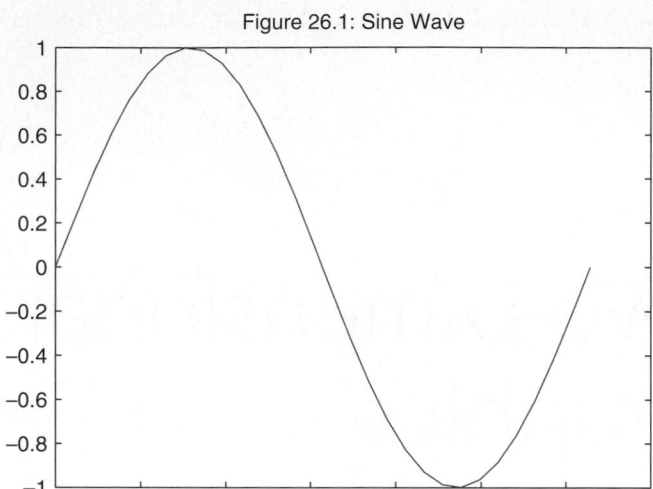

Figure 26.1: Sine Wave

axes to fit the data, plots the points, and then connects the points with straight lines. It also adds numerical scales and tick marks to the axes automatically. If a *Figure* window already exists, plot generally clears the current *Figure* window and draws a new plot.

The following code plots more than one curve or line:

```
>> z = cos(x);
>> plot(x,y,x,z)
>> title('Figure 26.2: Sine and Cosine')
```

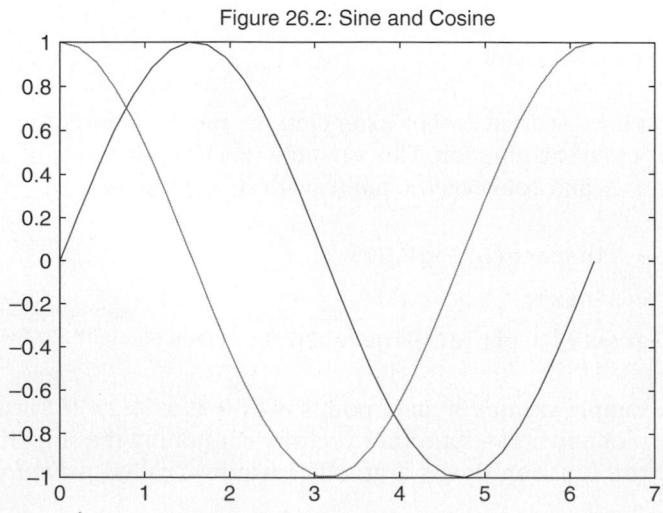

Figure 26.2: Sine and Cosine

Just giving plot another pair of arguments instructs it to generate a second line. This time, sin(x) versus x, and cos(x) versus x were plotted on the same plot. Although the figure doesn't show color, plot automatically draws the second curve in a different color. The function plot generates as many curves as it receives pairs of input arguments.

If one of the arguments is a matrix and the other a vector, the plot function plots each column of the matrix versus the vector. For example, the code

```
>> W = [y;z]; % create a matrix of the sine and cosine
>> plot(x,W)  % plot the columns of W vs. x
```

reproduces the preceding plot.

If you change the order of the arguments, the orientation of the plot changes accordingly:

```
>> plot(W,x) % plot x vs. the columns of W
>> title('Figure 26.3: Change Argument Order')
```

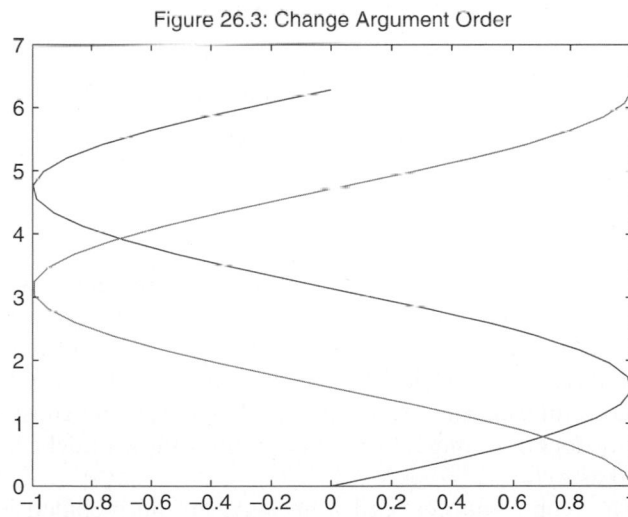

Figure 26.3: Change Argument Order

When the plot function is called with only one argument (e.g., plot(Y)) the plot function acts differently, depending on the data contained in Y. If Y is a *complex-valued* vector, plot(Y) is interpreted as plot(real(Y),imag(Y)). In all other cases, the imaginary components of the input vectors are *ignored*. On the other hand, if Y is *real-valued*, then plot(Y) is interpreted as plot(1:length(Y),Y); that is, Y is plotted versus an index of its values. When Y is a matrix, this interpretation is applied to each column of Y.

26.2 LINESTYLES, MARKERS, AND COLORS

In the previous examples, MATLAB chose the solid linestyle, and the colors blue and green, for the plots. You can specify your own colors, markers, and linestyles by giving `plot` a third argument after each pair of data arrays. This optional argument is a character string consisting of one or more characters from the following table:

Symbol	Color	Symbol	Marker	Symbol	Linestyle
b	Blue	.	Point	-	Solid line
g	Green	o	Circle	:	Dotted line
r	Red	x	Cross	-.	Dash-dot line
c	Cyan	+	Plus sign	--	Dashed line
m	Magenta	*	Asterisk		
y	Yellow	s	Square		
k	Black	d	Diamond		
w	White	v	Triangle (down)		
		^	Triangle (up)		
		<	Triangle (left)		
		>	Triangle (right)		
		p	Pentagram		
		h	Hexagram		

If you do not specify a color and you are using the default color scheme, MATLAB starts with blue and cycles through the first seven colors in the preceding table for each additional line. The default linestyle is a solid line, unless you explicitly specify a different linestyle. There is no default marker. If no marker is selected, no markers are drawn. The use of any marker places the chosen symbol at each data point, but does not connect the data points with a straight line unless a linestyle is specified as well.

If a color, a marker, and a linestyle are all included in the string, the color applies to both the marker and the line. To specify a different color for the marker, plot the same data with a different specification string:

```
>> plot(x,y,'b:p',x,z,'c-',x,1.2*z,'m+')
>> title('Figure 26.4: Linestyles and Markers')
```

As with many of the plots in this section, your computer displays color, but the figures shown here do not. If you are following along in MATLAB, just enter the commands listed in the examples to see the effects of color.

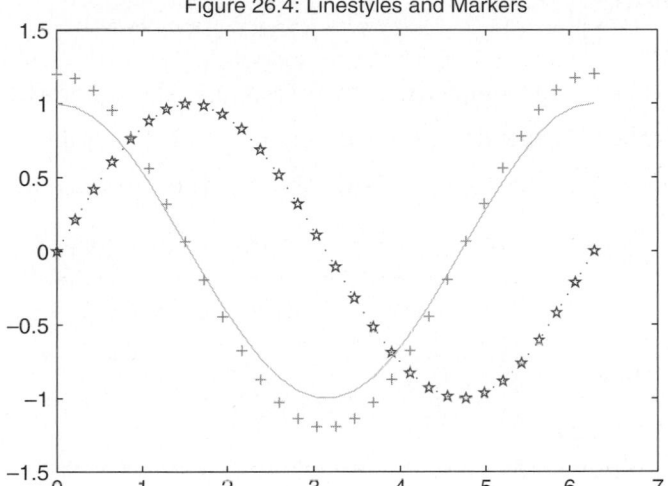

Figure 26.4: Linestyles and Markers

26.3 PLOT GRIDS, AXES BOX, AND LABELS

The grid on command adds grid lines to the current plot at the tick marks. The grid off command removes the grid. The command grid with no arguments alternately turns the grid lines on and off—that is, it *toggles* them. By default, MATLAB starts up with grid off for most plots. If you like to have grid lines on all of your plots by default, add the following lines to your startup.m file:

```
set(0,'DefaultAxesXgrid','on')
set(0,'DefaultAxesYgrid','on')
set(0,'DefaultAxesZgrid','on')
```

These lines illustrate the use of Handle Graphics features in MATLAB and the setting of default behavior. (More information on these topics can be found in Chapter 30.)

Normally, 2-D axes are fully enclosed by solid lines called an **axes box**. This box can be turned off with box off. The command box on restores the axes box. The box command toggles the state of the axes box. Horizontal and vertical axes can be labeled with the xlabel and ylabel functions, respectively. The title function adds a line of text at the top of the plot. The following example is illustrative:

```
>> x = linspace(0,2*pi,30);
>> y = sin(x);
>> z = cos(x);
```

```
>> plot(x,y,x,z)
>> box off                               % turn off the axes box
>> xlabel('Independent Variable X')      % label horizontal axis
>> ylabel('Dependent Variables Y and Z') % label vertical axis
>> title('Figure 26.5: Sine and Cosine Curves, No Box') % title
```

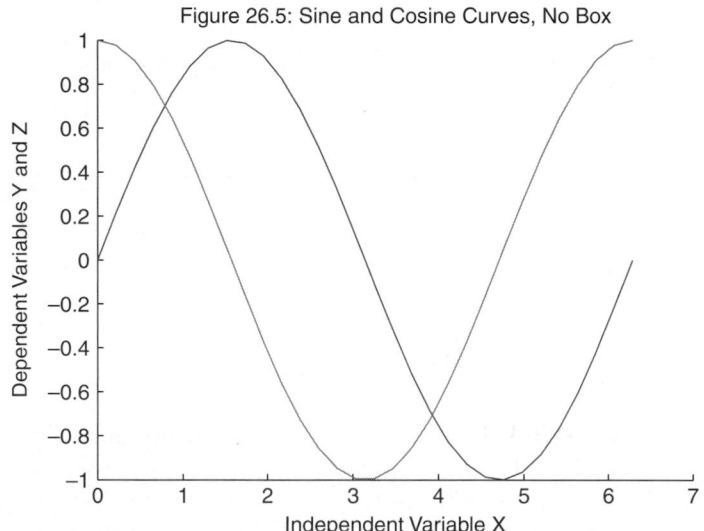

You can add a label or any other text string to any specific location on your plot with the text function. The syntax for text is text(x,y,'string'), where (x,y) represents the coordinates of the center left edge of the text string in units taken from the plot axes. For example, the following code segment places the text 'sin(x)' at the location $x = 2.5$, $y = 0.7$:

```
>> grid on, box on % turn axes box and grid lines on
>> text(2.5,0.7,'sin(x)')
>> title('Figure 26.6: Sine and Cosine Curves, Added Label')
```

If you want to add a label, but don't want to stop to figure out the coordinates to use, you can place a text string with the mouse. The gtext('text') function switches to the current *Figure* window, puts up a cross-hair that follows the mouse, and waits for a mouse click or key press. When either one occurs, the string argument to gtext is placed with the lower left corner of the first character at that location.

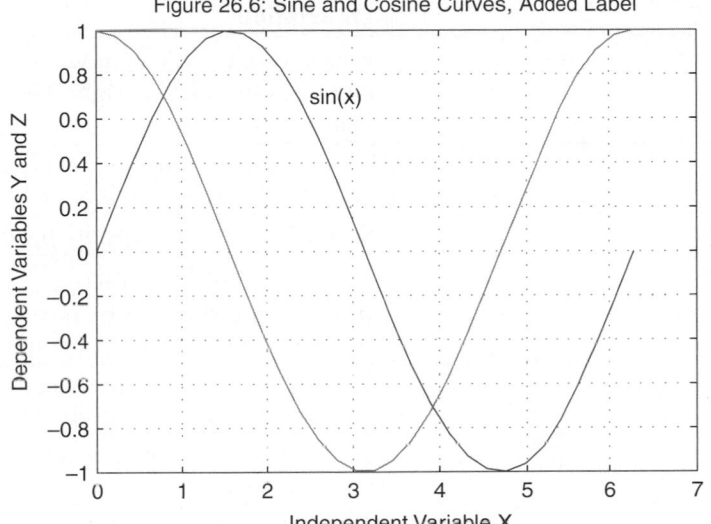

Figure 26.6: Sine and Cosine Curves, Added Label

26.4 CUSTOMIZING PLOT AXES

MATLAB gives you complete control over the scaling and appearance of both the horizontal and vertical axes of your plot with the `axis` command. Because this command has so many features, only the most useful are described here. The primary features of the `axis` command are given in the following table:

Command	Description
`axis([xmin xmax ymin ymax])`	Sets axis limits on the current plot
`V = axis`	Returns a row vector containing the current axis limits
`axis auto`	Returns axis scaling to automatic defaults
`axis manual`	Freezes axis scaling so that if `hold` is on, subsequent plots use the same axis limits
`axis tight`	Sets axis limits to the range of the plotted data
`axis fill`	Sets the axis limits and aspect ratio so that the axis fills the allotted space. This option has an effect only if `PlotBoxAspectRatio` or `DataAspect RatioMode` is `'manual'`.
`axis ij`	Puts axis in *matrix* mode. The horizontal axis increases from left to right. The vertical axis increases from top to bottom.

Command	Description
axis xy	Puts axis in Cartesian mode. The horizontal axis increases from left to right. The vertical axis increases from bottom to top.
axis equal	Sets the aspect ratio so that equal tick mark increments on each axis are equal in size
axis image	Sets axis limits appropriate for displaying an image
axis square	Makes the axis box square
axis normal	Restores the current axis box to full size and removes any restrictions on unit scaling
axis vis3d	Freezes the aspect ratio to enable rotation of 3-D objects without axis size changes
axis off	Turns off all axis labeling, tick marks, and background
axis on	Turns on all axis labeling, tick marks, and background

Multiple commands to axis can be given at once. For example, axis auto on xy is the default axis scaling. The axis command affects only the current plot. Therefore, it is issued after the plot command, just as grid, xlabel, ylabel, title, text, and so on, are issued after the plot is on the screen:

```
x = linspace(0,2*pi,30);
y = sin(x);
plot(x,y)
title('Figure 26.7: Fixed Axis Scaling')
axis([0 2*pi -1.5 2])  % change axis limits
```

Note that by specifying the maximum x-axis value to be 2*pi, the plot axis ends at exactly 2*pi rather than rounding the axis limit up to 7. The simplest way to see what the various axis command arguments do is to generate a simple plot, then issue multiple axis commands, and, finally view the resulting changes.

When you simply want to change the axis limits on a single axis, the axis command is cumbersome because it requires you to enter limits for all axes. To solve this problem, MATLAB provides the functions xlim, ylim, and zlim, which are described by the help text for xlim, with the obvious change for ylim and zlim:

```
>> help xlim
  XLIM X limits.
     XL = XLIM              gets the x limits of the current axes.
     XLIM([XMIN XMAX])      sets the x limits.
```

```
XLMODE = XLIM('mode') gets the x limits mode.
XLIM(mode)              sets the x limits mode.
                        (mode can be 'auto' or 'manual')
XLIM(AX,...)            uses axes AX instead of current axes.
XLIM sets or gets the XLim or XLimMode property of an axes.
See also PBASPECT, DASPECT, YLIM, ZLIM.
```

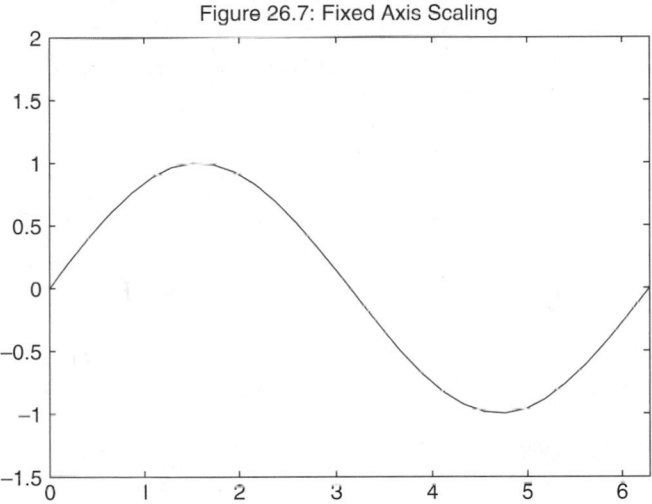

Figure 26.7: Fixed Axis Scaling

26.5 MULTIPLE PLOTS

You can add new plots to an existing plot by using the hold command. When you enter hold on, MATLAB does not remove the existing axes when new plot functions are issued. Instead, it adds new curves to the current axes. However, if the new data does not fit within the current axes limits, the axes are rescaled. Entering hold off releases the current *Figure* window for new plots. The hold command, without arguments, toggles the hold setting:

```
>> x = linspace(0,2*pi,30);
>> y = sin(x);
>> z = cos(x);
>> plot(x,y)
>> hold on
```

```
>> ishold  % return 1 (True) if hold is ON
ans =
    1
>> plot(x,z,'m')
>> hold off
>> ishold  % hold is no longer ON
ans =
     0
>> title 'Figure 26.8: Use of hold command'
```

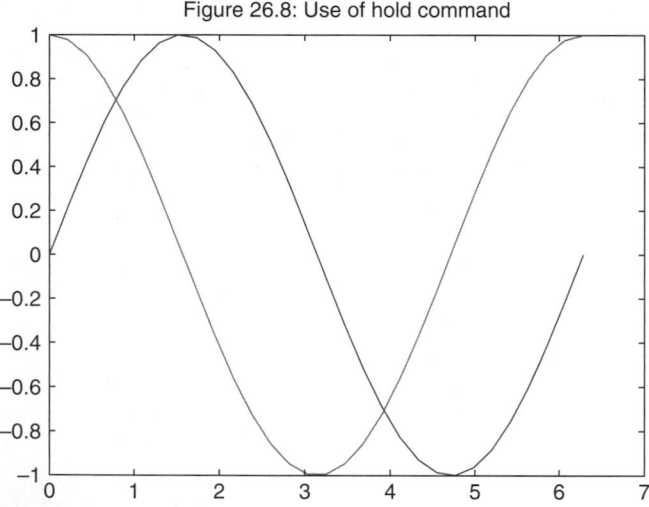

Figure 26.8: Use of hold command

Notice that this example specifies the color of the second curve. Since there is only one set of data arrays in each `plot` function, the line color for each `plot` function would otherwise default to the first color in the color order list, resulting in two lines plotted in the same color on the plot. Note also that the title text is not enclosed in parentheses, but the effect remains unchanged. In the alternative form shown, `title` is interpreted as a command rather than a function. In addition, in MATLAB 7, the command `hold all` holds both the current plot and the next line color to be used. Therefore, if `hold on` in the previous example is replaced by `hold all`, and no explicit color is specified for the second plot, then, by default, `plot(x,z)` would have produced a green line.

26.6 MULTIPLE FIGURES

It is possible to create multiple *Figure* windows and plot different data sets in different ways in each one. To create new *Figure* windows, use the `figure` command in the *Command* window or the **New Figure** selection from the **File** menu in the *Command* or *Figure* window. You can choose a specific *Figure* window to be the active, or current, figure by clicking on it with the mouse or by using `figure(n)`, where n is the number of the window. The current *Figure* window is the window that is active for subsequent plotting functions.

Every time a new *Figure* window is created, a number identifying it—that is, its **handle**—is returned and stored for future use. The figure handle is also displayed in the *Figure* window title bar. When a new *Figure* window is created, it is placed in the default figure position on the screen. As a result, when more than one *Figure* window is created, each new window covers all preceding *Figure* windows. To see the windows simultaneously, simply drag them around by using the mouse on the *Figure* window title bar.

To reuse a *Figure* window for a new plot, it must be made the active, or current, figure. Clicking on the figure of choice with the mouse makes it the current figure. From within MATLAB, `figure(h)` where h is the figure handle, makes the corresponding figure active or current. Only the current figure is responsive to the `axis`, `hold`, `xlabel`, `ylabel`, `title`, and `grid` commands.

Figure windows can be deleted by closing them with the mouse, similar to the way you may close windows on your computer. Alternatively, the command `close` can be issued. For example,

```
>> close
```

closes the current *Figure* window,

```
>> close(h)
```

closes the *Figure* window having handle h, and

```
>> close all
```

closes all *Figure* windows.

If you simply want to erase the contents of a *Figure* window without closing it, use the command `clf`. For example,

```
>> clf
```

clears the current *Figure* window, and

```
>> clf reset
```

clears the current *Figure* window and resets all properties, such as `hold`, to their default states.

26.7 SUBPLOTS

One *Figure* window can hold more than one set of axes. The subplot(m,n,p) command subdivides the current *Figure* window into an m-by-n matrix of plotting areas and chooses the pth area to be active. The subplots are numbered left to right along the top row, then along the second row, and so on.

The following code is illustrative:

```
x = linspace(0,2*pi,30);
y = sin(x);
z = cos(x);
a = 2*sin(x).*cos(x);
b = sin(x)./(cos(x)+eps);

subplot(2,2,1) % pick the upper left of a 2-by-2 grid of subplots
plot(x,y), axis([0 2*pi -1 1]), title('Figure 26.9a: sin(x)')

subplot(2,2,2) % pick the upper right of the 4 subplots
plot(x,z), axis([0 2*pi -1 1]), title('Figure 26.9b: cos(x)')

subplot(2,2,3) % pick the lower left of the 4 subplots
plot(x,a), axis([0 2*pi -1 1]), title('Figure 26.9c: 2sin(x)cos(x)')

subplot(2,2,4) % pick the lower right of the 4 subplots
plot(x,b), axis([0 2*pi -20 20]), title('Figure 26.9d: sin(x)/cos(x)')
```

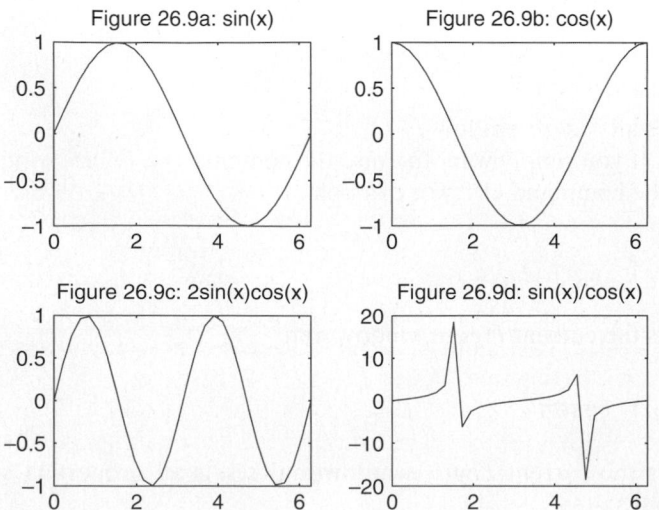

Note that when a particular subplot is active, it is the only subplot or axis that is responsive to the `axis`, `hold`, `xlabel`, `ylabel`, `title`, `grid`, and `box` commands. The other subplots are not affected. In addition, the active subplot remains active until another `subplot` or `figure` command is issued. When a new `subplot` command changes the number of subplots in the *Figure* window, previous subplots are erased to make room for the new orientation. To return to the default mode and use the entire *Figure* window for a single set of axes, use the command `subplot(1,1,1)`. When you print a *Figure* window containing multiple plots, all of them are printed on the same page. For example, when the current *Figure* window contains four subplots and the orientation is landscape mode, each of the plots uses one-quarter of the printed page.

26.8 INTERACTIVE PLOTTING TOOLS

Before the *Figure* window menu bar and toolbars existed, MATLAB offered several functions for annotating plots interactively. These functions are described in this section, and most are available from the *Figure* menu bar and toolbars.

Rather than individual text strings, a legend can be used to identify the data sets on your plot. The `legend` command creates a legend box on the plot, keying any text you supply to each line in the plot. If you wish to move the legend, simply click and hold down the mouse button near the lower left edge of the legend and drag the legend to the desired location. The command `legend off` deletes the legend. The following example is illustrative:

```
>> close % close figure containing subplots
>> x = linspace(0,2*pi,30);
>> y = sin(x);
>> z = cos(x);
>> plot(x,y,x,z)
>> legend('sin(x)','cos(x)')
>> title('Figure 26.10: Legend Example')
```

MATLAB provides an interactive tool to expand sections of a 2-D plot to see more detail or to ***zoom in*** on a region of interest. The command `zoom on` turns on the zoom mode. Clicking the left mouse button within the *Figure* window expands the plot by a factor of 2 centered around the point under the mouse pointer. Each time you click, the plot expands. Click the right mouse button to open a contextual menu to zoom out or select other properties. You can also click and drag a rectangular area to zoom in on a specific area. The `zoom out` command returns the plot to its initial state. The command `zoom off` turns off the zoom mode. Just `zoom` with no arguments toggles the zoom state of the active *Figure* window. The *Figure* toolbar and *Figure* window menus offer a GUI approach to implementing this feature as well.

Figure 26.10: Legend Example

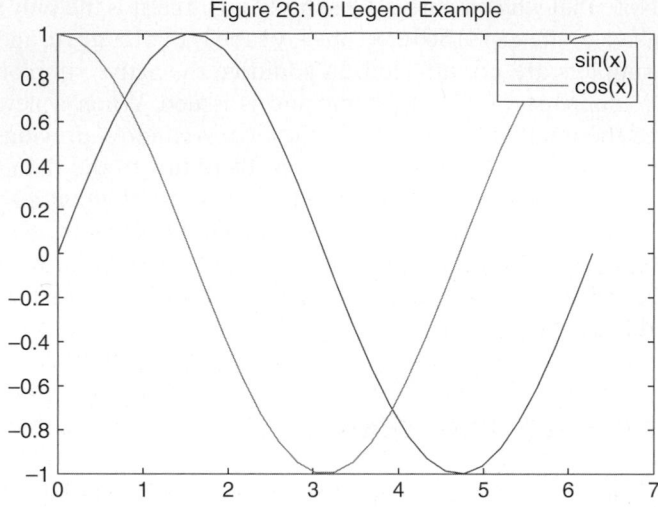

In some situations, it is convenient to select coordinate points from a plot in a *Figure* window. In MATLAB, this feature is embodied in the ginput function. The form [x,y]=ginput(n) gets n points from the current plot or subplot based on mouse click positions within the plot or subplot. If you press the **Return** or **Enter** key before all n points are selected, ginput terminates with fewer points. The points returned in the vectors x and y are the respective *x* and *y* data coordinate points selected. The returned data are not necessarily points from the plotted data, but rather the explicit *x*- and *y*-coordinate values where the mouse was clicked. If points are selected outside of the plot or subplot axes limits—for example, outside the plot box—the points returned are extrapolated values.

This function can be somewhat confusing when used in a *Figure* window containing subplots. The data returned is with respect to the current or active subplot. Thus, if ginput is issued after a subplot(2,2,3) command, the data returned is with respect to the axes of the data plotted in subplot(2,2,3). If points are selected from other subplots, the data is still with respect to the axes of the data in subplot(2,2,3). When an unspecified number of data points are desired, the form [x,y]=ginput without an input argument can be used. Here, data points are gathered until the **Return** key is pressed. The gtext function (described earlier in this chapter) uses the function ginput along with the function text for placing text with the mouse.

26.9 SCREEN UPDATES

Because screen rendering is relatively time-consuming, MATLAB does not always update the screen after each graphics command. For example, if the following commands are entered at the MATLAB prompt, MATLAB updates the screen after each graphics command (plot, axis, and grid):

```
>> x = linspace(0,2*pi); y = sin(x);
>> plot(x,y)
>> axis([0 2*pi -1.2 1.2])
>> grid
```

However, if the same graphics commands are entered on a single line, such as

```
>> plot(x,y), axis([0 2*pi -1.2 1.2]), grid
```

then MATLAB renders the figure only once—when the MATLAB prompt reappears. A similar procedure occurs when graphics commands appear as part of a script or function M-file. In this case, even if the commands appear on separate lines in the file, the screen is rendered only once—when all commands are completed and the MATLAB prompt reappears.

In general, six events cause MATLAB to render the screen:

1. A return to the MATLAB prompt
2. Encountering a function that temporarily stops execution, such as `pause`, `keyboard`, `input`, and `waitforbuttonpress`
3. Execution of a `getframe` command
4. Execution of a `drawnow` command
5. Execution of a `figure` command
6. Resizing of a *Figure* window

Of these, only the `drawnow` command specifically allows one to force MATLAB to redraw the screen at arbitrary times.

26.10 SPECIALIZED 2-D PLOTS

Up to this point, the basic plotting function `plot` has been illustrated. In many situations, plotting lines or points on linearly scaled axes does not convey the desired information. As a result, MATLAB offers other basic 2-D plotting functions, as well as specialized plotting functions that are embodied in function M-files.

In addition to `plot`, MATLAB provides the functions `semilogx` for plotting with a logarithmically scaled x-axis, `semilogy` for plotting with a logarithmically scaled y-axis, and `loglog` for plotting with both axes logarithmically spaced. All of the features discussed previously with respect to the function `plot` apply to these functions as well.

The `area` function is useful for building a stacked area plot. The function `area(x,y)` is the same as `plot(x,y)` for vectors x and y, except that the area under the plot is filled in with color. The lower limit for the filled area may be specified, but defaults to zero. To stack areas, use the form `area(X,Y)`, where Y is a matrix and X is a matrix or vector whose length equals the number of rows in Y. If X is omitted, as in the following for example, `area` uses the default value `X=1:size(Y,1)`;

```
>> z = -pi:pi/5:pi;
>> area([sin(z);cos(z)])
>> title('Figure 26.11: Stacked Area Plot')
```

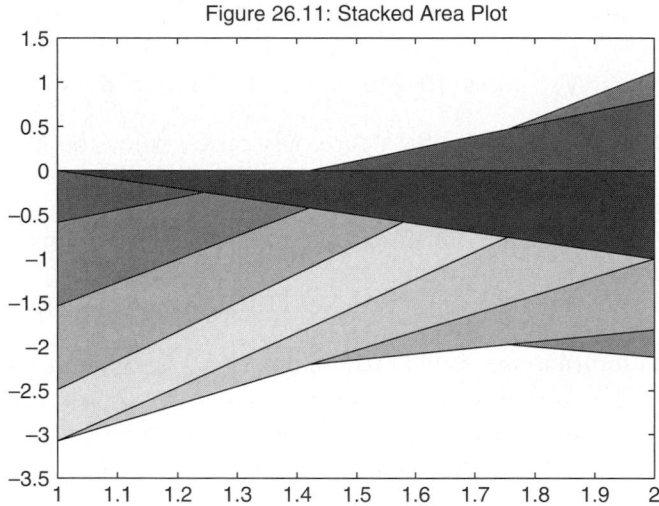

Figure 26.11: Stacked Area Plot

Filled polygons can be drawn by using the `fill` function. The function `fill(x,y,'c')` fills the 2-D polygon defined by the column vectors x and y with the color specified by c. The vertices of the polygon are specified by the pairs (xi,yi). If necessary, the polygon is closed by connecting the last vertex to the first. Like the `plot` function, `fill` can have any number of pairs of vertices and associated colors. Moreover, when x and y are matrices of the same dimension, the columns of x and y are assumed to describe separate polygons. In the code

```
>> t = (1:2:15)'*pi/8;
>> x = sin(t);
>> y = cos(t);
>> fill(x,y,'r') % a filled red circle using only 8 data points
>> axis square off
>> text(0,0,'STOP', ...
   'Color',[1 1 1], ...
   'FontSize',80, ...
   'FontWeight','bold', ...
   'HorizontalAlignment','center')
>> title('Figure 26.12: Stop Sign')
```

Figure 26.12: Stop Sign

the text(x,y,'string') function is used with extra arguments. The Color, Font-Size, FontWeight, and HorizontalAlignment arguments tell MATLAB to use Handle Graphics to modify the text. Handle Graphics is the name of MATLAB's underlying graphics functions. You can access this rich set of powerful, versatile graphics functions yourself. (See Chapter 31 for more information on these features.)

Standard pie charts can be created by using the pie(a,b) function, where a is a vector of values and b is an optional logical vector describing a slice or slices to be pulled out of the pie chart. The pie3 function renders the pie chart with a 3-D appearance:

```
>> a = [.5 1 1.6 1.2 .8 2.1];
>> pie(a,a==max(a));   % chart a and pull out the biggest slice
>> title('Figure 26.13: Example Pie Chart')
```

Figure 26.13: Example Pie Chart

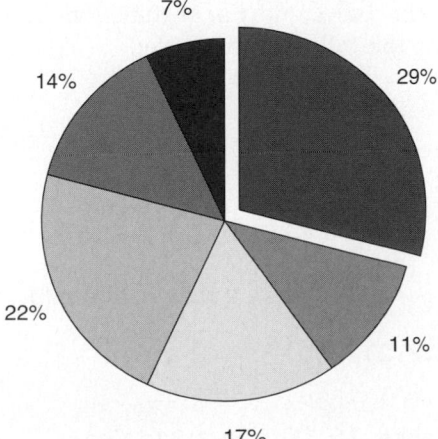

Sometimes it is desirable to plot two different functions on the same axes by using different y-axis scales. The function `plotyy` does just that:

```
>> x = -2*pi:pi/10:2*pi;
>> y = sin(x);
>> z = 3*cos(x);
>> subplot(2,1,1), plot(x,y,x,z)
>> title('Figure 26.14a: Two plots on the same scale.');
>> subplot(2,1,2), plotyy(x,y,x,z)
>> title('Figure 26.14b: Two plots on different scales.');
```

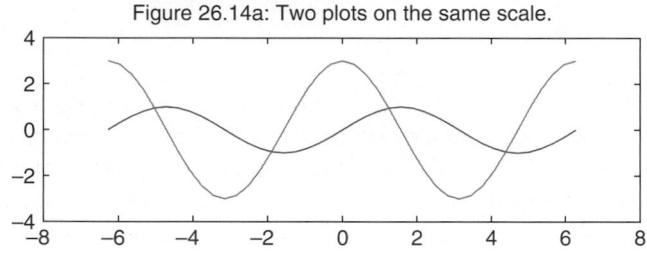
Figure 26.14a: Two plots on the same scale.

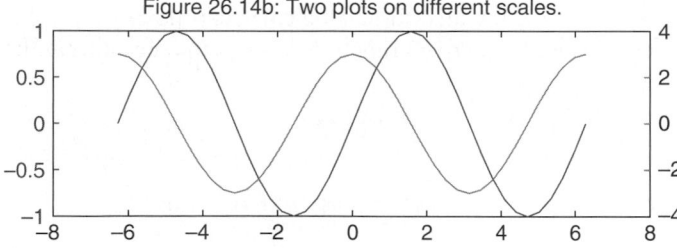
Figure 26.14b: Two plots on different scales.

Bar and stair plots can be generated by using the `bar`, `barh`, and `stairs` plotting functions. The `bar3` and `bar3h` functions render the bar charts with a 3-D appearance, as in the following example:

```
>> x = -2.9:0.2:2.9;
>> y = exp(-x.*x);

>> subplot(2,2,1)
>> bar(x,y)
>> title('Figure 26.15a: 2-D Bar Chart')

>> subplot(2,2,2)
>> bar3(x,y,'r')
>> title('Figure 26.15b: 3-D Bar Chart')
```

```
>> subplot(2,2,3)
>> stairs(x,y)
>> title('Figure 26.15c: Stair Chart')

>> subplot(2,2,4)
>> barh(x,y)
>> title('Figure 26.15d: Horizontal Bar Chart')
```

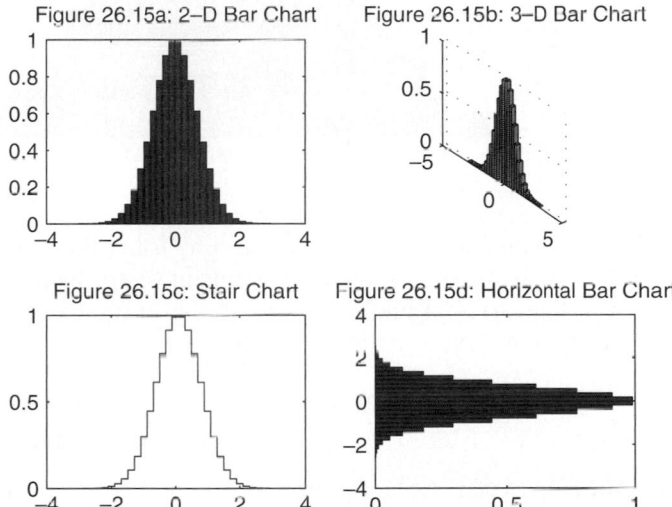

The various bar functions accept a single-color argument for all bars. Bars can be grouped or stacked as well. The form bar(x,Y) for vector x and matrix Y draws groups of bars corresponding to the columns of Y. The form bar(x,Y,'stacked') draws the bars stacked vertically. The functions barh, bar3, and bar3h have similar options.

Histograms illustrate the distribution of values in a vector. The function hist(y) draws a 10-bin histogram for the data in vector y. The function hist(y,n), where n is a scalar, draws a histogram with n bins. The function hist(y,x), where x is a vector, draws a histogram using the bins specified in x:

```
>> x = -2.9:0.2:2.9;  % specify the bins to use
>> y = randn(5000,1); % generate 5000 random data points

>> hist(y,x)          % draw the histogram
>> title('Figure 26.16: Histogram of Gaussian Data')
```

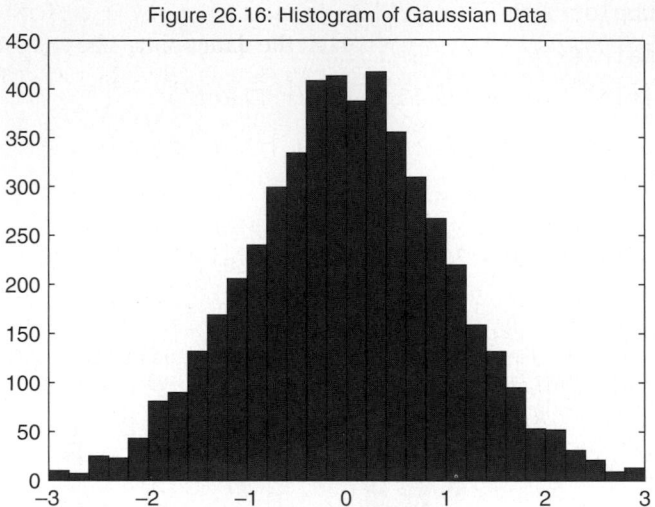

Figure 26.16: Histogram of Gaussian Data

Discrete sequence data can be plotted by using the `stem` function. The function `stem(z)` creates a plot of the data points in vector z connected to the horizontal axis by a line. An optional character-string argument can be used to specify linestyle, as in the following code:

```
>> z = randn(30,1);      % create some random data
>> stem(z,'--')          % draw a stem plot using dashed linestyle
>> set(gca,'YGrid','on') % turn grid on Y-axis only
>> title('Figure 26.17: Stem Plot of Random Data')
```

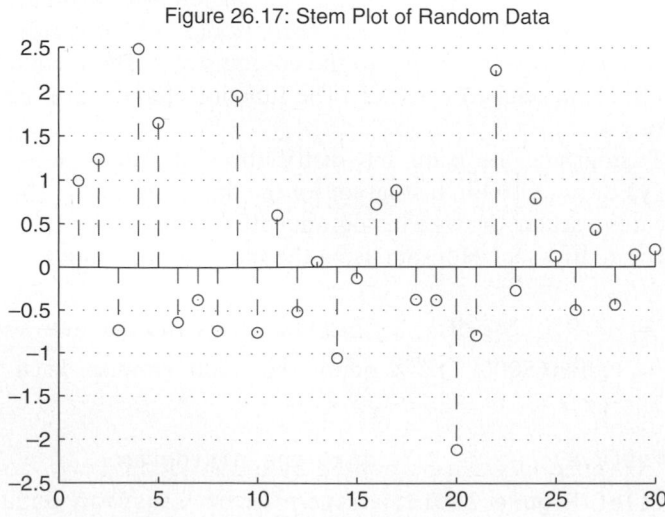

Figure 26.17: Stem Plot of Random Data

The function stem(x,z) plots the data points in z at the values specified in x.
A plot can include error bars at the data points. The function errorbar(x,y,e)
plots the graph of vector x versus vector y with error bars specified by vector e:

```
>> x = linspace(0,2,21); % create a vector
>> y = erf(x);           % y is the error function of x
>> e = rand(size(x))/10; % e contains random error values

>> errorbar(x,y,e)       % create the plot
>> title('Figure 26.18: Errorbar Plot')
```

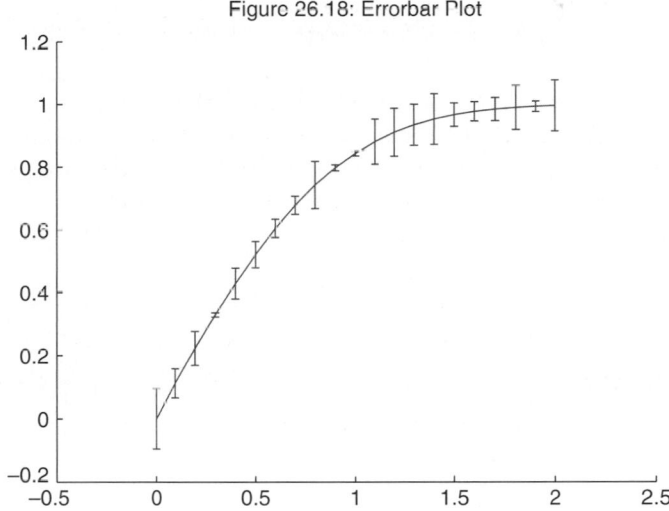

Figure 26.18: Errorbar Plot

Note that all vectors must be the same length. For each data point (xi,yi), an error
bar is drawn a distance ei above and ei below the point:
Plots in polar coordinates can be created by using the polar(t,r,S) function,
where t is the angle vector in radians, r is the radius vector, and S is an optional
character string describing color, marker symbol, and/or linestyle:

```
>> t = linspace(0,2*pi);
>> r = sin(2*t).*cos(2*t);
>> subplot(2,2,1)
>> polar(t,r), title('Figure 26.19a: Polar Plot')
```

Complex data can be plotted by using compass and feather. The function
compass(z) draws a plot that displays the angle and magnitude of the complex

elements of z as arrows emanating from the origin. The function feather(z) plots the
same data by using arrows emanating from equally spaced points on a horizontal line.
The functions compass(x,y) and feather(x,y) are equivalent to compass(x+i*y)
and feather(x+i*y), respectively. The following example is illustrative:

```
>> z = eig(randn(20));

>> subplot(2,2,2)

>> compass(z)

>> title('Figure 26.19b: Compass Plot')

>> subplot(2,2,3)

>> feather(z)

>> title('Figure 26.19c: Feather Plot')
```

The function rose(v) draws a 20-bin polar histogram for the angles in vector v.
The function rose(v,n), where n is a scalar, draws a histogram with n bins. The func-
tion rose(v,x), where x is a vector, draws a histogram using the bins specified in x.
The following code illustrates the first of these options:

```
>> subplot(2,2,4)

>> v = randn(1000,1)*pi;

>> rose(v)

>> title('Figure 26.19d: Angle Histogram')
```

Figure 26.19a: Polar Plot

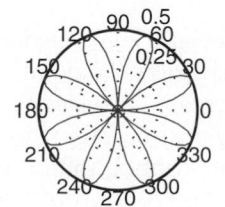

Figure 26.19b: Compass Plot

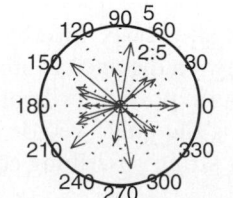

Figure 26.19c: Feather Plot

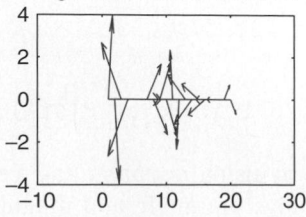

Figure 26.19d: Angle Histogram

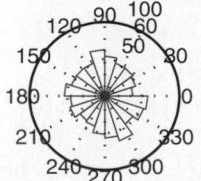

The function scatter generates a scatter plot—that is, a plot of circles at data points, where the circle size or color can vary, point by point:

```
>> x = rand(40,1);
>> y = randn(40,1);
>> area = 20+(1:40);
>> scatter(x,y,area)
>> box on
>> title('Figure 26.20: A scatter plot')
```

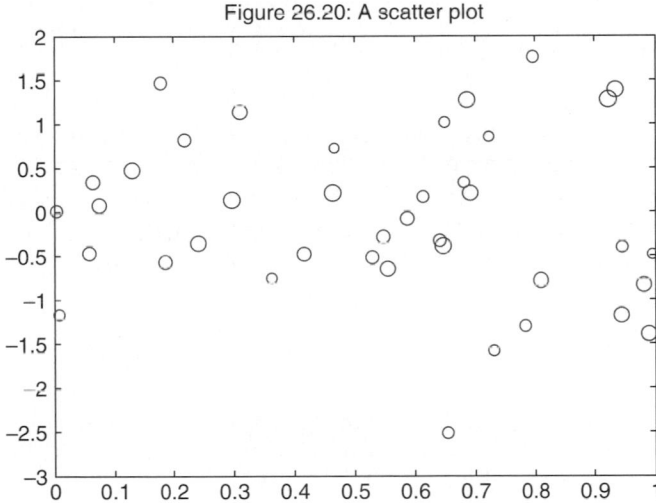

Figure 26.20: A scatter plot

26.11 EASY PLOTTING

When you don't want to take the time to specify the data points explicitly for a plot, MATLAB provides the functions fplot, ezplot, and ezpolar. The function fplot plots functions defined by M-file names or function handles. The functions ezplot and ezpolar plot functions defined by function handles, string expressions, or symbolic math objects, with the obvious difference in plot type. These functions simply spare the user from having to define the data for the independent variable, as the following code indicates:

```
>> subplot(2,2,1)
>> fplot(@humps,[-.5 3])
>> title('Figure 26.21a: Fplot of the Humps Function')
>> xlabel('x')
>> ylabel('humps(x)')
```

```
>> subplot(2,2,2)
>> f_hdl=@(x) sin(x)/x;
>> ezplot(f_hdl,[-15,15])
>> title(['Figure 26.21b: ' sin(x)/x])

>> subplot(2,2,3)
>> istr = '(x-2)^2/(2^2) + (y+1)^2/(3^2) - 1';
>> ezplot(istr,[-2 6 -5 3])
>> axis square
>> grid
>> title(['Figure 26.21c: ' istr])
```

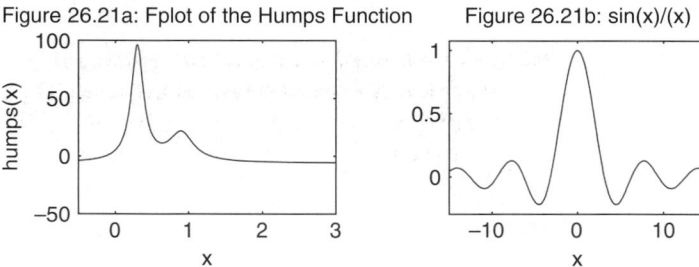

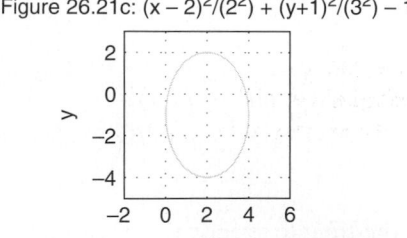

The last example shows that `ezplot` can be used to plot implicit functions. In this case, the string expression is that of an ellipse centered at $(2, -1)$.

26.12 TEXT FORMATTING

Multiline text can be used in any text string, including titles and axis labels, as well as the `text` and `gtext` functions. Simply use string arrays or cell arrays for multiline text. For example,

```
>> xlabel({'This is the first line','and this is the second.'});
```

labels the *x*-axis with two lines of text. Note that the string separator can be a space, a comma, or a semicolon; each style produces the same result. (See Chapter 9 for more details on cell arrays of strings.)

A selection of more than 75 symbols, including Greek letters and other special characters, can be included in MATLAB text strings by embedding commands within the string. The available symbols and the character strings used to define them are listed in the following table (this information can also be found by viewing the `string` property of the *text* Handle Graphics object in the online documentation):

Character Sequence	Symbol	Character Sequence	Symbol	Character Sequence	Symbol
\alpha	α	\rfloor	$\rfloor$	\cong	$\cong$
\beta	β	\lfloor	$\lfloor$	\approx	$\approx$
\gamma	γ	\perp	$\perp$	\Re	$\Re$
\delta	δ	\wedge	$\wedge$	\oplus	$\oplus$
\epsilon	ε	\rceil	$\rceil$	\cup	$\cup$
\zeta	ζ	\vee	$\vee$	\subseteq	$\subseteq$
\eta	η	\langle	$\langle$	\in	$\in$
\theta	θ	\upsilon	υ	\lceil	$\lceil$
\vartheta	ϑ	\phi	ϕ	\cdot	$\cdot$
\iota	ι	\chi	χ	\neg	$\neg$
\kappa	κ	\psi	ψ	\times	$\times$
\lambda	λ	\omega	ω	\surd	$\surd$
\mu	μ	\Gamma	Γ	\varpi	ϖ
\nu	ν	\Delta	Δ	\rangle	$\rangle$
\xi	ξ	\Theta	Θ	\sim	$\sim$
\pi	π	\Lambda	Λ	\leq	$\leq$
\rho	ρ	\Xi	Ξ	\infty	∞
\sigma	σ	\Pi	Π	\clubsuit	$\clubsuit$
\varsigma	ς	\Sigma	Σ	\diamondsuit	$\diamondsuit$
\tau	τ	\Upsilon	Υ	\heartsuit	$\heartsuit$
\equiv	$\equiv$	\Phi	Φ	\spadesuit	$\spadesuit$
\Im	$\Im$	\Psi	Ψ	\leftrightarrow	$\leftrightarrow$
\otimes	$\otimes$	\Omega	Ω	\leftarrow	$\leftarrow$
\cap	$\cap$	\forall	$\forall$	\uparrow	$\uparrow$
\supset	$\supset$	\exists	$\exists$	\rightarrow	$\rightarrow$
\int	$\int$	\ni	$\ni$	\downarrow	$\downarrow$

Character Sequence	Symbol	Character Sequence	Symbol	Character Sequence	Symbol
\circ	°	\neq	≠	\nabla	∇
\pm	±	\aleph	ℵ	\ldots	...
\geq	≥	\wp	℘	\prime	′
\propto	∝	\oslash	∅	\0	∅
\partial	∂	\supseteq	⊇	\mid	|
\bullet	•	\subset	⊂	\copyright	©
\div	÷	\o	o		

In several previous versions of MATLAB, only a limited subset of TEX formatting commands was available. In MATLAB 7, all TEX formatting commands are available. Complete information regarding these commands can be found online. Some of the common and important commands include the following: Superscripts and subscripts are specified by ∧ and _, respectively; text font and size are chosen by using the \fontname and \fontsize commands; and a font style are specified by using the \bf, \it, \sl, or \rm command to select a boldface, italic, oblique (or slant), or normal Roman font, respectively. To print the special characters used to define TeX strings, prefix them with the backslash (\) character. The characters affected are the backslash (\), left and right curly braces { }, underscore (_), and carat (∧). The following example illustrates the use of TeX formatting commands:

```
>> close % close last Figure window and start over
>> axis([0 1 0 0.5])
>> text(0.2,0.1,'\itE = M\cdotC^{\rm2}')
>> text(0.2,0.2,'\fontsize{16} \nabla \times H = J + \partialD/\partialt')
>> text(0.2,0.3,'\fontname{courier}\fontsize{16}\bf x_{\alpha}+y^{2\pi}')
>> fsstr='f(t) = A_o + \fontsize{30}_\Sigma\fontsize{10}';
>> text(0.2,0.4,[fsstr '[A_ncos(n\omega_ot) + B_nsin(n\omega_ot)]'])
>> title('Figure 26.22: TeX Formatting Examples')
```

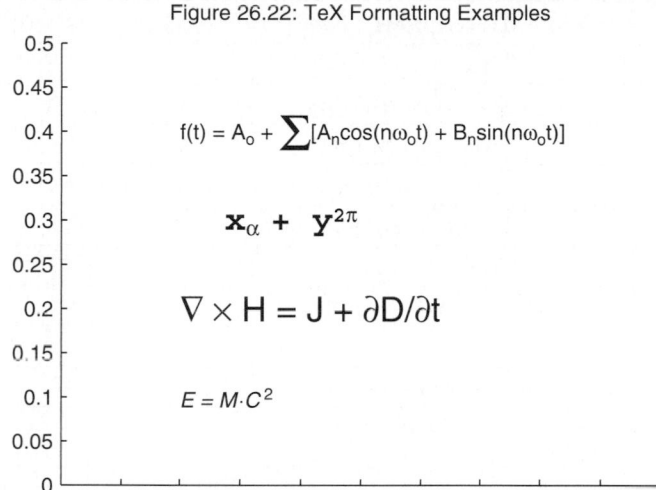

Figure 26.22: TeX Formatting Examples

26.13 SUMMARY

The following table lists MATLAB functions for 2-D plotting:

Function	Description
plot	Linear plot
loglog	Log-log plot
semilogx	Semilog x-axis plot
semilogy	Semilog y-axis plot
polar	Polar coordinate plot
plotyy	Linear plot with two y-axes
axis	Controls axis scaling and appearance
xlim	x-axis limits
ylim	y-axis limits
zlim	z-axis limits
daspect	Set and get data aspect ratio, i.e., axis equal
pbaspect	Set and get plot box aspect ratio, i.e., axis square
zoom	Zoom in and out

Function	Description
grid	Grid line visibility
box	Axis box visibility
hold	Hold current plot
subplot	Create multiple axes in *Figure* window
figure	Create *Figure* windows
legend	Add legend
title	Title at top of plot
xlabel	*x*-axis label
ylabel	*y*-axis label
text	Place text on plot
gtext	Place text with mouse
ginput	Get coordinates at cursor
area	Filled area plot
bar	Bar graph
barh	Horizontal bar graph
bar3	3-D bar graph
bar3h	3-D horizontal bar graph
compass	Compass graph
errorbar	Linear plot with error bars
ezplot	Easy line plot of string expression
ezpolar	Easy polar plot of string expression
feather	Feather plot
fill	Filled 2-D polygons
fplot	Plot function
hist	Histogram
pareto	Pareto chart
pie	Pie chart
pie3	3-D pie chart
plotmatrix	Scatter plot matrix
ribbon	Linear plot with 2-D lines as ribbons
scatter	Scatter plot
stem	Discrete sequence or stem plot
stairs	Stairstep plot

27

Three-Dimensional Graphics

MATLAB provides a variety of functions to display 3-D data. Some functions plot lines in three dimensions, while others draw surfaces and wire frames. In addition, color can be used to represent a fourth dimension. When color is used in this manner, it is called **pseudocolor**, since color is not an inherent or natural property of the underlying data in the way that color in a photograph is a natural characteristic of the image. To simplify the discussion of 3-D graphics, the use of color is postponed until the next chapter. In this chapter, the fundamental concepts of producing useful 3-D plots are discussed.

27.1 LINE PLOTS

The plot function from the 2-D world is extended into three dimensions with plot3. The format is the same as the 2-D plot, except that the data is supplied in triplets rather than pairs. The general format of the function call is plot3(x1,y1,z1,S1,x2,y2,z2,S2,...), where xn, yn, and zn are vectors or matrices and Sn are optional character strings specifying color, marker symbol, and or linestyle. The function plot3 is commonly used to plot a 3-D function of a single variable, as in the following example:

```
>> t = linspace(0,10*pi);
>> plot3(sin(t),cos(t),t)
>> xlabel('sin(t)'), ylabel('cos(t)'), zlabel('t')
>> text(0,0,0,'Origin')
>> grid on
```

```
>> title('Figure 27.1: Helix')
>> v = axis
v =
      -1      1     -1      1      0     35
```

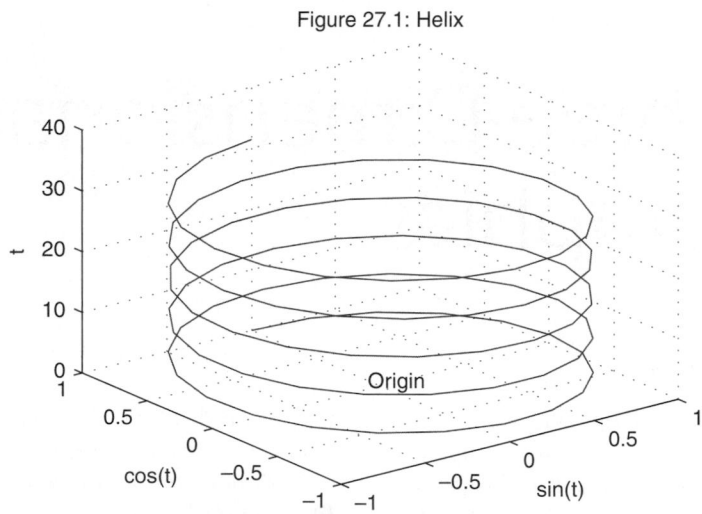

Figure 27.1: Helix

From this simple example, it is apparent that all of the basic features of 2-D graphics exist in 3-D graphics also. The `axis` command extends to 3-D by returning the z-axis limits (0 and 35) as two additional elements in the axis vector. There is a `zlabel` function for labeling the z-axis. The `grid` command toggles a 3-D grid underneath the plot, and the `box` command creates a 3-D box around the plot. The defaults for `plot3` are `grid off` and `box off`. The function `text(x,y,z,'string')` places a character string at the position identified by the triplet `x,y,z`. In addition, subplots and multiple *Figure* windows apply directly to 3-D graphics functions.

   In the last chapter, multiple lines or curves were plotted on top of one another by specifying multiple arguments to the `plot` function or by using the `hold` command. The function `plot3` and the other 3-D graphics functions offer the same capabilities. For example, the added dimension of `plot3` allows multiple 2-D plots to be stacked next to one another along one dimension, rather than directly on top of one another:

```
>> x = linspace(0,3*pi);   % x-axis data
>> z1 = sin(x);            % plot in x-z plane
>> z2 = sin(2*x);
>> z3 = sin(3*x);
>> y1 = zeros(size(x));    % spread out along y-axes
>> y3 = ones(size(x));     % by giving each curve different y-axis values
```

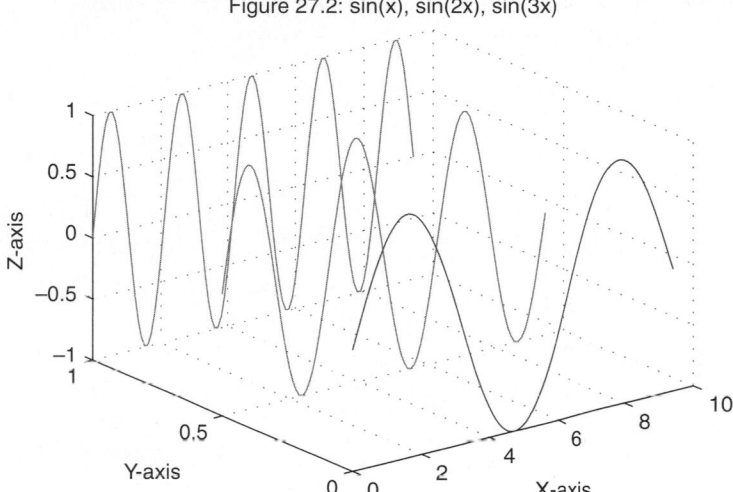

Figure 27.2: sin(x), sin(2x), sin(3x)

```
>> y2 = y3/2;
>> plot3(x,y1,z1,x,y2,z2,x,y3,z3)
>> grid on
>> xlabel('X-axis'), ylabel('Y-axis'), zlabel('Z-axis')
>> title('Figure 27.2: sin(x), sin(2x), sin(3x)')
>> pause(5)
>> plot3(x,z1,y1,x,z2,y2,x,z3,y3)
>> grid on
>> xlabel('X-axis'), ylabel('Y-axis'), zlabel('Z-axis')
>> title('Figure 27.3: sin(x), sin(2x), sin(3x)')
```

27.2 SCALAR FUNCTIONS OF TWO VARIABLES

As opposed to generating line plots with `plot3`, it is often desirable to visualize a scalar function of two variables—that is,

$$z = f(x, y)$$

Here, each pair of values for x and y produces a value for z. A plot of z as a function of x and y is a surface in three dimensions. To plot this surface in MATLAB, the values for z are stored in a matrix. As described in the section on 2-D interpolation, given that x and y are the independent variables, z is a matrix of the dependent variable and the association of x and y with z is

```
z(i,:) = f(x,y(i))     and     z(:,j) = f(x(j),y)
```

Figure 27.3: sin(x), sin(2x), sin(3x)

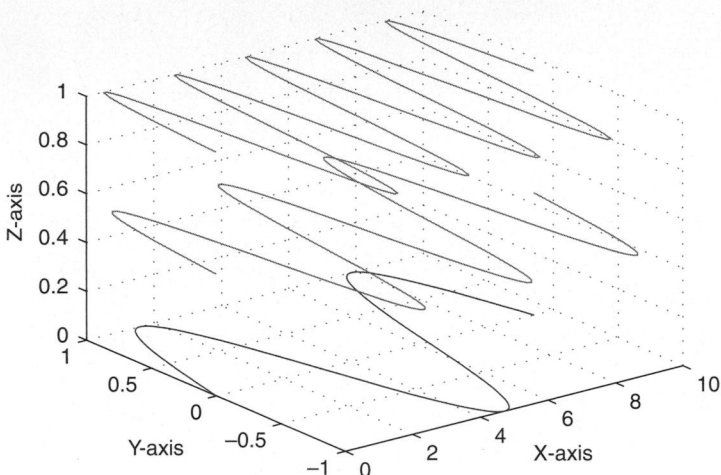

That is, the *i*th row of z is associated with the *i*th element of y, and the *j*th column of z is associated with the *j*th element of x.

When z=f(x,y) can be expressed simply, it is convenient to use array operations to compute all of the values of z in a single statement. To do so requires that we create matrices of all *x*- and *y*-values in the proper orientation. This orientation is sometimes called ***plaid*** by *The Mathworks Inc.* MATLAB provides the function meshgrid to perform this step:

```
>> x = -3:3;   % choose x-axis values
>> y = 1:5;    % y-axis values
>> [X,Y] = meshgrid(x,y)
X =
     -3    -2    -1     0     1     2     3
     -3    -2    -1     0     1     2     3
     -3    -2    -1     0     1     2     3
     -3    -2    -1     0     1     2     3
     -3    -2    -1     0     1     2     3
Y =
      1     1     1     1     1     1     1
      2     2     2     2     2     2     2
      3     3     3     3     3     3     3
      4     4     4     4     4     4     4
      5     5     5     5     5     5     5
```

As you can see, meshgrid duplicated x for each of the five rows in y. Similarly, it duplicated y as a column for each of the seven columns in x.

An easy way to remember which variable is duplicated which way by meshgrid is to think about 2-D plots. The *x*-axis varies from left to right, just as the X output of meshgrid does. Similarly the *y*-axis varies from bottom to top, just as the Y output of meshgrid does.

Given X and Y, if $z = f(x,y) = (x + y)^2$, then the matrix of data values defining the 3-D surface is given simply as

```
>> Z = (X+Y).^2
Z =
```

4	1	0	1	4	9	16
1	0	1	4	9	16	25
0	1	4	9	16	25	36
1	4	9	16	25	36	49
4	9	16	25	36	49	64

When a function cannot be expressed simply, you must use For Loops or While Loops to compute the elements of Z. In many cases, it may be possible to compute the elements of Z rowwise or columnwise. For example, if it is possible to compute Z rowwise, the following script file fragment can be helpful:

```
x= ???  % ostatement defining vector of x axis values
y= ???  % statement defining vector of y axis values

nx = length(x);   % length of x is no. of rows in Z
ny = length(y);   % length of y is no. of columns in Z
Z = zeros(nx,ny); % initialize Z matrix for speed

for r=1:nx
  (preliminary commands)
  Z(r,:)= {a function of y and x(r) defining r-th row of Z}
end
```

On the other hand, if Z can be computed columnwise, the following script file fragment can be helpful:

```
x= ???   % statement defining vector of x axis values
y= ???   % statement defining vector of y axis values


nx = length(x);     % length of x is no. of rows in Z
ny = length(y);     % length of y is no. of columns in Z
Z = zeros(nx,ny);   % initialize Z matrix for speed


for c=1:ny
  (preliminary commands)
  Z(:,c)= {a function of y(c) and x defining c-th column of Z}
end
```

Only when the elements of Z must be computed element by element does the computation usually require a nested For Loop, such as in the following script file fragment:

```
x= ???   % statement defining vector of x axis values
y= ???   % statement defining vector of y axis values
nx = length(x);     % length of x is no. of rows in Z
ny = length(y);     % length of y is no. of columns in Z
Z = zeros(nx,ny);   % initialize Z matrix for speed
for r=1:nx
  for c=1:ny
    (preliminary commands)
    Z(r,c)= {a function of y(c) and x(r) defining (r,c)-th element}
  end
end
```

27.3 MESH PLOTS

MATLAB defines a ***mesh*** surface by the z-coordinates of points above a rectangular grid in the x-y plane. It forms a mesh plot by joining adjacent points with straight lines. The result looks like a fishing net with knots at the data points. The following example is illustrative:

```
>> [X,Y,Z] = peaks(30);
>> mesh(X,Y,Z)
>> xlabel('X-axis'), ylabel('Y-axis'), zlabel('Z-axis')
>> title('Figure 27.4: Mesh Plot of Peaks')
```

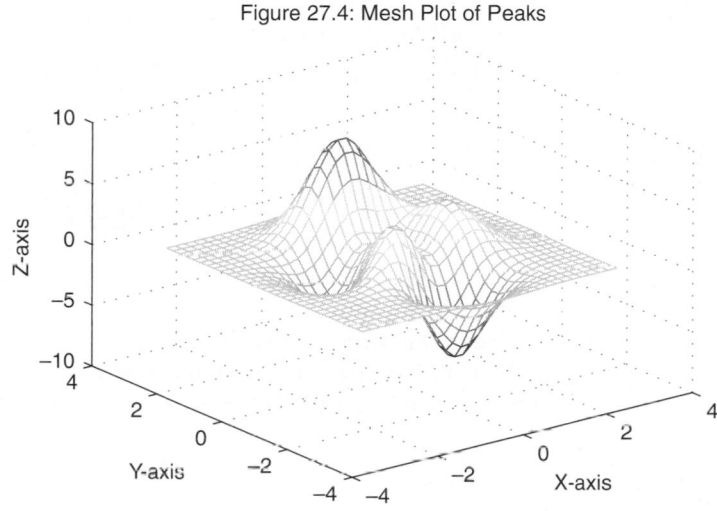

Figure 27.4: Mesh Plot of Peaks

Note on your monitor how the line colors are related to the height of the mesh. In general, mesh accepts optional arguments to control color use in the plot. (This ability to change how MATLAB uses color is discussed in the next chapter.) The use of color is called pseudocolor, since color is used to add a fourth effective dimension to the graph. Note also that the plot was drawn with a grid. ***Most 3-D plots, other than plot3 and a few other exceptions, default to grid on.***

In addition to these input arguments, mesh and most 3-D plot functions can also be called with a variety of input arguments. The syntax used here is the most specific, in that information is supplied for all three axes. The function mesh(Z) plots the matrix z versus its row and column indices. The most common variation is to use the vectors that were passed to meshgrid for the *x*- and *y*- axes—for example, mesh(x,y,Z).

As shown in the previous figure, the areas between the mesh lines are opaque rather than transparent. The MATLAB command hidden controls this aspect of mesh plots:

```
>> [X,Y,Z]=sphere(12);
>> subplot(1,2,1)
>> mesh(X,Y,Z), title('Figure 27.5a: Opaque')
>> hidden on
```

```
>> axis square off
>> subplot(1,2,2)
>> mesh(X,Y,Z), title('Figure 27.5b: Transparent')
>> hidden off
>> axis square off
```

Figure 27.5a: Opaque   Figure 27.5b: Transparent

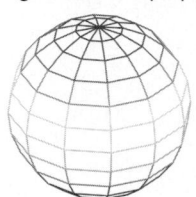 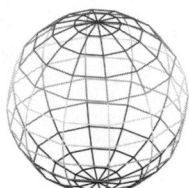

The sphere on the left is opaque (the lines are hidden), whereas the one on the right is transparent (the lines are not hidden).

    The MATLAB mesh function has two siblings: meshc, which is a mesh plot and underlying contour plot, and meshz, which is a mesh plot that includes a zero plane. The following example uses both of these forms:

```
>> [X,Y,Z] = peaks(30);
>> meshc(X,Y,Z)  % mesh plot with underlying contour plot
>> title('Figure 27.6: Mesh Plot with Contours')
>> pause(5)
>> meshz(X,Y,Z)  % mesh plot with zero plane
>> title('Figure 27.7: Mesh Plot with Zero Plane')
```

Figure 27.6: Mesh Plot with Contours

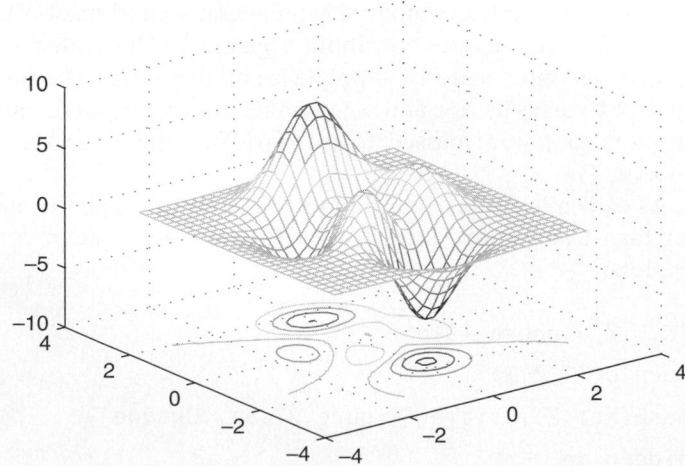

Figure 27.7: Mesh Plot with Zero Plane

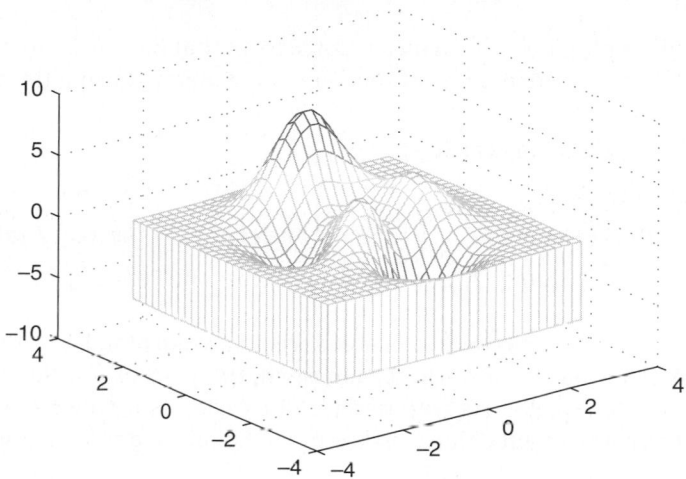

The function `waterfall` is identical to `mesh`, except that the mesh lines appear only in the *x*-direction:

```
>> waterfall(X,Y,Z)
>> xlabel('X-axis'), ylabel('Y-axis'), zlabel('Z-axis')
>> title('Figure 27.8: Waterfall Plot')
```

Figure 27.8: Waterfall Plot

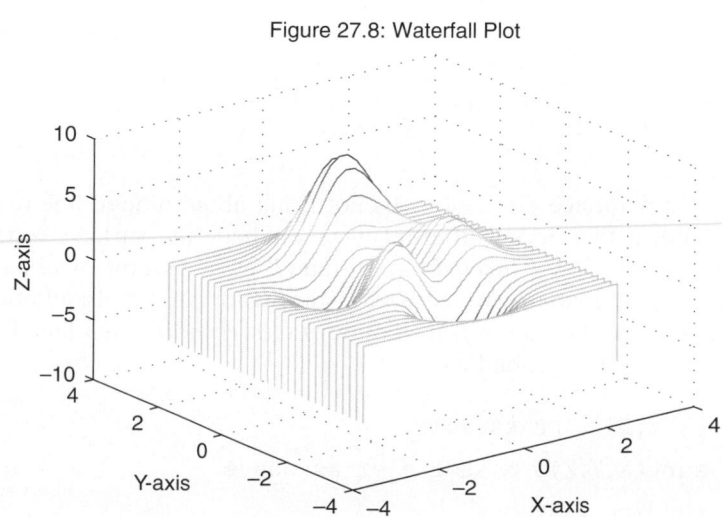

27.4 SURFACE PLOTS

A *surface* plot looks like a mesh plot, except that the spaces between the lines, called *patches*, are filled in. Plots of this type are generated using the surf function:

```
>> [X,Y,Z] = peaks(30);
>> surf(X,Y,Z)
>> xlabel('X-axis'), ylabel('Y-axis'), zlabel('Z-axis')
>> title('Figure 27.9: Surface Plot of Peaks')
```

Note how this plot type is a *dual* of sorts to a mesh plot. Here, the lines are black and the patches have color, whereas in mesh, the patches are the color of the axes and the lines have color. As with mesh, color varies along the *z*-axis, with each patch or line having constant color. Surface plots default to grid on also.

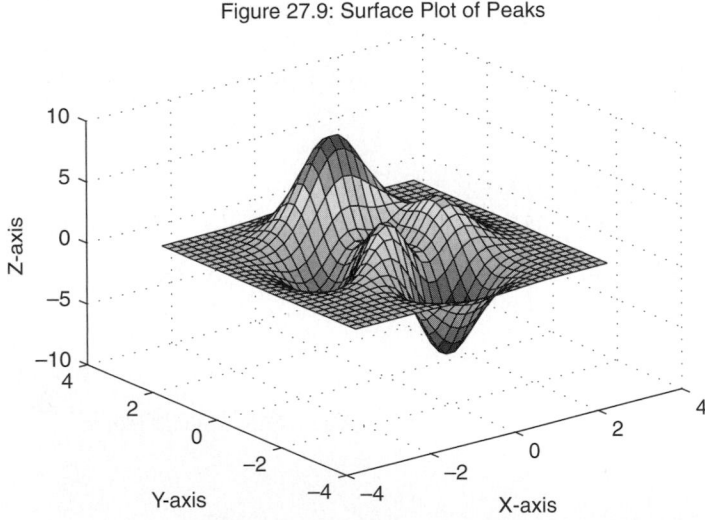

Figure 27.9: Surface Plot of Peaks

In a surface plot, one does not think about hidden line removal as in a mesh plot, but rather about different ways to shade the surface. In the preceding surf plot, the shading is *faceted* like a stained-glass window or object, where the black lines are the joints between the constant-color patches. In addition to faceted shading, MATLAB provides *flat* shading and *interpolated* shading. These are applied by using the function shading:

```
>> [X,Y,Z] = peaks(30);
>> surf(X,Y,Z)  % same plot as above
>> shading flat
```

Figure 27.10: Surface Plot with Flat Shading

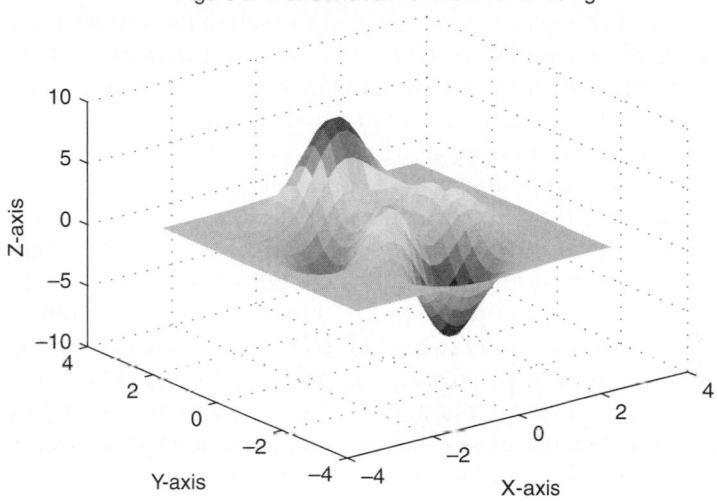

```
>> xlabel('X-axis'), ylabel('Y-axis'), zlabel('Z-axis')
>> title('Figure 27.10: Surface Plot with Flat Shading')
>> pause(5)
>> shading interp
>> title('Figure 27.11: Surface Plot with Interpolated Shading')
```

In flat shading, the black lines are removed and each patch retains its single color, whereas in interpolated shading, the lines are also removed but each patch is given

Figure 27.11: Surface Plot with Interpolated Shading

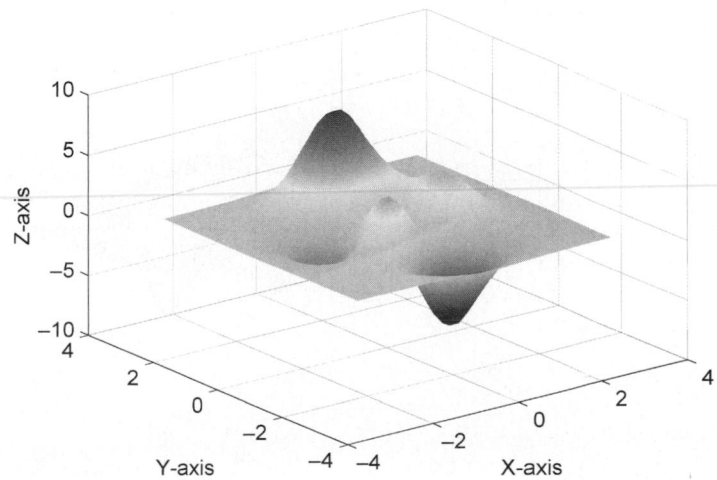

interpolated shading. That is, the color of each patch is interpolated over its area on the basis of the color values assigned to each of its vertices. Needless to say, interpolated shading requires much more computation than faceted and flat shading. While shading has a significant visual impact on surf plots, it also applies to mesh plots; although in this case the visual impact is relatively minor, since only the lines have color. Shading also affects pcolor and fill plots.

On some computer systems, interpolated shading creates extremely long printing delays or, at worst, printing errors. These problems are not due to the size of the PostScript data file, but rather to the enormous amount of computation required in the printer to generate shading that continually changes over the surface of the plot. Often, the easiest solution to this problem is to use flat shading for printouts.

In some situations, it may be convenient to remove part of a surface so that underlying parts of the surface can be seen. In MATLAB, this is accomplished by setting the data values where holes are desired to the special value NaN. *Since NaNs have no value, all MATLAB plotting functions simply ignore NaN data points, leaving a hole in the plot where they appear.*

```
>> [X,Y,Z] = peaks(30);
>> x = X(1,:);              % vector of x axis
>> y = Y(:,1);              % vector of y axis
>> i = find(y>.8 & y<1.2);  % find y axis indices of hole
>> j = find(x>-.6 & x<.5);  % find x axis indices of hole
>> Z(i,j) = nan;            % set values at hole indices to NaNs
>> surf(X,Y,Z)
>> xlabel('X-axis'), ylabel('Y-axis'), zlabel('Z-axis')
>> title('Figure 27.12: Surface Plot with a Hole')
```

Figure 27.12: Surface Plot with a Hole

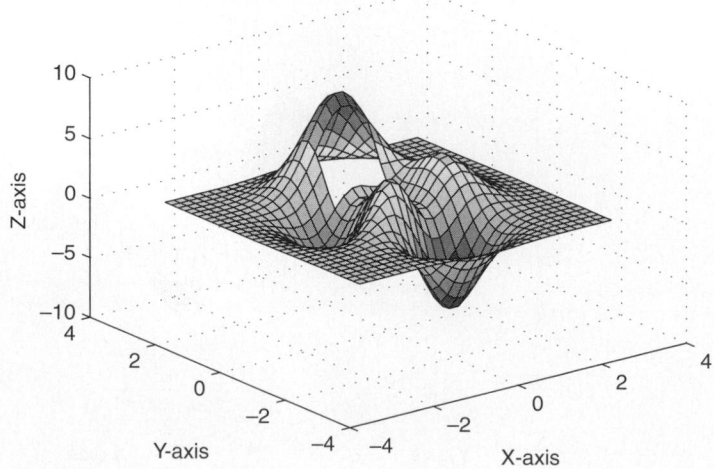

The MATLAB surf function also has two siblings: surfc, which is a surface plot and underlying contour plot, and surfl, which is a surface plot with lighting. Both are used in the following code:

```
>> [X,Y,Z] = peaks(30);
>> surfc(X,Y,Z)  % surf plot with contour plot
>> xlabel('X-axis'), ylabel('Y-axis'), zlabel('Z-axis')
>> title('Figure 27.13: Surface Plot with Contours')
>> pause(5)
>> surfl(X,Y,Z)     % surf plot with lighting
>> shading interp  % surfl plots look best with interp shading
>> colormap pink    % they also look better with shades of a single color
>> xlabel('X-axis'), ylabel('Y-axis'), zlabel('Z-axis')
>> title('Figure 27.14: Surface Plot with Lighting')
```

The function surfl makes a number of assumptions regarding the light applied to the surface. It does not use the *light* object. Rather, it simply modifies the color of the surface to give the appearance of lighting. (The next chapter provides more rigorous information regarding light properties.) Also, in the preceding commands, colormap is a MATLAB function for applying a different set of colors to a figure. (This function is discussed in the next chapter as well.)

The surfnorm(X,Y,Z) function computes surface normals for the surface defined by X, Y, and Z; plots the surface; and plots vectors normal to the surface at the data points:

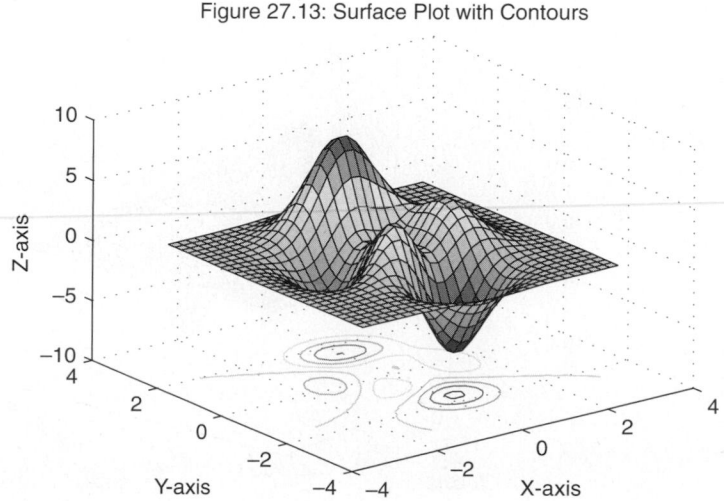

Figure 27.13: Surface Plot with Contours

Figure 27.14: Surface Plot with Lighting

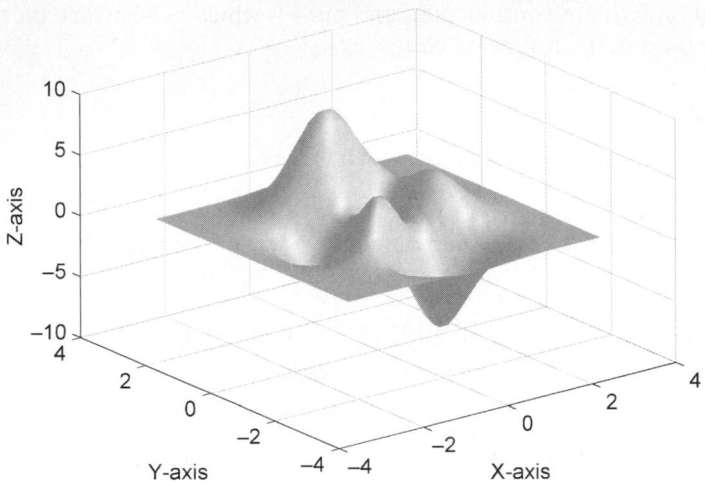

```
>> [X,Y,Z] = peaks(15);
>> surfnorm(X,Y,Z)
>> xlabel('X-axis'), ylabel('Y-axis'), zlabel('Z-axis')
>> title('Figure 27.15: Surface Plot with Normals')
```

Note that the surface normals are unnormalized and valid at each vertex. The form `[Nx,Ny,Nz]=surfnorm(X,Y,Z)` computes the 3-D surface normals and returns their components, but does not plot the surface.

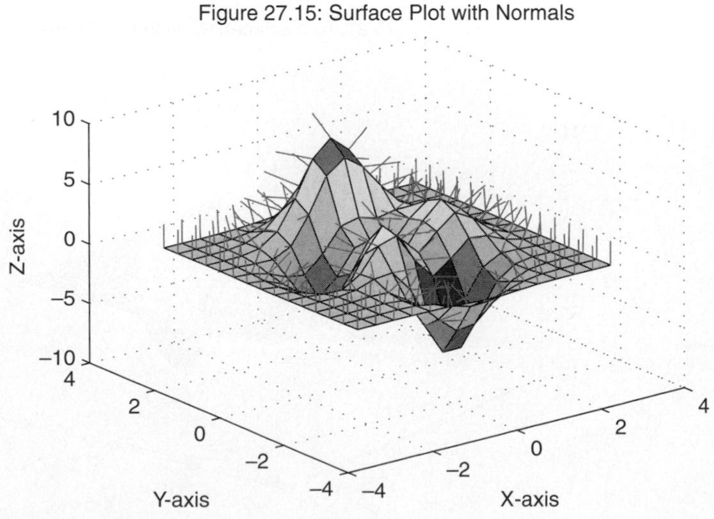

Figure 27.15: Surface Plot with Normals

27.5 MESH AND SURFACE PLOTS OF IRREGULAR DATA

Irregular or nonuniformly spaced data can be visualized using the functions trimesh, trisurf, and voronoi:

```
>> x = rand(1,50);
>> y = rand(1,50);
>> z = peaks(x,y*pi);
>> t = delaunay(x,y);
>> trimesh(t,x,y,z)
>> hidden off
>> title('Figure 27.16: Triangular Mesh Plot')
>> pause(5)
>> trisurf(t,x,y,z)
>> title('Figure 27.17: Triangular Surface Plot')
>> pause(5)
>> voronoi(x,y)
>> title('Figure 27.18: Voronoi Plot')
```

(See Chapter 19 for more information on Delaunay triangulation and Voronoi diagrams.)

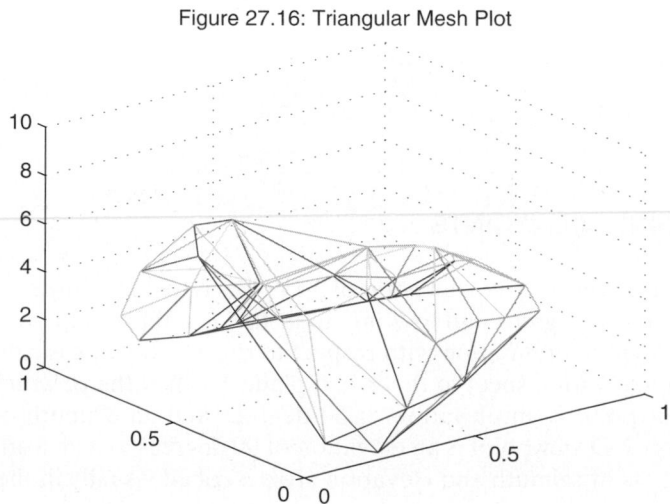

Figure 27.16: Triangular Mesh Plot

Figure 27.17: Triangular Surface Plot

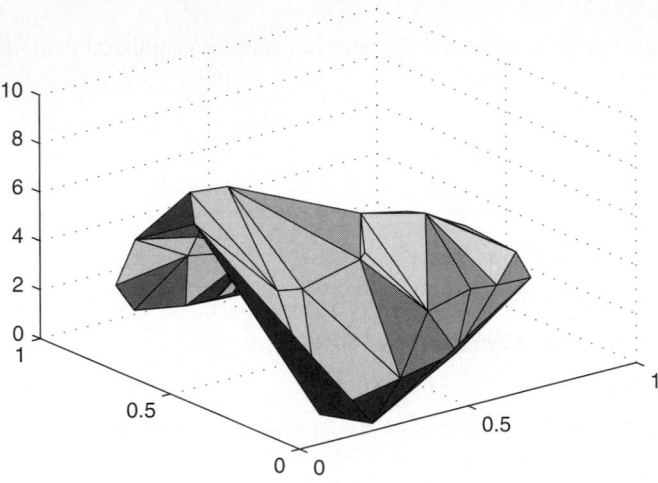

Figure 27.18: Voronoi Plot

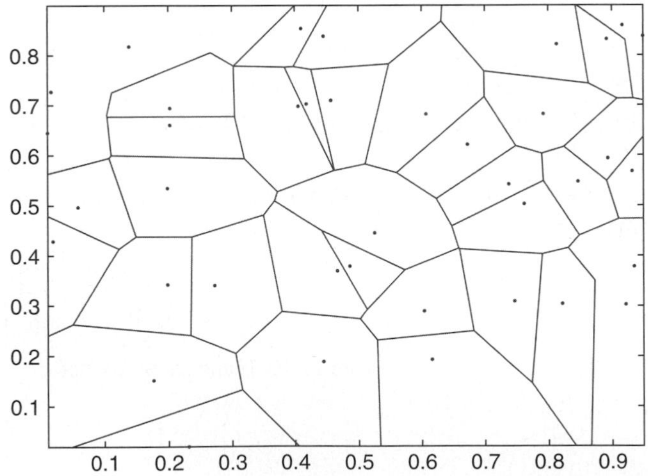

27.6 CHANGING VIEWPOINTS

Note that the default viewpoint of 3-D plots is looking down at the $z = 0$ plane at an angle of 30 degrees and looking up at the $x = 0$ plane at an angle of 37.5 degrees. The angle of orientation with respect to the $z = 0$ plane is called the ***elevation***, and the angle with respect to the $x = 0$ plane is called the ***azimuth***. Thus, the default 3-D viewpoint is an elevation of 30 degrees and an azimuth of -37.5 degrees. The default 2-D viewpoint is an elevation of 90 degrees and an azimuth of 0 degrees. The concepts of azimuth and elevation are described visually in the following figure.

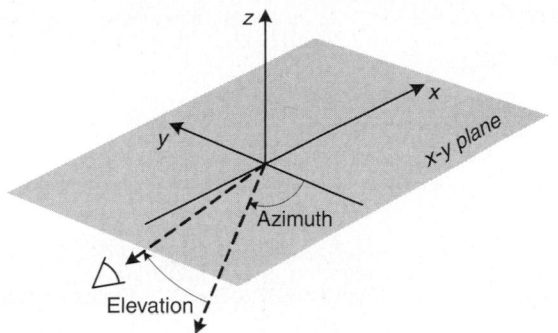

In MATLAB, the function view changes the graphical viewpoint for all types of 2-D and 3-D plots. The forms view(az,el) and view([az,el]) change the viewpoint to the specified azimuth az and elevation el:

```
>> x = -7.5:.5:7.5; y = x;          % create a data set
>> [X,Y] = meshgrid(x,y);
>> R = sqrt(X.^2+Y.^2)+eps;
>> Z = sin(R)./R;
>> subplot(2,2,1)
>> surf(X,Y,Z)
>> view(-37.5,30)
>> xlabel('X-axis'), ylabel('Y-axis'), zlabel('Z-axis')
>> title('Figure 27.19a: Default Az = -37.5, El = 30')

>> subplot(2,2,2)
>> surf(X,Y,Z)
>> view(-37.5+90,30)
>> xlabel('X-axis'), ylabel('Y-axis'), zlabel('Z-axis')
>> title('Figure 27.19b: Az Rotated to 52.5')

>> subplot(2,2,3)
>> surf(X,Y,Z)
>> view(-37.5,60)
>> xlabel('X-axis'), ylabel('Y-axis'), zlabel('Z-axis')
>> title('Figure 27.19c: El Increased to 60')
```

```
>> subplot(2,2,4)
>> surf(X,Y,Z)
>> view(0,90)
>> xlabel('X-axis'), ylabel('Y-axis')
>> title('Figure 27.19d: Az = 0, El = 90')
```

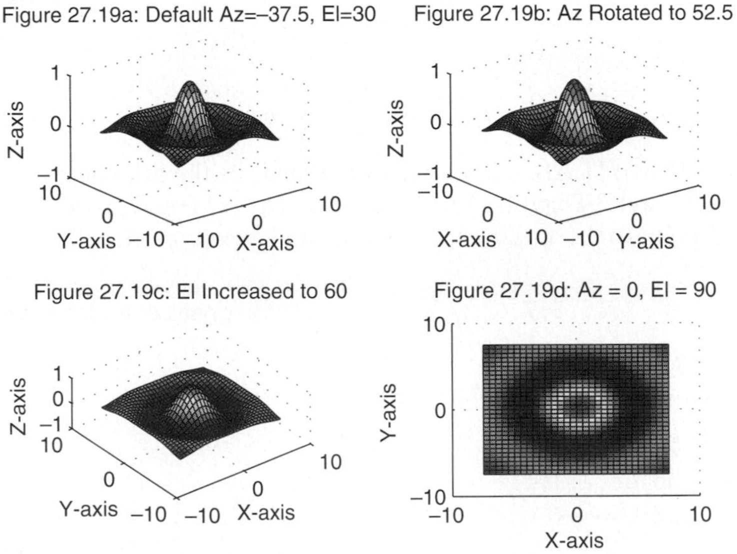

Figure 27.19a: Default Az=–37.5, El=30   Figure 27.19b: Az Rotated to 52.5

Figure 27.19c: El Increased to 60   Figure 27.19d: Az = 0, El = 90

In addition to these forms, view offers additional features that are summarized in its online documentation and in its help text:

```
>> help view
 VIEW    3-D graph viewpoint specification.
   VIEW(AZ,EL) and VIEW([AZ,EL]) set the angle of the view from which an
   observer sees the current 3-D plot.  AZ is the azimuth or horizontal
   rotation and EL is the vertical elevation (both in degrees). Azimuth
   revolves about the z-axis, with positive values indicating counter-
   clockwise rotation of the viewpoint. Positive values of elevation
   correspond to moving above the object; negative values move below.
   VIEW([X Y Z]) sets the view angle in cartesian coordinates. The
   magnitude of vector X,Y,Z is ignored.

Here are some examples:
```

AZ = -37.5, EL = 30 is the default 3-D view.

AZ = 0, EL = 90 is directly overhead and the default 2-D view.

AZ = EL = 0 looks directly up the first column of the matrix.

AZ = 180 is behind the matrix.

VIEW(2) sets the default 2-D view, AZ = 0, EL = 90.

VIEW(3) sets the default 3-D view, AZ = -37.5, EL = 30.

[AZ,EL] = VIEW returns the current azimuth and elevation.

VIEW(T) accepts a 4-by-4 transformation matrix, such as

the perspective transformations generated by VIEWMTX.

T = VIEW returns the current general 4-by-4 transformation matrix.

See also VIEWMTX, the AXES properties View, Xform.

In addition to the view function, the viewpoint can be set interactively with the mouse by using the function rotate3d. The function rotate3d on turns on mouse-based view rotation, rotate3d off turns it off, and rotate3d with no arguments toggles the state. This functionality is also available on the **Tools** menu in a *Figure* window, as well as on a button on the *Figure* toolbar. In reality, the **Rotate 3D** menu item and toolbar button both call rotate3d to do the requested work.

27.7 CAMERA CONTROL

The viewpoint control provided by the view function is convenient, but limited in capabilities. To provide complete control of a 3-D scene, camera capabilities are needed. That is, one must have all of the capabilities available when filming a movie with a camera. Alternatively, one must have all of the capabilities available in a 3-D computer or console game environment. In this environment, there are two 3-D coordinate systems to manage: one at the camera and one at what the camera is pointed at (i.e., the camera target). The camera functions in MATLAB manage and manipulate the relationships between these two coordinate systems and provide control over the camera lens.

Use of the camera functions in MATLAB is generally not easy for a novice. Therefore, to simplify the use of these functions, most are made available interactively from the **Tools** menu or from the *Camera* toolbar in a *Figure* window. To view the *Camera* toolbar, choose it from the **View** menu in the *Figure* window. By using the interactive camera tools, you avoid dealing with the input and output arguments that the *Command* window functions require. Given the ease with which the interactive tools can be used, the complexity involved in using and describing the camera functions, and the relatively small number of potential users, camera functions are not described in this text. The MATLAB documentation contains a rigorous discussion of these functions and their use. The camera functions available in MATLAB are listed at the end of this chapter.

27.8 CONTOUR PLOTS

Contour plots show lines of constant elevation or height. If you've ever seen a topographical map, you know what a contour plot looks like. In MATLAB, contour plots in 2-D and 3-D are generated by using the contour and contour3 functions, respectively:

```
>> [X,Y,Z] = peaks;
>> subplot(1,2,1)
>> contour(X,Y,Z,20)           % generate 20 2-D contour lines
>> axis square
>> xlabel('X-axis'), ylabel('Y-axis')
>> title('Figure 27.20a: 2-D Contour Plot')
>> subplot(1,2,2)
>> contour3(X,Y,Z,20)          % the same contour plot in 3-D
>> xlabel('X-axis'), ylabel('Y-axis'), zlabel('Z-axis')
>> title('Figure 27.20b: 3-D Contour Plot')
```

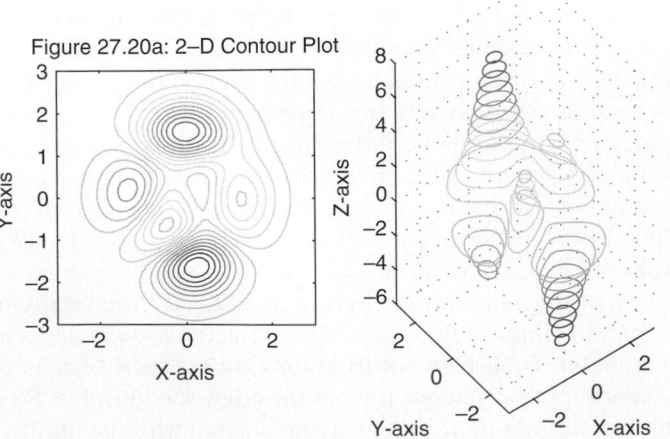

Figure 27.20b: 3–D Contour Plot

Figure 27.20a: 2–D Contour Plot

The pcolor function maps height to a set of colors and presents the same information as the contour plot, at the same scale:

```
>> subplot(1,2,1)
>> pcolor(X,Y,Z)
>> shading interp  % remove the grid lines
>> axis square
>> title('Figure 27.21a: Pseudocolor Plot')
```

Combining the idea of a pseudocolor plot with a 2-D contour produces a filled contour plot. In MATLAB, this plot is generated by the function contourf:

```
>> subplot(1,2,2)
>> contourf(X,Y,Z,12)          % filled contour plot with 12 contours
>> axis square
>> xlabel('X-axis'), ylabel('Y-axis')
>> title('Figure 27.21b: Filled Contour Plot')
```

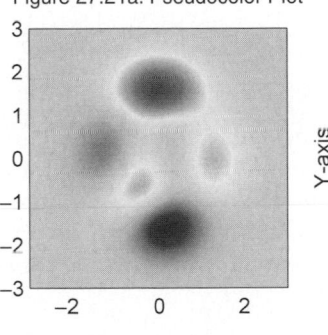 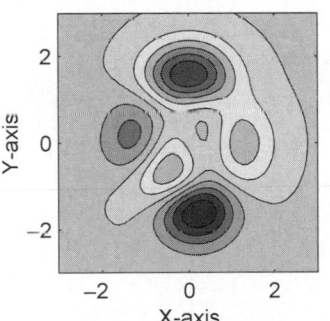

Contour lines can be labeled by using the clabel function, which requires a matrix of lines and optional text strings that are returned by contour, contourf, and contour3:

```
>> C = contour(X,Y,Z,12);
>> clabel(C)
>> title('Figure 27.22: Contour Plot With Labels')
```

Alternatively, inline labels are generated by using two arguments returned by the contour functions. For example, modifying the preceding MATLAB code to [C,h]=contour(X,Y,Z,12); clabel(C,h) produces inline labels that follow the

Figure 27.22: Contour Plot With Labels

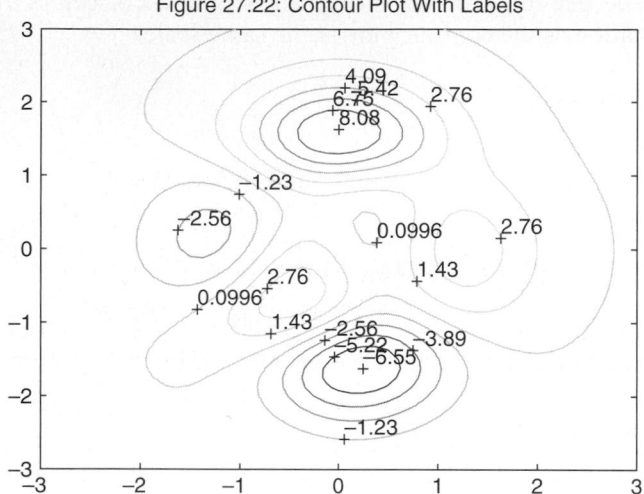

contours. Finally, by using the mouse, you can select which contours to label by providing `'manual'` as the last input argument to `clabel`. Inline or horizontal labels
are used, depending on the presence of the second argument `h`, as illustrated
previously.

27.9 SPECIALIZED 3-D PLOTS

MATLAB provides a number of specialized plotting functions in addition to those
already discussed. The function `ribbon(Y)` plots the columns of Y as separate ribbons. The function `ribbon(x,Y)` plots x versus the columns of Y. The width of the
ribbons can also be specified by using the syntax `ribbon(x,Y,width)`, where the
default width is 0.75. The following code uses the first of these forms:

```
>> Z = peaks;
>> ribbon(Z)
>> title('Figure 27.23: Ribbon Plot of Peaks')
```

The function `quiver(x,y,dx,dy)` draws directional or velocity vectors
(`dx,dy`) at the points (`x,y`):

```
>> [X,Y,Z] = peaks(16);
>> [DX,DY] = gradient(Z,.5,.5);
>> contour(X,Y,Z,10)
>> hold on
>> quiver(X,Y,DX,DY)
```

Figure 27.23: Ribbon Plot of Peaks

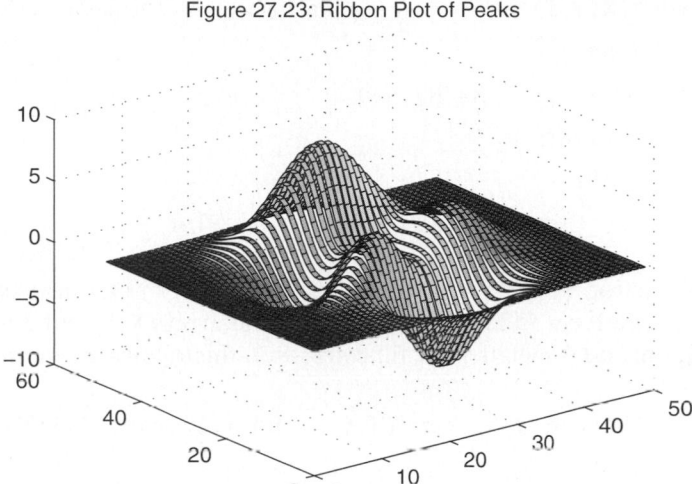

```
>> hold off
>> title('Figure 27.24: 2-D Quiver Plot')
```

Three-dimensional quiver plots of the form quiver3(x,y,z,Nx,Ny,Nz) display the vectors (Nx,Ny,Nz) at the points (x,y,z):

```
>> [X,Y,Z] = peaks(20);
>> [Nx,Ny,Nz] = surfnorm(X,Y,Z);
```

Figure 27.24: 2–D Quiver Plot

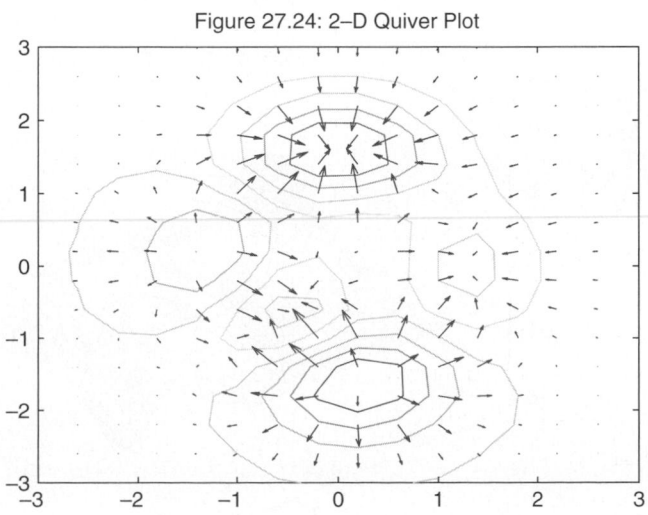

```
>> surf(X,Y,Z)
>> hold on
>> quiver3(X,Y,Z,Nx,Ny,Nz)
>> axis tight
>> hold off
>> title('Figure 27.25: 3-D Quiver Plot')
```

The function `fill3`, being the 3-D equivalent of `fill`, draws filled polygons in 3-D space. The form `fill3(X,Y,Z,C)` uses the arrays X, Y, and Z as the vertices of the polygon; and C specifies the fill color, as in the following example:

```
>> fill3(rand(3,5),rand(3,5),rand(3,5),rand(3,5))
>> grid on
>> title('Figure 27.26: Five Random Filled Triangles')
```

The 3-D equivalent of `stem` plots discrete sequence data in 3-D space. The form `stem3(X,Y,Z,C,'filled')` plots the data points in (X,Y,Z) with lines extending to the *x–y* plane. The optional argument C specifies the marker style or color, and the optional `'filled'` argument causes the marker to be filled in. The `stem3(Z)` function plots the points in Z and automatically generates X and Y values. The following code uses `stem3` with three arguments:

```
>> Z = rand(5);
>> stem3(Z,'ro','filled');
>> grid on
>> title('Figure 27.27: Stem Plot of Random Data')
```

Figure 27.25: 3–D Quiver Plot

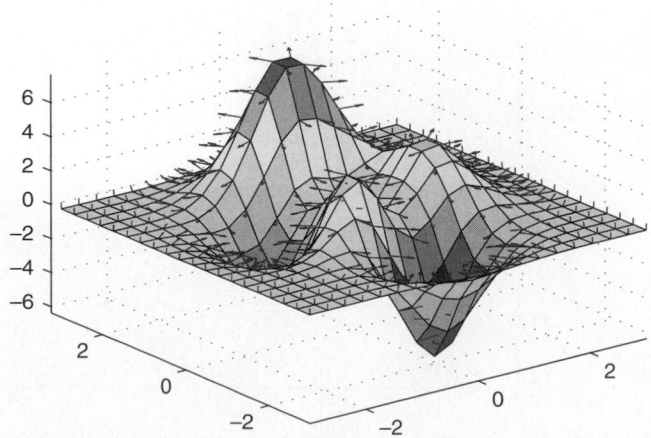

Figure 27.26: Five Random Filled Triangles

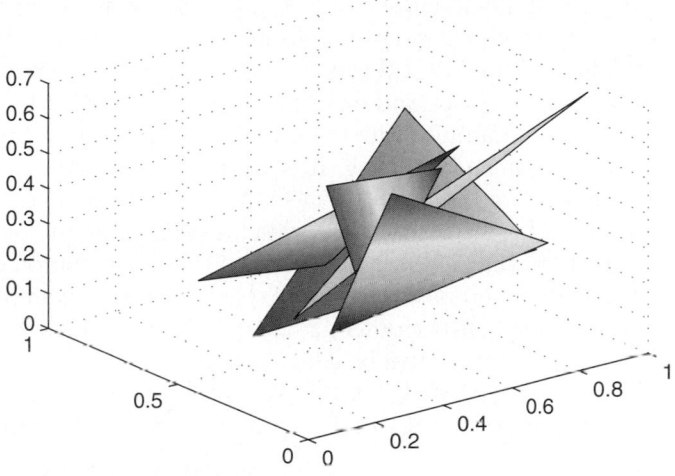

Figure 27.27: Stem Plot of Random Data

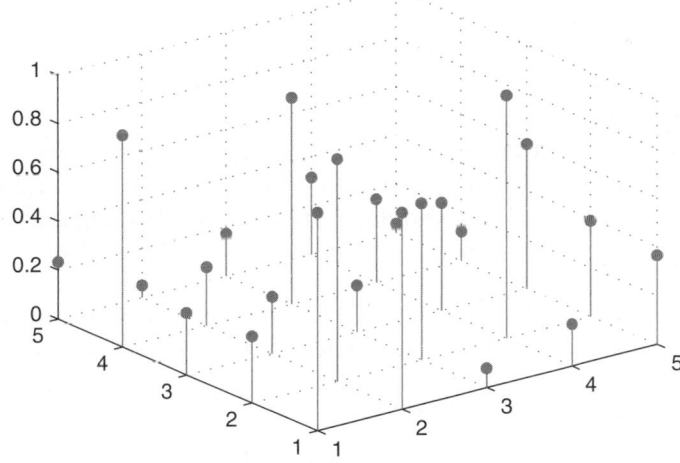

27.10 VOLUME VISUALIZATION

In addition to the common mesh, surface, and contour plots, MATLAB offers a variety of more complex volume and vector visualization functions. These functions construct plots of scalar and vector quantities in 3-D space. Because they generally construct volumes rather than surfaces, their input arguments are 3-D arrays, one for each axis x, y, and z. The points in each 3-D array define a grid of coordinates or data at the coordinates. For scalar functions, four 3-D arrays are required, one for each of the three coordinate axes, plus one for the scalar data at the coordinate points. These arrays are commonly identified as X, Y, Z, and V, respectively. For vector functions, six 3-D arrays are required, one for each of the three coordinate axes, plus one for each axis component of the vector at the coordinate points. These arrays are commonly identified as X, Y, Z, U, V, and W, respectively.

Use of the volume and vector visualization functions in MATLAB requires an understanding of volume and vector terminology. For example, *divergence* and *curl* describe vector processes, and *isosurfaces* and *isocaps* describe visual aspects of volumes. If you are unfamiliar with these terms, using MATLAB volume and visualization functions can be confusing. It is beyond the scope of this text to cover the terminology required to rigorously use these volume and vector visualization functions. However, the structure of the data arrays and the use of several functions are demonstrated in what follows. The MATLAB documentation contains a more thorough introduction and more rigorous explanations and examples.

Consider the construction of a scalar function defined over a volume. First, the volume coordinate axes must be constructed:

```
>> x = linspace(-3,3,13);      % x coordinate points
>> y = 1:20;                    % y coordinate points
>> z = -5:5;                    % z coordinate points
>> [X,Y,Z] = meshgrid(x,y,z); % meshgrid works here too!
>> size(X)
ans =
        20    13    11
```

Here, X, Y, Z are 3-D arrays defining the grid. The X array contains x duplicated for as many rows as length(y) and as many pages as length(z). Similarly, Y contains y transposed to a column and duplicated for as many columns as length(x) and as many pages as length(z). And Z contains z permuted to a 1-by-1-by-length(z) vector and duplicated for as many rows as length(y) and as many columns as length(x). As described, this is a direct extension of meshgrid to 3-D.

Next, we need to define a function of these data, such as

```
>> V = sqrt(X.^2 + cos(Y).^2 + Z.^2);
```

Now, the 3-D arrays X, Y, Z, and V define a scalar function $v = f(x,y,z)$ defined over a volume. To visualize what this looks like, we can look at slices along planes:

```
>> slice(X,Y,Z,V,[0 3],[5 15],[-3 5])
>> xlabel('X-axis')
>> ylabel('Y-axis')
>> zlabel('Z-axis')
>> title('Figure 27.28: Slice Plot Through a Volume')
```

This plot shows slices on planes defined by $x = 0$, $x = 3$, $y = 5$, $y = 15$, $z = -3$, and $z = 5$, as shown by the last three arguments to the function slice. The color of the plot is mapped to the values in V on the slices.

Figure 27.28: Slice Plot Through a Volume

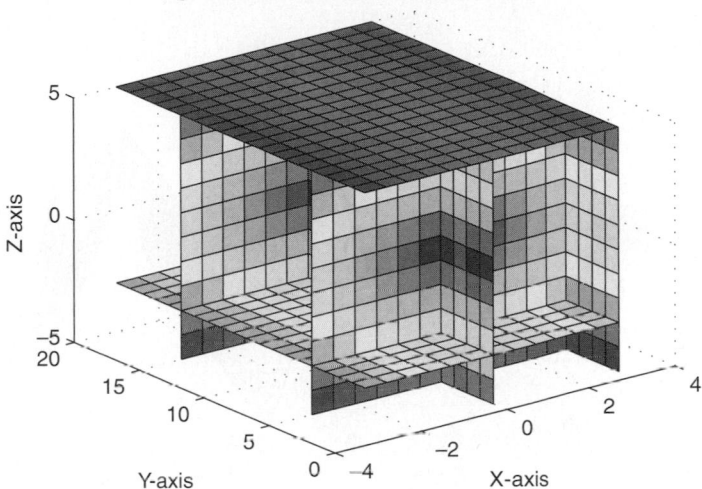

The slices displayed need not be planes. They can be any surface, as generated by the following code:

```
>> [xs,ys] = meshgrid(x,y);
>> zs = sin(-xs+ys/2); % a surface to use
>> slice(X,Y,Z,V,xs,ys,zs)
>> xlabel('X-axis')
>> ylabel('Y-axis')
>> zlabel('Z-axis')
>> title('Figure 27.29: Slice Plot Using a Surface')
```

Here, xs, ys, and zs define a surface to slice through the volume.

Going back to the original slice plot, it is possible to add contour lines to selected planes by using the contourslice function:

```
>> slice(X,Y,Z,V,[0 3],[5 15],[-3 5])
>> hold on
>> h = contourslice(X,Y,Z,V,3,[5 15],[]);
>> set(h,'EdgeColor','k','Linewidth',1.5)
>> xlabel('X-axis')
>> ylabel('Y-axis')
>> zlabel('Z-axis')
>> title('Figure 27.30: Slice Plot with Selected Contours')
>> hold off
```

Figure 27.29: Slice Plot Using a Surface

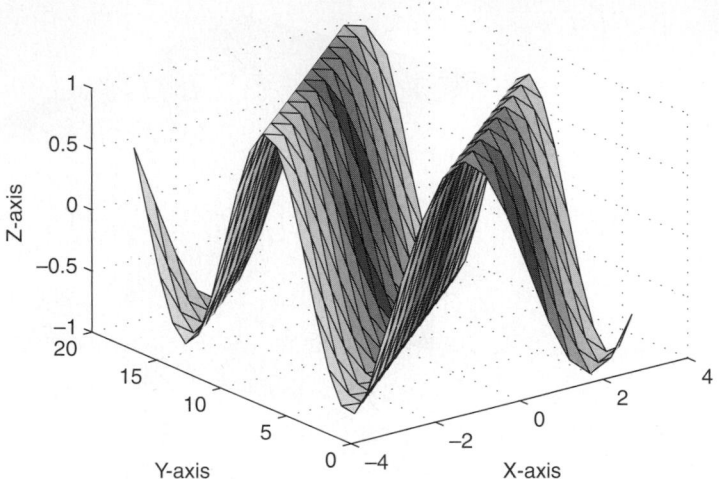

Here, contour lines are added to the $x = 3$, $y = 5$, and $y = 15$ planes. With the use of Handle Graphics features, the contour lines are set to black and their width is set to 1.5 points.

In addition to looking at slices through a volume, surfaces where the scalar volume data V has a specified value can be plotted by using the `isosurface` function. This function returns triangle vertices in a manner similar to Delaunay triangulation, the results being in the form required by the `patch` function, which plots the triangles:

Figure 27.30: Slice Plot with Selected Contours

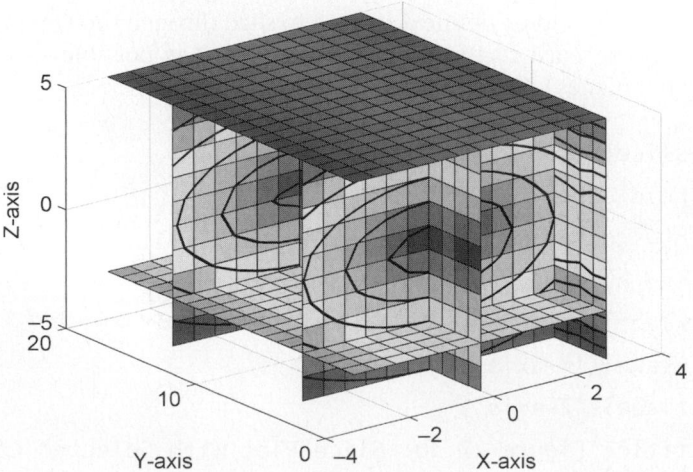

```
>> [X,Y,Z,V] = flow(13);              % get flow data
>> fv = isosurface(X,Y,Z,V,-2);       % find surface of value -2

>> subplot(1,2,1)
>> p = patch(fv);                     % plot V = -2 surface
>> set(p,'FaceColor',[.5 .5 .5],'EdgeColor','Black'); % modify patches
>> view(3), axis equal tight, grid on % pretty it up
>> title({'Figure 27.31a:' 'Isosurface Plot, V = 2'})

>> subplot(1,2,2)
>> p = patch(shrinkfaces(fv,.3));        % shrink faces to 30% of original
>> set(p,'Facecolor',[.5 .5 .5],'EdgeColor','Black'); % modify patches
>> view(3), axis equal tight, grid on % pretty it up
>> title({'Figure 27.31b:' 'Shrunken Face Isosurface Plot, V = 2'})
```

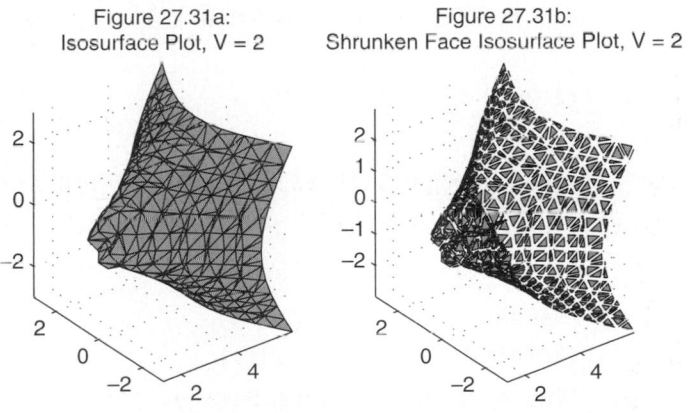

Figure 27.31a:
Isosurface Plot, V = 2

Figure 27.31b:
Shrunken Face Isosurface Plot, V = 2

The preceding plots also demonstrate the use of the function shrinkfaces, which does exactly what its name suggests.

Sometimes volume data contain too many points for efficient display. The functions reducevolume and reducepatch represent two ways to improve the display of an isosurface. The function reducevolume eliminates data before the isosurface is formed, whereas reducepatch seeks to eliminate patches while minimizing distortion in the underlying surface. The following code is illustrative:

```
>> [X,Y,Z,V] = flow;
>> fv = isosurface(X,Y,Z,V,-2);
>> subplot(2,2,1) % Original
```

```
>> p = patch(fv);
>> Np = size(get(p,'Faces'),1);
>> set(p,'FaceColor',[.5 .5 .5],'EdgeColor','Black');
>> view(3), axis equal tight, grid on  % pretty it up
>> zlabel(sprintf('%d Patches',Np))
>> title('Figure 27.32a: Original')

>> subplot(2,2,2) % Reduce Volume
>> [Xr,Yr,Zr,Vr] = reducevolume(X,Y,Z,V,[3 2 2]);
>> fvr = isosurface(Xr,Yr,Zr,Vr,-2);
>> p = patch(fvr);
>> Np = size(get(p,'Faces'),1);
>> set(p,'FaceColor',[.5 .5 .5],'EdgeColor','Black');
>> view(3), axis equal tight, grid on  % pretty it up
>> zlabel(sprintf('%d Patches',Np))
>> title('Figure 27.32b: Reduce Volume')

>> subplot(2,2,3) % Reduce Patch
>> p = patch(fv);
>> set(p,'FaceColor',[.5 .5 .5],'EdgeColor','Black');
>> view(3), axis equal tight, grid on  % pretty it up
>> reducepatch(p,.15) % keep 15 percent of the faces
>> Np = size(get(p,'Faces'),1);
>> zlabel(sprintf('%d Patches',Np))
>> title('Figure 27.32c: Reduce Patches')

>> subplot(2,2,4) % Reduce Volume and Patch
>> p = patch(fvr);
>> set(p,'FaceColor',[.5 .5 .5],'EdgeColor','Black');
>> view(3), axis equal tight, grid on  % pretty it up
>> reducepatch(p,.15) % keep 15 percent of the faces
>> Np = size(get(p,'Faces'),1);
>> zlabel(sprintf('%d Patches',Np))
>> title('Figure 27.32d: Reduce Both')
```

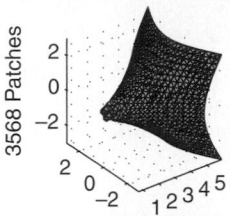

Figure 27.32a: Original

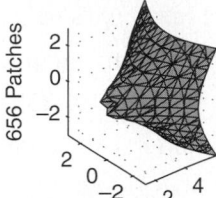

Figure 27.32b: Reduce Volume

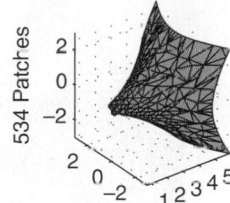

Figure 27.32c: Reduce Patches

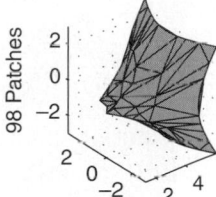

Figure 27.32d: Reduce Both

Three-dimensional data can also be smoothed by filtering it with the smooth3 function, as in the following code:

```
>> data = rand(10,10,10);          % random data
>> datas = smooth3(data,'box',3); % smoothed data

>> subplot(1,2,1) % random data
>> p = patch(isosurface(data,.5), ...
     'FaceColor','Blue','EdgeColor','none');
>> patch(isocaps(data,.5), ...
     'FaceColor', 'interp', 'EdgeColor', 'none');
>> isonormals(data,p)
>> view(3); axis vis3d tight off
>> camlight; lighting phong
>> title({'Figure 27.33a:' 'Random Data'})

>> subplot(1,2,2) % smoothed random data
>> p = patch(isosurface(datas,.5), ...
     'FaceColor','Blue','EdgeColor','none');
>> patch(isocaps(datas,.5), ...
     'FaceColor', 'interp', 'EdgeColor', 'none');
```

```
>> isonormals(datas,p)
>> view(3); axis vis3d tight off
>> camlight; lighting phong
>> title({'Figure 27.33b:' 'Smoothed Data'})
```

Figure 27.33a:
Random Data

Figure 27.33b:
Smoothed Data

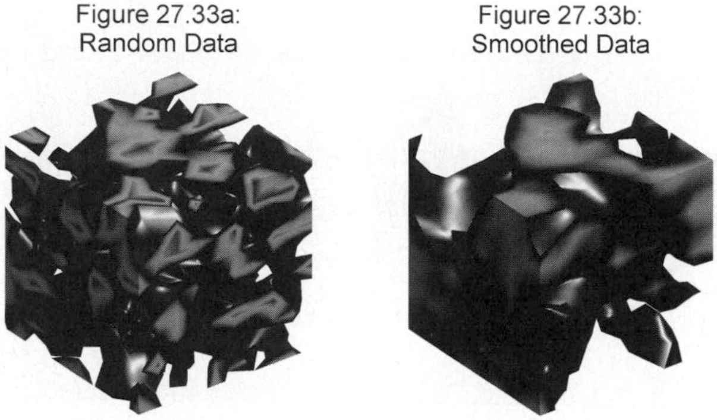

This example demonstrates the use of the functions isocaps and isonormals. The function isocaps creates the faces on the outer surfaces of the block. The function isonormals modifies properties of the drawn patches, so that lighting works correctly.

27.11 EASY PLOTTING

For those occasions when you don't want to take the time to specify the data points explicitly for a 3-D plot, MATLAB provides the functions ezcontour, ezcontour3, ezmesh, ezmeshc, ezplot3, ezsurf, and ezsurfc. These functions construct plots like their equivalents without the ez prefix. However, the input arguments are functions defined by function handles, string expressions, or symbolic math objects and, optionally, the axis limits over which the plot is to be generated. Internally, the functions compute the data and generate the desired plot, as in the following example:

```
>> fstr = ['3*(1-x).^2.*exp(-(x.^2) - (y+1).^2)' ...
   ' - 10*(x/5 - x.^3 - y.^5).*exp(-x.^2-y.^2)' ...
   ' - 1/3*exp(-(x+1).^2 - y.^2)'];

>> subplot(2,2,1)
>> ezmesh(fstr)
>> title('Figure 27.34a: Mesh of peaks(x,y)')
>> subplot(2,2,2)
```

```
>> ezsurf(fstr)
>> title('Figure 27.34b: Surf of peaks(x,y)')
>> subplot(2,2,3)
>> ezcontour(fstr)
>> title('Figure 27.34c: Contour of peaks(x,y)')
>> subplot(2,2,4)
>> ezcontourf(fstr)
>> title('Figure 27.34d: Contourf of peaks(x,y)')
```

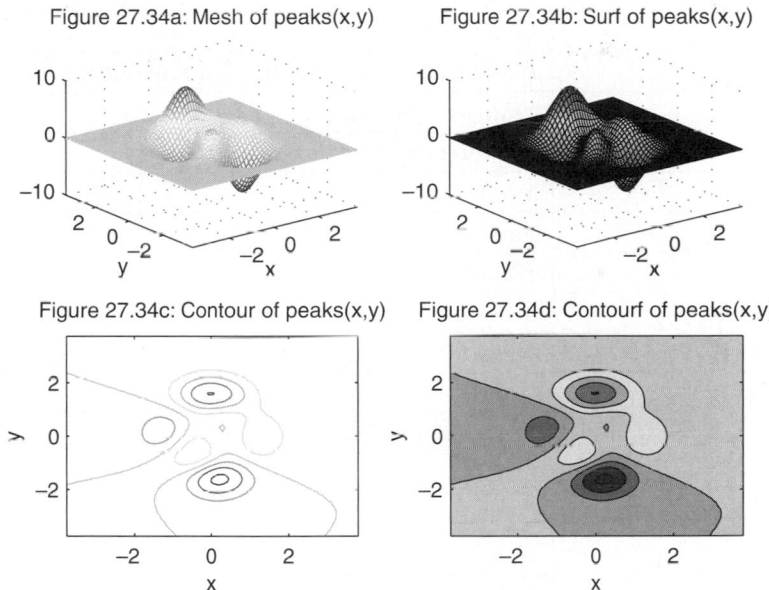

Figure 27.34a: Mesh of peaks(x,y)   Figure 27.34b: Surf of peaks(x,y)

Figure 27.34c: Contour of peaks(x,y)   Figure 27.34d: Contourf of peaks(x,y)

27.12 SUMMARY

The following table documents MATLAB functions for 3-D plotting:

Function	Description
plot3	Plot lines and points in 3-D space
mesh	Mesh surface
meshc	Mesh with underlying contour plot
meshz	Mesh with zero plane
surf	Surface plot

Function	Description
surfc	Surface plot with underlying contour plot
surfl	Surface plot with basic lighting
fill3	Filled 3-D polygons
shading	Color shading mode
hidden	Mesh hidden line removal
surfnorm	Surface normals
axis	Control axis scaling and appearance
grid	Grid line visibility
box	Axis box visibility
hold	Hold current plot
subplot	Creates multiple axes in *Figure* window
daspect	Data aspect ratio
pbaspect	Plot box aspect ratio
xlim	x-axis limits
ylim	y-axis limits
zlim	z-axis limits
view	3-D viewpoint specification
viewmtx	View transformation matrix
rotate3d	Interactive axes rotation
campos	Camera position
camtarget	Camera target
camva	Camera view angle
camup	Camera up vector
camproj	Camera projection
camorbit	Camera orbit
campan	Pan camera
camdolly	Dolly camera
camzoom	Zoom camera
camroll	Roll camera
camlookat	Look at specific object
camlight	Camera lighting creation and placement
title	Plot title
xlabel	x-axis label
ylabel	y-axis label
zlabel	z-axis label

Function	Description
text	Places text on plot
gtext	Places text with mouse
contour	Contour plot
contourf	Filled contour plot
contour3	3-D contour plot
clabel	Contour labeling
pcolor	Pseudocolor plot
voronoi	Voronoi diagram
trimesh	Triangular mesh plot
trisurf	Triangular surface plot
scatter3	3-D scatter plot
stem3	3-D stem plot
waterfall	Waterfall plot
ezmesh	Easy mesh plot of string expression
ezmeshc	Easy mesh plot with contour plot of string expression
ezplot3	Easy 3-D linear plot of string expression
ezsurf	Easy surface plot of string expression
ezsurfc	Easy surface plot with contour plot of string expression
ezcontour	Easy contour plot of string expression
ezcontourf	Easy filled contour plot of string expression
vissuite	Help for visualization suite
isosurface	Isosurface extractor
isonormals	Isosurface normals
isocaps	Isosurface end caps
isocolors	Isosurface and patch colors
contourslice	Contours in slice planes
slice	Volumetric slice plot
streamline	Streamlines from data
stream3	3-D streamlines
stream2	2-D streamlines
quiver3	3-D quiver plot
quiver	2-D quiver plot
divergence	Divergence of a vector field
curl	Curl and angular velocity of a vector field
coneplot	Cone plot

Function	Description
streamtube	Stream tube
streamribbon	Stream ribbon
streamslice	Streamlines in slice planes
streamparticles	Display stream particles
interpstreamspeed	Interpolate streamline vertices from speed
subvolume	Extract subset of volume data set
reducevolume	Reduce volume data set
volumebounds	Return volume and color limits
smooth3	Smooth 3-D data
reducepatch	Reduce number of patch faces
shrinkfaces	Reduce size of patch faces

28

Using Color and Light

MATLAB provides a number of tools for displaying information visually in two and three dimensions. For example, the plot of a sine curve presents more information at a glance than a set of data points could. The technique of using plots and graphs to present data sets is known as ***data visualization***. In addition to being a powerful computational engine, MATLAB excels in presenting data visually in interesting and informative ways.

Often, however, a simple 2-D or 3-D plot cannot display all of the information you would like to present at one time. Color can provide an additional dimension. Many of the plotting functions discussed in previous chapters accept a *color* argument that can be used to add that additional dimension.

This discussion begins with an investigation of colormaps—how to use them, display them, alter them, and create them. Next, techniques for simulating more than one colormap in a *Figure* window or for using only a portion of a colormap are illustrated. Finally, lighting models are discussed, and examples are presented. As in the preceding chapters, the figures in this chapter do not exhibit color, although they do have color on a computer screen. As a result, if you are not following along in MATLAB, it may take some imagination to understand the concepts covered in this chapter.

28.1 UNDERSTANDING COLORMAPS

MATLAB uses a numerical array with three columns to represent color values. This array is called a ***colormap***, and each row in the matrix represents an individual color by using numbers in the range 0 to 1. The numbers in each row indicate the intensity of red, green, and blue that make up a specific color. The following table illustrates the correspondence between numerical values in a colormap and colors:

Red	Green	Blue	Color
1	0	0	Red
0	1	0	Green
0	0	1	Blue
1	1	0	Yellow
1	0	1	Magenta
0	1	1	Cyan
0	0	0	Black
1	1	1	White
0.5	0.5	0.5	Medium gray
0.67	0	1	Violet
1	0.4	0	Orange
0.5	0	0	Dark red
0	0.5	0	Dark green

The first column in a colormap is the intensity of red, the second column is the intensity of green, and the third column is the intensity of blue. Colormap values are restricted to the range from 0 to 1.

A colormap is a sequence of rows containing red-green-blue (RGB) values that vary in some prescribed way from the first row to the last. MATLAB provides a number of predefined colormaps, as shown in the following table:

Colormap Function	Description
hsv	Hue-saturation-value colormap, begins and ends with red
jet	Variant of hsv that starts with blue and ends with red
hot	Black to red to yellow to white
cool	Shades of cyan and magenta
summer	Shades of green and yellow
autumn	Shades of red and yellow
winter	Shades of blue and green
spring	Shades of magenta and yellow
white	All white
gray	Linear gray scale
bone	Gray with a tinge of blue
pink	Pastel shades of pink
copper	Linear copper tone
prism	Alternating red, orange, yellow, green, blue, and violet
flag	Alternating red, white, blue, and black
lines	Alternating plot line colors
colorcube	Enhanced color cube

By default, each of these colormaps generates a 64-by-3 array specifying the RGB descriptions of 64 colors. Each of these functions accepts an argument specifying the number of rows to be generated. For example, hot(m) generates an m-by-3 matrix containing the RGB values of colors ranging from black, through shades of red, orange, and yellow, to white.

28.2 USING COLORMAPS

The statement colormap(M) installs the matrix M as the colormap to be used in the current *Figure* window. For example, colormap(cool) installs a 64-entry version of the cool colormap. The form colormap default installs the default colormap, usually hsv or jet.

Line plotting functions, such as plot and plot3, do not use colormaps; they use the colors listed in the plot color and linestyle table. The sequence of colors used by these functions varies, depending on the plotting style you have chosen. Most other plotting functions, such as mesh, surf, contour, fill, pcolor, and their variations, use the current colormap to determine color sequences.

Plotting functions that accept a *color* argument usually accept the argument in one of three forms:

(**1**) a character string representing one of the colors in the plot color and linestyle table—for example, 'r' or 'red' for red;

(**2**) a three-entry row vector representing a single RGB value—for example, [.25 .50 .75];

(**3**) an array. If the color argument is an array, the elements are scaled and used as indices into the current colormap.

28.3 DISPLAYING COLORMAPS

Colormaps can be viewed in a number of ways. One way is to view the elements in a colormap matrix directly:

```
>> hot(8)
ans =
        0.33333                0            0
        0.66667                0            0
              1                0            0
              1          0.33333            0
              1          0.66667            0
              1                1            0
```

```
              1                1              0.5
              1                1                1
>> gray(5)
ans =
              0                0                0
           0.25             0.25             0.25
            0.5              0.5              0.5
           0.75             0.75             0.75
              1                1                1
```

This example shows two standard colormaps. The first is an eight-element hot colormap, and the second is a five-element gray colormap. The gray colormap increments all three components equally, thereby producing various shades of gray.

A colormap is best visualized graphically. The pcolor and rgbplot functions are useful in this case, as in the following example:

```
>> n = 21;
>> map = copper(n);
>> colormap(map)
>> subplot(2,1,1)
>> [xx,yy] = meshgrid(0:n,[0 1]);
>> c = [1:n+1;1:n+1];
>> pcolor(xx,yy,c)
>> set(gca,'Yticklabel',")
>> title('Figure 28.1a: Pcolor of Copper')
>> subplot(2,1,2)
>> rgbplot(map)
>> xlim([0,n])
>> title('Figure 28.1b: RGBplot of Copper')
```

These figures show how a copper colormap varies, from its first row (shown on the left) to its last row (shown on the right). The function rgbplot simply plots the three columns of the colormap in red, green, and blue, respectively, thereby dissecting the three components visually.

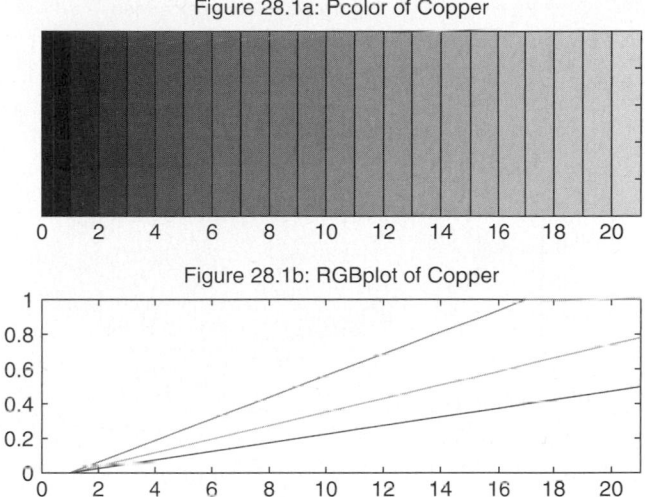

Figure 28.1a: Pcolor of Copper

Figure 28.1b: RGBplot of Copper

When a 3-D plot is made, the color information in the plot can be displayed as auxiliary information in a colorbar by using the colorbar function. For example, consider the following:

```
>> mesh(peaks)
>> axis tight
>> colorbar
>> title('Figure 28.2: Colorbar Added')
```

In this plot, color is associated with the z-axis, and the colorbar associates z-coordinate values with the colors in the colormap.

28.4 CREATING AND ALTERING COLORMAPS

The fact that colormaps are arrays means that you can manipulate them exactly like other arrays. The function brighten takes advantage of this feature to adjust a given colormap to increase or decrease the intensity of the colors. The form brighten(beta) brightens (0 < beta $\leq$ 1) or darkens (-1 $\leq$ beta < 0) the current colormap. The form brighten(beta) followed by brighten(−beta) restores the original colormap. The command newmap=brighten(beta) creates a brighter or darker version of the current colormap without changing the current map. The command mymap=brighten(cmap,beta) creates an adjusted version of the specified colormap without affecting either the current colormap or the specified colormap cmap.

Colormaps can be created by generating an m-by-3 array mymap and installing it with colormap(mymap). Each value in a colormap matrix must be between 0 and 1.

Figure 28.2: Colorbar Added

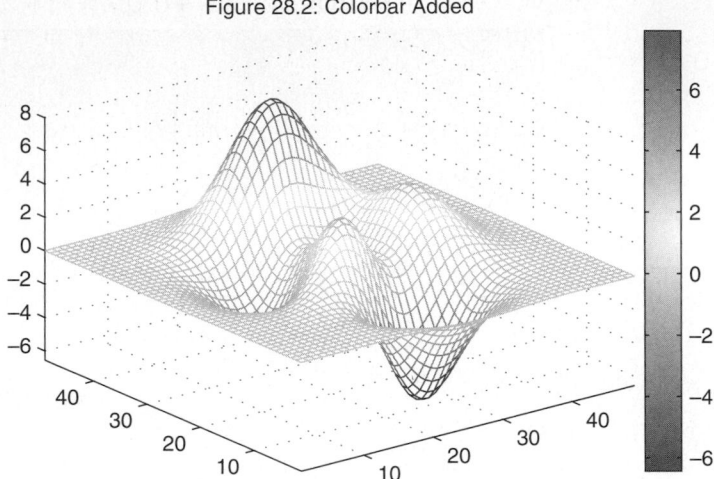

If you try to use a matrix with more or fewer than three columns, or one containing any values less than 0 or greater than 1, colormap will report an error.

Colormaps can be converted between the red-green-blue (RGB) standard and the hue-saturation-value (HSV) standard by using the rgb2hsv and hsv2rgb functions. MATLAB, however, always interprets colormaps as RGB values. Colormaps can be combined as well, as long as the result satisfies the size and value constraints. For example, the colormap called pink is simply

```
pinkmap = sqrt(2/3*gray + 1/3*hot);
```

Again, the result is a valid colormap only if all elements of the m-by-3 matrix are between 0 and 1 inclusive.

Normally, a colormap is scaled to extend from the minimum to the maximum values of your data—that is, the entire colormap is used to render your plot. You may occasionally wish to change the way these colors are used. The caxis function, which stands for *color axis*, allows you to use the entire colormap for a subset of your data range or to use only a portion of the current colormap for your entire data set.

The statement [cmin, cmax] = caxis returns the minimum and maximum data values mapped to the first and last entries of the colormap, respectively. These are normally set to the minimum and maximum values of your data. For example, mesh(peaks) creates a mesh plot of the peaks function and sets caxis to [−6.5466, 8.0752], the minimum and maximum z values. Data points between these values use colors interpolated from the colormap.

The function caxis([cmin,cmax]) uses the entire colormap for data in the range between cmin and cmax. Data points greater than cmax are rendered with the color associated with cmax, and data points less than cmin are rendered with the color associated with cmin. If cmin is less than min(data), or cmax is greater than max(data), the colors associated with cmin or cmax will never be used. Only the portion of the colormap associated with data will be used. The function caxis('auto') or the command form caxis auto restores the default values of cmin and cmax. The following example illustrates color axis settings:

```
>> N = 17;
>> data = [1:N+1;1:N+1]';

>> subplot(1,3,1)
>> colormap(hsv(N))
>> pcolor(data)
>> set(gca,'XtickLabel','')
>> title('Figure 28.3: Auto Limits')
>> caxis auto       % automatic limits (default)

>> subplot(1,3,2)
>> pcolor(data)
>> axis off
>> title('Extended Limits')
>> caxis([-5,N+5]) % extend the color limits

>> subplot(1,3,3)
>> pcolor(data)
>> axis off
>> title('Restricted Limits')
>> caxis([5,N-5]) % restrict the color limits
```

The first plot on the left is the default plot. This plot covers the complete colormap. In the center plot, the color axis is extended, forcing all plotted values to use a subset of the colormap. In the plot on the right, the color axis is restricted, forcing the colormap to cover a region in the center of the plot. The plot extremes simply use the colormap extremes.

28.5 USING COLOR TO DESCRIBE A FOURTH DIMENSION

Surface plots such as mesh and surf vary color along the z-axis, unless a color argument is given—for example, surf(X,Y,Z) is equivalent to surf(X,Y,Z,Z).

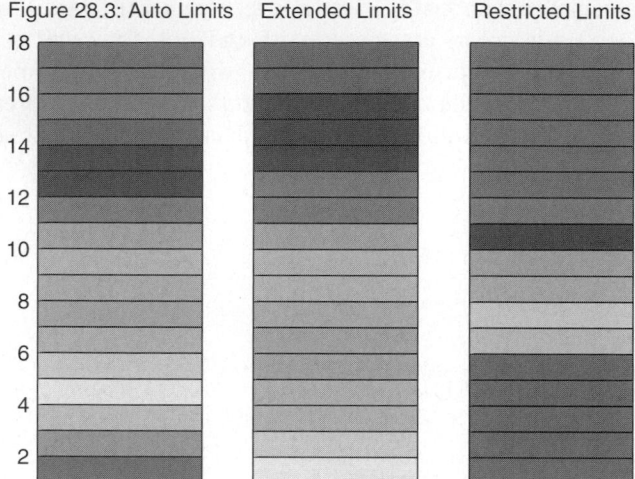

Figure 28.3: Auto Limits   Extended Limits   Restricted Limits

Applying color to the *z*-axis produces a colorful plot, but does not provide additional information, since the *z*-axis already exists. To make better use of color, it is suggested that color be used to describe some property of the data not reflected by the three axes. To do so requires specifying different data for the color argument to 3-D plotting functions.

If the color argument to a plotting function is a vector or a matrix, it is scaled and used as an index into the colormap. This argument can be any real vector or matrix that is the same size as the other arguments. The following code is illustrative:

```
>> x = -7.5:.5:7.5;            % data
>> [X Y] = meshgrid(x);        % create plaid data
>> R = sqrt(X.^2 + Y.^2)+eps; % create sombrero
>> Z = sin(R)./R;


>> subplot(2,2,1)
>> surf(X,Y,Z,Z)      % default color order
>> colormap(gray)
>> shading interp
>> axis tight off
>> title('Figure 28.4a: Default, Z')


>> subplot(2,2,2)
>> surf(X,Y,Z,Y)     % Y axis color order
```

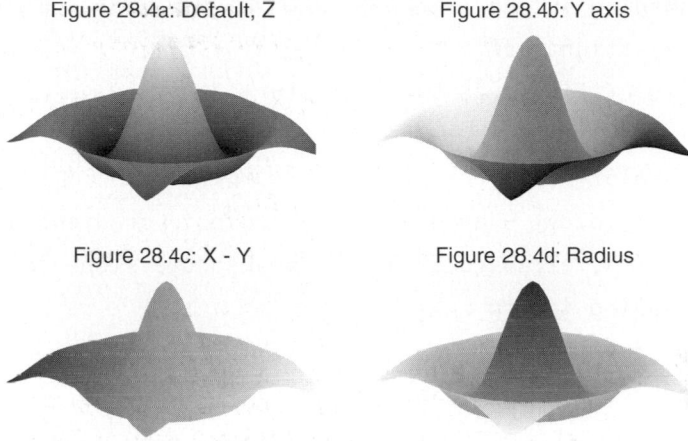

Figure 28.4a: Default, Z   Figure 28.4b: Y axis

Figure 28.4c: X - Y   Figure 28.4d: Radius

```
>> shading interp
>> axis tight off
>> title('Figure 28.4b: Y axis')

>> subplot(2,2,3)
>> surf(X,Y,Z,X-Y) % diagonal color order
>> shading interp
>> axis tight off
>> title('Figure 28.4c: X - Y')

>> subplot(2,2,4)
>> surf(X,Y,Z,R)    % radius color order
>> shading interp
>> axis tight off
>> title('Figure 28.4d: Radius')
```

These subplots demonstrate four simple ways to use color as a fourth dimension. Whatever data are provided as the fourth argument to surf is used to interpolate the colormap. Any function of the first three arguments can be provided, or some completely independent variable works as well. Using the functions del2 and gradient allows one to apply color with respect to curvature and slope, respectively:

```
>> subplot(2,2,1)
>> surf(X,Y,Z,abs(del2(Z)))    % absolute Laplacian
>> colormap(gray)
```

```
>> shading interp
>> axis tight off
>> title('Figure 28.5a: |Curvature|')

>> subplot(2,2,2)
>> [dZdx,dZdy] = gradient(Z); % compute gradient of surface
>> surf(X,Y,Z,abs(dZdx))      % absolute slope in x-direction
>> shading interp
>> axis tight off
>> title('Figure 28.5b: |dZ/dx|')

>> subplot(2,2,3)
>> surf(X,Y,Z,abs(dZdy))      % absolute slope in y-direction
>> shading interp
>> axis tight off
>> title('Figure 28.5c: |dZ/dy|')

>> subplot(2,2,4)
>> dR = sqrt(dZdx.^2 + dZdy.^2);
>> surf(X,Y,Z,abs(dR))        % absolute slope in radius
>> shading interp
>> axis tight off
>> title('Figure 28.5d: |dR|')
```

Figure 28.5a: |Curvature| Figure 28.5b: |dZ/dx|

Figure 28.5c: |dZ/dy| Figure 28.5d: |dR|

Note how color in these subplots provides an additional dimension to the plotted surface. The function del2 is the discrete Laplacian function that applies color based on the curvature of the surface. The function gradient approximates the gradient, or slope, of the surface with respect to the two coordinate directions.

28.6 LIGHTING MODELS

The graphics functions pcolor, fill, fill3, mesh, and surf discussed in the previous chapters render objects that appear to be well lit from all sides by very diffuse light. This technique emphasizes the characteristics of the objects in the *Figure* window and enhances the user's ability to visualize the data being analyzed. Although the data can be visualized quite clearly, the realism of the scene can be enhanced or diminished by creating different lighting effects.

The shading function selects faceted, flat, or interpolated shading. (Examples of each of these were illustrated in the chapter on 3-D graphics.) Although requiring more computational power and subsequently more time to render, interpolated shading of the objects in a scene can enhance the realism of the scene being rendered.

Once or more light sources can be added to simulate the highlights and shadows associated with directional lighting. The function light creates a white light source infinitely far away along the vector [1 0 1]. Once the light source has been created, the lighting function allows you to select from four different lighting models: none (which ignores any light source), flat (the default when a light source is created), phong, and gouraud. Each of these models uses a different algorithm to change the appearance of the object. Flat lighting uses a uniform color for each face of the object. Gouraud lighting interpolates the face colors from the vertices. Phong lighting interpolates the normals at the vertices across each face and calculates the reflectance at each pixel. While colormaps are properties of *Figure* windows, light is a property, or child, of the axes. Therefore, each axis in a *Figure* window can be separately lit, as in the following example:

```
>> subplot(2,2,1)
>> sphere
>> light
>> shading interp
>> axis square off
>> lighting none
>> title('Figure 28.6a: No Lighting')

>> subplot(2,2,2)
>> sphere
>> light
```

```
>> shading interp
>> axis square off
>> lighting flat
>> title('Figure 28.6b: Flat Lighting')


>> subplot(2,2,3)
>> sphere
>> light
>> shading interp
>> axis square off
>> lighting gouraud
>> title('Figure 28.6c: Gouraud Lighting')


>> subplot(2,2,4)
>> sphere
>> light
>> shading interp
>> axis square off
>> lighting phong
>> title('Figure 28.6d: Phong Lighting')
```

Figure 28.6a: No Lighting

Figure 28.6b: Flat Lighting

Figure 28.6c: Gouraud Lighting

Figure 28.6d: Phong Lighting

In addition to lighting, the appearance of objects in an axis can be changed by modifying the apparent reflective characteristics, or ***reflectance***, of surfaces. Reflectance is made up of a number of components:

- ***Ambient light***—strength of the uniform directionless light in the figure
- ***Diffuse reflection***—intensity of the soft directionless reflected light
- ***Specular reflection***—intensity of the hard directional reflected light
- ***Specular exponent***—controls the size of the specular "hot spot" or spread
- ***Specular color reflectance***—determines the contribution of the surface color to the reflectance

Some predefined surface reflectance properties are available by using the `material` function. Options include `shiny`, `dull`, `metal`, and `default` for restoring the default surface reflectance properties. The form `material([ka kd ks n sc])`, where `n` and `sc` are optional, sets the ambient strength, diffuse reflectance, specular reflectance, specular exponent, and specular color reflectance of objects in an axis. The following code illustrates the first four options:

```
>> subplot(2,2,1)
>> sphere
>> colormap(gray)
>> light
>> shading interp
>> axis square off
>> material default
>> title('Figure 28.7a: Default Material')

>> subplot(2,2,2)
>> sphere
>> light
>> shading interp
>> axis square off
>> material shiny
>> title('Figure 28.7b: Shiny Material')

>> subplot(2,2,3)
>> sphere
>> light
>> shading interp
```

```
>> axis square off
>> material dull
>> title('Figure 28.7c: Dull Material')


>> subplot(2,2,4)
>> sphere
>> light
>> shading interp
>> axis square off
>> material metal
>> title('Figure 28.7d: Metal Material')
```

Figure 28.7a: Default Material

Figure 28.7b: Shiny Material

Figure 28.7c: Dull Material

Figure 28.7d: Metal Material

Note that the function light is fairly limited. It creates white light emanating from an infinite distance away along a given direction. In reality, light is a Handle Graphics object creation function. The function light offers a variety of properties that can be set. For example, the light color, position, and style can be set, with style meaning that the light can be a point source at a given position or can be infinitely far away along some vector:

```
>> Hl = light('Position',[x,y,z],'Color',[r,g,b],'Style','local');
```

This command creates a light source at position (x,y,z) using light color [r,g,b] and specifies that the position is a location ('local') rather than a vector ('infinite'). It also saves the handle of the *light* object (H1), which can be used to change the properties of the light source at a later time. For example,

```
>> set(H1,'Position',[1 0 1],'Color',[1 1 1],'Style','infinite');
```

sets the light source defined by the handle H1 back to its original default characteristics. (For more information about Handle Graphics, see Chapter 31.)

28.7 SUMMARY

The following table documents MATLAB functions for color and lighting:

Function	Description
light	Light object creation function
lighting	Set lighting mode (flat, gouraud, phong, or none)
lightangle	Position light object in spherical coordinates
material	Set material reflectance (default, shiny, dull, or metal)
camlight	Set light object with respect to camera
brighten	Brighten or darken colormap
caxis	Set or get color axis limits
diffuse	Find surface diffuse reflectance
specular	Find surface specular reflectance
surfnorm	Compute surface normals
colorbar	Create colorbar
colordef	Define default color properties
colormap	Set or get *Figure* window colormap
colormapeditor	GUI for creating colormaps
hsv2rgb	Convert hue-saturation-value color values to red-green-blue model
rgb2hsv	Convert red-green-blue color values to hue-saturation-value model
rgbplot	Plot colormap
shading	Surface shading (flat, faceted, or interp)
spinmap	Spin colormap
whitebg	Change plot axes background color
graymon	Set graphics defaults for grayscale monitors
autumn	Colormap with shades of red and yellow

Function	Description
bone	Gray-scale colormap with a tinge of blue
cool	Colormap with shades of cyan and magenta
copper	Colormap with linear copper tone
flag	Colormap with alternating red, white, blue, and black
gray	Colormap with linear gray scale
hot	Colormap with black, red, yellow, and white
hsv	Colormap based on hue, saturation, and value progression
jet	Colormap variant of hsv that starts with blue and ends with red
lines	Colormap based on line colors
prism	Colormap with alternating red, orange, yellow, green, blue, and violet
spring	Colormap with shades of magenta and yellow
summer	Colormap with shades of green and yellow
winter	Colormap with shades of blue and green

29

Images, Movies, and Sound

MATLAB provides commands for displaying several types of images. Images can be created and stored as standard double-precision floating-point numbers (`double`) and, optionally, as 8-bit (`uint8`) or 16-bit (`uint16`) unsigned integers. MATLAB can read and write image files in a number of standard graphics file formats, as well as use `load` and `save` to save image data in MAT-files. MATLAB provides commands for creating and playing animations as movies (sequences of frames). Sound functions are available as well for computers that support sound.

29.1 IMAGES

Images in MATLAB consist of a data matrix and usually an associated colormap matrix. There are three types of image data matrices, each interpreted differently: indexed images, intensity images, and truecolor, or RGB, images.

An *indexed image* requires a colormap and interprets the image data as indices into the colormap matrix. The colormap matrix is a standard colormap—any m-by-3 array containing valid RGB data. Given an image data array $X(i,j)$ and a colormap array cmap, the color of each image pixel P_{ij} is cmap($X(i,j)$,:). This implies that the data values in X are integers within the range [1 length(cmap)]. This image can be displayed by using

```
>> image(X); colormap(cmap)
```

An *intensity image* scales the image data over a range of intensities. This form is normally used with images to be displayed in grayscale or one of the other monochromatic colormaps, but other colormaps can be used if desired. The image data are not required to be in the range [1 length(cmap)], as is the case with indexed images. The data are scaled over a given range, and the result is used to index into the colormap. For example,

```
>> imagesc(X,[0 1]); colormap(gray)
```

associates the value 0 with the first colormap entry, and the value 1 with the last colormap entry. Values in X between 0 and 1 are scaled and used as indices into the colormap. If the scale is omitted, it defaults to [min(min(X)) max(max(X))].

A *truecolor*, or **RGB image** is created from an m-by-n-by-3 data array containing valid RGB triples. The row and column dimensions specify the pixel location, and the page of third dimension specifies each color component. For example, pixel P_{ij} is rendered in the color specified by X(i,j,:). A colormap is not required, because the color data are stored within the image data array itself. For example,

```
>> image(X)
```

where X is an m-by-n-by-3 truecolor, or RGB, image, displays the image. The image X can contain uint8, unit16, or double data.

If images are displayed on default axes, the aspect ratio will often be incorrect and the image will be distorted. Issuing

```
>> axis image off
```

sets the axis properties so that the aspect ratio matches the image and the axis labels and ticks are hidden. To force each pixel in the image to occupy one pixel on the display requires setting *figure* and *axes* object properties, as in the following example:

```
>> load clown        % sample image
>> [r,c] = size(X); % pixel dimensions
>> figure('Units','Pixels','Position',[100 100 c r])
>> image(X)
>> set(gca,'Position',[0 0 1 1])
>> colormap(map)
```

Here, the *figure* is set to display exactly the same number of pixels as the image by setting its width and height equal to that of the image. Then, the *axes* position is set to occupy the entire figure in normalized units.

MATLAB installations have a number of sample images in addition to clown.mat used in the preceding example. The demos subdirectory on the MATLAB path contains cape.mat, clown.mat, detail.mat, durer.mat, flujet.mat, gatlin.mat, mandrill.mat, and spine.mat. Each of these images can be displayed by issuing

```
>> load filename
>> image(X), colormap(map)
```

```
>> title(caption)
>> axis image off
```

29.2 IMAGE FORMATS

The `image` and `imagesc` commands can display 8- and 16-bit images without first converting them to the `double` format. However, the range of data values for `uint8` data is [0 255], as supported in standard graphics file formats, and the range of data values for `uint16` is [0 65535].

For indexed images, `image` maps the value 0 to the first entry in a 256-entry colormap and maps the value 255 to the last entry, by automatically supplying the proper offset. Since the normal range of `double` data for indexed images is [1 length(cmap)], converting between `uint8` and `double` or `uint16` and `double` requires shifting the values by 1. Before MATLAB 7, mathematical operations on `uint8` and `uint16` arrays were not defined. As a result, to perform mathematical operations on unsigned integers in previous versions of MATLAB required that they must be converted to the `double` format. For example, the code

```
>> Xdouble = double(Xuint8) + 1;
>> Xuint8 = uint8(Xdouble - 1);
```

converts the `uint8` data in `Xuint8` to `double` and back, taking into account the offset of 1. With the introduction of arithmetic for integer data types in MATLAB 7, it is now possible to perform operations without conversion to double precision. This facilitates easier image manipulation in MATLAB.

For 8-bit intensity and RGB images, the range of values is normally [0 255] rather than [0 1]. To display 8-bit intensity and RGB images, use the following commands:

```
>> imagesc(Xuint8,[0 255]); colormap(cmap)
>> image(Xuint8)
```

Conversions to `double` can also be normalized, as in the following code:

```
>> Xdouble = double(Xuint8)/255;
>> Xuint8 = uint8(round(Xdouble*255));
```

The 8-bit color data contained in an RGB image is automatically scaled when it is displayed. For example, the color white is normally [1 1 1] when doubles are used. If the same color is stored as 8-bit data, the color white is represented as [255 255 255].

The optional *Image Processing Toolbox* available for MATLAB contains many functions for manipulating images. This Toolbox is valuable if you regularly manipulate images.

29.3 IMAGE FILES

Image data can be saved to files and reloaded into MATLAB by using many different file formats. The normal MATLAB save and load functions support image data in double, uint8, or uint16 format in the same way that they support any other MATLAB variable and data type. When saving indexed images or intensity images with nonstandard colormaps, be sure to save the colormap as well as the image data, using, for example, the command

```
>> save myimage.mat X map
```

MATLAB also supports several industry-standard image file formats using the imread and imwrite functions. Information about the contents of a graphics file can be obtained with the imfinfo function. The help text for imread gives extensive information regarding image read formats and features. Following is part of this help text:

```
Supported file types
--------------------

JPEG   Any baseline JPEG image; JPEG images with some
       commonly used extensions; 8-bit and 12-bit lossy
       compressed RGB and grayscale images; 8-bit and 12-bit
       lossless compressed RGB images; 8-bit, 12-bit, and
       16-bit lossless compressed grayscale images

TIFF   Any baseline TIFF image, including 1-bit, 8-bit, and
       24-bit uncompressed images; 1-bit, 8-bit, and 24-bit
       images with packbits compression; 1-bit images with
       CCITT compression; 16-bit grayscale, 16-bit indexed,
       and 48-bit RGB images; 24-bit and 48-bit ICCLAB
       and CIELAB images; 32-bit and 64-bit CMYK images; and
       8-bit tiled TIFF images with any compression and colorspace
       combination listed above.

GIF    Any 1-bit to 8-bit GIF image

BMP    1-bit, 4-bit, 8-bit, 16-bit, 24-bit, and 32-bit uncompressed
       images; 4-bit and 8-bit run-length encoded (RLE) images

PNG    Any PNG image, including 1-bit, 2-bit, 4-bit, 8-bit,
       and 16-bit grayscale images; 8-bit and 16-bit
       indexed images; 24-bit and 48-bit RGB images
```

HDF   8-bit raster image datasets, with or without an
      associated colormap; 24-bit raster image datasets

PCX   1-bit, 8-bit, and 24-bit images

XWD   1-bit and 8-bit ZPixmaps; XYBitmaps; 1-bit XYPixmaps

ICO   1-bit, 4-bit, and 8-bit uncompressed images

CUR   1-bit, 4-bit, and 8-bit uncompressed images

RAS   Any RAS image, including 1-bit bitmap, 8-bit indexed,
      24-bit truecolor and 32-bit truecolor with alpha.

PBM   Any 1-bit PBM image. Raw (binary) or ASCII (plain) encoded.

PGM   Any standard PGM image. ASCII (plain) encoded with
      arbitrary color depth. Raw (binary) encoded with up
      to 16 bits per gray value.

PPM   Any standard PPM image. ASCII (plain) encoded with
      arbitrary color depth. Raw (binary) encoded with up
      to 16 bits per color component.

The calling syntax for imwrite varies, depending on the image type and file format.
The help text for imwrite, shown in part as follows, gives extensive information
regarding image save formats and features:

```
Table: summary of supported image types
---------------------------------------
```

BMP   1-bit, 8-bit and 24-bit uncompressed images

TIFF Baseline TIFF images, including 1-bit, 8-bit, 16-bit,
      and 24-bit uncompressed images; 1-bit, 8-bit, 16-bit,
      and 24-bit images with packbits compression; 1-bit
      images with CCITT 1D, Group 3, and Group 4 compression;
      CIELAB, ICCLAB, and CMYK images

JPEG Baseline JPEG images

PNG   1-bit, 2-bit, 4-bit, 8-bit, and 16-bit grayscale
        images; 8-bit and 16-bit grayscale images with alpha
        channels; 1-bit, 2-bit, 4-bit, and 8-bit indexed
        images; 24-bit and 48-bit truecolor images; 24-bit
        and 48-bit truecolor images with alpha channels

HDF   8-bit raster image datasets, with or without associated
        colormap; 24-bit raster image datasets; uncompressed or
        with RLE or JPEG compression

PCX   8-bit images

XWD   8-bit ZPixmaps

RAS   Any RAS image, including 1-bit bitmap, 8-bit indexed,
        24-bit truecolor and 32-bit truecolor with alpha.

PBM   Any 1-bit PBM image, ASCII (plain) or raw (binary) encoding.

PGM   Any standard PGM image. ASCII (plain) encoded with
        arbitrary color depth. Raw (binary) encoded with up
        to 16 bits per gray value.

PPM   Any standard PPM image. ASCII (plain) encoded with
        arbitrary color depth. Raw (binary) encoded with up
        to 16 bits per color component.

PNM   Any of PPM/PGM/PBM (see above) chosen automatically.

29.4 MOVIES

Animation in MATLAB takes one of two forms. First, if the computations needed
to create a sequence of images can be performed quickly enough, *figure* and *axes*
properties can be set so that screen rendering occurs sufficiently quickly so that animation is visually smooth. On the other hand, if computations require significant
time or the resulting images are complex enough, you must create a movie.

In MATLAB, the functions getframe and movie provide the tools required to
capture and play movies. The getframe command takes a snapshot of the current

figure, and movie plays back the sequence of frames after they have been captured. The output of getframe is a structure containing all of the information needed by movie. Capturing multiple frames is simply a matter of adding elements to the structure. The following code is illustrative:

```
% moviemaking example: rotate a 3-D surface plot
[X,Y,Z]=peaks(50);          % create data
surfl(X,Y,Z)                % plot surface with lighting
axis([-3 3 -3 3 -10 10])    % fix axes so that scaling does not change
axis vis3d off              % fix axes for 3D and turn off axes ticks etc.
shading interp              % make it pretty with interpolated shading
colormap(copper)            % choose a good colormap for lighting
for i=1:15                  % rotate and capture each frame
    view(-37.5+15*(i-1),30)% change the viewpoint for this frame
    m(i)=getframe;          % add this figure to the frame structure
end
cla         % clear axis for movie
movie(m)    % play the movie
```

The preceding script file creates a movie by incrementally rotating the peaks surface and capturing a frame at every increment. Finally, the movie is played after clearing the *axes*. The variable m contains a structure array, with each array element containing a single frame:

```
>> m

m =

1x15 struct array with fields:
    cdata
    colormap

>> size(m(1).cdata)
ans =
    342    306      3
```

The color data holding the image cdata make up a truecolor, or RGB, bitmap image. As a result, the complexity of contents of the *axes* does not influence the bytes required to store a movie. The size of the *axes* in pixels determines the size of the image and therefore the number of bytes required to store a movie.

29.5 IMAGE UTILITIES

Conversion between indexed images and movie frames is possible with the im2frame and frame2im functions. For example,

```
>> [X,cmap] = frame2im(M(n))
```

converts the nth frame of the movie matrix M into an indexed image X and associated colormap cmap. Similarly,

```
>> M(n) = im2frame(X,cmap)
```

converts the indexed image X and colormap cmap into the nth frame of the movie matrix M. Note that im2frame can be used to convert a series of images into a movie in the same way that getframe converts a series of *figures* or *axes* into a movie.

29.6 SOUND

In addition to the high level audiorecorder and audioplayer functions now available, MATLAB supports sound by using a variety of lower-level functions. The function sound(y,f,b) sends the signal in vector y to the computer's speaker at sample frequency f. Values in y outside of the range [-1 1] are clipped. If f is omitted, the default sample frequency of 8192 Hz is used. MATLAB plays the sound by using b bits persecond, if possible. Most platforms support b=8 or b=16. If b is omitted, b=16 is used.

The soundsc function is the same as sound, except that the values in y are *scaled* to the range [-1 1] rather than clipped. This results in a sound that is as loud as possible without clipping. An additional argument is available that permits mapping a range of values in y to the full sound range. The format is soundsc(y,...,[smin smax]). If omitted, the default range is [min(y) max(y)].

Two industry-standard sound file formats are supported in MATLAB. NeXT/Sun audio format (*file.au*) files and Microsoft WAVE format (*file.wav*) files can be written and read.

The NeXT/Sun Audio sound storage format supports multichannel data for 8-bit mu-law, 8-bit linear, and 16-bit linear formats. The most general form of auwrite is auwrite(y,f,n,'*method*','*filename*'), where y is the sample data, f is the sample rate in hertz, b specifies the number of bits in the encoder, '*method*' is a string specifying the encoding method, and '*filename*' is a string specifying the name of the output file. Each column of y represents a single channel. Any value in y outside of the range [-1 1] is clipped prior to writing the file. The f, n, and '*method*' arguments are optional. If omitted, f=8000, n=8, and method='mu'. The method argument must be either 'linear' or 'mu'. If the file name string contains no extension, '.au' is appended.

Conversion between mu-law and linear formats can be performed by using the mu2lin and lin2mu functions. (More information about the exact conversion processes involved with these two functions can be found by using the on-line help.)

Multichannel 8-bit or 16-bit WAVE sound storage format sound files can be created with the wavwrite function. The most general form is wavwrite (y,f,n,'filename'), where y is the sample data, f is the sample rate in hertz, b specifies the number of bits in the encoder, and 'filename' is a string specifying the name of the output file. Each column of y represents a single channel. Any value in y outside of the range [-1 1] is clipped prior to writing the file. The f and n arguments are optional. If omitted, f=8000 and n=16. If the file name string contains no extension, '.wav' is appended.

Both auread and wavread have the same syntax and options. The most general form is [y,f,b]=auread('filename',n), which loads a sound file specified by the string 'filename' and returns the sampled data into y. The appropriate extension (.au or .wav) is appended to the file name if no extension is given. Values in y are in the range $[-1\ 1]$. If three outputs are requested, as previously illustrated, the sample rate in hertz and the number of bits per sample are returned in f and b, respectively. If n is given, only the first n samples are returned from each channel in the file. If n=[n1 n2], only samples from n1 through n2 are returned from each channel. The form [samples,channels]=wavread('filename','size') returns the size of the audio data in the file, rather than the data itself. This form is useful for preallocating storage or estimating resource use.

29.7 SUMMARY

The following table summarizes the image, movie, and sound capabilities in MATLAB:

Function	Description
image	Create indexed or truecolor (RGB) *image* object
imagesc	Create intensity *image* object
colormap	Apply colormap to *image*
axis image	Adjust axis scaling for *image*
uint8	Conversion to unsigned 8-bit integer
uint16	Conversion to unsigned 16-bit integer
double	Conversion to double precision
imread	Read image file
imwrite	Write image file
imfinfo	Image file information
getframe	Place movie frame in structure

Function	Description
movie	Play movie from movie structure
frame2im	Convert movie frame to image
im2frame	Convert image to movie frame
avifile	Create avi movie file
addframe	Add frame to avi movie file
close	Close avi movie file
aviread	Read avi movie file
aviinfo	Information about an avi movie file
movie2avi	Convert movie in MATLAB format to avi format
audiorecorder	Audio recorder object
audioplayer	Audio player object
audiodevinfo	Audio device information
sound	Play vector as sound
soundsc	Autoscale and play vector as sound
wavplay	Play WAVE format sound file
wavrecord	Record sound using Windows audio input device
wavread	Read WAVE format sound file
wavwrite	Write WAVE format sound file
auread	Read NeXT/SUN sound file
auwrite	Write NeXT/SUN sound file
lin2mu	Convert linear audio to mu-law
mu2lin	Convert mu-law audio to linear

30

Printing and Exporting Graphics

MATLAB graphics are very effective tools for data visualization and analysis. Therefore, it is often desirable to create hard-copy output or to use these graphics in other applications. MATLAB provides a very flexible system for printing graphics and creating output in many different graphics formats, including EPS and TIFF. Most other applications can import one or more of the graphics file formats supported by MATLAB.

Perhaps the most important issue to recognize when printing or exporting graphics is that the *figure* is ***rerendered*** in the process. That is, what you see on the screen is *not* what you get on a printed page or in an exported file. By default, MATLAB can choose a different renderer (`painters`, `zbuffer`, or `OpenGL`), can change the *axes* tick-mark distribution, and can change the size of the *figure* being printed and exported. Naturally, MATLAB provides the capability to enforce what-you-see-is-what-you-get (WYSIWYG), but this is not the default action taken.

In general, printing and exporting are not simple, because there is an almost uncountable number of possible combinations involving printer drivers, printer protocols, graphics file types, renderers, bit-mapped versus vector graphics descriptions, dots-per-inch selection, color versus black and white, compression, platform limitations, and so on. Most of this complexity is hidden when default output is requested. However, when very specific printed characteristics are required, or when exporting a *figure* for insertion into a word processing or presentation document, the complexities of printing and exporting must be understood. In these cases, it is not uncommon to spend a significant amount of time tweaking output until it matches the characteristics desired.

This chapter introduces the printing and exporting capabilities of MATLAB. As in other areas, MATLAB provides menu items, as well as *Command* window functions, for printing and exporting. The menu approach offers convenience, but limited flexibility, whereas the function approach offers complete flexibility, but requires much more knowledge on the part of the user.

> Accessing the on-line documentation for the function `print` by issuing `>> doc print` provides a great deal of information about printing and exporting.

30.1 PRINTING AND EXPORTING USING MENUS

When graphics are displayed in a *Figure* window, the top of the window contains a menu bar and possibly one or more toolbars. The menu bar has a number of menus, including **File, Edit, View, Insert, Tools, Desktop, Window**, and **Help**. Of these, the **File** menu lists menu items for printing and exporting the current *figure*. In addition, the **Edit** menu on the Windows platform contains menu items for exporting the current *figure* to the system clipboard. The **File** menu includes the menu items **Export Setup, Page Setup, Print Setup, Print Preview**, and **Print**. Each of these items offers dialog boxes for setting various aspects of printing and exporting.

The **Export Setup** menu item makes it possible to save the current *figure* in one of many graphics formats. The **Export Setup** dialog box presents an **Export. . .** dialog, where the user chooses a directory, file name, and file type for the current *figure*. The size and features of the saved *figure* can be set from the dialog box.

Choosing the **Page Setup** menu item (or the `pagesetupdlg` command) opens a tabbed dialog box for setting a number of features that can be used when a *figure* is rendered for printing or exporting. This dialog box is the primary place to specify the size, orientation, placement, and many of the other features of the rendered *figure*.

Choosing the **Print Setup** menu item (or the `print -dsetup` command) on the Windows platform opens a standard dialog box with selections for a printer and options such as paper size, paper source, and orientation. On UNIX systems, the **Print Setup** menu item (or the `printdlg -setup` command) opens a dialog box identical to the **Print** dialog box, except for the title of the box and the fact that the **OK** button closes the dialog box, but does not send the *figure* to the printer or to a file. The UNIX **Print Setup** dialog box includes options for selecting a printer (or a filename) and a printer driver, *figure* size, axes and tick options. This dialog box also offers an **Options** button providing access to a dialog box for setting many additional options, including choosing a renderer and setting the output resolution. On both platforms, options chosen in this dialog box persist during a MATLAB session and override any like settings made in the **Page Setup** dialog box.

The **Print Preview** menu item (or the `printpreview` command) opens a window showing how the *figure* will be rendered on the printed page. Buttons are provided to call the **Page Setup** dialog box to make any necessary changes and to call the **Print** dialog box to print the *figure*.

The **Print** menu item (or the `printdlg` command) opens a standard **Print** dialog box on PC systems. From this dialog box, the printer and the number of copies can be chosen. The option to print to a file is available here as well. On UNIX systems, the printer or the filename, the printer driver, and options to freeze the axes and tick marks can be chosen. An **Options** button provides access to a dialog box that is useful for setting a variety of additional options, all of which also appear in the **Page Setup** dialog box.

> **Page Setup** options apply to the current *figure* and are saved with the current *figure*. **Print Setup** options persist through the current MATLAB session and override any like **Page Setup** options for the current *figure*. **Print** options apply only to the current *figure* and override any like options in **Page Setup** or **Print Setup** for the current *figure*.

On the Windows platform, the **Edit** menu provides export capabilities through the system clipboard. The **Copy Figure** menu item places the current *figure* on the clipboard based on the options set in the preferences dialog box opened by selecting the **Copy Options** menu item. The current *figure* can be copied in either bitmap (BMP) or enhanced meta file (EMF) format. A template can be used to change line widths and font sizes, as well as to set other default options for copy operations. These options apply only when a *figure* is copied to the clipboard by using the **Copy Figure** menu item. They have no effect when printing or exporting a *figure* to a graphics file.

30.2 COMMAND LINE PRINTING AND EXPORTING

The function `print` handles all printing and exporting from the *Command* window. This single function offers numerous options that are specified as additional input arguments. The syntax for the command form of `print` is

```
>> print -device -option -option filename
```

where all parameters or arguments are optional. The parameter *-device* specifies the device driver to be used, *-option* specifies one or more options, and `filename` specifies the name of an output file if the output is not sent directly to a printer. Because of command-function duality, `print` can be called as a function as well. For example,

```
>> print('-device','-option','-option','filename')
```

is equivalent to the preceding `print` command statement.

In general, print options including *-device* can be specified in any order. Recognized print *-option* strings are described in the following table:

-option	Description
-adobecset	Select PostScript default character set encoding (early PostScript printer drivers and EPS file formats only)
-append	Append *figure* to existing file (PostScript printer drivers only)
-cmyk	Print with CMYK colors instead of RGB colors (PostScript printer drivers and EPS file format only)
-device	Printer driver to be used
-dsetup	Display the **Print Setup** dialog box (Windows only)
-fhandle	Specify numerical handle of *figure* to print or export
-loose	Use *loose* PostScript bounding box (PostScript, EPS, and Ghost-Script only)
-noui	Suppress printing of *uicontrol* objects
-opengl	Render using OpenGL algorithm (bitmap format)
-painters	Render using Painter's algorithm (vector format)
-Pprinter	Specify name of printer to use (UNIX only)
-rnumber	Specify resolution in dots per inch (dpi). Settable for most devices for printing. Settable for built-in MATLAB file formats except EMF and ILL. Not settable for many GhostScript export formats. Default export resolution is 150 dpi for Z-buffer and OpenGL renderers, and 864 dpi for Painter's renderer.
-swindowtitle	Specify name of SIMULINK system window to print or export
-v	Verbose. Display the **Print** dialog box (Windows only)
-zbuffer	Render using Z-buffer algorithm (bitmap format)

30.3 PRINTERS AND EXPORT FILE FORMATS

The print function supports a number of output devices (printers and file types). Many of these printers are supported through the use of GhostScript printer drivers, which convert PostScript printer code to native printer code. This conversion process is transparent to the user, but limits printable fonts to those supported in PostScript.

The function print also supports exporting to a file or to the clipboard in a number of graphics formats. Export file formats are also specified by the *-device* option to print. Some of the supported file formats are shown in the following table.

-device	Description
-dbmp16m	24-bit BMP (Windows only) (uses GhostScript)
-dbmp256	8-bit BMP with fixed colormap (Windows only) (uses GhostScript)
-dbmp	24-bit BMP (Windows only)

`-dmeta`	EMF (Windows only)
`-deps`	EPS level 1, black and white, including grayscale
`-depsc`	EPS level 1, color
`-deps2`	EPS level 2, black and white, including grayscale
`-depsc2`	EPS level 2, color
`-hdf`	HDF, 24-bit (Windows only)
`-dill`	Adobe Illustrator
`-djpeg`	JPEG, 24-bit, quality setting of 75 (rendered using Z-buffer)
`-djpegNN`	JPEG, 24-bit, quality setting of *NN*
`-dpbm`	PBM plain format (UNIX only) (uses GhostScript)
`-dpbmraw`	PBM raw format (UNIX only) (uses GhostScript)
`-dpcxmono`	PCX, 1-bit (Windows only) (uses GhostScript)
`-dpcx24b`	PCX, 24-bit color (Windows only) (uses GhostScript)
`-dpcx256`	PCX, 8-bit color (Windows only) (uses GhostScript)
`-dpcx16`	PCX, 16 colors (Windows only) (uses GhostScript)
`-dpcx`	PCX, 8-bit color (Windows only)
`-dpgm`	PGM portable graymap, plain (UNIX only) (uses GhostScript)
`-dpgmraw`	PGM portable graymap, raw (UNIX only) (uses GhostScript)
`-dpng`	PNG 24 bit
`-dppm`	PPM portable pixmap, plain (UNIX only) (uses GhostScript)
`-dppmraw`	PPM portable pixmap, raw (UNIX only) (uses GhostScript)
`-dtiff`	TIFF (rendered using Z-buffer)
`-tiff`	Add TIFF preview to EPS formats only—must be used in addition to an EPS device specification

MATLAB also provides methods for exporting images to graphics files. The functions `getframe`, `imwrite`, `avifile`, and `addframe` provide the capability to create and save image files from *figures*. (See Chapter 29 for more information.)

30.4 POSTSCRIPT SUPPORT

All PostScript devices as well as devices that use GhostScript, offer limited font support. This includes both devices that print and those that save images. The following table shows the supported fonts, in addition to common Windows fonts that map into standard PostScript fonts (Note that fonts not listed in the table are mapped to `Courier` and that the Windows devices `-dwin` and `-dwinc` use standard Windows printer drivers and therefore support all installed fonts):

PostScript Font	Windows Equivalent
AvantGarde	
Bookman	
Courier	Courier New
Helvetica	Arial
Helvetica-Narrow	
NewCenturySchlBk	New Century Schoolbook
Palatino	
Symbol	
Times-Roman	Times New Roman
ZapfChancery	
ZapfDingbats	

If your printer supports PostScript, a built-in PostScript driver should be used. Level 1 PostScript is an older specification and is required for some printers. Level 2 PostScript produces smaller and faster code and should be used if possible.

If you are using a color printer, a color driver should be used. Black-and-white or grayscale printers can use either driver. However, when a color driver is used for a black-and-white printer, the file is larger and colors are dithered, making lines and text less clear in some cases. When colored lines are printed by using a black-and-white driver, they are converted to black. When colored lines are printed by using a color driver on a black-and-white printer, the lines are printed in grayscale. In this case, unless the lines have sufficient width, they often do not have sufficient contrast with the printed page.

As implemented in MATLAB, PostScript supports *surfaces* and *patches* (only with triangular faces) that have interpolated shading. When printed, the corresponding PostScript files contain color information at the *surface* or *patch* vertices, requiring the printer to perform the shading interpolation. Depending on the printer characteristics, this may take an excessive amount of time, possibly leading to a printer error. One way to solve this problem is to use flat shading along with a finer meshed surface. Another alternative that ensures that the printed output matches the screen image is to print by using either the Z-buffer or OpenGL renderer with a sufficiently high resolution. In this case, the output is in bitmap format and may result in a large output file, but no interpolation is required by the printer.

30.5 CHOOSING A RENDERER

A renderer processes graphics data such as arrays of vertices and color data into a format suitable for display, printing, or export. There are two major categories of graphics formats: bitmap (or raster) graphics and vector graphics.

Bitmap graphics formats contain information such as color at each point in a grid. As the divisions between points decrease, and the total number of points increases, the resolution of the resulting graphic increases, and the size of the resulting file increases. Increasing the number of bits used to specify the color of each point in the grid increases the total number of possible colors in the resulting graphic, and also increases the size of the resulting file. Increasing the complexity of the graphic has no effect on the size of the resulting output file.

Vector graphics formats contain instructions for recreating the graphic by using points, lines, and other geometric objects. Vector graphics are easily resized and usually produce higher-resolution results than bitmap graphics. However, as the number of objects in the graphic increases, the number of instructions required to recreate the graphic increases, and the size of the resulting file increases. At some point, the complexity of the instructions can become too much for an output device to handle, and the graphic simply cannot be output on the specific output device.

MATLAB supports three rendering methods: OpenGL, Z-buffer, and Painter's. Painter's uses vector format, while OpenGL and Z-buffer produce bitmaps. By default, MATLAB automatically selects a renderer based on the characteristics of the *figure* and the printer driver or file format used.

The renderer MATLAB selects for printing or exporting is not necessarily the same renderer used to display a *figure* on the screen.

MATLAB's default selection can be overridden by the user, and this is often done to make the printed or exported *figure* look the same as it does on the screen or to avoid imbedding a bitmap in a vector-format output file, such as PostScript or EPS.

Examples of some situations that require specific renderers are as follows:

- If the *figure* uses RGB color rather than a single color for surface or patch objects, the graphic must be rendered by using a bitmap method to properly capture color.
- HPGL (-dhpgl) and Adobe Illustrator (-dill) output formats use the Painter's renderer.
- JPEG (-djpeg) and TIFF (-dtiff) output formats always use Z-buffer rendering.
- Lighting effects cannot be reproduced by using the vector-format Painter's renderer, and so a bitmap method must be used in this case as well.
- The OpenGL renderer is the only method that supports transparency.

The renderer used for printing and exporting can be chosen by using dialog boxes or options to the print command, or by setting Handle Graphics properties. The dialog box opened by selecting the **Options** button in the **Print Setup** and **Print** dialog boxes on UNIX platforms and the **Axes and Figures** tab in the **Page Setup** dialog box on all platforms can be used to select a renderer. The -zbuffer, -opengl, and -painters options to the print command select a specific renderer when printing or exporting that overrides any other selections made.

30.6 HANDLE GRAPHICS PROPERTIES

A number of Handle Graphics properties influence the way graphics are printed or exported. Many of the options selected from printing and exporting dialog boxes make changes to these properties for the current *figure*. The following table lists the *figure* properties that influence printing and exporting:

'PropertyName'	'PropertyValue' choices, {default}	Description
Color	[RGB vector]	Sets *figure* background color
InvertHardcopy	[{on} | off]	Determines whether *figure* background color is printed or exported. When set to on, forces a white *figure* background independent of the Color property.
PaperUnits	[{inches} | centimeters | normalized | points]	Units used to measure the size of a printed or exported *figure*
PaperOrientation	[{portrait} | lanscape | rotated]	Orientation of *figure* with respect to paper
PaperPosition	[left bottom width height] vector	Position of *figure* on paper or in exported file. width and height determine the size of the *figure* printed or exported.
PaperPositionMode	[auto | {manual}]	Determines whether PaperPosition width and height are used. When set to auto, the *figure* is printed or exported using the *Figure* window displayed width and height (i.e., output is WYSIWYG)
PaperSize	[width height] vector	Paper size measured in PaperUnits
PaperType	[{usletter} | uslegal | A0 | A1 | A2 | A3 | A4 | A5 | B0 | B1 | B2 | B3 | B4 | B5 | arch-A | arch-B | arch-C | arch-D | arch-E | A | B | C | D | E | tabloid | <custom	Type of paper used. Selecting a paperType sets the PaperSize accordingly

Renderer	[{painters} | zbuffer | OpenGL]	Displays renderer
RendererMode	[{auto} manual]	Determines how the renderer is chosen. When set to auto, MATLAB chooses the renderer automatically and independently for display, printing, and export. When set to manual, the renderer set by the Renderer property is used for display, printing, and export.

Certain *axes* properties also influence printing and exporting. Particularly important are the *axes* tick mode properties:

```
XTickMode [ {auto} | manual ]
YTickMode [ {auto} | manual ]
ZtickMode [ {auto} | manual ]
```

When a *figure* is printed or exported, the resulting graphic is normally rendered at a different size than the *figure* on the screen. Since the width and height are different, MATLAB can rescale the number and placement of tick marks on each axis to reflect the new size. Setting the tick mode properties to 'manual' prevents MAT-LAB from changing the tick marks on the *axes* when the *figure* is printed or exported.

Some *line* properties can be used to advantage for output as well:

```
Color: [ 3-element RGB vector ]
LineStyle: [ {-} | – | : | -. | none ]
LineWidth: [ scalar ]
Marker: [ + | o | * | . | x | square | diamond | v | ^ | > | < | pentagram
         | hexagram | {none} ]
MarkerSize: [ scalar ]
MarkerEdgeColor: [ none | {auto} ] –or– a ColorSpec.
MarkerFaceColor: [ {none} | auto ] –or– a ColorSpec.
```

When colored lines are printed, the result depends on the capabilities of the output device and the printer driver. If a color printer and a color printer driver are used, the result is in color, as expected. If a black-and-white printer driver is used, the result is black and white. If a color printer driver is used with a black-and-white printer, the result is grayscale. This can be a problem, since grayscale is printed by using dithering, which may lead to lines that are not distinct. A different, but related, problem can occur when printing in black and white. When multiple solid lines are printed in black, they lose the distinction that color provides on the screen. In this case, the line properties `'Color'`, `'LineStyle'`, `'LineWidth'`, and `'Marker'` can be used to add distinction to plotted lines.

Finally, it is often advantageous to modify *text* when printing or exporting *figures*. Useful *text* properties include the following:

```
Color: [ 3-element RGB vector ]
FontAngle: [ {normal} | italic | oblique ]
FontName: [ font name ]
FontSize: [ scalar ]
FontUnits: [ inches | centimeters | normalized | {points} | pixels ]
FontWeight: [ light | {normal} | demi | bold ]
```

Increasing the size of *text* strings such as titles and *axes* labels when printing or exporting to a smaller-sized graphic can often make the text easier to read. If the font used in the *figure* is not one of the 11 previously listed fonts supported by MATLAB for PostScript output, changing the font to one of these 11 can prevent unwanted font substitution. Sometimes a bold font shows up better in printed output than a normal-weight font. Changes such as these can often improve the appearance of the printed or exported output. Font characteristics are particularly important when exporting graphics for inclusion in presentation software, where fonts must be large and distinct enough to be readable from across a large room.

30.7 SETTING DEFAULTS

MATLAB sets the following factory default options for printing and exporting:

- *Figure* size is 8-by-6 inches (may be different outside of the United States).
- Orientation is portrait.
- *Figure* and *axes* background colors are inverted to white.
- U.S. letter (8.5-by-11 inch) paper is used if available (may be different outside of the United States).
- *Figure* is centered on the paper.
- *Figure* is cropped.
- Output is RGB (not CYMK).
- Tick marks are recalculated.
- MATLAB chooses the renderer.
- Uicontrols print.
- Print device is `-dps2` on UNIX and `-dwin` on the PC platform.

The default print device is set in the `$TOOLBOX/local/printopt.m` file. Edit this file to change the default print device across MATLAB sessions. If you do not have write access to this file, edit the file by using the command `edit printopt.m`, make your changes, save the `printopt.m` file into a local directory, and make sure the local directory is in the MATLABPATH before `$TOOLBOX/local`. For example, if you use a color PostScript printer for printed output on a UNIX platform, edit `printopt.m` and add the line `dev = '-dpsc2';` to the file at the location specified in the file.

Other option defaults can be changed by setting default properties in the `startup.m` file. For example, `set(0,'DefaultFigurePaperType','A4');` changes the default paper type to A4. The handle 0 addresses the *root* object. The `'DefaultFigurePaperType'` property sets the default value for the *figure* property `'PaperType'`. Adding the prefix `'Default'` to any Handle Graphics property name specifies that the accompanying property value should be used as the default. Default properties for a given object must be set at a higher level in the Handle Graphics hierarchy. For example, default *figure* properties must be set at the *root* object level, and default *axes* properties must be set at the *figure* object or *root* object level. Therefore, defaults are usually set at the *root* level. For example, the code

```
set(0,'DefaultFigurePaperOrientation','landscape');

set(0,'DefaultFigurePaperPosition',[0.25 0.25 10.5 8]);

set(0,'DefaultAxesXGrid','on','DefaultAxesYGrid','on');

set(0,'DefaultAxesLineWidth','1');
```

tells MATLAB to use the landscape mode, fill the page with the plot, print (and display) *x*- and *y*-axis grids, and display and print lines that are 1 point wide, as defaults in all *figures* and *axes*.

Many other options can be set using Handle Graphics properties. Most printing and exporting properties are *figure* and *axes* properties. (For more information on Handle Graphics properties and default values, see the next chapter.)

30.8 PUBLISHING

MATLAB provides additional options for publishing your formatted results, including graphics, to a number of different formats. Publishing in MATLAB provides the ability to execute M-files and capture the output in a formatted document. Document formats supported by MATLAB include HTML, XML, and LaTeX. On PC platforms, publishing to Microsoft Word or Powerpoint (if installed) is also supported. The MATLAB Notebook is another option for publishing on PC platforms. (See Chapter 36 for an example using the MATLAB Notebook.)

Publishing requires that the user define *Cells* (sections of code) in the M-file to be executed. (See the documentation on the MATLAB Editor for additional information on defining Cells.) When an M-file is published, the M-file is executed and the output—including the source code in each Cell, along with any *Command Window* output and graphical output—is formatted and published into document files in the selected format. Text markup is also supported. Additions can include an overall title, section titles, descriptive text, and equations and symbols using TeX formatting. Text may be formatted by using bold text, indented or preformatted text, bulleted lists, monospaced text, or even HTML links.

In addition to the **Publish** entries in the **File** menu of the MATLAB Editor, the publish function is available in the *Command* Window. Publishing preferences can be set in the Editor/Debugger section of the MATLAB **Preferences** menu. Specific instructions and examples of publishing are available in the on-line documentation.

30.9 SUMMARY

MATLAB provides a very flexible system for printing graphics and creating output in many different graphics formats. Most other applications can import one or more of the graphics file formats supported by MATLAB, but many have limited ability to edit the resulting graphic. The best results are achieved if the *figure* is edited and appropriate options are set before the *figure* is printed or exported. The most widely used, flexible output formats are PostScript, EPS, EMF, and TIFF. MATLAB contains native support for all of these formats and uses Ghostscript to translate them into many others.

MATLAB also provides the ability to publish a document that includes portions of an M-file—along with *Command* Window output, graphics output, descriptive text, and comments—in a number of document formats, including HTML, XML, LaTeX, Microsoft Word, and Powerpoint.

31

Handle Graphics

Handle Graphics is the name given to a large collection of low-level graphics features that specify how graphics behave and are displayed in MATLAB. Through interaction with Handle Graphics *objects* and their associated properties, you can gain almost infinite control over the graphical features available in MATLAB. As MATLAB continues to evolve, many Handle Graphics features can be manipulated interactively by using the numerous menus, contextual menus, toolbars, palettes, browsers, and editors available within *Figure* windows. These interactive tools make it possible to customize graphics while requiring minimal knowledge of Handle Graphics. As a result, detailed knowledge of Handle Graphics is required only when noninteractive graphics customization is desired. In this case, Handle Graphics function calls are commonly placed in an M-file.

Because of the diversity and breadth of the graphics features in MATLAB, there are well over one thousand pages of MATLAB-supplied documentation on graphics. As a result, it is simply not possible to provide a comprehensive survey of Handle Graphics in this text. Doing so would at least double the size of this text.

Given the space limitations and broader goals of the text, this chapter develops a basic understanding of Handle Graphics features that provides a basis for learning the many more specific aspects of graphics in MATLAB.

31.1 OBJECTS

Handle Graphics is based on the idea that every component of a MATLAB graphic is an *object*, that each object has a unique identifier, or *handle*, associated with it, and that each object has *properties* that can be modified as desired. Here, the term *object* is defined as a closely related collection of data and functions that form a unique whole. In MATLAB, a graphics object is a distinct component that can be manipulated individually.

All plotting and graphics in MATLAB create graphics objects. Some produce **composite objects**, and some produce **core objects**. Graphical objects in MATLAB are arranged in a hierarchy of parent objects and associated child objects, as shown in the following diagram:

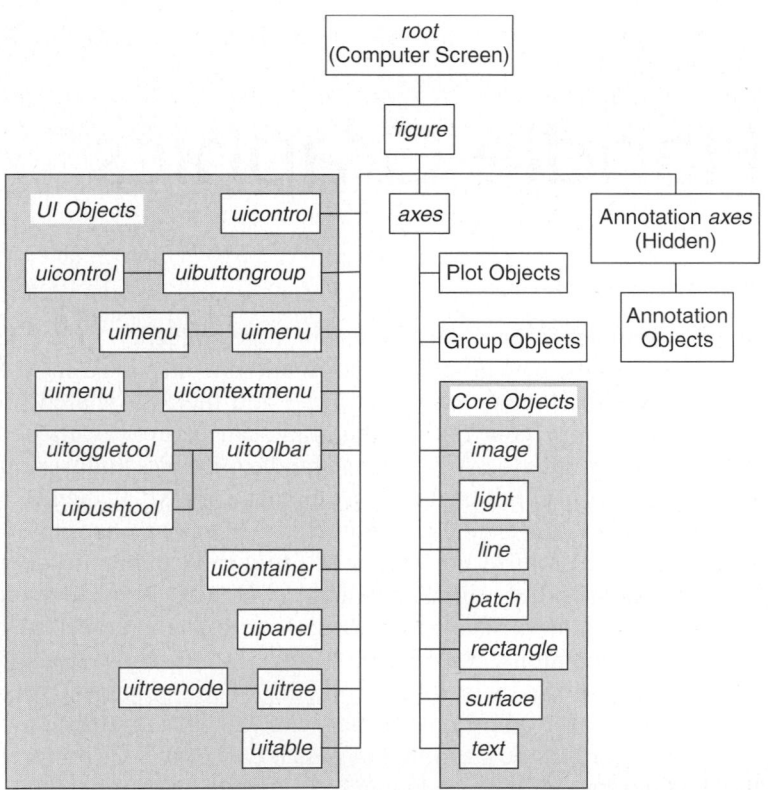

The computer screen itself is the *root* object and the parent of everything else. One or more *figure* objects are children of the *root*. The objects *axes*, annotation *axes*, and the collection of UI objects are children of *figures*. Plot objects, group objects, and the collection of core objects are children of *axes*. All objects that are children of the *figure* object in the hierarchy, except for the core objects, are called composite objects.

This hierarchy is new to MATLAB 7. In MATLAB 6, the graphics hierarchy did not document the hidden annotation *axes* and did not have **plot** objects or **group** objects. Core objects and some UI objects existed, but were simply not grouped and named. UI objects and core objects behave in the same way in MATLAB 7 as they did in previous MATLAB versions, whereas plot objects and group objects are new in MATLAB 7.

31.2 OBJECT HANDLES

In MATLAB, each object has an associated identifier, called a *handle*, which is a double-precision number. Each time an object is created, a unique handle is created for it. The handle of the *root* object, the computer screen, is always zero. The Hf_fig=figure command creates a new *figure* and returns its handle in the Hf_fig variable. *Figure* handles are normally integers and are usually displayed in the *Figure* window title bar. Other object handles are typically noninteger numbers in full MATLAB precision. All object creation functions return the handles of the objects they create.

Before MATLAB 7, the high-level graphics functions (discussed in Chapters 26 and 27) returned a column vector of handles to the core objects they created. For example, Hl=plot(...) returned handles to all *line* objects created by the plot function. Similarly, Hs=surf(...) returned a handle to a *surface* object. Now, in MATLAB 7, many high-level graphics creation functions return handles to plot objects. That is, Hls=plot(...) returns handles to *lineseries* plot objects. In the presence of these changes in the graphics hierarchy, fundamental operations using Handle Graphics remain unchanged. As a result, before these new features are addressed, the fundamental creation and manipulation of Handle Graphics objects through use of their handles is discussed.

It is suggested that variable names used to store handles start with an upper case H, followed by one or more letters identifying the object type, then an underscore, and finally one or more descriptive characters. Thus, Hf_fig is a handle to a *figure*, Ha_ax1 is a handle to an *axes* object, and Ht_title is a handle to a *text* object. When an object type is unknown, the letter x is used, as in Hx_obj. While handles can be given any name, following this convention makes it easy to spot object handle variables in an M-file.

31.3 OBJECT PROPERTIES

All objects have a set of ***properties*** that define their characteristics. It is by setting these properties that you modify how graphics are displayed. The properties associated with each object type (e.g., *axes*, *line*, and *surface*) are unique, although a number of property names are common to all objects. Object properties include characteristics such as an object's position, color, object type, parent object handle, child object handles, and many others. Each distinct object has properties associated with it that can be changed without affecting other objects of the same type.

Object properties consist of property ***names*** and their associated property ***values***. Property names are character strings. They are typically displayed in mixed case, with the initial letter of each word capitalized. For example, the property name 'LineStyle' identifies the line style for a *line* object. When identifying a property name in an M-file or at the *Command* window prompt, MATLAB recognizes properties regardless of case. In addition, you need use only enough characters to uniquely identify the property name. For example, the position property of an *axes* object can be called `'Position'`, `'position'`, or even `'pos'`.

When an object is created, it is initialized with a full set of default property values that can be changed in either of two ways. The object-creation function can be issued with (`'Property-name'`, `Property-value`) pairs, or property values can be changed after the object is created. For example,

```
>> Hf_1 = figure('Color','yellow')
```

creates a new *figure* with default properties, except that the background color is set to yellow rather than to the default color.

In addition to the *Figure* window menu and toolbar features in MATLAB, the function `inspect` provides a GUI for inspection and modification of object properties. To use this function, simply issue `inspect(H)`, where H is the handle of the desired object.

31.4 get AND set

The two functions `get` and `set` are used to obtain or change Handle Graphics object properties. The function `get` returns the current value of one or more properties of an object. The most common syntax is `get(handle,'PropertyName')`. For example,

```
>> p = get(Hf_1,'Position')
```

returns the position vector of the *figure* having the handle Hf_1. Similarly,

```
>> c = get(Hl_a,'Color')
```

returns the color of an object identified by the handle Hl_a.

The function `set` changes the values of Handle Graphics object properties and uses the syntax `set(handle,'PropertyName',PropertyValue)`. For example,

```
>> set(Hf_1,'Position',p_vect)
```

sets the position of the *figure* having the handle Hf_1 to that specified by the vector p_vect. Likewise,

```
>> set(Hl_a,'Color','r')
```

sets the color of the object having the handle Hl_a to red. In general, the `set` function can have any number of (`'PropertyName'`, `PropertyValue`) pairs. For instance,

```
>> set(Hl_a,'Color',[1 0 0],'LineWidth',2,'LineStyle','--')
```

changes the color of the *line* having the handle Hl_a to red, its line width to 2 points, and its line style to dashed.

In addition to these primary purposes, the get and set functions provide feed-back about object properties. For example, set(handle,'PropertyName') returns a list of values that can be assigned to the object described by handle. For example,

```
>> set(Hf_1,'Units')
[ inches | centimeters | normalized | points | {pixels} | characters ]
```

shows that there are six allowable character-string values for the 'Units' property of the *figure* referenced by Hf_1 and that 'pixels' is the default value.

If you specify a property that does not have a fixed set of values, MATLAB informs you of that fact:

```
>> set(Hf_1,'Position')
A figure's 'Position' property does not have a fixed set of property values.
```

In addition to the set command, Handle Graphics object-creation functions accept multiple pairs of property names and values. For example,

```
>> figure('Color','blue','NumberTitle','off','Name','My Figure')
```

creates a new *figure* with a blue background entitled 'My Figure', rather than the default window title 'Figure 1'.

As an illustration of the preceding concepts, consider the following example:

```
>> Hf_fig = figure  % create a figure
Hf_fig =
     1
>> Hl_light = light % add default light to an axes in the figure
Hl_light =
        107
>> set(Hl_light)    % find settable properties of light
    Position
    Color
    Style: [ {infinite} | local ]

    ButtonDownFcn: string -or- function handle -or- cell array
    Children
    Clipping: [ {on} | off ]
    CreateFcn: string -or- function handle -or- cell array
    DeleteFcn: string -or- function handle -or- cell array
```

```
      BusyAction: [ {queue} | cancel ]
      HandleVisibility: [ {on} | callback | off ]
      HitTest: [ {on} | off ]
      Interruptible: [ {on} | off ]
      Parent
      Selected: [ on | off ]
      SelectionHighlight: [ {on} | off ]
      Tag
      UIContextMenu
      UserData
      Visible: [ {on} | off ]

>> get(Hl_light)    % get all properties and names for light
      Position = [1 0 1]
      Color = [1 1 1]
      Style = infinite

      BeingDeleted = off
      ButtonDownFcn =
      Children = []
      Clipping = on
      CreateFcn =
      DeleteFcn =
      BusyAction = queue
      HandleVisibility = on
      HitTest = on
      Interruptible = on
      Parent = [108]
      Selected = off
      SelectionHighlight = on
      Tag =
      Type = light
      UIContextMenu = []
      UserData = []
      Visible = on
```

The *light* object was used because it contains the fewest properties of any MATLAB object. A *figure* was created and returned a handle. A *light* object was then created that returned a handle of its own. An *axes* object was also created, since *light* objects are children of *axes*; the *axes* handle is available in the 'Parent' property of the *light*.

Note that the property lists for each object are divided into two groups. The first group lists properties that are unique to the particular object type, and the second group lists properties common to all object types. Note also that the set and get functions return slightly different property lists. The function set lists only properties that can be changed with the set command, while get lists all visible object properties. In the previous example, get listed the 'Type' property, while set did not. This property can be read but not changed; that is, it is a read-only property.

The number of properties associated with each object type is fixed in each release of MATLAB, but the number varies among object types. As already shown, a *light* object lists 3 unique and 18 common properties, or 21 properties in all. On the other hand, an *axes* object lists 103 properties. Clearly, it is beyond the scope of this text to thoroughly describe and illustrate all of the properties of all object types!

As an example of the use of object handles, consider the problem of plotting a line in a nonstandard color. In this case, the line color is specified by using an RGB value of [1 .5 0], a medium orange color:

```
>> x = -2*pi:pi/40:2*pi;  % create data
>> y = sin(x);            % find sine of x
>> Hls_sin = plot(x,y)    % plot sine and save lineseries handle
Hls_sin -
   59.0002
>> set(Hls_sin,'Color',[1 .5 0],'LineWidth',3) % Change color and width
```

Now add a cosine curve in light blue:

```
>> z = cos(x);            % find cosine of x
>> hold on                % keep sine curve
>> Hls_cos = plot(x,z);   % plot cosine and save lineseries handle
>> set(Hls_cos,'Color',[.75 .75 1])  % color it light blue
>> hold off
```

It's also possible to do the same thing with fewer steps:

```
>> Hls_line = plot(x,y, x,z);       % plot both curves and save handles
>> set(Hls_line(1),'Color',[1 .5 0],'LineWidth', 3)
>> set(Hls_line(2),'Color',[.75 .75 1])
```

How about adding a title and making the font size larger than normal?

```
>> title('Handle Graphics Example') % add a title
>> Ht_text = get(gca,'Title')       % get handle to title
>> set(Ht_text,'FontSize',16)       % customize font size
```

The last example illustrates an interesting point about *axes* objects. Every object has a `'Parent'` property, as well as a `'Children'` property, which contains handles to descendant objects. A *lineseries* object on a set of axes has the handle of the *axes* object in its `'Parent'` property and the empty array in the `'Children'` property. At the same time, the *axes* object has the handle of its *figure* in its `'Parent'` property and the handles of *lineseries* objects in the `'Children'` property. Even though *text* objects created with the `text` and `gtext` commands are children of *axes* and their handles are included in the `'Children'` property, the handles associated with the title string and axis labels are not. These handles are kept in the `'Title'`, `'XLabel'`, `'YLabel'`, and `'ZLabel'` properties of the *axes*. These *text* objects are always created when an *axes* object is created. The `title` command simply sets the `'String'` property of the title *text* object within the current *axes*. Finally, the standard MATLAB functions `title`, `xlabel`, `ylabel`, and `zlabel` return handles and accept property and value arguments. For example, the following command adds a 24-point green title to the current plot and returns the handle of the title *text* object:

```
>> Ht_title = title('This is a title.','FontSize',24,'Color','green')
```

In addition to `set` and `get`, MATLAB provides several other functions for manipulating objects and their properties. Objects can be ***copied*** from one parent to another using the `copyobj` function. For instance,

```
>> Ha_new = copyobj(Ha_ax1,Hf_fig2)
```

makes copies of the *axes* object with handle `Ha_ax1` and all of its children, assigns new handles, and places the objects in the *figure* with handle `Hf_fig2`. A handle to the new *axes* object is returned in `Ha_new`. Any object can be copied into any valid parent object based on the hierarchy described earlier. Either one or both arguments to `copyobj` can be vectors of handles.

Note that any object can be ***moved*** from one parent to another simply by changing the `'Parent'` property value to the handle of another valid parent object. For example,

```
>> figure(1)
```

```
>> set(gca,'Parent',2)
```

moves the current *axes* and all of its children from the *figure* having handle 1 to the *figure* having handle 2. Any existing objects in *figure* 2 are not affected, except that they may become obscured by the relocated objects.

Any object and all of its children can be **deleted** using the delete(handle) function. Similarly, reset(handle) resets all object properties associated with handle, except for the 'Position' property, to the defaults for that object type. If handle is a column vector of object handles, all referenced objects are affected by set, reset, copyobj, and delete.

The get and set functions return a structure when an output is assigned. Consider the following example:

```
>> lprop = get(H1_light)
lprop =
          BeingDeleted: 'off'
            BusyAction: 'queue'
         ButtonDownFcn: ''
              Children: [0x1 double]
              Clipping: 'on'
                 Color: [1 1 1]
             CreateFcn: ''
             DeleteFcn: ''
      HandleVisibility: 'on'
               HitTest: 'on'
         Interruptible: 'on'
                Parent: 108
              Position: [1 0 1]
              Selected: 'off'
    SelectionHighlight: 'on'
                 Style: 'infinite'
                   Tag: ''
                  Type: 'light'
         UIContextMenu: []
              UserData: []
               Visible: 'on'
>> class(lprop) % class of get(H1_light)
ans =
struct

>> lopt = set(H1_light)
lopt =
```

```
          BusyAction: {2x1 cell}
       ButtonDownFcn: {}
            Children: {}
            Clipping: {2x1 cell}
               Color: {}
           CreateFcn: {}
           DeleteFcn: {}
    HandleVisibility: {3x1 cell}
             HitTest: {2x1 cell}
       Interruptible: {2x1 cell}
              Parent: {}
            Position: {}
            Selected: {2x1 cell}
    SelectionHighlight: {2x1 cell}
               Style: {2x1 cell}
                 Tag: {}
       UIContextMenu: {}
            UserData: {}
             Visible: {2x1 cell}
>> class(lopt) % class of set(Hl_light)
ans =
struct
```

The field names of the resulting structures are the object property-name strings and are assigned alphabetically. Note that even though property names are not case sensitive, these field names are:

```
>> lopt.BusyAction
ans =
    'queue'
    'cancel'
>> lopt.busyaction
??? Reference to non-existent field 'busyaction'.
```

Combinations of property values can be set by using structures as well. For example,

```
>> newprop.Color = [1 0 0];
>> newprop.Position = [-10 0 10];
```

```
>> newprop.Style = 'local';
>> set(Hl_light,newprop)
```

changes the 'Color', 'Position', and 'Style' properties, but has no effect on any other properties of the *light* object. Note that you cannot simply obtain a structure of property values using get and use the same structure to set values:

```
>> light_prop = get(Hl_light);
>> light_prop.Color = [1 0 0];   % change the light color to red
>> set(Hl_light,light_prop);     % reapply the property values
??? Error using ==> set
Attempt to modify read-only light property: 'Type'.
```

Since 'Type' is the only read-only property of a *light* object, you can work around the problem by removing the 'Type' field from the structure:

```
>> light_prop = rmfield(light_prop,'Type');
>> set(Hl_light,light_prop)
```

For objects with more read-only properties, all read-only properties must be removed from the structure before using it to set property values.

A cell array can also be used to query a selection of property values. To do so, create a cell array containing the desired property names in the desired order and pass the cell array to get. The result is returned as a cell array as well:

```
>> plist = {'Color','Position','Style'}
plist =
    'Color'     'Position'    Style'
>> get(Hl_light,plist)
ans =
    [ double]    [1x3 double]    'local'
>> class(ans) % cell array in, cell array out
ans =
cell
```

One more point about the get function should be noted. If H is a vector of handles, get(H, 'Property') returns a cell array rather than a vector. Consider the following example, given a *Figure* window with four subplots:

```
>> Ha = get(gcf,'Children')      % get axes handles
Ha =
      15.0002
      13.0002
```

```
        11.0002
         9.0002
>> Ha_kids = get(Ha,'Children')   % get handles of axes children
Ha_kids =
    [    16.0002]
    [4x1 double]
    [    12.0002]
    [2x1 double]]
>> class(Ha_kids)
ans =
    cell
>> Hx = cat(1,Ha_kids{:})          % convert to column vector
Hx =
        16.0002
        26.0002
        24.0002
        18.0002
        22.0002
        12.0002
        14.0002
        10.0002
>> class(Hx)
ans =
    double
```

Now Hx can be used as an argument to Handle Graphics functions expecting a vector of object handles.

31.5 FINDING OBJECTS

As has been shown, Handle Graphics provides access to objects in a *figure* and allows the user to customize graphics by using the get and set commands. The use of these functions requires that you know the handles of the objects to be manipulated. In cases where handles are unknown, MATLAB provides a number of functions for finding object handles. Two of these functions, gcf and gca, were introduced earlier. For example,

```
>> Hf_fig = gcf
```

returns the handle of the current *figure*, and

```
>> Ha_ax = gca
```

returns the handle of the current *axes* in the current *figure*.

In addition to the above, MATLAB includes gco, a function to obtain the handle of the current object. For example,

```
>> Hx_obj = gco
```

returns the handle of the current object in the current *figure*; alternatively,

```
>> Hx_obj = gco(Hf_fig)
```

returns the handle of the current object in the *figure* associated with the handle Hf_fig.

The current object is defined as the last object clicked on with the mouse within a *figure*. This object can be any graphics object except the *root*. When a *figure* is initially created, no current object exists, and gco returns an empty array. The mouse button must be clicked while the pointer is within a *figure* before gco can return an object handle.

Once an object handle has been obtained, the object type can be found by querying an object's 'Type' property, which is a character-string object name such as 'figure', 'axes', or 'text'. The 'Type' property is common to all objects. For example,

```
>> x_type = get(Hx_obj,'Type')
```

is guaranteed to return a valid object string for all objects.

When something other than the 'CurrentFigure', 'CurrentAxes', or 'CurrentObject' is desired, the function get can be used to obtain a vector of handles to the children of an object. For example,

```
>> Hx_kids = get(gcf,'Children');
```

returns a vector containing handles of the children of the current *figure*.

To simplify the process of finding object handles, MATLAB contains the built-in function findobj, which returns the handles of objects with specified property values. The form Hx=findobj(Handles,'flat','PropertyName',PropertyValue) returns the handles of all objects in Handles whose 'PropertyName' property contains the value PropertyValue. Multiple ('PropertyName', PropertyValue) pairs are allowed, and all must match. When Handles is omitted, the *root* object is assumed. When no ('PropertyName',PropertyValue) pairs are given, all objects match and all Handles are returned. When 'flat' is omitted, all objects in Handles **and all descendants of these objects**, including axes titles and labels, are searched. When no objects are found to match the specified criteria, findobj returns an empty array. As an example, finding all green *line* objects is easily accomplished with the following statement:

```
>> Hl_green = findobj(0,'Type','line','Color',[0 1 0]);
```

It is possible to hide specific handles by using the `'HandleVisibility'` property common to all objects. This property is convenient, because it keeps the user from inadvertently deleting or changing the properties of an object. When an object has its `'HandleVisibility'` property set to `'off'` or to `'callback'`, `findobj` does not return handles to these objects when called from the *Command* window. Hidden handles do not appear in lists of children or as the output of `gcf`, `gca`, or `gco`. However, when the property is set to `'callback'`, these handles can be found during the execution of a callback (which is discussed in the next chapter).

31.6 SELECTING OBJECTS WITH THE MOUSE

The `gco` command returns the handle of the current object, which is the last object clicked on with the mouse. When a mouse click is made near the intersection of more than one object, MATLAB uses rules to determine which object becomes the current object. Each object has a selection region associated with it. A mouse click within this region selects the object. For *line* objects, the selection region includes the *line* and all of the area within a 5-pixel distance from the *line*. The selection region of a *surface*, *patch*, or *text* object is the smallest rectangle that contains the object. The selection region of an *axes* object is the *axes* box itself, plus the areas where labels and titles appear. Objects within *axes*, such as *lines* and *surfaces*, are higher in the stacking order, and clicking on them selects the associated object rather than the *axes*. Selecting an area outside of the *axes* selection region selects the *figure* itself.

When a mouse click is within the border of two or more objects, the ***stacking order*** determines which object becomes the current object. The stacking order determines which overlapping object is on *top* of the others. Initially, the stacking order is determined when the object is created, with the newest object at the top of the stack. For example, when you issue two `figure` commands, two *figures* are created. The second *figure* is drawn on top of the first. The resulting stacking order has *figure* 2 on top of *figure* 1, and the handle returned by `gcf` is 2. If the `figure(1)` command is issued or if *figure* 1 is clicked on, the stacking order changes. *Figure* 1 moves to the top of the stack and becomes the current *figure*.

In the preceding example, the stacking order was apparent from the window overlap on the computer screen. However, this is not always the case. When two *lines* are plotted, the second *line* drawn is on top of the first at the points where they intersect. If the first *line* is clicked on with the mouse at some other point, the first *line* becomes the current object, but the stacking order does not change. A click on the intersecting point continues to select the second *line* until the stacking order is explicitly changed.

The stacking order is given by the order in which `'Children'` handles appear for a given object. That is, `Hx_kids=get(handle,'Children')` returns handles of child objects, in the stacking order. The first element in the vector `Hx_kids` is at the top of the stack, and the last element is at the bottom of the stack. The stacking order can be changed by changing the order of the `'Children'` property value of the parent object. For example,

```
>> Hf = get(0,'Children');
>> if length(Hf) > 1
     set(0,'Children',Hf([end 1:end-1]));
   end
```

moves the bottom *figure* to the top of the stack, where it becomes the new current *figure*.

31.7 POSITION AND UNITS

The `'Position'` property of *figure* objects and of most other Handle Graphics objects is a four-element row vector. As shown in the following figure, the values in this vector are [`left, bottom, width, height`], where [`left, bottom`] is the position of the lower left corner of the object relative to its parent and [`width, height`] is the width and height of the object.

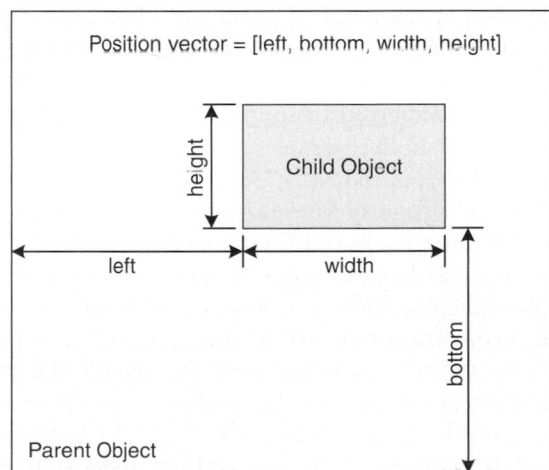

These position vector values are in units specified by the `'Units'` property of the object. For example, the code

```
>> get(gcf,'Position')
ans =
```

```
     920    620    672    504
>> get(gcf,'units')
ans =
pixels
```

shows that the lower left corner of the current *figure* object is 920 pixels to the right and 620 pixels above the lower left corner of the screen and that the *figure* object is 672 pixels wide and 504 pixels high. ***Note that the 'Position' vector for a figure gives the drawable area within the figure object itself and does not include window borders, scroll bars, menus, or the title bar of the Figure window.***

Alternatively, *figures* have an undocumented 'OuterPosition' property that does include the window borders:

```
>> get(gcf,'Position')      % drawable position
ans =
     920    620    672    504
>> get(gcf,'OuterPosition') % outside position
ans =
     916    616    680    531
```

Here, the outer position contains the left, bottom, width, and height of the outer boundary of the *Figure* window. When the *Figure* or *Camera* toolbar is displayed or hidden in response to a choice from the **View** menu in a *Figure* window, either the drawable position or the outer position parameters can be maintained by setting the 'ActivePositionProperty' of the *figure*. Setting this property to 'Position' gives priority to the drawable position, thereby keeping the 'Position' values unchanged when toolbars are displayed or hidden. Setting this property to 'OuterPosition' gives similar priority to the 'OuterPosition' values, leaving them unchanged when toolbars are displayed or hidden.

The 'Units' property for *figures* defaults to pixels, but can be inches, centimeters, points, characters, or normalized coordinates. Pixels represent screen pixels, the smallest rectangular object that can be represented on a computer screen. For example, a computer display set to a resolution of 800 by 600 is 800 pixels wide and 600 pixels high. Points are a typesetting standard, where 1 point is equal to 1/72 of an inch. Character units are units relative to the width of a character in the default system font. A value of 1 is equal to the width of the letter x in the default system font. Normalized coordinates are in the range 0 to 1. In normalized coordinates, the lower left corner of the parent is at [0, 0] and the upper right corner is at [1, 1]. Inches and centimeters are self-explanatory.

To illustrate various 'Units' property values, consider again the preceding example:

```
>> set(gcf,'units','inches')    % INCHES
>> get(gcf,'position')
```

```
ans =

        7.9224          5.3362          5.7931          4.3448
>> set(gcf,'units','cent')       % CENTIMETERS
>> get(gcf,'position')
ans =

        20.108          13.544          14.703          11.027
>> set(gcf,'units','normalized')% NORMALIZED
>> get(gcf,'position')
ans =

        0.57438         0.51583         0.42            0.42
>> set(gcf,'units','points')     % POINTS
>> get(gcf,'position')
ans =

        570.41          384.21          417.1           312.83
>> set(gcf,'units','char')       % CHARACTERS
>> get(gcf,'position')
ans =

        153.17          38.688          112             31.5
```

All of these values represent the same *figure* position relative to the computer screen for a particular monitor and screen resolution.

The positions of *axes* objects are also four-element vectors having the same form, [left, bottom, width, height], but specifying the object position relative to the lower left corner of the parent *figure*. **In general, the 'Position' property of a child is relative to the position of its parent.**

To be more descriptive, the computer screen, or *root* object, position property is not called 'Position', but rather 'ScreenSize'. In this case, [left, bottom] is always [0, 0], and [width, height] are the dimensions of the computer screen in units specified by the 'Units' property of the *root* object.

31.8 DEFAULT PROPERTIES

MATLAB assigns default properties to each object as it is created. The built-in defaults are referred to as *factory* defaults. To override these defaults, you must set or get the values using set and get. In cases where you want to change the same properties every time, MATLAB allows you to set your own default properties. It lets you change the default properties for individual objects and for object types at any point in the object hierarchy. When an object is created, MATLAB looks for a default value at the ***parent*** level. If no default is found, the search continues up the object hierarchy until a default value is found or until it reaches the built-in factory default.

You can set your own default values at any level of the object hierarchy by using a special property-name string consisting of `'Default'`, followed by the object type and the property name. The handle you use in the `set` command determines the point in the object parent-child hierarchy at which the default is applied. For example,

```
>> set(0,'DefaultFigureColor',[.5 .5 .5])
```

sets the default background color for all new *figure* objects to medium gray rather than the MATLAB default. This property applies to the *root* object (whose handle is always zero), and so all new *figures* will have a gray background. Other examples include the following:

```
>> set(0,'DefaultAxesFontSize',14)      % larger axes fonts - all figures
>> set(gcf,'DefaultAxesLineWidth',2)    % thick axis lines - this figure only
>> set(gcf,'DefaultAxesXColor','y')     % yellow X axis lines and labels
>> set(gcf,'DefaultAxesYGrid','on')     % Y axis grid lines - this figure
>> set(0,'DefaultAxesBox','on')         % enclose axes - all figures
>> set(gca,'DefaultLineLineStyle',':') % dotted linestyle - these axes only
```

When a default property is changed, only objects created *after* the change is made are affected. Existing objects already have property values assigned and do not change.

When working with existing objects, it is always a good idea to restore them to their original state after they are used. If you change the default properties of objects in an M-file, save the previous settings and restore them when exiting the routine:

```
oldunits = get(0,'DefaultFigureUnits');

set(0,'DefaultFigureUnits','normalized');

    <MATLAB statements>

set(0,'DefaultFigureUnits',oldunits);
```

To customize MATLAB to use user-defined default values at all times, simply include the desired `set` commands in your `startup.m` file. For example, the code

```
set(0,'DefaultAxesXGrid','on')

set(0,'DefaultAxesYGrid','on')
```

```
set(0,'DefaultAxesZGrid','on')
set(0,'DefaultAxesBox','on')
set(0,'DefaultFigurePaperType','A4')
```

creates all axes with grids, and an enclosing box is turned on and sets the default paper size to A4. Defaults set at the *root* level affect every object in every *Figure* window.

There are three special property value strings that reverse, override, or query user-defined default properties. They are 'remove', 'factory', and 'default'. If you've changed a default property, you can reverse the change, thereby resetting it to the original defaults using 'remove':

```
>> set(0,'DefaultFigureColor',[.5 .5 .5])  % set a new default
>> set(0,'DefaultFigureColor','remove')    % return to MATLAB defaults
```

To temporarily override a default and use the original MATLAB default value for a particular object, use the special property value 'factory':

```
>> set(0,'DefaultFigureColor',[.5 .5 .5])   % set a new user default
>> figure('Color','factory')   % figure using default color
```

The third special property value string is 'default'. This value forces MATLAB to search up the object hierarchy until it encounters a default value for the desired property. If one is found, MATLAB uses this default value. If the *root* object is reached and no user-defined default is found, the MATLAB factory default value is used. This feature is useful when you want to set an object property to a default property value after it was created with a different property value, as in the following code:

```
>> set(0,'DefaultLineColor','r')  % set default at the root level
>> set(gcf,'DefaultLineColor','g') % current figure level default
>> Hl_rand = plot(rand(1,10));     % plot a line using 'ColorOrder' color
>> set(Hl_rand,'Color','default') % the line becomes green
>> close(gcf)                      % close the window
>> Hl_rand = plot(rand(1,10));     % plot a line using 'ColorOrder' color again
>> set(Hl_rand,'Color','default') % the line becomes red
```

Note that the plot command does not use *line* object defaults for the line color. If a color argument is not specified, the plot command uses the *axes* 'ColorOrder' property to specify the color of each *line* it generates.

A list of all of the factory defaults can be obtained by issuing

```
>> get(0,'factory')
```

Default properties that have been set at any level in the object hierarchy can be listed by issuing

```
>> get(handle,'default')
```

The *root* object contains default values for a number of color properties and the *figure* position at startup:

```
>> get(0,'default')
ans =
                    defaultTextColor: [0 0 0]
                   defaultAxesXColor: [0 0 0]
                   defaultAxesYColor: [0 0 0]
                   defaultAxesZColor: [0 0 0]
              defaultPatchFaceColor: [0 0 0]
              defaultPatchEdgeColor: [0 0 0]
                    defaultLineColor: [0 0 0]
        defaultFigureInvertHardcopy: 'on'
                  defaultFigureColor: [0.8 0.8 0.8]
                    defaultAxesColor: [1 1 1]
               defaultAxesColorOrder: [7x3 double]
               defaultFigureColormap: [64x3 double]
             defaultSurfaceEdgeColor: [0 0 0]
               defaultFigurePosition: [920 620 672 504]
```

Other defaults are not listed until they have been created by the user:

```
>> get(gcf,'default')
ans =
0x0 struct array with fields:
>> set(gcf,'DefaultLineMarkerSize',10)
>> get(gcf,'default')
ans =
    defaultLineMarkerSize: 10
```

31.9 COMMON PROPERTIES

All Handle Graphics objects share the common set of object properties shown in the following table:

Property	Description
BeingDeleted	Set to 'on' when object is about to be deleted
BusyAction	Controls how MATLAB handles callback interuptions
ButtonDownFcn	Callback code executed when the mouse button is pressed over the object
Children	Handles of visible children
Clipping	Enable or disable clipping of *axes* children
CreateFcn	Callback code executed immediately *after* an object is created
DeleteFcn	Callback code executed immediately *before* an object is deleted
BusyAction	Determines how callbacks to this object are interrupted by other callbacks
HandleVisibility	Determines whether the object handle is visible in the *Command* window or while executing callbacks
HitTest	Determines whether the object can be selected with the mouse and then become the current object
Interruptible	Determines whether callbacks to this object are interruptible
Parent	Handle of parent object
Selected	Determines whether an object has been selected as the current object
SelectionHighlight	Determines whether a selected object shows visible selection handles or not
Tag	User-defined character string used to identify or *tag* the object. Often useful in association with findobj. For example, findobj(0,'tag','mytagstring').
Type	Character string identifying object type
UIContextMenu	Handle of contextual menu associated with an object
UserData	Storage of any user-defined variable associated with an object
Visible	Visibility of object

Three of these properties contain *callbacks*: 'ButtonDownFcn', 'CreateFcn', and 'DeleteFcn'. Callbacks identify MATLAB code to be executed when the action described by the property occurs. In many cases, the code is a function call. The 'Parent' and 'Children' properties contain handles of other objects in the hierarchy. Objects drawn on the *axes* are clipped at the *axes* limits if 'Clipping' is 'on', which is the default for all axes children except *text* objects. 'Interruptible' and 'BusyAction' control the behavior of callbacks if a later callback is triggered while a previous callback is currently executing. `'Type'` is a string specifying the object type. `'Selected'` is `'on'` if this object is the `'CurrentObject'` of the *figure*, and `'SelectionHighlight'` determines whether the object changes appearance when selected. `'HandleVisibility'` specifies whether the object handle is visible, invisible, or visible only to callbacks. The *root* `'ShowHiddenHandles'` property overrides the `'HandleVisibility'` property of all objects, if needed. If `'Visible'` is set to `'off'`, the object disappears from view. It is still there, and the object handle is still valid, but it is not rendered on the computer screen. Setting `'Visible'` to `'on'` restores the object to view. The `'Tag'` and `'UserData'` properties are reserved for the user. The `'Tag'` property is typically used to tag an object with a user-defined character string. For example,

```
>> set(gca,'Tag','My Axes')
```

attaches the string `'My Axes'` to the current *axes* in the current *figure*. This string does not display in the *axes* or in the *figure*, but you can query the `'Tag'` property to identify the object. For instance, when there are numerous *axes*, you can find the handle to the above *axes* object by issuing the command

```
>> Ha_myaxes = findobj(0,'Tag','My Axes');
```

The `'UserData'` property can contain any variable you wish to put into it. A character string, a number, a structure, or even a multidimensional cell array can be stored in any object's `'UserData'` property. No MATLAB function changes, or makes assumptions about, the values contained in these properties.

The properties listed for each object using the `get` and `set` commands are the documented properties. There are also undocumented, or hidden, properties used by MATLAB developers. Some of them can be modified, but others are read-only. Undocumented properties are simply hidden from view. These properties still exist and can be modified. The undocumented *root* property `'HideUndocumented'` controls whether `get` returns all properties or only documented properties. For example,

```
>> set(0,'HideUndocumented','off')
```

makes undocumented object properties visible within MATLAB.

Since undocumented properties have been purposely left undocumented, one must be cautious when using them. They are sometimes less robust than documented properties and are always subject to change. Undocumented properties may appear, disappear, change functionality, or even become documented in future versions of MATLAB.

31.10 PLOT OBJECTS

As stated earlier, MATLAB 7 introduces plot objects and group objects. Plot objects are objects associated with the high-level graphics functions discussed in Chapters 26 and 27. Plot objects were created in MATLAB 7 as a way to group the core objects created by high-level graphics functions, so that the overall properties of the graphics are more easily identified and modified. That is, plot objects have additional properties that do not exist in the core objects used in their creation. These additional properties specify properties specific to the type of graphic created. For example, consider the creation of a bar graph:

```
>> hbs_mm7 = bar(randn(1,6));
>> title('Figure 31.1 Random Bar Graph')
>> get(hbs_mm7,'type')
ans =
hggroup
>> hx_hbs = get(hbs_mm7,'children');
>> get(hx_hbs,'type')
ans =
patch
```

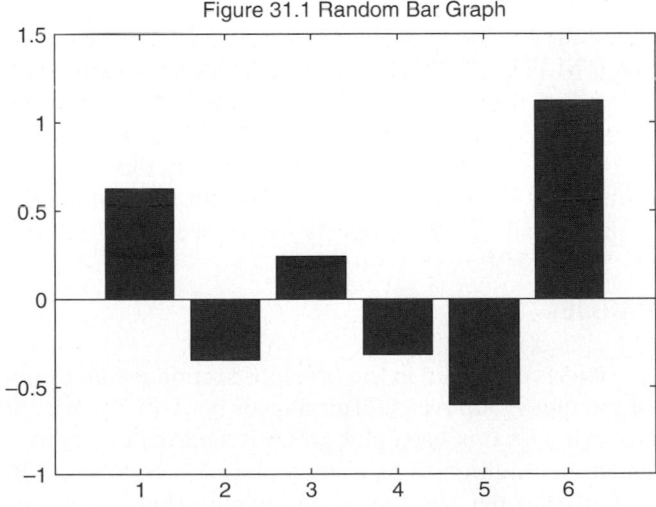

Figure 31.1 Random Bar Graph

This bar graph contains a composite *barseries* plot object whose handle type is designated hggroup. The children of this object are single *patch* objects. Rather than have all of the properties of the *patch* object, the *barseries* object has properties such as 'BaseLine', 'BarLayout', and 'BarWidth', which directly specify aspects of the created bar chart. Changing these properties in turn changes the appropriate properties of the underlying *patch* object. By adding this layer on top of the *patch* object, manipulation of the bar chart is much easier. You do not need to know how to manipulate *patch* object properties to make changes in the bar chart.

The plot objects in MATLAB 7 are shown in the following table (also included is the `'type'` property returned when `get(H,'type')` is called, where H is a handle of the corresponding plot object):

Plot Object	**Get(H,'type')**	**Created by the graphics functions**
areasseries	hggroup	area
barseries	hggroup	bar, bar3, bar3h.
contourgroup	hggroup	contour, contour3, contourf
errorbarseries	hggroup	errorbar
lineseries	line	plot, plot3, semilogx, semilogy, loglog
quivergroup	hggroup	quiver, quiver3
scattergroup	hggroup	scatter, scatter3
stairseries	hggroup	stairs
stemseries	hggroup	stem, stem3
surfaceplot	surface	surf, mesh

Since the functions in this table returned handles to core objects in previous versions of MATLAB, MATLAB 7 provides backward compatibility when `'v6'` is used as the first argument to these functions. For example, `H1 = plot('v6', ...)` returns handles to *line* objects rather than to *lineseries* objects.

As with the core objects, there are simply too many plot object properties to discuss each one here. (For further information about all object properties and the action taken by all of their property values, see the MATLAB documentation.)

31.11 GROUP OBJECTS

The plot objects discussed in the previous section are an example of grouped objects. Each of the plot group types forms a layer between the user and the underlying core objects created. That is, each plot group is a parent object to one or more underlying core objects. In addition to furnishing plot objects, MATLAB 7 provides this grouping capability through the function `hggroup`. That is, you can group any number of *axes* object children into a group. These children can include any of the core objects, plot objects, or other objects grouped with the use of the `hggroup` function. As such, a group object has an *axes* object, or another group object, as its parent object.

By grouping Handle Graphics objects and giving them a single group handle, the visibility and selectability of the group are determined globally. That is, setting the `'visible'` property of the group sets the individual `'visible'` properties of all underlying group children. In addition, it is possible to select all members in a group by clicking on any group member with the mouse. While the plot objects created as part of MATLAB have different or additional property names and property values than their underlying core objects, MATLAB does not provide this capability for groups created by means of the hggroup function. Group objects have all of the common object properties listed in section 31.9. In addition, they have an 'EraseMode' property that controls the 'EraseMode' property of all core objects in the group.

To illustrate the creation of a group object, consider the following code segment:

```
Ha = newplot; % create new axes object

Hl = line(1:6,[rand(1,6);1:6]);  % create 2 core line objects
Hp = patch([3 5 4],[2 2 5],'w'); % create a core patch object

Hg_mm7 = hggroup; % create a new group object and return handle

set(Hl,'Parent',Hg_mm7) % place lines in group
set(Hp,'Parent',Hg_mm7) % place patch in group

get(Hg_mm7) % look at group properties
     EraseMode = normal
     HitTestArea = off

     BeingDeleted = off
     ButtonDownFcn =
     Children = [ (3 by 1) double array]
     Clipping = on
     CreateFcn =
     DeleteFcn =
     BusyAction = queue
     HandleVisibility = on
     HitTest = on
     Interruptible = on
     Parent = [151.017]
     Selected = off
     SelectionHighlight = on
     Tag =
```

```
Type = hggroup
UIContextMenu = []
UserData = []
Visible = on
```

The children of the group object having handle Hg_mm7 are the two *line* objects and one *patch* object. The properties of the lines and patch are hidden under group object, and the group 'type' property is hggroup.

There are many cases where it is desirable to rotate, translate, or scale a group of objects within an axes. While the hggroup function does not provide this capability, the group function hgtransform does. Like the hggroup function, hgtransform creates a group object that becomes the parent of any number of *axes* object children. These children can include any of the core objects, plot objects, or other objects grouped by means of the hgtransform function. As such, a group object has an *axes* object, or another group object, as its parent object. The construction of a hgtransform group object follows that of the hggroup function. Group objects created by using the hgtransform function have all of the common object properties listed in section 31.9. In addition, they have a 'Matrix' property that specifies a graphical transformation matrix to be applied to all members of the group object. The content of this transformation matrix is constructed following standard graphics transformation practice. The function makehgtform facilitates the construction of this matrix. (For further information see the MATLAB documentation and other resources on graphics transformations.)

31.12 ANNOTATION *AXES*

The annotation *axes* shown in the Handle Graphics hierarchy at the beginning of this chapter is simply a core object *axes* with its handle visibility turned Off. This object contains all of the annotations added to a *figure*, using the numerous menus, toolbars, palettes, browsers, and editors available within *Figure* windows. The 'units' property of this axes is set to 'normalized', and its position is set to [0 0 1 1], so that it covers the entire drawing area of the *figure*. Lines, arrows, rectangles, text boxes, etc., can be created by the use of the interactive aspects of *Figure* windows; in addition, they can be created with the annotation function. Because annotations added to a *figure* exist in this separate *axes*, they do not change in response to additions or modifications to any underlying data plots. Therefore, it is important to create annotations after the underlying plots are finalized.

31.13 LINKING OBJECTS

In some situations it is desirable to link identical properties of a number of objects, so that changing a linked property of any one of the objects makes the same change to all linked objects. To support this capability, MATLAB provides the functions linkprop and linkaxes. The function linkprop provides general object-property linking, whereas linkaxes is an M-file function that uses linkprop to specifically link axis limits on two or more *axes* objects.

To demonstrate the use of these functions, consider the following code:

```
>> H_a1 = subplot(2,1,1);      % create axes object
>> plot(rand(1,30))            % plot some random data
>> H_a2 = subplot(2,1,2);      % create a second axes
>> plot(randn(1,30))           % plot random data on second axes

>> linkaxes([H_a1 H_a2],'xy') % link x and y axis limits
```

After this code is executed, the axis limits on the two *axes* are linked. Using the pan or zoom features on the *Figure* window toolbar, panning or zooming one *axes* performs the identical function on the other *axes*. Alternatively, issuing an axis command on one *axes* applies the command to all linked *axes* objects.

The previous example can also be implemented with the use of linkprop by executing the following code:

```
>> close                       % close current figure to start over
>> H_a1 = subplot(2,1,1);      % create axes object
>> plot(rand(1,30))            % plot some random data
>> H_a2 = subplot(2,1,2);      % create a second axes
>> plot(randn(1,30))           % plot random data on second axes

>> H_link = linkprop([H_a1 H_a2],{'Xlim','Ylim'})
H_link =
       graphics.linkprop

>> class(H_link)
ans =
graphics.linkprop

>> methods(H_link)
Methods for class graphics.linkprop:
addprop        addtarget       linkprop       removeprop      removetarget
```

The output H_link of the linkprop function is a handle to a *linkprop* object named graphics.linkprop. This object is an object-oriented class created by using

MATLAB's object-oriented programming features (as discussed in Chapter 33). As shown, this class has five methods that facilitate manipulation of *linkprop* objects. Once properties are linked via the function linkprop, the returned handle H_link must remain in existence. In the preceding example, H_link exists in the base workspace. If H_link is deleted or destroyed by being overwritten, the linked properties are unlinked. Modifying a *linkprop* object requires passing the associated object handle to one of its methods. For example,

```
>> addprop(H_link,'Zlim')
```

adds z-axis limit linking to the object in H_link. (See the MATLAB documentation for more information on linking Handle Graphics object linking.)

31.14 NEW PLOTS

When a new graphics object is created by using a low-level command such as line or text, the object appears on the current *axes* in the current *figure* by default. High-level graphics functions like mesh and plot, however, clear the current *axes* and reset most *axes* properties to their defaults before displaying a plot. As discussed earlier, the hold command can be used to change this default behavior. Both *figures* and *axes* have a 'NextPlot' property used to control how MATLAB reuses existing *figures* and *axes*. The hold, newplot, reset, clf, and cla functions all affect the 'NextPlot' property of *figures* and *axes*. 'NextPlot' has three possible values:

```
>> set(gcf,'NextPlot')
    [ {add} | replace | replacechildren ]
>> set(gca,'NextPlot')
    [ add | {replace} | replacechildren ]
```

The default setting for *figures* is 'add'; the default for *axes* is 'replace'. When 'NextPlot' is set to 'add', a new plot is added without clearing or resetting the current *figure* or *axes*. When the value is 'replace', a new object causes the *figure* or *axes* to remove all child objects and reset all properties except 'Position' and 'Units' to their defaults before drawing the new plot. The default settings clear and reset the current *axes* and reuse the current *figure*.

The third possible setting for 'NextPlot' is 'replacechildren'. This setting removes all child objects, but does not change the current *figure* or *axes* properties. The command hold on sets both *figure* and *axes* 'NextPlot' properties to 'add'. The command hold off sets the *axes* 'NextPlot' property to 'replace'.

The newplot function creates a new *axes* by following the 'NextPlot' guidelines. This function is meant to be called to create an *axes* that will contain children created by a core object-creation function such as line, patch, and so on, as opposed to a high-level graphics function such as plot or surf. The newplot function contains code similar to the following code segment:

```
function Ha = newplot
Hf = gcf;                          % get current figure or create one
next = lower(get(Hf,'NextPlot'));
switch next
  case 'replacechildren', clf;     % delete figure children
  case 'replace', clf('reset');    % delete children and reset properties
end
Ha = gca;                          % get current axes or create one
next = lower(get(Ha,'NextPlot'));
switch next
  case 'replacechildren', cla;     % delete axes children
  case 'replace', cla('reset');    % delete children and reset properties
end
```

31.15 RENDERING SPEED

There are many applications in which you either wish to show the evolution of a graphical solution as one or more properties of the solution change or to dynamically modify a graphical image in response to mouse feedback. In both of these cases, it is desirable to maximize the screen rendering speed and eliminate flicker. MATLAB *figure, axes, line, patch, rectangle,* and *surface* objects offer properties that influence rendering speed and flicker.

MATLAB *figures* have three properties that influence rendering speed: 'Renderer', 'DoubleBuffer', and 'BackingStore'. 'Renderer' refers to the underlying algorithm used to create or render a graphical image on the screen. This property has three possible values—'painters', 'zbuffer', and 'openGL'. Normally, MATLAB selects among the three renderers automatically on the basis of the complexity of the graphics objects to be rendered and on the capabilities of the computer being used. For the most part, MATLAB makes the optimum choice for each *figure*. However, in special cases, the renderer can be specified by setting this property as desired.

'BackingStore' is normally set to 'on' and controls the existence of a duplicate copy of the *Figure* window data. When multiple overlapping windows appear on a computer screen, having 'BackingStore' set to 'on' allows obscured or partially obscured *Figure* windows to be refreshed quickly from the off-screen copy when uncovered. When 'BackingStore' is set to 'off', no off-screen copy is created, thereby increasing rendering speed, but forcing a recomputation of graphical data when a *figure* is uncovered.

The *figure* `'DoubleBuffer'` property is normally set to `'off'`. Setting this property to `'on'` instructs MATLAB to draw graphics objects into an off-screen buffer and then dump the data to the screen when the buffer is fully updated. By drawing into an off-screen buffer rather than directly updating the screen, the rendering progress is hidden from the user. MATLAB sets this property to `'off'` because it is applicable only to graphics rendered with the Painter's algorithm. In addition, it produces flicker-free rendering only for simple graphics, such as those produced by the function `plot`. Graphics containing *surface* or *patch* objects generally do not benefit from double buffering.

When an *axes* is rendered with the Painter's algorithm, the amount of behind-the-scenes work done to render objects such as *lines* can be controlled by the `'DrawMode'` *axes* property. By default, this property is set to `'normal'`. However, setting it to `'fast'` sacrifices rendering accuracy for added speed. For Z-buffer and OpenGL rendering, the property has no effect.

The graphics drawing objects *line*, *patch*, *rectangle*, and *surface* all have an `'EraseMode'` property that allows the user to specify how they are erased and rerendered. The default value for this property is `'normal'`, with `'none'`, `'background'`, and `'xor'` being alternatives. This property determines how these objects are rendered when they are redrawn because of a change in their own properties or the placement of other objects on top of them. For example, when `'EraseMode'` is set to `'none'`, the original object is not erased when its properties are changed, the modified object is simply rendered. In most cases, this leaves the *axes* with undesirable ghosts. Setting `'EraseMode'` to `'background'` erases the original object by setting the pixels it occupies to the *axes* background color. This damages objects appearing beneath the original object. Setting `'EraseMode'` to `'xor'` erases and redraws the *line* by applying an exclusive OR operation on the color underneath the object. This does not damage objects beneath the changed object, but can affect the color of the changed object. Given these property value descriptions, rendering speed is increased most dramatically with `'EraseMode'` set to `'none'`, but this results in ghosts that are undesirable in most cases. Setting `'EraseMode'` to `'xor'` or `'background'` offers a compromise between speed and rendering accuracy. In most cases, `'xor'` is the best choice for creating animations.

31.16 CALLBACKS

All Handle Graphics objects have the properties `'ButtonDownFcn'`, `'CreateFcn'`, and `'DeleteFcn'`. In addition, *figures* have the properties `'CloseRequestFcn'`, `'KeyPressFcn'`, `'WindowButtonDownFcn'`, `'WindowButtonMotionFcn'`, and `'WindowButtonUpFcn'`, and the user interface functions have a property `'CallBack'`. The property names associated with each of these properties are called **callbacks**. These callbacks identify code to be executed when the specific user actions associated with a property name are taken. In the simplest case, callbacks are character strings that are evaluated by `eval` in the *Command* window workspace. These strings can contain any sequence of valid MATLAB statements. In most cases, callbacks are function calls, often to the same function where the callbacks are defined. The setting of callbacks tells MATLAB to perform certain tasks in response to an action taken by the user. These callbacks form the basis for MATLAB's GUI features to be discussed in the next chapter.

The simplest callback is the close request callback, which, by default, is not empty:

```
>> get(gcf,'CloseRequestFcn')
ans =
closereq
>> class(ans)
ans =
char
```

By default, when a *Figure* window is closed by clicking the close box in the *figure* title bar, the string `closereq` is passed to `eval`. This string is the name of a function in MATLAB that simply deletes the current *figure*. So, by default, clicking the close box deletes the associated *Figure* window. This behavior can be changed simply by replacing the preceding string with another to be evaluated on a close request—for example,

```
>> set(gcf,'CloseRequestFcn','')
```

This replacement disables closure via the close box. The close request function is an empty string, and so no action is taken. This callback string can be any valid sequence of MATLAB statements. Therefore, the string could prompt the user for confirmation of the close request before actually doing it.

31.17 M-FILE EXAMPLES

There are many, many examples of Handle Graphics usage in MATLAB itself. Almost all of the specialized plotting functions in the `specgraph` directory (`>> doc specgraph`) are composed of Handle Graphics functions. Even the M-file function `axis` is implemented by using Handle Graphics function calls. This section provides further illustrations of Handle Graphics usage.

This function `mmis2d` returns logical True if an axis is a 2-D view of the *x-y* plane:

```
function [tf,xa,ya]=mmis2d(H)
%MMIS2D True for Axes that are 2D.
% MMIS2D(H) returns True if the axes having handle H displays
% a 2D viewpoint of the X-Y plane where the X- and Y-axes are
% parallel to the sides of the associated figure window.
```

```
%
% [TF,Xa,Ya]=MMIS2D(H) in addition returns the angles of x- and y-axes
%
% e.g., if the x-axis increases from right-to-left Xa=180
% e.g., if the y-axis increases from left-to-right Ya=0
% e.g., if the x-axis increases from bottom-to-top Xa=90


if ~ishandle(H)
        error('H Must be a Handle.')
end
if ~strcmp(get(H,'Type'),'axes')
        error('H Must be a Handle to an Axes Object.')
end
v = get(H,'view');
az = v(1);
el = v(2);
tf = rem(az,90)==0 && abs(el)==90;


if nargout==3
        xdir = strcmp(get(H,'Xdir'),'reverse');
        ydir = strcmp(get(H,'Ydir'),'reverse');
        s = sign(el);


        xa = mod(-s*az - xdir*180,360);
        ya = mod(s*(90-az) - ydir*180,360);
end
```

This function makes use of the function ishandle, which returns logical True for arguments that are valid object handles. It checks to see if the supplied handle is that of an *axes* by getting the 'Type' property. If successful, it then gets the 'View' property to determine the requested output.

The function mmgetpos finds the position of an object in a specific set of units. This function does the right thing, in that it gets the current 'Units' property, sets the units to the 'Units' of the desired output, gets the 'Position' property in the desired units, and then resets the 'Units' property:

```
function p=mmgetpos(H,u,cp)
%MMGETPOS Get Object Position Vector in Specified Units.
% MMGETPOS(H,'Units') returns the position vector associated with the
% graphics object having handle H in the units specified by 'Units'.
% 'Units' is one of: 'pixels', 'normalized', 'points', 'inches', 'cent',
% or 'character'.
% 'Units' equal to 'data' is valid for text objects only.
%
% MMGETPOS does the "right thing", i.e., it: (1) saves the current units,
% (2) sets the units to those requested, (3) gets the position, then
% (4) restores the original units.
%
% MMGETPOS(H,'Units','CurrentPoint') returns the 'CurrentPoint' position
% of the figure having handle H in the units specified by 'Units'.
%
% MMGETPOS(H,'Units','Extent') returns the 'Extent' rectangle of the text
% object having handle H.
%
% 'Uimenu', 'Uicontextmenu', 'image', 'line', 'patch', 'surface',
% 'rectangle' and 'light' objects do NOT have position properties.

if ~ischar(u)
   error('Units Must be a Valid String.')
end
if ~ishandle(H)
        error('H is Not a Valid Handle.')
end
Htype = get(H,'Type');

if nargin==3 && ~isempty(cp) && ischar(cp)

      if strcmp(Htype,'figure') && lower(cp(1))=='c'
            pname = 'CurrentPoint';
      elseif strcmp(Htype,'text') && lower(cp(1))=='e'
            pname = 'Extent';
```

```
      else
            error('Unknown Input Syntax.')
      end

elseif H~=0
      pname = 'Position';

elseif H==0  % root object
      pname = 'ScreenSize';
else
      error('Unknown Input Syntax.')
end

hu = get(H,'units'); % get original units
set(H,'units',u)      % set to desired units

p = get(H,pname);     % get position in desired units
set(H,'units',hu)     % set units back to original units
```

The `mmzap` function illustrates a technique that is very useful when writing Handle Graphics function M-files: It uses a combination of `waitforbuttonpress` and `gco` to get the handle to an object selected using the mouse. The command `waitforbuttonpress` is a built-in MATLAB function that waits for a mouse click or key press. Following is its help text:

```
>> help waitforbuttonpress

WAITFORBUTTONPRESS Wait for key/buttonpress over figure.
    T = WAITFORBUTTONPRESS stops program execution until a key or
    mouse button is pressed over a figure window.  Returns 0
    when terminated by a mouse buttonpress, or 1 when terminated
    by a keypress.  Additional information about the terminating
    event is available from the current figure.
```

After a mouse button is pressed with the mouse pointer over a *figure*, `gco` returns the handle of the selected object. This handle is then used to manipulate the selected object:

```
function mmzap(arg)
%MMZAP Delete Graphics Object Using Mouse.
% MMZAP waits for a mouse click on an object in
% a figure window and deletes the object.
% MMZAP or MMZAP text erases text objects.
% MMZAP axes  erases axes objects.
% MMZAP line  erases line objects.
% MMZAP surf  erases surface objects.
% MMZAP patch  erases patch objects.
%
% Clicking on an object other than the selected type, or striking
% a key on the keyboard aborts the command.

if nargin<1
   arg='text';
end
Hf = get(0,'CurrentFigure');
if isempty(Hf)
   error('No Figure Window Exists.')
end
if length(findobj(0,'Type','figure'))==1
      figure(Hf) % bring only figure forward
end
key = waitforbuttonpress;

if key  % key on keyboard pressed
      return
      else    % object selected
            object = gco;
            type = get(object,'Type');
            if strncmpi(type,arg,4)
                  delete(object)
            end
end
```

The functions xlim, ylim, and zlim in MATLAB allow you to set and get axis limits for the three plot axes independently. There are no equivalent functions for grid lines. The function grid turns grid lines on and off along all three axes. To add this feature the function mmgrid allows you to turn grid lines on and off along any individual axis:

```
function mmgrid(varargin)
%MMGRID  Individual Axis Grid Lines on the Current Axes.
% V is a single character X, Y, or Z denoting the desired axis.
% MMGRID V toggles the major grid lines on the V-axis.
% MMGRID V ON adds major grid lines to the V-axis.
% MMGRID V ON MINOR adds minor grid lines to the V-axis.
% MMGRID V OFF removes major grid lines from the V-axis.
% MMGRID V OFF MINOR removes minor grid lines from the V-axis.
%
% See also GRID

ni = nargin;
if ni==0
   error('At Least One Input Argument Required.')
end
if ~iscellstr(varargin)
   error('Input Arguments Must be Strings.')
end
Hf = get(0,'CurrentFigure'); % get current figure if it exists
if isempty(Hf) % no figure so do nothing
   return
end
Ha = get(Hf,'CurrentAxes'); % get current axes if it exists
if isempty(Ha) % no axes so do nothing
   return
end
% parse input and do work
V = varargin{1};
idx = strfind('xXyYzZ',V(1));
```

```
if isempty(idx)
   error('Unknown Axis Selected.')
end
VGrid = [upper(V(1)) 'Grid']; % XGrid, YGrid, or ZGrid

if ni==1  % MMGRID V        Toggle Grid
   Gstate = get(Ha,VGrid);
   if strcmpi(Gstate,'on')
      set(Ha,VGrid,'off')
   else
      set(Ila,VGrid,'on')
   end
elseif ni==2  % MMGRID V ON   or MMGRID V OFF
   OnOff = varargin{2};
   if strcmpi(OnOff,'on')
      set(Ha,VGrid,'on')
   elseif strcmpi(OnOff,'off')
      set(Ha,VGrid,'off')
   else
      error('Second Argument Must be On or OFF.')
   end
elseif ni==3  % MMGRID V ON MINOR   or MMGRID V OFF MINOR
   if ~strcmpi(varargin{3},'minor')
      error('Unknown Third Argument.')
   end
   VGrid = [upper(V(1)) 'MinorGrid'];
   OnOff = varargin{2};
   if strcmpi(OnOff,'on')
      set(Ha,VGrid,'on')
   elseif strcmpi(OnOff,'off')
      set(Ha,VGrid,'off')
   else
      error('Second Argument Must be On or OFF.')
   end
end
```

Rather than use the functions gcf and gca to get the current *figure* and *axes*, respectively, Handle Graphics calls are used to get the 'CurrentFigure' and 'CurrentAxes' properties. By doing so, no automatic creation of a *figure* or *axes* is performed. This allows the function to terminate without performing any action, if no *figure* or *axes* exists. The rest of the function simply modifies the 'XGrid', 'YGrid', 'ZGrid', 'XMinorGrid', 'YMinorGrid', or 'ZMinorGrid' properties of the current *axes*, as directed by the input arguments.

31.18 SUMMARY

Handle Graphics functions provide the ability to fine-tune the appearance of visual aspects of MATLAB. Each graphics object has a handle associated with it that can be used to manipulate the object. The following table documents pertinent Handle Graphics functions in MATLAB:

Function	Description
get	Get object properties
set	Set object properties
gcf	Get current *figure*
gca	Get current *axes*
gco	Get current object
findobj	Find visible objects having specified properties
findall	Find visible and invisible objects having specified properties
findfigs	Return visible *figure* windows to the computer screen
allchild	Get visible and invisible children handles for an object
copyobj	Copy object to new parent
inspect	Open Handle Graphics property inspector GUI
root	Root computer object, handle = 0
figure	*figure* object creation
axes	*axes* object creation
image	*image* object creation
light	*light* object creation
line	*line* object creation
patch	*patch* object creation
rectangle	*rectangle* object creation
surface	*surface* object creation
text	*text* object creation

Function	Description
linkaxes	Create a link object for linking axis limits on two or more *axes* objects
linkprop	Create a link object for linking listed properties of two or more Handle Graphics objects
uibuttongroup	User interface container object for managing 'radiobutton' and 'togglebutton' style *uicontrol* objects
uicontainer	User interface *container* object creation
uicontrol	User interface *uicontrol* object creation
uimenu	User interface *uimenu* object creation
uicontextmenu	User interface *uicontextual menu* object creation
uipanel	User interface *uipanel* object creation
uitoolbar	User interface *uitoolbar* object creation
uipushtool	Momentary contact pushbutton for a *uitoolbar*
uitoggletool	On/Off pushbutton for a *uitoolbar*
uitable	User interface *uitable* object creation
uitree	User interface *uitree* object creation
uitreenode	User interface *uitreenode* object creation
areasseries	Plot object created by the function area
barseries	Plot object created by the functions bar, bar3, bar3h
contourgroup contourf	Plot object created by the functions contour, contour3,
errorbarseries	Plot object created by the function errorbar
lineseries	Plot object created by the functions plot, plot3, semilogx, semilogy, loglog
quivergroup	Plot object created by the functions quiver, quiver3
scattergroup	Plot object created by the functions scatter, scatter3
stairseries	Plot object created by the function stairs
stemseries	Plot object created by the functions stem, stem3
surfaceplot	Plot object created by the functions surf, mesh
hggroup	Create group object
hgtransform	Create group object
makehgtform	Create transformation matrix for hgtransform group object
annotation	Create annotation *axes* object
reset	Reset object properties to default values
clf	Clear current *figure*
cla	Clear current *axes*

Function	Description
`ishandle`	True for arguments that are object handles
`delete`	Delete object
`close`	Close *figure* using close request function
`refresh`	Refresh *figure*
`gcbo`	Get current callback object
`gcbf`	Get current callback *figure*
`closereq`	Default *figure* `'CloseRequestFcn'` callback
`newplot`	Create *axes* with knowledge of `'NextPlot'` properties

32

Graphical User Interfaces

This chapter introduces the graphical user interface features available in MATLAB. These features include predefined dialog boxes, as well as window menus, contextual menus, pushbuttons, scroll bars, radio buttons, toggle buttons, popup menus, list boxes, and toolbars that can be placed in *Figure* windows. In addition to these features, it is possible to dynamically track mouse position and motion, and perform tasks based on this position and motion. These features are so extensive and detailed that entire texts have been devoted exclusively to them. Given the broader goals of this text, the best that can be offered here is a useful introduction and several illustrative examples.

32.1 WHAT'S A GUI?

A user interface is the point of contact or method of interaction between a person and a computer or computer program. It is the method used by the computer and the user to exchange information. The computer displays text and graphics on the computer screen and may generate sounds with a speaker. The user communicates with the computer by using an input device such as a keyboard, mouse, trackball, drawing pad, or microphone. The user interface defines the look and feel of the computer, operating system, or application. Often a computer or program is chosen on the basis of pleasing design and the functional efficiency of its user interface.

A graphical user interface, or GUI (pronounced *goo' ey*), is a user interface incorporating graphics objects such as windows, icons, buttons, menus, and text. Selecting or activating these objects in some way usually causes an action or change to occur. The most common activation method is to use a mouse or other pointing device to control the movement of a pointer on the screen and to press a mouse button to signal object selection or some other action.

In the same way that the Handle Graphics capabilities of MATLAB discussed in the previous chapter let you customize the way MATLAB displays information, the Handle Graphics user interface functions described in this chapter let you customize the way you interact with MATLAB.

This chapter illustrates the use of predefined dialog boxes and the use of Handle Graphics *uicontrol*, *uimenu*, *uicontextmenu*, and *uitoolbar* objects to add graphical user interfaces to MATLAB functions and M-files. *uimenu* objects create drop-down menus and submenus in *Figure* windows. *uicontrol* objects create objects such as buttons, sliders, popup menus, and text boxes. *uicontextmenu* objects create contextual menus that open over objects. *uitoolbar* objects contain pushbuttons (*uipushtool* objects) and toggle buttons (*uitoggletool* objects).

MATLAB includes an excellent example of its GUI capabilities in the demo command. To explore these demonstrations, simply issue the command

```
>> demo
```

32.2 PREDEFINED DIALOG BOXES

Essentially, every computer application offers the capability to open and save data files. To navigate the directory structure, operating systems or window managers provide dialog boxes to perform the tasks of identifying file names and directory paths. In MATLAB, these tasks are performed by the functions uigetfile, uiputfile, and uigetdir. None of these functions actually opens or saves a file. All they do is return a file name or a directory path that can be used, along with the low-level file I/O functions discussed in Chapter 14, to perform directory and file manipulation.

When possible, uigetfile, uiputfile, and uigetdir use standard dialog boxes available on your computer platform. As a result, their layout varies from platform to platform, but their functionality does not. These dialog boxes match the operation of standard operating system dialog boxes that identify files for opening and saving and for choosing directories. For example, using uigetfile to find the startup.m file on the author's computer produces the following results:

```
>> [fname,dirpath] = uigetfile('*.m')
fname =
startup.m
dirpath =
C:\matlabR14\work\
```

Appending the file name to the directory path gives the unique file specfication for startup.m:

```
>> myfile = [dirpath fname]
myfile =
C:\matlabR14\work\startup.m
```

It is important to note that uigetfile is not limited to directories on the MATLAB search path. The directory path returned will be for whatever directory the dialog box points to just before it is closed.

Using uiputfile to save a MAT-file in the work directory on the author's computer produces the following results:

```
>> [fname,dirpath] = uiputfile('*.mat')

fname =

mydata

dirpath =

C:\matlabR14\work\
```

Here, the string mydata was entered as the file name in the dialog box. Even though '*.mat' specified the file type, '.mat' was not automatically appended to the returned fname. It is up to the programmer to append the file extension.

Specifying colors in MATLAB based on RGB triples is straightforward, but lacks visual feedback about the actual color specified by a given triple. In addition, most graphics objects have a 'Color' property that can be set. The function uisetcolor provides a GUI for choosing RGB triples or setting the color of a particular graphics object.

As with uigetfile, the function uisetcolor uses the standard color dialog box available on your computer platform (when one is available). As a result, the layout and features of uisetcolor vary from platform to platform, but the functionality does not.

Another built-in function, named uisetfont, provides a GUI for selecting fonts and font attributes. Using the function to choose Palatino 12-point italics produces the following results:

```
>> fc = uisetfont

fc =

      FontName: 'Palatino'
     FontUnits: 'points'
      FontSize: 12
    FontWeight: 'normal'
     FontAngle: 'italic'
```

Here, fc is a structure containing the features chosen in the uisetfont GUI. The results are in the proper form to set the font properties of any *text* object.

32.3 M-FILE DIALOG BOXES

In addition to the built in dialog functions discussed earlier, MATLAB offers a number of dialog boxes implemented as M-file functions. Some of these functions prompt the user for input of some kind. The functions available include axlimdlg, dialog, inputdlg, menu, and msgbox.

The function `axlimdlg` provides a rudimentary dialog box for setting plot axes limits. The *Figure* window editing features in MATLAB are much more comprehensive. The function `inputdlg` creates a set of text edit box *uicontrols* and prompts the user for input into each. This function is useful for prompting the user for a few values. The function `Menu` offers the user a selection of alternatives using *uicontrol* pushbuttons.

The functions `dialog` and `msgbox` are generic dialog-box creation functions. Of these functions, `dialog` simply creates a *Figure* window with default properties set appropriate for *Figure* windows that do not contain any *axes* objects. *Figure* windows created by using `dialog` use minimal system resources. For example, they do not allocate memory for a colormap. The function `msgbox` is used by all message dialog boxes and manages all of the underlying activities required to put up a dialog box and accept input.

The standard message dialog boxes in MATLAB include `errordlg`, `helpdlg`, `questdlg`, and `warndlg`. These boxes display an appropriate icon, along with at least one pushbutton to acknowledge and dismiss the dialog box. In addition to being helpful in application development, these functions are good examples of GUI construction to study.

32.4 DIALOG BOX SUMMARY

The dialog box functions available in MATLAB are summarized in the following table:

Function	Description
`axlimdlg`	Axes limits dialog box
`dialog`	Creates figure for dialog box or GUI
`errordlg`	Error dialog box
`helpdlg`	Help dialog box
`inputdlg`	Input dialog box
`listdlg`	List selection dialog box
`menu`	Menu choice selection dialog box
`msgbox`	Generic message dialog box
`pagedlg`	Page position dialog box
`pagesetupdlg`	Page setup dialog box
`printdlg`	Print dialog box
`printpreview`	Print preview dialog box
`questdlg`	Question dialog box
`uigetfile`	Standard open file dialog box
`uiputfile`	Standard save file dialog box
`uisetcolor`	Color selection dialog box
`uisetfont`	Font and font attributes dialog box
`waitbar`	Displays wait bar
`warndlg`	Warning dialog box

32.5 GUI OBJECT HIERARCHY

Graphical user interface creation utilizes a number of MATLAB functions. These functions create Handle Graphics objects that are shown in the UI Objects group in the following figure:

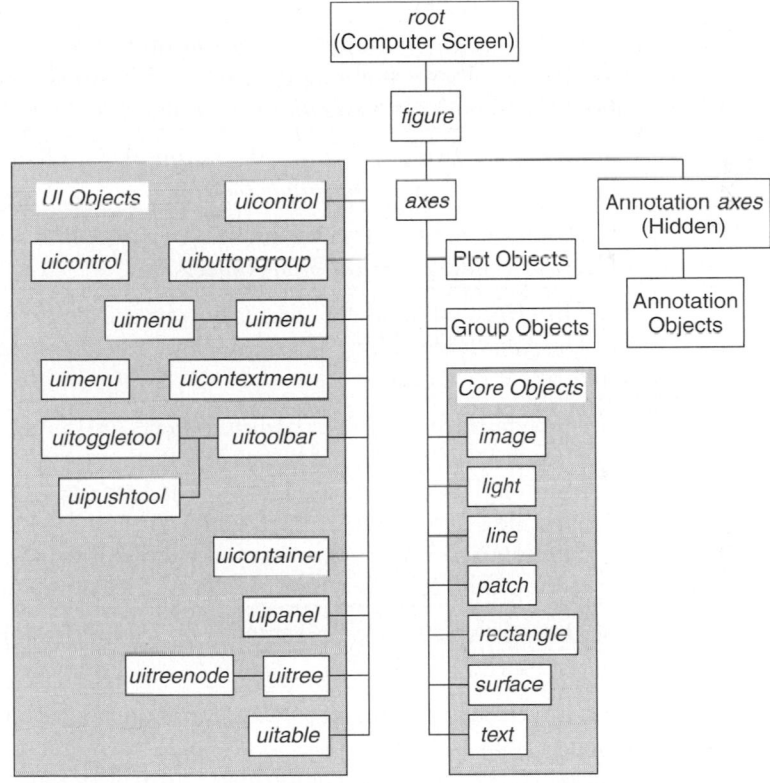

The *UI Objects* group contains numerous types of GUI objects. These features are summarized in the following table:

UI Object	Description
uicontrol	General user interface control object having one of the following styles: `'checkbox'`, `'edit'`, `'frame'`, `'listbox'`, `'popupmenu'`, `'pushbutton'`, `'radiobutton'`, `'slider'`, `'text'`, and `'togglebutton'`. All *uicontrol* objects are the children of *figure* objects.
uimenu	User interface menu object. *uimenu* objects that are children of *figure* objects are menu headings at the top of a *Figure* window. *uimenu* objects that are children of other *uimenu* objects are menu items or submenu items.

UI Object	Description
uicontextmenu	User interface contextual menu object. All *uicontextmenu* objects are children of *figure* objects. The handle of a *uicontextmenu* is placed in the `'UIContextMenu'` property of the Handle Graphics object for which they are to appear. All *uicontextmenu* objects have child *uimenu* objects that define their menu items.
uitoolbar	User interface toolbar object. *uitoolbar* objects are children of *figure* objects. A toolbar is populated with toggle buttons (*uitoggletool* objects) and push buttons (*uipushtool* objects).
uitoggletool	User interface two position toggle button on a toolbar. *uitoggletool* objects are children of *uitoolbar* objects.
uipushtool	User interface momentary push button on a toolbar. *uipushtool* objects are children of *uitoolbar* objects.
uitable	User interface table object. *uitable* objects are children of *figure* objects. All *uitable* objects have rows and columns similar to a spreadsheet program.
uitree	User interface tree object. All *uitree* objects are children of *figure* objects. *uitree* objects create visual representations of hierarchical information. The *Content* tab in the *Help Navigator* section of the *Help* window shows a tree that visually organizes MATLAB help information. The visual presentation of a file directory tree showing parent and child folders is another example of a tree object.
uitreenode	User interface tree node object. *uitreenode* objects define nodes on a *uitree* object.
uibuttongroup	User interface container object. All *uibuttongroup* objects are children of *figure* objects. They have *uicontrol* and other Handle Graphics objects as their children. The primary purpose of *uibuttongroup* objects is to manage the exclusive selection behavior of radio buttons and toggle buttons contained within it.
uicontainer	User interface container object. *uicontainer* objects contain groups of other user interface objects and become their parent. The position of *uicontainer* child objects is relative to the container and therefore move relative to the container.
uipanel	User interface container object. Like *uicontainer* objects, *uipanel* objects contain groups of other user interface objects and become their parent. However, *uipanel* objects have borders and titles that can be set.

There are 10 different styles of *uicontrol* objects. Each of these styles creates a different GUI component. Because these components are fundamental parts of a graphical user interface, they are described in the following table:

uicontrol **Style**	**Description**
`'checkbox'`	Check boxes consist of buttons with a label and a small square box to the left of the label text. When activated, the control is toggled between checked and cleared. When checked, the box is filled or contains an x, depending on the platform; when cleared, the box becomes empty. Check boxes are typically used to indicate the state of an option or attribute.
`'edit'`	Editable text boxes display text in a box so that you can modify or replace the text string dynamically, just as you would by using a text editor or word processor. Editable text boxes can contain one or more lines of text. A single-line editable text box accepts one line of text from the user, while a multiline text box accepts more than one line of text. Multiline editable text boxes have a vertical scroll bar. Single-line text entry is terminated by pressing the **Return** key. Multiline text entry is terminated with **Control-Return**.
`'frame'`	Frame *uicontrol* objects are simply opaque, shaded, rectangular regions with a border. Frames are analogous to the `'Separator'` property of *uimenu* objects, in the sense that they provide visual separation. Frames are typically used to logically group radio buttons or other *uicontrol* objects. Frames should be defined before other objects are placed within them. Otherwise, the frame may cover the other *uicontrols*. In MATLAB 7, frame style *uicontrol* objects have been superseded by *uicontainer* and *uipanel* objects, which are much more versatile.
`'listbox'`	List box *uicontrol* objects look like multiline text boxes. However, in this case one or more lines of text text are selectable, but they are not editable.
`'popupmenu'`	Popup menus are used to present a list of mutually exclusive choices to the user. When closed, a popup menu appears as a rectangle or button containing the label of the current selection with a small raised rectangle, or downward-pointing arrow, beside the label to indicate that the object is a popup menu. When the pointer is positioned over a popup control and the mouse button is pressed, other choices appear. Moving the pointer to a different choice and releasing the mouse button closes the popup menu and displays the new selection. MS Windows and some X-Window System platforms allow the user to click on a popup menu to open it and then click again on a menu choice to select it.
`'pushbutton'`	Pushbuttons, sometimes called *command buttons* or just *buttons*, are small, rectangular screen objects that usually contain a text label. Selecting a pushbutton with the mouse by moving the pointer over the object and clicking the mouse button causes MATLAB to perform the action defined by the object's callback string. Immediately after a pushbutton is pressed, it returns to its default up state. Pushbuttons are typically used to perform an action, rather than change a state or set an attribute.

uicontrol Style	Description
'radiobutton'	Radio buttons consist of buttons containing a label and a small circle or diamond to the left of the label text. When selected, the circle or diamond is filled; when unselected, the indicator is cleared. Radio buttons are typically used to select one of a group of mutually exclusive options.
'slider'	Sliders, or ***scroll bars***, consist of three distinct parts: the trough, or the rectangular area representing the range of valid object values; the indicator within the trough representing the current value of the slider; and arrows at each end of the trough. Sliders are typically used to select a value from a range of values. Slider values can be set in three ways: First, the indicator can be moved by positioning the mouse pointer over the indicator, holding down the mouse button while moving the mouse, and releasing the mouse button when the indicator is in the desired location. The second method is to click the mouse button while the pointer is in the trough, but to one side of the indicator. The indicator moves in that direction over a default distance equal to about 10 percent of the total range of the slider. Finally, clicking on one of the arrows at the ends of the slider moves the indicator by default about 1 percent of the slider range in the direction of the arrow. Sliders are often accompanied by separate text objects used to display labels, the current slider value, and range limits.
'text'	Static text boxes are controls that simply display a text string. Static text boxes are typically used to display labels, user information, or current values. Static text boxes are static in the sense that the user cannot dynamically change the text displayed.
'togglebutton'	Toggle buttons are identical to pushbuttons, except that they toggle between two states, up and down, when repeatedly pushed by clicking on them with the mouse pointer.

Given the 12 GUI objects and the 10 styles of *uicontrol* objects described in the preceding tables, there are 21 distinct GUI objects that can be used to create graphical user interfaces in MATLAB. Each of these objects has numerous Handle Graphics properties that can be accessed by using the functions set and get. Taken as a whole, these objects, their property names, and associated property values constitute hundreds of features and options. Discussion of all of these features and options is beyond the scope of this text. The remainder of this chapter introduces some of the main features of GUI creation and ends with several illustrative examples.

32.6 GUI CREATION FUNDAMENTALS

The fundamental steps of GUI creation are as follows:

1. Understand what you want the GUI to do. This is often the most difficult step. In many cases, what you want the GUI to do changes as new ideas and options are discovered during GUI construction.

2. Sketch the GUI layout on a piece of paper. This often skipped step saves time in the long run, because iterating hand sketches of potential GUI layouts on paper is much faster than building and rebuilding GUI prototypes in MATLAB.

3. On the basis of the final sketch, create the GUI using UI objects. This step can be done by directly entering the creation code in to an M-file. Alternatively, the MATLAB graphical user interface development environment (GUIDE), can be used.

4. Create the MATLAB code that is executed when a user interacts with the GUI. This callback code is most often contained in the same M-file as the code that creates the GUI.

5. Test and debug the GUI. In particular, if the GUI will be used by others, try interacting with the GUI in ways that you never intended, yet in ways a less knowledgeable user may actually use it.

32.7 GUI OBJECT SIZE AND POSITION

The `'Position'` and `'Units'` properties of UI objects are used to locate objects within their parent object, which in many cases is a *Figure* window. Given a *Figure* window or container object of some size, GUI layout is a problem in 2-D geometry. When designing a GUI, you must make objects large enough to be seen and to display the text or image that is part of the object. You also must decide if the GUI *figure* is to be resizable.

Often, sizing and placing UI objects is a process of trial and error. Even when you are satisfied with the result, the appearance of the *figure* on another platform may be sufficiently different to require more adjustments. Many times, it is desirable to make objects a bit larger than appears necessary, simply to ensure usability on all computer platforms.

Just because a *figure* has a default size, there is no guarantee that all *figures* on all platforms are that size. If you add UI objects to an existing *figure*, you may find that the *figure* is smaller or larger than the default. In addition, you can resize any *figure* at any time, unless you are prevented from doing so because the *figure* object's `'Resize'` property has been set to `'off'`. Therefore, in general it is a good idea to specifically set the size of a *Figure* window to be used for a GUI.

When you are adding UI objects to a *figure* that may be resized, the `'Units'` property and the restrictions imposed by fixed font-size character strings can be a problem. When the position of each UI object is specified in absolute units such as pixels, inches, centimeters, or points, resizing the *Figure* window does not change the size or placement of the objects. The objects maintain the same absolute position relative to the bottom and left sides of the *figure*. If the *figure* is made smaller, some of the UI Objects objects may move outside of the *figure* and no longer be visible.

When the positions of UI objects are specified in normalized units, the objects maintain their relative relationships to each other and to the *figure* itself when the *figure* is resized. However, there is a drawback in this case as well. If the *figure* is made smaller and the objects are resized accordingly, label strings may become unreadable, since the font size is fixed. Any portion of a label outside of the dimensions of the resized object is clipped.

There is no way to avoid this conflict between readable character strings of fixed size and GUI size. Good GUI layout practice usually leaves sufficient white space around objects so that clipping does not occur in transferring from one computer platform to the next. If a GUI is meant to be resizable, the `'ResizeFcn'` callback property of the GUI *figure* can be used to move and reshape the objects in response to a change in *figure* size.

32.8 CAPTURING MOUSE ACTIONS

GUI functions use the location of the mouse pointer and the status of mouse buttons to control MATLAB actions. This section discusses the interaction between pointer and object locations and mouse button actions and how MATLAB responds to changes or events, such as a button press, a button release, or pointer movement.

All Handle Graphics objects have a `'ButtonDownFcn'` property. Most UI objects have a `'CallBack'` property, and *uicontrol* objects have a `'KeyPressFcn'` property. *figure* objects have `'WindowButtonDownFcn'`, `'WindowButtonUpFcn'`, and `'WindowButtonMotionFcn'` properties, as well as `'KeyPressFcn'`, `'CloseRequestFcn'`, and `'ResizeFcn'` properties. All Handle Graphics objects also have `'CreateFcn'` and `'DeleteFcn'` properties. The value associated with each of these properties is a callback that tells MATLAB what to do when the corresponding action occurs. The mouse pointer location determines which callbacks are involved and the order in which they are invoked when an event occurs.

The previous chapter contained a discussion of stacking order and object selection regions that is relevant to this discussion. MATLAB determines which callback is invoked on the basis of three regions within a *figure*. When the pointer is within a Handle Graphics object, as determined by its `'Position'` property, the pointer is considered to be on the object. If the pointer is not on an object, but is within an object's selection region, the pointer is near the object. Finally, if the pointer is within the *figure*, but not on or near another object, the pointer is off the other objects. When objects or their selection regions overlap, the stacking order determines which one is selected.

The selection region of Handle Graphics *line*, *surface*, *patch*, *text*, and *axes* objects was discussed in the previous chapter. *uimenu* objects have no external selection region. Either the pointer is on a *uimenu* object, or it is not. *uicontrols* have a selection region that extends about 5 pixels beyond the control's position in all directions. The pointer can be either on or near a control.

A **button click** is defined as the press and subsequent release of a mouse button while the mouse pointer is over an object. If the mouse pointer is over a *uimenu*, *uicontextmenu*, or *uicontrol* object, a button click triggers execution of the object's `'CallBack'` property string, as long as the `'Enable'` property of the object is set to `'on'`. The button press prepares the *uicontrol* and often changes the *uicontrol* or *uimenu* visually, and the button release triggers callback execution. If the mouse pointer is not on a *uicontrol* or *uimenu*, both button press and button release events are triggered as explained next.

When the mouse button is pressed, a ***button press*** event is created. When this happens with the pointer located within a *Figure* window, a number of different actions can occur, on the basis of the location of the pointer and the proximity of Handle Graphics objects. If an object is selected, it becomes the current object. If no object is selected, the *figure* itself becomes the current object. The *figure* object's `'CurrentPoint'` and `'SelectionType'` properties are also updated. The appropriate callbacks are then invoked.

The following table lists the pointer location options and the callbacks executed for a button press event:

Button Press Pointer Location	**Action Taken**
On a *uimenu* item if the `'Enable'` property is `'on'`.	Change *uimenu* appearance and prepare for a release event
On a *uicontrol* if the `'Enable'` property is `'on'`.	Change *uicontrol* appearance and prepare for a release event
On a *uimenu* item if the `'Enable'` property is `'off'`.	Ignore Button press
On a *uicontrol* if the `'Enable'` property is `'off'` or `'inactive'`	Evaluate *figure's* `'WindowButtonDownFcn'` callback and then *uicontrol's* `'ButtonDownFcn'` callback
On or near any Handle Graphics object except for *uimenus* and *uicontrols*	Evaluate *figure's* `'WindowButtonDownFcn'` callback and then the object's `'ButtonDownFcn'` callback
Within a *figure*, but not on or near any other object	Evaluate *figure's* `'WindowButtonDownFcn'` callback and then *figure's* `'ButtonDownFcn'` callback

Note that a button press event always invokes the *figure's* `'WindowButtonDownFcn'` callback before the selected object's `'ButtonDownFcn'` callback, except when the pointer is on a *uicontrol* or *uimenu* object. When the pointer is near a *uicontrol* or on a *uicontrol* with the `'Enable'` property `'off'` or `'inactive'`, the *uicontrol's* `'ButtonDownFcn'` callback is invoked (rather than the `'CallBack'` property) after the *figure's* `'WindowButtonDownFcn'` callback has finished. The `'ButtonDownFcn'` callback of a *uimenu* is never invoked.

When the mouse button is released, a ***button release*** event is created. When this happens, the *figure* object's `'CurrentPoint'` property is updated and the *figure* object's `'WindowButtonUpFcn'` callback is invoked. If the `'WindowButtonUpFcn'` callback is not defined, the `'CurrentPoint'` property is not updated when the button is released.

When the pointer is moved within a figure, ***pointer movement*** events are created. When this happens, the *figure* object's `'CurrentPoint'` property is updated and the *figure* object's `'WindowButtonMotionFcn'` callback is invoked. If the

'WindowButtonMotionFcn' callback is not defined, the 'CurrentPoint' property is not updated when the pointer moves.

32.9 THE EVENT QUEUE

Because GUIs rely on human actions, GUI execution is based on the concept of an *event queue*. That is, a GUI is presented to the user. The user in turn interacts with the GUI in an essentially arbitrary way over time, with each interaction creating an event that is placed in the event queue. In addition, other commands that involve *Figure* window input or output generate events. Events include pointer movements and mouse button actions that generate callbacks, waitfor and waitforbuttonpress functions, and commands that redraw graphics such as drawnow, figure, getframe, or pause. MATLAB acts on all of these events in the order in which they appear in the queue. MATLAB manages its event queue in a default manner that is appropriate in most cases. However, the common Handle Graphics object properties 'Interruptible' and 'BusyAction' can be used to specify particular actions.

In all cases, a callback executes until it reaches a waitfor, drawnow, waitforbuttonpress, getframe, pause, or figure command. *Callbacks that do not contain any of these commands cannot be interrupted.* When one of these special commands is reached, MATLAB suspends execution of the callback and examines each of the pending events in the event queue. If the 'Interruptible' property of the object with the suspended callback is set to 'on', which is its default value, all pending events are processed before the suspended callback is resumed. If the 'Interruptible' property is set to 'off', then only pending screen-update events are processed. At the same time, if the 'BusyAction' property of the object with the suspended callback is set to 'cancel', the interrupting callback events are discarded. If the 'BusyAction' property is set to 'queue', interrupting callback events are held in the event queue until the interrupted callback finishes. Even if an executing callback is not interruptible, pending screen-update events are still processed when the callback reaches a waitfor, waitforbuttonpress, drawnow, figure, getframe, or pause command.

32.10 CALLBACK PROGRAMMING

Handle Graphics and GUI functions make extensive use of callbacks to perform the tasks associated with user selections in a GUI. Prior to MATLAB 7, callbacks were always character strings evaluated in the *Command* window workspace. In this scenario, a callback string was passed to the function eval each time the callback was invoked. For example,

```
H_p1 = uicontrol('Style','PushButton',...
                 'Callback','myguifcn push1');
```

sets the callback for a pushbutton *uincontrol* object to the string 'myguifcn push1'. When this GUI pushbutton is pressed, this command syntax form is interpreted as the function call myguifcn('push1') and is executed in the *Command* window workspace. In most situations, myguifcn is the name of the function containing the preceding uicontrol statement. That is, callbacks are often directed to call the same function where they are defined. In this way, a single M-file creates the GUI and contains all of the code that is executed when the GUI is run. This approach of placing all GUI callbacks in the same M-file leads to a ***switchyard programming*** structure. For example, suppose a GUI function M-file myguifcn.m creates three pushbuttons—Apply, Revert, and Done—with the callbacks to these buttons being myguifcn('Apply'), myguifcn('Revert'), and myguifcn('Done'), respectively. Then the function myguifcn can implement the switchyard programming callback structure with a Switch-Case construction, shown as follows:

```
function myguifcn(arg)
%MYGUIFCN Sample Switchyard Programming Fxample.

persistent mydata
if nargin==0
   arg = 'Initialize';
end
switch arg
case 'Initialize'
   % code that creates the GUI and sets the callbacks
   uicontrol('Style','PushButton',...
             'String','Apply',...
             % In a real GUI other properties must be set also
             'Callback','myguifcn Apply')
   uicontrol('Style','PushButton',...
             'String','Revert',...
             % In a real GUI other properties must be set also
             'Callback','myguifcn Revert')
   uicontrol('Style','PushButton',...
             'String','Done',...
             % In a real GUI other properties must be set also
             'Callback','myguifcn Done')

   mydata = ... % data to be stored in the GUI for callback use,
              % most commonly a structure
```

```
case 'Apply'
    % code that performs Apply button callback actions

case 'Revert'
    % code that performs Revert button callback actions

case 'Done'
    % code that performs Done button callback actions

otherwise
    % report error?
end
```

The most difficult aspect of callback programming, as shown, is retrieving data. Even though switchyard programming callbacks call the same function repeatedly, each callback creates a separate instance of the function, each with its own separate workspace. As a result, callback instances do not have access to anything other than the function argument passed to it in the callback string. MATLAB provides a variety of ways to make data available to separate callback instances. The simplest approach is to declare one or more variables associated with needed data persistent. In doing so, the declared variables persist and are available whenever the function is called. In the preceding example, the variable mydata is declared persistent and is assigned data when the GUI is created; then the data stored in mydata are available when any switchyard callback is invoked. The only disadvantage of using the persistent declaration is that only one instance of the GUI can exist at a given time. If multiple GUIs existed simultaneously, they would all share the same persistent data.

As an alternative to using persistent variables, the 'Tag' property of Handle Graphics objects can be set to a unique character string. Then, the function findobj can be used to search for objects having the specified tag string. Once the object is found, the functions getappdata, setappdata, rmappdata, and isappdata can be used to manipulate data stored in the tagged object. The following example demonstrates this approach:

```
function myguifcn(arg)
%MYGUIFCN Sample Switchyard Programming Example.

if nargin==0
    arg = 'Initialize';
end
```

```
switch arg
case 'Initialize'
   % code that creates the GUI and sets the callbacks
   Hf = figure(... % create figure for gui
               % In a real GUI other properties must be set also
               'Tag','MyGuiFcnTag')

   uicontrol('Parent',Hf,...
             'Style','PushButton',...
             'String','Apply',...
             % In a real GUI other properties must be set also
             'Callback','myguifcn Apply')
   uicontrol('Parent',Hf,...
             'Style','PushButton',...
             'String','Revert',...
             % In a real GUI other properties must be set also
             'Callback','myguifcn Revert')
   uicontrol('Parent',Hf,...
             'Style','PushButton',...
             'String','Done',...
             % In a real GUI other properties must be set also
             'Callback','myguifcn Done')
   mydata = ... % data to be stored in the GUI for callback use,
                % most commonly a structure
   setappdata(Hf,'MyGuiFcnData',mydata)

case 'Apply'
   % code that performs Apply button callback actions
   Hf = findobj('Tag','MyGuiFcnTag');
   mydata = getappdata(Hf,'MyGuiFcnData');

case 'Revert'
   % code that performs Revert button callback actions
   Hf = findobj('Tag','MyGuiFcnTag');
   mydata = getappdata(Hf,'MyGuiFcnData');
```

```
case 'Done'
    % code that performs Done button callback actions
    Hf = findobj('Tag','MyGuiFcnTag');
    mydata = getappdata(Hf,'MyGuiFcnData');

otherwise
    % report error?
end
```

As shown setappdata is used to store a variable called mydata in the GUI *figure* application data property. The GUI figure is given the tag 'MyGuiFcnTag'. In the code executed by each callback, findobj is used to find the handle of the GUI figure. Then getappdata is used to retrieve the application data stored in the *figure*.

An alternative to the previous example uses the function guidata, which stores and retrieves data in the GUI *figure* 'ApplicationData' property. Using guidata, the previous example simplifies to the following:

```
function myguifcn(arg)
%MYGUIFCN Sample Switchyard Programming Example.

if nargin==0
    arg = 'Initialize';
end
switch arg
case 'Initialize'
    % code that creates the GUI and sets the callbacks
    Hf = figure(...) % create figure for gui

    uicontrol('Parent',Hf,...
              'Style','PushButton',...
              'String','Apply',...
              % In a real GUI other properties must be set also
              'Callback','myguifcn Apply')
```

```matlab
    uicontrol('Parent',Hf,...
              'Style','PushButton',...
              'String','Revert',...
              % In a real GUI other properties must be set also
              'Callback','myguifcn Revert')
    uicontrol('Parent',Hf,...
              'Style','PushButton',...
              'String','Done',...
              % In a real GUI other properties must be set also
              'Callback','myguifcn Done')
    mydata = ... % data to be stored in the GUI for callback use,
                 % most commonly a structure

    guidata(Hf,mydata)

case 'Apply'
    % code that performs Apply button callback actions
    mydata = guidata(gcbo);

case 'Revert'
    % code that performs Revert button callback actions
    mydata = guidata(gcbo);

case 'Done'
    % code that performs Done button callback actions
    mydata = guidata(gcbo);

otherwise
    % report error?
end
```

Object tagging is no longer required. Instead, the function gcbo **g**ets the handle of the **c**all**b**ack **o**bject (thus the reason for calling this function gcbo). That is, it returns the handle of the object that invoked the callback. While this handle is not that of the *figure* containing the data, guidata finds the *figure* by searching through the parent objects of the callback object, as returned by gcbo. The function guidata uses setappdata and getappdata behind the scenes and simplifies the syntax for storing and retrieving data between callback instances.

The definition of callbacks as character strings evaluated in the *Command* window workspace has been around since the introduction of Handle Graphics. This approach has a number of weaknesses, including the facts that callback strings are evaluated each time a callback is invoked and that it is not possible to directly pass data to callbacks. Because of these weaknesses, MATLAB 7 introduces an alternative approach to defining callbacks. MATLAB now supports defining callbacks as function handles rather than as character strings. For example,

```
H_p1 = uicontrol('Style','PushButton',...
                 'Callback',@mycbfcn);
```

defines the callback for this *uicontrol* as a function handle to the function mycbfcn. When this callback is invoked, MATLAB calls the function pointed to by the function handle as mycbfcn(cbo,eventdata). The first argument cbo contains the handle of the callback object—the same handle that is returned by the function gcbo. The second argument contains event data specific to the callback object and the type of callback invoked. As of this writing, the content of this variable is unspecified.

The strength of this function handle approach is that all of the information required to execute the function is created when the function handle is created, namely, at the time the GUI is created. That is, the function pointed to by @mycbfcn can be a subfunction or a nested function within the main GUI creation M-file. In accordance with this approach, the earlier callback examples can be rewritten as follows:

```
function myguifcn
%MYGUIFCN Sample GUI Funcition Using Function Callbacks

% code that creates the GUI and sets the callbacks
Hf = figure(...)% create figure for gui

uicontrol('Parent',Hf,...
          'Style','PushButton',...
          'String','Apply',...
          % In a real GUI other properties must be set also
          'Callback',@Applyfcn)
uicontrol('Parent',Hf,...
          'Style','PushButton',...
          'String','Revert',...
          % In a real GUI other properties must be set also
          'Callback',@Revertfcn)
```

```
uicontrol('Parent',Hf,...
         'Style','PushButton',...
         'String','Done',...
         % In a real GUI other properties must be set also
         'Callback',@Donefcn)

mydata = ... % data to be stored in the GUI for callback use,
             % most commonly a structure

guidata(Hf,mydata)

%-----------------------------
function Applyfcn(cbo,eventdata)
% subfunction code that performs Apply button callback actions
mydata = guidata(cbo);

%-----------------------------
function Revertfcn(cbo,eventdata)
% subfunction code that performs Revert button callback actions
mydata = guidata(cbo);

%-----------------------------
function Donefcn(cbo,eventdata)
% subfunction code that performs Done button callback actions
mydata = guidata(cbo);
```

The switchyard programming structure is no longer needed. Each callback calls its own subfunction. In addition, since the callback object handle is the first argument, the function guidata easily retrieves GUI data stored in the GUI creation *figure*.

In addition to this default use of function handles as callbacks, MATLAB supports function handle callbacks that pass additional data variables directly. For example,

```
H_p1 = uicontrol('Style','PushButton',...
                 'Callback',{@mycbfcn, mydata1, mydata2});
```

defines the callback for this *uicontrol* as a cell array containing a function handle to the function mycbfcn, plus additional input arguments mydata1 and mydata2.

When this callback is invoked, MATLAB calls the function pointed to by the function handle as

```
mycbfcn(cbo,eventdata,mydata1,mydata2)
```

where the content of the variables `mydata1` and `mydata2` are set at the time the GUI is created. Any number of additional variables can be passed to callback functions in this way.

In accordance with this approach, this example can be rewritten as follows:

```
function myguifcn
%MYGUIFCN Sample GUI Funcition Using Function Callbacks

% code that creates the GUI and sets the callbacks
Hf = figure(...)% create figure for gui

mydata = ... % data to be stored in the GUI for callback use,
             % most commonly a structure

uicontrol('Parent',Hf,...
          'Style','PushButton',...
          'String','Apply',...
          % In a real GUI other properties must be set also
          'Callback',{@Applyfcn, mydata})
uicontrol('Parent',Hf,...
          'Style','PushButton',...
          'String','Revert',...
          % In a real GUI other properties must be set also
          'Callback',{@Revertfcn, mydata})
uicontrol('Parent',Hf,...
          'Style','PushButton',...
          'String','Done',...
          % In a real GUI other properties must be set also
          'Callback',{@Donefcn, mydata})

%-------------------------------------------
function Applyfcn(cbo,eventdata,mydata)
% subfunction code that performs Apply button callback actions
```

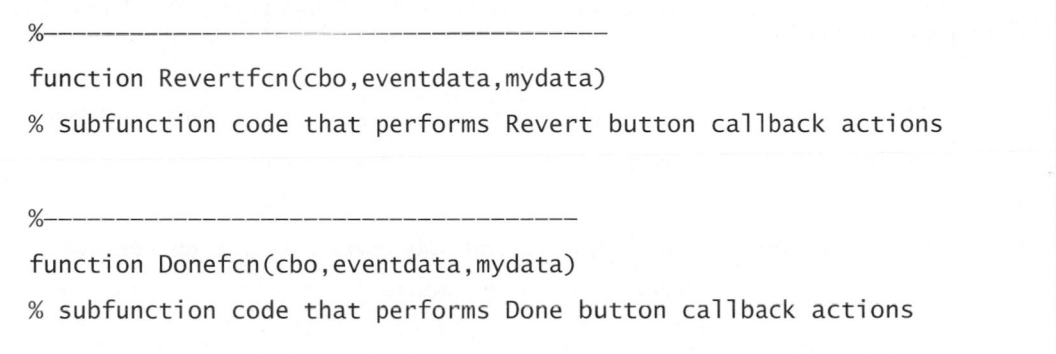

```
%--------------------------------------
function Revertfcn(cbo,eventdata,mydata)
% subfunction code that performs Revert button callback actions


%--------------------------------------
function Donefcn(cbo,eventdata,mydata)
% subfunction code that performs Done button callback actions
```

Now it is no longer necessary to retrieve data by using the guidata function. The only restriction is that the content of the variable mydata must be known before it appears in the callback cell array.

While the above examples demonstrate the use of function handles pointing to subfunctions as callbacks, function handle callbacks can point to any function type that is within scope (i.e., is visible) at the time the function handle is created. Therefore, in addition to subfunctions, function handle callbacks can be nested functions, anonymous functions, or private functions. Once the function handle is created, MATLAB can execute the underlying function even if it goes out of scope or would otherwise be invisible to MATLAB. Clearly, the versatility of function handles and the ability to use them as callbacks increase the productivity of MATLAB GUI programming.

32.11 M-FILE EXAMPLES

It is difficult to illustrate GUI programming with rigorous examples, because GUI M-files are typically very long and their user-interactive nature makes it difficult to illustrate their results on a printed page. The best way to become familiar with GUI programming is to find a preexisting one in MATLAB or on a website, open it in an editor to view the source code, and run it to see what the code produces. Comparing the visual operation of the GUI to the MATLAB code that produces it provides a wealth of information that cannot be gained in any other way.

Given that disclaimer, two GUI examples are illustrated in this section. In the first example, the functions mmtext6 and mmtext7 demonstrate the use of the *figure* object callback properties 'WindowButtonDownFcn', 'WindowButtonMotionFcn', and 'WindowButtonUpFcn'. These functions interactively place text on an *axes* similar to the MATLAB function gtext. The added feature in these functions is the ability to drag the text around the *axes* with the mouse after it has been placed. The function mmtext6 demonstrates the use of character-string callbacks that were available in MATLAB 6 and earlier, whereas mmtext7 demonstrates the use of function handle callbacks now available in MATLAB 7. Here is the code:

```
function h=mmtext6(arg)
%MMTEXT6 Place and Drag Text with Mouse.
% MMTEXT6 waits for a mouse click on a text object in the current figure
% then allows it to be dragged while the mouse button remains down.
%
% MMTEXT6('whatever') places the string 'whatever' on the current axes
% and allows it to be dragged with the mouse
%
% Ht = MMTEXT6('whatever') returns the handle to the text object.
%
% MMTEXT6 becomes inactive after the move is complete or no text
% object is selected.

if nargin==0, arg = 0; end

if ischar(arg)  % user entered text to be placed
   Ht = text('Units','normalized',...
             'Position',[.5 .5],...
             'String',arg,...
             'HorizontalAlignment','center',...
             'VerticalAlignment','middle');

   if nargout>0, h=Ht; end
   mmtext6(0)  % call mmtext6 again to drag it
elseif arg==0  % initial call, select text for dragging

   Hf = get(0,'CurrentFigure');
   if isempty(Hf)
      error('No Figure Window Exists.')
   end
   set(Hf,'BackingStore','off',... % speed up rendering
          'DoubleBuffer','on',...  % get rid of screen flicker
          'WindowButtonDownFcn','mmtext6(1)')
```

```
    figure(Hf)  % bring figure forward

elseif arg==1 & strcmp(get(gco,'type'),'text') % text object selected

    set(gco,'Units','data',...
            'HorizontalAlignment','left',...
            'VerticalAlignment','baseline');
    set(gcf,'Pointer','topr',...
            'WindowButtonMotionFcn','mmtext6(2)',...
            'WindowButtonUpFcn','mmtext6(99)')

elseif arg==2  % dragging text object

    cp = get(gca,'CurrentPoint'); % get current mouse point
    set(gco,'Position',cp(1,1:3)) % move text to the current point

else       % mouse button up or incorrect object selected, reset everything
    set(gcf,'WindowButtonDownFcn','',...
            'WindowButtonMotionFcn','',...
            'WindowButtonUpFcn','',...
            'Pointer','arrow',...
            'DoubleBuffer','off',...
            'BackingStore','on')
end
```

The procedure used in mmtext6 is to set the 'WindowButtonDownFcn' property of the *figure* to 'mmtext6(1)', which is a callback to mmtext6. Then, gco is used to identify the handle of the selected text. Next, the 'WindowButtonMotionFcn' property of the *figure* is set to 'mmtext6(2)', and 'WindowButtonUpFcn' is set to 'mmtext6(99)'. These three different numerical arguments to mmtext6 cause it to perform the desired operations. Finally, when the text move is completed, the callbacks are all reset to empty strings, thereby disabling text selection and dragging.

The MATLAB 7 equivalent of mmtext6 is shown as follows as the function mmtext7.

```matlab
function h=mmtext7(arg)
%MMTEXT7 Place and Drag Text with Mouse.
% MMTEXT7 waits for a mouse click on a text object in the current figure
% then allows it to be dragged while the mouse button remains down.
%
% MMTEXT7('whatever') places the string 'whatever' on the current axes
% and allows it to be dragged with the mouse
%
% Ht = MMTEXT7('whatever') returns the handle to the text object.
%
% MMTEXT7 becomes inactive after the move is complete or no text
% object is selected.

if nargin==0        % call subfunction to set up string drag
   local_mmtext_init

elseif ischar(arg) % user entered text to be placed

   Ht = text('Units','normalized',...
             'Position',[.5 .5],...
             'String',arg,...
             'HorizontalAlignment','center',...
             'VerticalAlignment','middle');

   if nargout>0, h=Ht; end
   local_mmtext_init     % call subfunction to set up string drag
else
   error('String Input Expected.')
end
%----------------------------------------------------------
function local_mmtext_init(cbo,eventdata)
% two input arguments required even if not used
```

```
   Hf = get(0,'CurrentFigure');
   if isempty(Hf)
      error('No Figure Window Exists.')
   end
   set(Hf,'BackingStore','off',... % speed up rendering
          'DoubleBuffer','on',...  % get rid of screen flicker
          'WindowButtonDownFcn',@local_mmtext_down)
   figure(Hf)  % bring figure forward


%-----------------------          ---------------------------------
function local_mmtext_down(cbo,eventdata)
% two input arguments required even if not used

if strcmp(get(gco,'type'),'text') % text object selected

   set(gco,'Units','data',...
           'HorizontalAlignment','left',...
           'VerticalAlignment','baseline');
   set(gcf,'Pointer','topr',...
           'WindowButtonMotionFcn',@local_mmtext_drag,...
           'WindowButtonUpFcn',@local_mmtext_up)
else
   local_mmtext_up % reset everything
end


%---------------------------------------------------------------
function local_mmtext_drag(cbo,eventdata) % dragging text object
% two input arguments required even if not used

cp = get(gca,'CurrentPoint'); % get current mouse point
set(gco,'Position',cp(1,1:3)) % move text to the current point
```

```
%------------------------------------------------------------
function local_mmtext_up(cbo,eventdata) % reset everything
% two input arguments required even if not used

set(gcf,'WindowButtonDownFcn','',...
        'WindowButtonMotionFcn','',...
        'WindowButtonUpFcn','',...
        'Pointer','arrow',...
        'DoubleBuffer','off',...
        'BackingStore','on')
```

This function handle callback case is not dramatically different from the character-string case in mmtext6, but it is generally easier to read. The process of setting and clearing callbacks at the appropriate times is the same in both functions. The primary difference is that the switchyard programming structure has been replaced by callbacks to subfunctions.

The next example creates the following GUI:

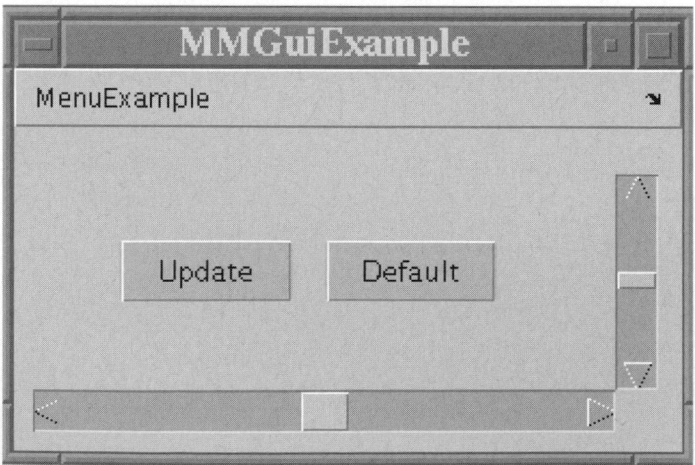

This GUI contains a *uimenu* object, two pushbutton style *uicontrol* objects, and two slider style *uicontrol* objects. Interacting with the horizontal slider moves the *figure* horizontally, and interacting with the vertical slider moves the *figure* vertically. The sliders or scroll bars are made so that the *figure* can be moved all around the current screen. The **Default** button moves the *figure* back to its default size and centered position. If the *figure* is manually dragged with the mouse, pressing the **Update** button updates the slider indicators to reflect the current *figure* position. The **MenuExample** menu contains a single selection named **Close**, which closes the GUI *figure*. The code that creates this GUI is shown as follows:

```
function mmguiexample
%MMGUIEXAMPLE Example GUI in Mastering MATLAB 7

% Build GUI

set(0,'Units','pixels');
Ssize = get(0,'ScreenSize'); % get screen size so gui can be centered

H.gui = dialog('WindowStyle','normal',... % Figure object with good
               'Resize','on',...          % properties for a gui
               'Name','MMGuiExample',...   % Add modifications
               'Units','pixels',...
               'Position',[(Ssize(3)-310)/2 (Ssize(4)-150)/2 310 150]);

DefOutPos = get(H.gui,'OuterPosition'); % undocumented figure property
set(H.gui,'UserData',DefOutPos) % store default outer position here

Hm = uimenu('Parent',H.gui,'Label','MenuExample'); % create top level menu
uimenu('Parent',Hm,... % add 'Close' menu item to top level menu
       'Label','Close',...
       'Callback','close(gcbf)'); % simple string callback

H.Hslider = uicontrol('Style','slider',... % horizontal slider
                      'Parent',H.gui,...
                      'Units','pixels',...
                      'Position',[10 10 270 20],...
                      'Min',20,'Max',Ssize(3)-DefOutPos(3)-20,...
                      'Value',DefOutPos(1),...
                      'Callback',{@local_Hslider,H});
```

```
H.Vslider = uicontrol('Style','slider',... % vertical slider
                      'Parent',H.gui,...
                      'Units','pixels',...
                      'Position',[280 30 20 100],...
                      'Min',20,'Max',Ssize(4)-DefOutPos(4)-20,...
                      'Value',DefOutPos(2),...
                      'Callback',{@local_Vslider,H});

H.Update = uicontrol('Style','pushbutton',... % Update pushbutton
                     'Parent',H.gui,...
                     'Units','pixels',...
                     'Position',[50 70 80 30],...
                     'String','Update',...
                     'Callback',{@local_Update,H});

H.Default = uicontrol('Style','pushbutton',... % Default pushbutton
                      'Parent',H.gui,...
                      'Units','pixels',...
                      'Position',[145 70 80 30],...
                      'String','Default',...
                      'Callback',{@local_Default,H});
%-------------------------------------------------------
% Subfunction callbacks
%-------------------------------------------------------
function local_Hslider(cbo,eventdata,h)
% Callback for horizontal slider
% Move gui figure horizontally
% Slider value contains desired outer left position

SliderValue = get(cbo,'Value');
pos = get(h.gui,'OuterPosition');
set(h.gui,'OuterPosition',[SliderValue pos(2:4)])
```

```
%-------------------------------------------------------
function local_Vslider(cbo,eventdata,h)
% Callback for vertical slider
% Move gui figure vertically
% Slider value contains desired outer bottom position

SliderValue = get(cbo,'Value');
pos = get(h.gui,'OuterPosition');
set(h.gui,'OuterPosition',[pos(1) SliderValue pos(3:4)])


%-------------------------------------------------------
function local_Update(cbo,eventdata,h)
% Callback for Update pushbutton
% Update slider values to reflect current GUI position
% This button is only needed if the user drags the GUI
%  window manually with the mouse

OutPos = get(h.gui,'OuterPosition');
set(h.Hslider,'Value',OutPos(1))
set(h.Vslider,'Value',OutPos(2))


%-------------------------------------------------------
function local_Default(cbo,eventdata,h)
% Callback for Default pushbutton
% Return GUI position vector to default value

defoutpos = get(h.gui,'UserData'); % retrieve default outer position
set(h.gui,'OuterPosition',defoutpos)
set(h.Hslider,'Value',defoutpos(1))
set(h.Vslider,'Value',defoutpos(2))
```

As shown in the code, with the exception of the *uimenu* for closing the GUI *figure*, callbacks are specified as function handles. Comments in the M-file describe the important aspects of the GUI. Given the power of function handle callbacks, no attempt was made to write this function by using only character-string callbacks. It is important to note that the above example reflects just one of many ways that this GUI could be created. While handles for the horizontal and vertical sliders are stored, they are never used, because the variable `cbo` in their callback functions contains their handles. In addition, the `'Userdata'` property of the GUI *figure* was used to store the default figure position vector. Alternatively, this variable could have been passed to the `local_Default` subfunction as an additional input argument.

32.12 GUIDE

GUIDE (Graphical User Interface Design Environment) is a GUI tool that has been designed to make building your own GUIs easier, faster, and more consistent. To create, place, align, and resize UI objects, the function `guide` provides a property editor and inspector that lists object properties and lets the user modify these properties interactively and a menu editor for interactive editing and rearranging user-defined pull-down and contextual menus. GUIDE provides an interactive approach to GUI development and performs well in laying out the geometry of a GUI. It also makes the structure of GUI code consistent from one GUI to the next. GUIDE stores the graphical user interface in a FIG-file, and creates an M-file to initiate the GUI and store the code used for callbacks. GUIDE has improved considerably with every MATLAB release. At this point, GUIDE works well and creates easily debugged and maintainable GUIs. The only weakness of GUIDE (which cannot be eliminated) is the wealth of knowledge required to understand all the nuances involved in creating a well-functioning GUI that does exactly what is desired. For example, Handle Graphics properties such as `'HitTest'`, `'Interruptible'`, and `'BusyAction'` must be understood, as must button-click sequencing, the event queue, and callbacks.

32.13 SUMMARY

Graphical user interface design is not for everyone. However, if you have a need for a GUI, it can be constructed in MATLAB. With the powerful numerical and graphical capabilities available in MATLAB, GUIs can become impressive tools for demonstrating and exploring a wide variety of technical data.

The following table summarizes the GUI functions available in MATLAB, including a number that have not been discussed elsewhere in this chapter (further information regarding GUI construction can be found in the MATLAB documentation):

Function	Description
uibuttongroup	User interface container object for managing 'radiobutton' and 'togglebutton' style *uicontrol* objects
uicontainer	User interface *container* object creation
uicontrol	User interface *uicontrol* object creation
uimenu	User interface *uimenu* object creation
uicontextmenu	User interface *uicontextual menu* object creation
uipanel	User interface *uipanel* object creation
uitoolbar	User interface *uitoolbar* object creation
uipushtool	Momentary contact pushbutton for a *uitoolbar*
uitoggletool	On/Off pushbutton for a *uitoolbar*
uitable	User interface *uitable* object creation
uitree	User interface *uitree* object creation
uitreenode	User interface *uitreenode* object creation
drawnow	Process all pending graphics events and update screen now
gcbf	Get callback *figure* handle
gcbo	Get callback object handle
dragrect	Drag rectangles with the mouse
rbbox	Capture the position of a rubberband box
selectmoveresize	Interactively select, move, and resize *axes* and *uicontrol* objects
waitforbuttonpress	Wait for a key press or a button press over a *figure*
waitfor	Block execution and wait for an event
uiwait	Block execution and wait for resume
uiresume	Resume execution of a blocked M-file
uistack	Control the stacking order of objects
uisuspend	Suspend the interactive state of a *figure*
uirestore	Restore the interactive state of a *figure*
uiclearmode	Clear the current interactive mode
guide	Graphical user interface design environment
inspect	Inspect object properties
ishandle	True if valid object handle
isprop	True if valid object property
align	Align *uicontrols* and *axes*
propedit	Open property editor GUI
makemenu	Create *uimenu* structure
umtoggle	Toggle checked status of a *uimenu* item

Function	Description
getpixelposition	Get position of object in pixels
setpixelposition	Set position of object in pixels
getptr	Get *figure* pointer
setptr	Set *figure* pointer
hidegui	Hide or unhide GUI
movegui	Move *figure* window to a specified part of screen
guidata	Store or retrieve application data
getappdata	Get application data associated with a GUI
setappdata	Set application data associated with a GUI
rmappdata	Remove application data associated with a GUI
isappdata	True if named application data exists
guihandles	Create structure of handles
overobj	Get handle of object that pointer is over
popupstr	Get popup menu selection string
remapfig	Transform positions of objects in a *figure*

33

MATLAB Classes and Object-Oriented Programming

MATLAB has a number of fundamental data types otherwise known as classes. For example, arrays of numbers are commonly double-precision arrays. A variable containing such an array has a class called `double`. Similarly, character strings are another data type or class. Variables containing character strings have a class called `char`. Consider the following example:

```
>> pi          % a simple double
ans =
     3.1416

>> class(pi)
ans =
double

>> s = 'pi'    % a simple string
s =
pi
```

```
>> class(s)
ans =
char
```

Data types or classes in basic MATLAB include `double`, `char`, `logical`, `cell`, and `struct`. These data types are the most commonly used classes in MATLAB. In addition, MATLAB includes the lesser-used classes `function`, `handle`, `inline`, `java`, `single`, and a variety of integer data types.

For each of these classes, MATLAB defines operations that can be performed. For example, addition is a defined operator for elements of the class `double`, but is not defined for elements of the class `char` or for elements of the class `cell`:

```
>> x = pi+2
x =
        5.1416


>> y = 'hello' + 'there'
y =
    220    205    209    222    212


>> {'hello' 'there'}+{'sunny' 'day'}
??? Function 'plus' is not defined for values of class 'cell'.
```

Here, adding two character strings created a numerical array rather than a character string. Rather than report an error, MATLAB chose to convert `'hello'` and `'there'` to their ASCII numerical equivalents, and then perform element-by-element numerical addition. Even though MATLAB produced a result for this character-string example, it did so only after converting the elements on the right-hand side to the class double. MATLAB does this implicit type or class conversion for convenience, not because addition is defined for character strings. On the other hand, trying to add two cell arrays produces an immediate error.

Starting with version 5, MATLAB added the ability to define new operations for the basic data types and, more importantly, added the ability to create user-defined data types or classes. Creating and using data types is called **object oriented programming** (OOP), in which variables in each data type or class are called **objects**. Operations on objects are defined by methods that encapsulate data and overload operators and functions. The vocabulary of OOP includes terms such as *operator* and *function overloading*, *data encapsulation*, *methods*, *inheritance*, and *aggregation*. These terms and the fundamental principles of object-oriented programming in MATLAB are discussed in this chapter.

33.1 OVERLOADING

Before getting involved in the details of OOP and creating new variable classes, consider the process of overloading standard classes in MATLAB. The techniques used to overload standard classes are identical to those used for user-created classes. Once overloading is understood for standard classes, it is straightforward for user-created classes.

When the MATLAB interpreter encounters an operator such as addition, or a function with one or more input arguments, it considers the data type or class of the arguments to the operator or function, and acts according to the rules it has defined internally. For example, addition means to compute the numerical sum of the arguments if the arguments are numerical values or can be converted to numerical values, such as character strings. When the internal rules for an operation or function are redefined, the operator or function is said to be **overloaded**.

> **Operator** and **function overloading** allow a user to redefine what actions MATLAB performs when it encounters an operator or function.

In MATLAB, the redefined rules for interpreting operators and functions are simply function M-files stored in **class directories** just off of the MATLAB search path.

> The collection of rules or M-files for redefining operators and functions define **methods**. The files themselves are commonly referred to as method functions.

That is, class directories themselves are not and cannot be on the MATLAB search path, but they are and must be subdirectories of directories that are on the MATLAB search path. To find class subdirectories, MATLAB requires that class directories be named as *@class*, where *class* is the variable class that the M-files in *@class* apply to. In addition, MATLAB supports multiple class directories. That is, there can be multiple *@class* directories for the same data type just off of the MATLAB path. When looking for functions in class directories, MATLAB follows the order given by the MATLAB search path and uses the first matching method function file found.

For example, if a directory @char appears just **off** of the MATLAB search path, the M-files contained in this directory can redefine operations and functions on character strings. To illustrate this, consider the function M-file plus.m:

```
function s=plus(s1,s2)
% Horizontal Concatenation for char Objects.

if ischar(s1)&ischar(s2)
```

```
    s=cat(2,s1(:).',s2(:).');
elseif isnumeric(s2)
    s=double(s1)+s2;
else
    error('Operator + Not Defined.')
end
```

If this M-file is stored in any @char directory just off of the MATLAB search path, addition of character strings is redefined as horizontal concatenation. For example, repeating the statement y = 'hello' + 'there' made earlier becomes

```
>> y = 'hello' + 'there'
y =
hellothere
```

MATLAB no longer converts the strings on the right-hand side to their ASCII equivalents and adds the numerical results! What MATLAB did was (1) construe character strings to appear on both sides of the addition symbol (+); (2) look down the MATLAB search path for an @char subdirectory; (3) find the one we created and look for a function M-file named plus.m; (4) find the previous plus.m function, pass the two arguments to the addition operator to the function, and let it determine what action to perform; and (5) finally, return the function output as the result of the addition operation.

> To speed operation, MATLAB *caches* class subdirectories at startup. So, for this example to work, you must create the subdirectory and M-file and then restart MATLAB or issue the rehash command to get MATLAB to cache the newly created class subdirectory and associated M-files.

When addition is performed between two different data types, such as char and double, MATLAB considers the precedence and order of the arguments. For variables of equal precedence, MATLAB gives precedence to the leftmost argument to an operator or function as in the following example:

```
>> z = 2 + 'hello'
z =
    106    103    110    110    113
```

The classes double and char have equal precedence. As a result, MATLAB considers addition to be numerical and applies its internal rules, converting 'hello' to its

ASCII equivalent and performing numerical addition. On the other hand, if the order of the preceding operands is reversed, as in the code

```
>> z = 'hello' + 2

z =

   106   103   110   110   113
```

then MATLAB considers addition to be a char class operation. In this case, the @char/plus.m function is called as plus('hello',2). As written, plus.m identifies this mixed class call with isnumeric(s2) and returns the same result as z = 2 + 'hello'.

As illustrated earlier, a function named plus.m defines addition in a class sub-directory. To support overloading of other operators, MATLAB assigns the function names shown in the following table to operators:

Operator	Function Name	Description
a + b	plus(a,b)	Numerical addition
a − b	minus(a,b)	Numerical subtraction
−a	uminus(a)	Unary minus
+a	uplus(a)	Unary plus
a .* b	times(a,b)	Element-by-element multiplication
a * b	mtimes(a,b)	Matrix multiplication
a ./ b	rdivide(a,b)	Element-by-element right division
a .\ b	ldivide(a,b)	Element-by-element left division
a / b	mrdivide(a,b)	Matrix right division
a \ b	mldivide(a,b)	Matrix left division
a .^ b	power(a,b)	Element-by-element exponentiation
a ^ b	mpower(a,b)	Matrix exponentiation
a < b	lt(a,b)	Less than
a > b	gt(a,b)	Greater than
a <= b	le(a,b)	Less than or equal to
a >= b	ge(a,b)	Greater than or equal to
a ~= b	ne(a,b)	Not equal
a == b	eq(a,b)	Equal
a & b	and(a,b)	Logical AND
a | b	or(a,b)	Logical OR
~a	not(a)	Logical NOT

Operator	Function Name	Description
`a:d:b`	`colon(a,d,b)`	Colon operator
`a:b`	`colon(a,b)`	
`a'`	`ctranspose(a)`	Conjugate transpose
`a.'`	`transpose(a)`	Transpose
`[a b]`	`horzcat(a,b)`	Horizontal concatenation
`[a; b]`	`vertcat(a,b)`	Vertical concatenation
`a(s1,s2, ...)`	`subsref(a,s)`	Subscripted reference
`a(s1,s2, ...) = b`	`subsasgn(a,s,b)`	Subscripted assignment
`b(a)`	`subsindex(a)`	Subscript index
	`display(a)`	*Command* window output
`end`	`end(a,k,n)`	Subscript interpretation of `end`

Continuing with the previous `char` class example, note that subtraction can be overloaded with the following function:

```
function s=minus(s1,s2)
% Subtraction for char Objects.
% Delete occurrences of s2 in s1.

if ischar(s1)&ischar(s2)
   s=strrep(s1,s2,'');
elseif isnumeric(s2)
   s=double(s1)-s2;
else
   error('Operator - Not Defined.')
end
```

As defined, subtraction is interpreted as the deletion of matching substrings:

```
>> z = 'hello' - 'e'
z =
hllo


>> a = 'hello' - 2
```

```
a =

    102      99     106     106     109
```

Again, this mixed-class case returns the MATLAB default action.

When multiple operators appear in a statement, MATLAB adheres to its usual order of precedence rules, working from left to right in an expression, as in the following example:

```
>> a = 'hello' + ' ' + 'there'
a =
hello there


>> a - 'e'
ans =
hllo thr


>> a = 'hello' + ' ' + ('there' - 'e')
a =
hello thr
```

When a statement assigns a character string to an output without a terminating semicolon, MATLAB displays the string in the *Command* window. What is displayed in the *Command* window can be overloaded by using the function display.m. While the default character display behavior of MATLAB is convenient, it can be overloaded with a function such as the following:

```
function display(s)
% Display for char objects.

isloose=strcmp(get(0,'FormatSpacing'),'loose');
ssiz=size(s);

if isloose, disp(' '), end
disp(['A Character Array of Size: ' mat2str(ssiz)])
if isloose, disp(' '), end
```

This function redefines how character strings are displayed in the *Command* window, as in the following code:

```
>> 'hello'
A Character Array of Size: [1 5]


>> a = 'hello' + ' ' + 'there'
A Character Array of Size: [1 11]


>> format loose
>> a

A Character Array of Size: [1 11]


>> format compact
>> s=char('hello','there')
A Character Array of Size: [2 5]
```

This discussion concludes how MATLAB overloads operators. Overloading functions follows the same procedure. In this case, the function stored in the class subdirectory has the same name as that of the standard MATLAB function:

```
function s=cat(varargin)
%CAT Concatenate Strings as a Row.

if length(varargin)>1 & ~ischar(varargin{2})
    error('CAT Not Defined for Mixed Classes.')
else
    s=cat(2,varargin{:});
end
```

This function overloads the function cat for character strings. It is called only if the first argument to the cat function is a character string; that is, it is of class char. If the first argument is numerical, the standard cat function is called:

```
>> cat('hello','there') % call overloaded cat
ans =
hellothere
```

```
>> cat('hello',2)          % call overloaded cat
??? Error using ==> char/cat
CAT Not Defined for Mixed Classes.

>> cat(2,'hello')          % call built in cat
ans =
hello
```

In addition to the operator overloading functions listed in this section, MATLAB provides several OOP utility functions. They include methods, isa, class, loadobj, and saveobj. The functions isa and class help to identify the data type or class of a variable or object:

```
>> a = 'hello';
>> class(a) % return class of argument
ans =
char
>> isa(a,'double') % logical class test
ans =
     0
>> isa(a,'char') % logical class test
ans =
     1
```

The function methods returns a listing of the methods, or overloading operators, and functions associated with a given class:

```
>> methods cell

Methods for class cell:
```

cell2struct	ismember	regexprep	str2func	strmatch	tril
ctranspose	lower	reshape	strcat	strncmp	triu
diag	permute	setdiff	strcmp	strncmpi	union
display	regexp	setxor	strcmpi	strtok	unique
intersect	regexpi	sort	strfind	transpose	upper

This result shows that MATLAB itself has overloading functions that are called when input arguments are cell arrays. These functions extend the functionality of basic MATLAB functions to cell arrays without requiring the basic functions themselves to be rewritten to accept cell array arguments.

Finally, the functions `loadobj` and `saveobj` are called (if they exist) whenever the functions `load` and `save` are called with user-defined classes, respectively. Adding these functions to a class subdirectory allows you to modify a user-defined variable after a load operation or before a save operation.

33.2 CLASS CREATION

Operator and function overloading are key aspects of OOP. MATLAB's implementation relies on a simple scheme whereby the methods associated with a variable class are stored in class directories just off of the MATLAB search path. The methods themselves are contained in standard M-files. (They can be P-file or MEX-file equivalents of M-files as well.) User-defined classes use the same scheme for method creation and storage. This section illustrates the creation of user-defined classes.

A new variable class is created when a class directory `@classname` is created and populated by at least two function M-files. The first M-file has the name `classname.m`, which is used to define the creation of variables in the new class. This M-file is called the ***constructor***. The second M-file has the name `display.m` and is used to display the new variable in the *Command* window. No variable class is useful without additional method M-files, but the constructor and `display.m` are a minimum requirement.

> In the vocabulary of OOP, the constructor creates an ***instance*** of the class. This instance is an object that utilizes the methods that overload how operators and functions act in the presence of the object.

The constructor function `classname.m` is a standard function call with input arguments containing the data needed to create an output variable of the desired class. For greatest flexibility, the constructor should handle three different sets of input arguments. Just as there are empty strings, arrays, cells, and so on, the constructor should produce an empty variable if no arguments are passed to it. On the other hand, if the constructor is passed a variable of the same class as that created by the constructor, the constructor should simply pass it back as an output argument. Finally, if creation data are provided, a new variable of the desired class should be created. In this last case, the input data can be checked for appropriateness. Inside of the constructor, the data used to create a variable of the desired class are stored in the fields of a structure. Once the structure fields are populated, the new variable is created by a call to the `class` function. For example, the following is the constructor function for a rational polynomial object:

```
function r=mmrp(varargin)
%MMRP Mastering MATLAB Rational Polynomial Object Constructor.
% MMRP(p) creates a polynomial object from the polynomial vector p
% with 'x' as the variable.
% MMRP(p,'s') creates the polynomial object using the letter 's' as
% the variable in the display of p.
% MMRP(n,d) creates a rational polynomial object from the numerator
% polynomial vector n and denominator polynomial d.
% MMRP(n,d,'s') creates the rational polynomial using the letter 's' as
% the variable in the display of p.
%
% All coefficients must be real.

[n,d,v,msg]=local_parse(varargin); % parse input arguments

if isempty(v) % input was mmrp so return it
   r=n;
else
   error(msg) % return error if it exists
   tol=100*eps;
   if length(d)==1 & abs(d)>tol % enforce scalar d=1
      r.n=n/d;
      r.d=1;
   elseif abs(d(1))>tol % make d monic if possible
      r.n=n/d(1);
      r.d=d/d(1);
   else                          % can't be made monic
      r.n=n;
      r.d=d;
   end
   r.v=v(1);


   r=class(r,'mmrp');   % create object from parts
   r=minreal(r);        % perform pole-zero cancellation
end
```

To simplify the function, parsing of the input arguments is handled by a subfunction entitled `local_parse`. This function is not shown in the preceding code, but it returns four outputs, n, d, v, and msg. The variables n and d are numerical row vectors containing the coefficients of the numerator and denominator of the rational polynomial, respectively. The variable v contains the string variable used to display the polynomial. Last, msg contains an error message if `local_parse` encounters invalid inputs.

Creation of the rational polynomial object is performed by the statement `r=class(r,'mmrp')`. This usage of the function `class` is valid only within the constructor itself. In other contexts, `class` returns the character-string name of the class of its single input argument. The previous constructor considers all three sets of input arguments. If no input arguments exist, n, d, and v are returned to create an empty rational polynomial. If the input argument is a rational polynomial object, it is simply returned as the output argument. Finally, if data are supplied, a rational polynomial object is created. In the simplest case, the denominator is simply equal to 1 and the display variable is `'x'`. The last assignment statement in `mmrp` passes the created rational polynomial to the overloaded but not shown function `minreal`, which returns a minimal realization of the object by canceling like poles and zeros.

Within the constructor M-file, it is important that the structure fields be created in the same order under all circumstances. Violation of this rule may cause the created object to behave erratically.

Given the `mmrp` constructor, we have the following associated `display.m` file:

```
function display(r)
%DISPLAY Command Window Display of Rational Polynomial Objects.

loose=strcmp(get(0,'FormatSpacing'),'loose');
if loose, disp(' '), end
var=inputname(1);
if isempty(var)
     disp('ans =')
else
     disp([var ' ='])
end
nstr=mmp2str(r.n,r.v); % convert polynomial to string
nlen=length(nstr);
if length(r.d)>1 | r.d~=1
     dstr=mmp2str(r.d,r.v);
```

```
else
     dstr=[];
end
dlen=length(dstr);
dash='-';
if loose, disp(' '), end
if dlen % denominator exists
     m=max(nlen,dlen);
     disp('MMRP Rational Polynomial Object:')
     disp([blanks(ceil((m-nlen)/2)) nstr]);
     disp(dash(ones(1,m)));
     disp([blanks(fix((m-dlen)/2)) dstr]);
else
     disp('MMRP Rational Polynomial Object:')
     disp(nstr);
end
if loose, disp(' '), end
```

This @mmrp/display.m function calls the function mmp2str to convert a numerical polynomial vector and a desired variable to a character-string representation. This mmp2str function must exist elsewhere on the MATLAB search path, not in the @mmrp directory. In this case, mmp2str is not part of the MATLAB installation. If it existed in the class directory, MATLAB would not find it, since the arguments to mmp2str are double and char, respectively, and not of class mmrp. As described in the last section, a method function is called only if the leftmost or highest-precedence input argument has a class that matches that of the method.

Within a method function, it is possible to act on objects as shown in the last assignment statement in the constructor r=minreal(r), where the variable r on the right-hand side is an object having class mmrp. There are two exceptions to this property. ***The overloading functions subsref and subsasgn are not called when subscripted reference and subscripted assignment appear within a method function.*** This allows the user to more freely access and manipulate a class variable within a method function.

It is also possible to act on the data contained in an object by simply addressing the structure fields of the object, as shown in numerous places within display.m. In this case, the class of the data determines how MATLAB acts.

> Outside of method functions (for example, in the *Command* window), it is not possible to gain access to the contents of fields of an object, nor is it possible to determine the number of or names of the fields. This property is called data encapsulation.

The following examples demonstrate the creation and display of rational polynomial objects:

```
>> p = mmrp([1 2 3])
p =
MMRP Rational Polynomial Object:
x^2 + 2x^1 + 3

>> q = mmrp([1 2 3],[4 5 6],'z')
q =
MMRP Rational Polynomial Object:
0.25z^2 + 0.5z^1 + 0.75
-----------------------
  z^2 + 1.25z^1 + 1.5

>> r = mmrp(conv([1 2],[1 4]),conv([1 2],[1 3]))
r =
MMRP Rational Polynomial Object:
x^1 + 4
-------
x^1 + 3
```

Rational polynomial objects have little value unless operators and functions are overloaded. In particular, it is convenient to define arithmetic operations on mmrp objects. The following M-files define addition, subtraction, multiplication, and division for mmrp objects (since multiplication and division offer multiple methods, they are all overloaded with the associated polynomial manipulation):

```
function r=plus(a,b)
%PLUS Addition for Rational Polynomial Objects.

if isnumeric(a)
    rn=mmpadd(a*b.d,b.n); % see chapter 20 for mmpadd
```

```
      rd=b.d;
      rv=b.v;
elseif isnumeric(b)
      rn=mmpadd(b*a.d,a.n);
      rd=a.d;
      rv=a.v;
else % both polynomial objects
      if ~isequal(a.d,b.d)
            rn=mmpadd(conv(a.n,b.d),conv(b.n,a.d));
            rd=conv(a.d,b.d);
      else
            rn=mmpadd(a.n,b.n);
            rd=b.d;
      end
      if ~strcmp(a.v,b.v)
            warning('Variables Not Identical')
      end
      rv=a.v;
end
r=mmrp(rn,rd,rv); % create new MMRP object from results
```

```
function r=minus(a,b)
%MINUS Subtraction for Rational Polynomial Objects.

r=a+(-b); % use plus and uminus to implement minus
```

```
function r=uminus(a)
%UMINUS Unary Minus for Rational Polynomial Objects.

r=mmrp(-a.n,a.d,a.v);
```

```
function r=times(a,b)
%TIMES Dot Times for Rational Polynomial Objects.

a=mmrp(a); % convert inputs to mmrp if necessary
b=mmrp(b);
rn=conv(a.n,b.n);
rd=conv(a.d,b.d);
if ~strcmp(a.v,b.v)
    warning('Variables Not Identical')
end
rv=a.v;
r=mmrp(rn,rd,rv); % create new MMRP object from results
```

```
function r=mtimes(a,b)
%MTIMES Times for Rational Polynomial Objects.
r=a.*b; % simply call times.m
```

```
function r=rdivide(a,b)
%RDIVIDE Right Dot Division for Rational Polynomial Objects.

a=mmrp(a); % convert inputs to mmrp if necessary
b=mmrp(b);
rn=conv(a.n,b.d);
rd=conv(a.d,b.n);
if ~strcmp(a.v,b.v)
    warning('Variables Not Identical')
end
rv=a.v;
r=mmrp(rn,rd,rv); % create new MMRP object from results
```

```
function r=mrdivide(a,b)
%MRDIVIDE Right Division for Rational Polynomial Objects.

r=a./b; % simply call rdivide.m
```

```
function r=ldivide(a,b)
%LDIVIDE Left Dot Division for Rational Polynomial Objects.

r=b./a; % simply call rdivide.m
```

```
function r=mldivide(a,b)
%MLDIVIDE Left Division for Rational Polynomial Objects.

r=b./a; % simply call rdivide.m
```

The preceding method functions are self-explanatory, in that they implement simply polynomial arithmetic. Examples of their use include the following:

```
>> a = mmrp([1 2 3])
a =
MMRP Rational Polynomial Object:
x^2 + 2x^1 + 3

>> b = a + 2                    % addition
b =
MMRP Rational Polynomial Object:
x^2 + 2x^1 + 5

>> a - b                        % subtraction
ans =
MMRP Rational Polynomial Object:
-2
```

```
>> a + b                      % addition
ans =
MMRP Rational Polynomial Object:
2x^2 + 4x^1 + 8


>> 2*b                        % multiplication
ans =
MMRP Rational Polynomial Object:
2x^2 + 4x^1 + 10


>> a * b                      % multiplication
ans =
MMRP Rational Polynomial Object:
x^4 + 4x^3 + 12x^2 + 16x^1 + 15


>> b/2                        % division
ans =
MMRP Rational Polynomial Object:
0.5x^2 + x^1 + 2.5


>> 2/b                        % division
ans =
MMRP Rational Polynomial Object:
        2
--------------
x^2 + 2x^1 + 5


>> c = a/b                    % division
c =
MMRP Rational Polynomial Object:
x^2 + 2x^1 + 3
---------------
x^2 + 2x^1 + 5


>> d = c/(1+c)                % mixed
```

```
d =
MMRP Rational Polynomial Object:
0.5x^2 + x^1 + 1.5
------------------

  x^2 + 2x^1 + 4


>> (a/b)*(b/a)                    % mixed
ans =
MMRP Rational Polynomial Object:
1
```

Given the polynomial functions available in MATLAB and the ease with which they can be manipulated, there are many functions that can be overloaded. For example, the basic MATLAB functions roots and zeros can be overloaded by the following method M-files:

```
function [z,p]=roots(r)
%ROOTS Find Roots of Rational Polynomial Objects.
% ROOTS(R) returns the roots of the numerator of R.
% [Z,P]=ROOTS(R) returns the zeros and poles of R in
% Z and P respectively.

z=roots(r.n);
if nargout==2
      p=roots(r.d);
end
```

```
function z=zeros(r)
%ZEROS Zeros of a Rational Polynomial Object.

z=roots(r.n);
```

The method function roots calls the basic MATLAB function roots, because the arguments within the method are of class double. With the creation of the zeros method, the function zeros has two entirely different meanings, depending on what its arguments are. The beauty of OOP is that functions can have multiple meanings or contexts without having to imbed them all in a single M-file. The class of the input arguments dictates which function is called into action.

Mimicking their Handle Graphics usage, it is common to use them to `set` or `get` individual class structure fields, as in the following example:

```matlab
function set(r,varargin)
%SET Set Rational Polynomial Object Parameters.
% SET(R,Name,Value, ...) sets MMRP object parameters of R
% described by the Name/Value pairs:
%
% Name            Value
% 'Numerator'     Numeric row vector of numerator coefficients
% 'Denominator'   Numeric row vector of denominator coefficients
% 'Variable'      Character Variable used to display polynomial

if rem(nargin,2)~=1
   error('Parameter Name/Values Must Appear in Pairs.')
end
for i=2:2:nargin-1
   name=varargin{i-1};
   if ~ischar(name), error('Parameter Names Must be Strings.'), end
   name=lower(name(isletter(name)));
   value=varargin{i};
   switch name(1)
   case 'n'
      if ~isnumeric(value) | size(value,1)>1
         error('Numerator Must be a Numeric Row Vector.')
      end
      r.n=value;
   case 'd'
      if ~isnumeric(value) | size(value,1)>1
         error('Denominator Must be a Numeric Row Vector.')
      end
      r.d=value;
```

```matlab
   case 'v'
      if ~ischar(value) | length(value)>1
         error('Variable Must be a Single Character.')
      end
      r.v=value;
   otherwise
      warning('Unknown Parameter Name')
   end
end
vname=inputname(1);
if isempty(vname)
   vname='ans';
end
r=mmrp(r.n,r.d,r.v);
assignin('caller',vname,r);
```

```matlab
function varargout=get(r,varargin)
%GET Get Rational Polynomial Object Parameters.
% GET(R,Name) gets the MMRP object parameter of R described by
% one of the following names:
%
% Name              Description
% 'Numerator'       Numeric row vector of numerator coefficients
% 'Denominator'     Numeric row vector of denominator coefficients
% 'Variable'        Character Variable used to display polynomial
%
% [A,B,...]=get(R,NameA,NameB,...) returns multiple parameters
% in the corresponding output arguments.

if (nargout+(nargout==0))~=nargin-1
   error('No. of Outputs Must Equal No. of Names.')
end
```

```
for i=1:nargin-1
   name=varargin{i};
   if ~ischar(name), error('Parameter Names Must be Strings.'), end
   name=lower(name(isletter(name)));
     switch name(1)
     case 'n'
       varargout{i}=r.n;
     case 'd'
       varargout{i}=r.d;
     case 'v'
       varargout{i}=r.v;
     otherwise
       warning('Unknown Parameter Name')
     end
end
```

These functions allow you to modify an mmrp object or to get data out of one, as the following example shows:

```
>> c  % recall data
c =
MMRP Rational Polynomial Object:
x^2 + 2x^1 + 3
--------------
x^2 + 2x^1 + 5

>> n = get(c,'n') % get numerator vector
n =
             1           2           3

>> set(c,'Numerator',[3 1]) % change numerator
>> c
c =
MMRP Rational Polynomial Object:
```

$$3x^1 + 1$$

$$----------$$

$$x^2 + 2x^1 + 5$$

```
>> class(c)  % class and isa know about mmrp objects
ans =
mmrp
>> isa(c,'mmrp')
ans =
      1
```

33.3 SUBSCRIPTS

Because of MATLAB's array orientation, user-defined classes can also make use of subscripts. In particular, $V(...)$, $V\{...\}$, and $V.field$ are all supported. In addition, these constructions can appear on either side of an assignment statement. When they appear on the right-hand side of an assignment statement, they are referencing the variable V; and when they appear on the left-hand side, they are assigning data to some part of the variable V. These indexing processes are called subscripted reference and subscripted assignment, respectively. The method functions that control how they are interpreted when applied to an object are subsref and subasgn, respectively. These functions are not as straightforward to understand as other operator and function overloading methods. As a result, this section specifically addresses them. To facilitate this discussion, the mmrp object created in the preceding section is used in the examples.

When dealing with rational polynomials, there are two obvious interpretations for subscripted reference. For a rational polynomial object R, R(x), where x is a data array, can return the results of evaluating R at the points in x. Alternatively, R('v'), where 'v' is a single character, could change the variable used to display the object to the letter provided.

The part of the help text for subsref that describes how to write a subsref method is as follows:

```
B = SUBSREF(A,S) is called for the syntax A(I), A{I}, or A.I
when A is an object.  S is a structure array with the fields:
    type -- string containing '()', '{}', or '.' specifying the
            subscript type.
    subs -- Cell array or string containing the actual subscripts.
```

For instance, the syntax A(1:2,:) invokes SUBSREF(A,S) where S is a
1-by-1 structure with S.type='()' and S.subs = {1:2,':'}. A colon
used as a subscript is passed as the string ':'.

Similarly, the syntax A{1:2} invokes SUBSREF(A,S) where S.type='{}'
and the syntax A.field invokes SUBSREF(A,S) where S.type='.' and
S.subs='field'.

These simple calls are combined in a straightforward way for
more complicated subscripting expressions. In such cases
length(S) is the number of subscripting levels. For instance,
A(1,2).name(3:5) invokes SUBSREF(A,S) where S is 3-by-1 structure
array with the following values:

S(1).type='()'	S(2).type='.'	S(3).type='()'
S(1).subs={1,2}	S(2).subs='name'	S(3).subs={3:5}

According to the help text, if R is an mmrp object, R(x) creates S.type='()' and
S.subs=x, where x contains the values where R is to be evaluated, not indices into
an array. Similarly, R('v') creates S.type='()' and S.subs='v'. All other possibilities should produce an error.

Using this information leads to the subsref method function:

```
function y=subsref(r,s)
%SUBSREF(R,S) Subscripted Reference for Rational Polynomial Objects.
% R('z') returns a new rational polynomial object having the same numerator
% and denominator, but using the variable 'z'.
%
% R(x) where x is a numerical array, evaluates the rational polynomial R
% at the points in x, returning an array the same size as x.

if length(s)>1
        error('MMRP Objects Support Single Arguments Only.')
end
```

```
if strcmp(s.type,'()') % R(x) or R('v')
   arg=s.subs{1};
   argc=class(arg);
   if strcmp(argc,'char')
      if strcmp(arg(1),':')
         error('MMRP Objects Do Not Support R(:).')
      else
          y=mmrp(r.n,r.d,arg(1)); % change variables
      end
   elseif strcmp(argc,'double')
      if length(r.d)>1
         y=polyval(r.n,arg)./polyval(r.d,arg);
      else
         y=polyval(r.n,arg);
      end
   else
      error('Unknown Subscripts.')
   end
else % R{ } or R.field
   error('Cell and Structure Addressing Not Supported.')
end
```

Examples using this method include the following:

```
>> c  % recall data
c =
MMRP Rational Polynomial Object:
    3x^1 + 1

    --------------

x^2 + 2x^1 + 5


>> c = c('t') % change variable
```

```
c =
MMRP Rational Polynomial Object:
    3t^1 + 1
    --------------
t^2 + 2t^1 + 5


>> x = -2:2
x =
    -2    -1    0    1    2
>> c(x)         % evaluate c(x)
ans =
          -1          -0.5          0.2          0.5          0.53846


>> c{3}         % try cell addressing
??? Error using ==> mmrp/subsref
Cell and Structure Addressing Not Supported.


>> c.n          % Try field addressing
??? Error using ==> mmrp/subsref
Cell and Structure Addressing Not Supported.
```

As stated earlier, outside of method functions, the field structure of objects is hidden from view. If it were not, issuing c.n above would have returned the numerator row vector from the object c. To enable this feature, it must be explicitly included in the subsref method:

```
function y=subsref(r,s)
%SUBSREF(R,S) Subscripted Reference for Rational Polynomial Objects.
% R('z') returns a new rational polynomial object having the same numerator
% and denominator, but using the variable 'z'.
%
% R(x) where x is a numerical array, evaluates the rational polynomial R
% at the points in x, returning an array the same size as x.
%
% R.n returns the numerator row vector of R.
```

```
% R.d returns the denominator row vector of R.
% R.v returns the variable associated with R.


if length(s)>1
   error('MMRP Objects Support Single Arguments Only.')
end
if strcmp(s.type,'()') % R( )
   arg=s.subs{1};
   argc=class(arg);
   if strcmp(argc,'char')
      if strcmp(arg(1),':')
         error('MMRP Objects Do Not Support R(:).')
      else
         y=mmrp(r.n,r.d,arg(1));
      end
   elseif strcmp(argc,'double')
      if length(r.d)>1
         y=polyval(r.n,arg)./polyval(r.d,arg);
      else
         y=polyval(r.n,arg);
      end
   else
      error('Unknown Subscripts.')
   end
elseif strcmp(s.type,'.') % R.field
   arg=lower(s.subs);
   switch arg(1)
   case 'n'
      y=r.n;
   case 'd'
      y=r.d;
```

```
   case 'v'
      y=r.v;
   otherwise
      error('Unknown Data Requested.')
   end
else % R{ }
   error('Cell Addressing Not Supported.')
end
```

Examples using this method include the following:

```
>> c.n  % return numerator
ans =
      3     1

>> c.v  % return variable
ans =
x

>> c.nadfdf % only first letter is checked
ans =
      3     1

>> c.t       % not n, d, or v
??? Error using ==> mmrp/subsref
Unknown Data Requested.

>> c.d(1:2) % we didn't include subaddressing in subsref
??? Error using ==> mmrp/subsref
MMRP Objects Support Single Arguments Only.
```

As stated earlier, the overloading functions subsref and subsasgn are *not* implicitly called when subscripted reference and subscripted assignment appear within a method function. Overloading the MATLAB polynomial evaluation function polyval demonstrates this fact, as shown by the following method:

```
function y=polyval(r,x)
%POLYVAL Evaluate Rational Polynomial Object.
% POLYVAL(R,X) evaluates the rational polynomial R at the
% values in X.

if isnumeric(x)
   %y=r(x);          % what we'd like to do, but can't
   S.type='()';
   S.subs={x};
   y=subsref(r,S); % must call subsref explicitly
else
   error('Second Input Argument Must be Numeric.')
end
```

Because of how subsref is written for the mmrp object, polynomial evaluation is simply a matter issuing R(x), where R is an mmrp object and x contains the values where R is to be evaluated. Since this matches the expected operation of polyval, simply issuing y=r(x) within polyval should cause MATLAB to call the subsref method to evaluate the rational polynomial. This does not happen, because MATLAB does not call subsref or subsasgn within methods. However, to force this to happen, one can explicitly call subsref with the desired arguments, as previously shown.

When dealing with rational polynomials, there is one obvious interpretation for subscripted assignment. For a rational polynomial object R, R(1,p) = v, where p is a numerical vector identifying variable powers and v is a numerical vector of the same length, the elements of v become the coefficients of the numerator polynomial associated with the powers in p. Likewise, R(2,q) = w changes the denominator coefficients for those powers identified in q.

The part of the help text for subsasgn that describes how to write a subsasgn method is as follows:

```
A = SUBSASGN(A,S,B) is called for the syntax A(I)=B, A{I}=B, or
A.I=B when A is an object.  S is a structure array with the fields:
    type -- string containing '()', '{}', or '.' specifying the
              subscript type.
    subs -- Cell array or string containing the actual subscripts.
```

For instance, the syntax A(1:2,:)=B calls A=SUBSASGN(A,S,B) where
S is a 1-by-1 structure with S.type='()' and S.subs = {1:2,':'}. A
colon used as a subscript is passed as the string ':'.

Similarly, the syntax A{1:2}=B invokes A=SUBSASGN(A,S,B) where
S.type='{}' and the syntax A.field=B invokes SUBSASGN(A,S,B) where
S.type='.' and S.subs='field'.

These simple calls are combined in a straightforward way for
more complicated subscripting expressions. In such cases
length(S) is the number of subscripting levels. For instance,
A(1,2).name(3:5)=B invokes A=SUBSASGN(A,S,B) where S is 3-by-1
structure array with the following values:
```
    S(1).type='()'       S(2).type='.'       S(3).type='()'
    S(1).subs={1,2}      S(2).subs='name'    S(3).subs={3:5}
```

On the basis of this help text and the desired subscripted assignment, the following
subsasgn method is created:

```
function a=subsasgn(a,s,b)
%SUBSASGN Subscripted assignment for Rational Polynomial Objects.
%
% R(1,p)=C sets the coefficients of the Numerator of R identified
% by the powers in p to the values in the vector C.
%
% R(2,p)=C sets the coefficients of the Denominator of R identified
% by the powers in p to the values in the vector C.
%
% R(1,:) or R(2,:) simply replaces the corresponding polynomial
% data vector.
%
% For example, for the rational polynomial object
```

```
%              2x^2 + 3x + 4
% R(x) = --------------------
%            x^3 + 4x^2 + 5x + 6
%
% R(1,2)=5            changes the coefficient 2x^2 to 5x^2
% R(2,[3 2])=[7 8] changes x^3 + 4x^2 to 7x^3 + 8x^2
% R(1,:)=[1 2 3]    changes the numerator to x^2 + 2x + 3

if length(s)>1
    error('MMRP Objects Support Single Arguments Only.')
end
if strcmp(s.type,'()') % R(1,p) or R(2,p)
   if length(s.subs)~=2
      error('Two Subscripts Required.')
   end
   nd=s.subs{1}; % numerator or denominator
   p=s.subs{2};  % powers to modify
   if ndims(nd)~=2 | length(nd)~=1 | (nd~=1 & nd~=2)
      error('First Subscript Must be 1 or 2.')
   end
   if isnumeric(p) & ...
      (ndims(p)~=2 | any(p<0) | any(fix(p)~=p))
      error('Second Subscript Must Contain Nonnegative Integers.')
   end
   if ndims(b)~=2 | length(b)~=prod(size(b))
      error('Right Hand Side Must be a Vector.')
   end
   b=b(:).';    % make sure b is a row
   p=p(:)';     % make sure p is a row
   if ischar(p) & length(p)==1 & strcmp(p,':') % R(1,:) or R(2,:)
      if nd==1 % replace numerator
         r.n=b;
         r.d=a.d;
      else     % replace denominator
```

```
            r.n=a.n;
            r.d=b;
        end
    elseif isnumeric(p) % R(1,p) or R(2,p)
        plen=length(p);
        blen=length(b);
        nlen=length(a.n);
        dlen=length(a.d);
        if plen~=blen
            error('Sizes Do Not Match.')
        end
        if nd==1  % modify numerator
            r.d=a.d;
            rlen=max(max(p)+1,nlen);
            r.n=zeros(1,rlen);
            r.n=mmpadd(r.n,a.n);
            r.n(rlen-p)=b;
        else      % modify denominator
            r.n=a.n;
            rlen=max(max(p)+1,dlen);
            r.d=zeros(1,rlen);
            r.d=mmpadd(r.d,a.d);
            r.d(rlen-p)=b;
        end
    else
        error('Unknown Subscripts.')
    end
else % R{ } or R.field
    error('Cell and Structure Addressing Not Supported.')
end
a=mmrp(r.n,r.d,a.v);
```

Examples using this method include the following:

```
>> a = mmrp([3 1],[1 2 5 10]) % create test object
a =
MMRP Rational Polynomial Object:
      3x^1 + 1
---------------------
x^3 + 2x^2 + 5x^1 + 10

>> a(1,:) = [1 2 4]      % replace entire numerator
a =
MMRP Rational Polynomial Object:
    x^2 + 2x^1 + 4
---------------------
x^3 + 2x^2 + 5x^1 + 10

>> a(2,2) = 12           % replace x^2 coef in denominator
a =
MMRP Rational Polynomial Object:
    x^2 + 2x^1 + 4
----------------------
x^3 + 12x^2 + 5x^1 + 10


>> a(1,0) = 0            % replace 4x^0 with 0x^0
a =
MMRP Rational Polynomial Object:
    x^2 + 2x^1
----------------------
x^3 + 12x^2 + 5x^1 + 10


>> a(1,:)=a.d            % subsref and subsasn! (a.n and a.d cancel)
a =
MMRP Rational Polynomial Object:
1
```

33.4 CONVERTER FUNCTIONS

As demonstrated in earlier chapters, the functions double, char, and logical convert their inputs to the data type matching their name. For example, double('hello') converts the character string 'hello' to its numerical ASCII equivalent. Whenever possible, methods for these converter functions should be included in a class directory. For mmrp objects, double and char have obvious interpretations. The double method should extract the numerator and denominator polynomials, and the char method should create a string representation such as that displayed by display.m, as in the following code:

```
function [n,d]=double(r)
%DOUBLE Convert Rational Polynomial Object to Double.
% DOUBLE(R) returns a matrix with the numerator of R in
% the first row and the denominator in the second row.
% [N,D]=DOUBLE(R) extracts the numerator N and denominator D
% from the rational polynomial object R.

if nargout<=1 & length(r.d)>1
   nlen=length(r.n);
   dlen=length(r.d);
   n=zeros(1,max(nlen,dlen));
   n=[mmpadd(n,r.n);mmpadd(n,r.d)];
elseif nargout<=1
      n=r.n;
else % nargout==2
      n=r.n;
      d=r.d;
end
```

```
function [n,d,v]=char(r)
%CHAR Convert Rational Polynomial Object to Char.
% CHAR(R) returns a 3 row string array containing R in the
% format used by DISPLAY.M
% [N,D]=CHAR(R) extracts the numerator N and denominator D
```

```
% as character strings from the rational polynomial object R.
% [N,D,V]=CHAR(R) in addition returns the variable V.

if nargout<=1
   nstr=mmp2str(r.n,r.v);
   nlen=length(nstr);
   if length(r.d)>1
      dash='-';
      dstr=mmp2str(r.d,r.v);
      dlen=length(dstr);
      m=max(nlen,dlen);
      n=char([blanks(ceil((m-nlen)/2)) nstr],...
         dash(ones(1,m)),...
         [blanks(fix((m-dlen)/2)) dstr]);
   else
      n=nstr;
   end
elseif nargout>1
   n=mmp2str(r.n); % converts polynomial to string
   d=mmp2str(r.d);
end
if nargout>2
   v=r.v;
end
```

33.5 PRECEDENCE, INHERITANCE, AND AGGREGATION

MATLAB automatically gives user-defined classes ***higher precedence*** than the built-in classes in MATLAB. Therefore, operators and functions containing a mixture of built-in classes and a user-defined class always call the methods of the user-defined class. While this default precedence is usually sufficient for simple classes, the presence of multiple user-defined classes requires that some mechanism exist to allow the user to control the precedence of classes with respect to one another, or perhaps to force a user-defined class to have lower precedence than a built-in class. The functions inferiorto and superiorto provide this capability within MATLAB. If you are using them, calls to ***these functions must appear within the constructor*** *function for a class*. The arguments to both functions contain a list of character strings identifying the classes that have lower or higher precedence than the object created by the constructor. For example, `superiorto('double')` dictates that the object has higher precedence than double-precision variables, and `inferiorto('mmrp','char')` dictates that an object has lower precedence than `mmrp` and `char` objects.

For large programming projects, it may be convenient to create a hierarchy of object types. In this case, it may be beneficial to let one object type inherit methods from another type. In doing so, fewer methods need to be written, and method modifications are more centralized. In the vocabulary of OOP, an object that inherits the properties of another is called a *child* class, and the class it inherits from is called the *parent* class. In the simplest case, a child class inherits methods from a single parent class. This is called *simple* or *single inheritance*. It is also possible for a child class to inherit methods from multiple classes, which is called *multiple inheritance*.

In single inheritance, the child class is given all of the fields of the parent class, plus one or more unique fields of its own. As a result, methods associated with the parent can be directly applied to objects of the child class. Quite naturally, methods of the parent class have no knowledge of the fields unique to the child and therefore cannot use them in any way. Similarly, fields of the parent class cannot be accessed by methods of the child. The child must use the methods it inherited from the parent to gain access to the parent fields. The `lti` object in the *Control Toolbox* is an example of a parent class having child classes `tf`, `zpk`, `ss`, `dss`, `frd`.

In multiple inheritance, a child class is given all of the fields of all parent classes, plus one or more unique fields of its own. As in single inheritance, the parents and child do not have direct access to each other's fields. With multiple parents, the complexity of determining which parent methods are called under what circumstances is more difficult to describe. (Information and detailed examples regarding inheritance can be found in the MATLAB documentation.)

In the `mmrp` class used as an example earlier, the object fields contained data that were elements of the MATLAB classes `double` and `char`. In reality, there is no reason object fields cannot contain other data types, including user-defined classes. In the vocabulary of OOP, this is called *containment* or *aggregation*. The rules for operator and function overloading do not change. Within method functions of one class, the methods of the classes of other classes are called as needed to operate on the fields of the original class.

34

MATLAB Programming Interfaces

MATLAB provides a number of ways to interface with external programs. C functions and FORTRAN subroutines can be called from MATLAB by using MEX-files. Furthermore, MATLAB can perform computations and return the results to C or FORTRAN programs by using the *MATLAB **Engine***. MATLAB provides header files and libraries for creating and accessing standard MATLAB MAT-files. Data can be collected and brought directly into MATLAB, using the built-in serial port interface. In addition, MATLAB can use Java classes, objects, and methods and can exchange data with PC applications by using Component Object Model (COM) objects and Dynamic Data Exchange (DDE). MATLAB can also act as a COM Automation server to communicate with Visual Basic (VB) applications or Visual Basic for Applications (VBA)-enabled PC applications, such as Microsoft Excel, PowerPoint, and Word. The use of Java, COM, DDE, and ActiveX are covered in the following chapters. This chapter focuses on interfacing with C and FORTRAN and serial port communications.

MATLAB supplies a rich set of external programming interfaces, many more than can be covered in this text. A comprehensive treatment of the MATLAB application programming interface (API) could easily fill a book by itself. This chapter and the next two chapters provide a constructive introduction, with examples, to the API features in MATLAB. (The MATLAB documentation provides a more comprehensive treatment of this subject.)

The discussions and examples in this chapter are based on the use of an ANSI C compiler on a UNIX platform, with additional discussion of platform-specific differences. The examples were developed and tested by using the GCC compiler version 3.3.3 on a Linux i386 platform, with the 2.6.6 kernel (glnx86 architecture). The GCC compiler version 3.3 was also used as a test on a Macintosh platform running OS X version 10.3.4 (mac architecture). The LCC compiler supplied with MATLAB was used to test the C language examples on a Windows 2000 Pentium PC (win32 architecture), and the G77 FORTRAN compiler version 3.3.3 was used to test FORTRAN code on the Linux platform.

34.1 ACCESSING MATLAB ARRAYS

Anyone writing programs to interface with MATLAB should have a basic understanding of the structure of MATLAB arrays. Any MEX, MAT, or Engine program must access MATLAB arrays to perform useful work. This section explains a little about how MATLAB stores arrays and lists some of the functions available for use in programs to access MATLAB arrays.

The MATLAB Array

MATLAB supports only one type of object: the MATLAB array. All MATLAB variables are stored as MATLAB arrays. MATLAB array elements are composed of one of the fundamental data types (double, sparse, char, cell, object, single, or 8-, 16-, 32-, or 64-bit signed or unsigned integers) or another MATLAB array. All scalars and composite elements such as strings, vectors, matrices, cell arrays, and structures are MATLAB arrays.

MATLAB stores array data in columnwise order. This fact is used later in the mmcellstr MEX-file (MEX Example 4) when data is extracted from a string buffer and reassembled into individual strings. The following example illustrates this data organization:

```
>> x = [1 2 3; 4 5 6; 7 8 9]
x =
        1       2       3
        4       5       6
        7       8       9

>> size(x)
ans =
        3       3

>> x(:)'
ans =
        1       4       7       2       5       8       3       6       9
```

Accessing MATLAB Arrays: The mxArray

A MATLAB array in C can be considered a new C data type—an object called an mxArray built from a C structure. The mxArray structure stores information about the type of array (double, sparse, cell, and so on), the array dimensions, and references to the data. The `mxArray` structure also stores type-specific information, such as the complexity of a numerical array (real or complex), the number and names of fields of a structure or an object, and the indices and maximum number of nonzero elements of a sparse array. The names of the functions used to access the characteristics and the elements of an `mxArray` all begin with the prefix `mx`.

Numerical matrices are stored as two vectors of doubles (or any of the other numerical data types); one vector contains the real elements, and the second vector contains the imaginary elements. Two pointers are used to access the data—a pointer to the real vector (`pr`) and another pointer to the vector of imaginary numbers (`pi`). The `pi` pointer is `NULL` if the array is real. Strings are stored as 16-bit integers (16 bits per character) with no imaginary component. C strings are null terminated, but MATLAB strings are not. The length of a MATLAB string is always available from the array dimensions.

Cell arrays are collections of arrays. The data vector contains pointers to other arrays. There is no imaginary component. Each pointer in the vector of pointers is called a cell. Structures are stored as folded cell arrays in which each element of the data vector is a field. Each field is associated with a name stored in another structure element of the `mxArray`. Objects are stored as named structures (the class name) with registered collections of methods. There are no imaginary components of cell arrays, structures, or objects.

One of the elements of an `mxArray` is the size of the array, implemented as a vector of integers. Each integer corresponds to the length of the associated dimension. If the number of elements in the vector is greater than 2, the `mxArray` is a multidimensional array. If any one of these dimensions is zero, the array is considered to be an empty array. Another `mxArray` structure element can be used to flag any noncomplex array as a logical array.

Sparse matrices consist of the elements described earlier—two numerical vectors, along with pointers to the vectors. These vectors contain the nonzero elements of the matrix. Sparse arrays also contain the parameters `nzmax`, `ir`, and `jc`, where `ir` is a pointer to an array of the row indices of the corresponding nonzero elements and `jc` is a pointer to a vector of column indices. The maximum number of nonzero elements in the sparse matrix is stored in `nzmax`.

MX Functions

MATLAB provides more than 100 functions and subroutines to access and manipulate `mxArrays` from C or FORTRAN programs. The logical subroutines (`mxIs...`) return one (1) for True and zero (0) for False. The `mxCreate...` functions normally come in two versions: an `mxCreate...Matrix` function for 2-D `mxArrays` and an `mxCreate...Array` version for n-D `mxArrays`. FORTRAN `mxCopyPtrTo...` functions are used to copy data from an `mxArray` into a FORTRAN array, and `mxCopy...ToPtr` subroutines are used to copy data from a FORTRAN array into an `mxArray`. These subroutines are often used to copy `mxArrays` into FORTRAN arrays, send the data to subroutines, and copy the resulting FORTRAN arrays back into `mxArrays`. Examples later in this chapter will illustrate the use of many of these functions and subroutines. Complete lists of MX functions and subroutines, including parameters and return values, are included in the MATLAB documentation.

34.2 CALLING C OR FORTRAN FROM MATLAB

MEX-files are compiled C or FORTRAN functions that can be called from the MATLAB environment, just as if they were M-file functions. Preexisting C functions or FORTRAN subroutines can be modified by adding a few lines of code to manage access to MATLAB data and functions. Compiling the modified code with the `mex` command produces a MEX-file that can be called from within MATLAB. Most computational functions perform faster and more efficiently by using MATLAB M-files. However, a few operations, such as certain For Loops, may be more efficiently implemented in C or FORTRAN. If you cannot eliminate iteration through vectorization, or improve performance sufficiently through loop optimization techniques, MEX-file creation may provide a viable solution.

Compiled MEX-files have platform-specific file name extensions. For example, `dll` is the MEX-file extension on the Windows platform, and `mexsol` is the MEX-file extension for Sun Solaris platforms. On all platforms, the `mexext` function returns the appropriate MEX-file extension.

Each MEX function you create should have an associated M-file (*myfunc.m*) to provide help text for the MEX function. The MATLAB documentation provides additional information about using other architectures and other compilers.

Preparing the MEX Environment

The MEX environment must be initialized to access an installed C or FORTRAN compiler. MATLAB supports a number of compilers, including those available for no charge. For example, the GCC compiler (including FORTRAN support) is available by download from many sources on the Internet, including the GCC home page at `http://gcc.gnu.org/`, and is supplied with many operating system distributions. Another option is to use the LCC compiler supplied with MATLAB on the PC platform. Standard third-party ANSI C compilers and FORTRAN compilers are supported, such as the Microsoft and Borland compilers on the PC platform, the GCC compiler on the Linux and BSD platforms, and the vendor-supplied optional ANSI C or FORTRAN compilers on Unix workstations.

Certain compilers do not mask floating-point exceptions by default. MATLAB arrays allow nonfinite values, such as `Inf` and `NaN`, which generate floating-point exceptions if they are not masked. The MATLAB documentation and the compiler documentation contain information describing ways to mask floating-point exceptions when compiling MEX, Engine, and MAT-file programs that use these compilers.

The command » `mex -setup` selects the appropriate initialization or options file for the compiler and computer platform (operating system and architecture). The options file sets certain environment variables and specifies the locations of the appropriate header files and libraries for a platform and compiler. Options files are located in `$MATLAB/bin` on Unix, Linux, and Macintosh platforms and in `$MATLAB\bin\win32\mexopts` on Windows PC platforms. Once the MEX environment has been initialized, the command » `mex myprog.c` compiles the MEX source file `myprog.c` into the compiled MEX-file `myprog.dll` (or `myprog.mexglx`, and so on). The `mex` command line option `-f` can be used to temporarily select a different initialization file if needed. The -f option is used later to compile MATLAB Engine and MAT programs. The -v option can be used to list the compiler settings and to observe the compile and link stages. The command » mex -help provides a list of mex command-line options. Other standard compiler options, such as -c, -g, and -D, are also supported by mex. Debuggers (such as dbx, ddd, or gdb) can be used to debug MEX programs if the -g option is used. The mex command is implemented both as a function to be called from the MATLAB *Command* window and as a batch or script file for use outside of the MATLAB environment. The MATLAB documentation contains more information about the mex command, command-line options, supported compilers, and debugging procedures.

MEX-files must be designed to operate as standard MATLAB functions using a variable number of input arguments and a variable number of output arguments. Normal MATLAB functions pass data by value rather than by reference. The inputs to a MEX function should never be modified. If the purpose of a function is to change the input, the changed value should be returned as an output. For example, if the function inc increments the value of an input variable b, the function should be called by using the syntax b=inc(b).

MEX Source File Organization

MEX source code files can be divided into ***interface*** and ***subroutine*** sections. The interface section (also known as the ***gateway routine*** in the MATLAB documentation) contains any necessary #include and #define directives, along with code to access MATLAB data and functions from your program. The subroutine section (also known as the ***computational routine*** in the MATLAB documentation) actually performs operations on the data. This section can be written as a separate function or subroutine, or the code can be incorporated into the interface section.

In addition to any other necessary include files, the interface section must contain an #include "mex.h" directive to provide support for MEX functions. The mex prefix designates functions that can be used only within MEX-files and that can operate in the MATLAB workspace. The "mex.h" header file also includes "matrix.h" to provide MX functions to support MATLAB data types, along with the standard header files <stdio.h>, <stdlib.h>, and <stddef.h>. The mx prefix designates functions that operate on MATLAB data types.

The interface section is used to communicate with the MATLAB environment. Every MEX-file requires a mexFunction statement. This is the entry point to the program and corresponds to the main function in a stand-alone C program. The mexFunction function definition is

```
void mexFunction( int nlhs,        mxArray *plhs[],
                  int nrhs, const mxArray *prhs[] )
```

The variable nlhs is an integer containing the number of left-hand-side (output) arguments corresponding to the number returned from the MATLAB nargout function. Similarly, nrhs contains the number of right-hand-side (input) arguments. prhs[] is an array of pointers to the input arguments. prhs[0] is a pointer to the first input argument, prhs[1] is a pointer to the second input argument, and prhs[nrhs-1] is a pointer to the last input argument. All MATLAB data types are MATLAB arrays, and so all input and output arguments to a MEX-file are the mxArray data type.

At the time the compiled MEX function is called, prhs contains pointers to the input arguments, while plhs contains null pointers. It is the responsibility of the programmer to create any output arrays and assign array pointers to the plhs pointer array. An output array can be created even if nlhs contains 0, indicating that no output arguments have been requested. If called from the *Command* window, any output is assigned to the ans variable in the MATLAB base workspace.

Workspace Issues

When a MATLAB M-file function is called, the function operates in a private workspace that is distinct from the MATLAB base workspace or the caller function workspace. Variables in the base or caller workspace are passed by value rather than by reference and are not affected by changes within the called function. The evalin and assignin functions are exceptions; these functions affect variables outside of the current workspace.

MEX functions operate in their own environment, in that the scope of local variables is limited to the MEX function itself. However, this environment is not a MATLAB workspace as such. MX functions (functions with the mx prefix) operate on MATLAB data types. MEX functions (functions with the mex prefix) operate within the MATLAB workspace itself. For example, the mexEvalString and mexCallMATLAB functions evaluate their string arguments in the caller workspace. The mexGetVariable function copies a variable from a specified MATLAB workspace ('base', 'caller', or 'global') into an mxArray. The function mexGetVariablePtr gets a read-only pointer to a MATLAB variable. The function mexPutVariable copies an mxArray into a specified MATLAB workspace. All other mex functions operate in the caller workspace.

Another difference is the use of the mexPrintf function for printing formatted strings. The function mexPrintf must be used, rather than the standard C printf, within MEX-files. This function calls the MATLAB printf function in the base workspace and prints in the *Command* window (and in the diary file if it is in use).

The following is a list of MEX functions available in C programs, FORTRAN programs, or both:

Function	**C**	**F**	**Purpose**
mexAtExit	C	F	Register a function to be called when the MEX-file is cleared
mexCallMATLAB	C	F	Call a MATLAB function, M-file, or MEX-file
mexFunction	C	F	Entry point to the MEX-file
mexFunctionName	C	F	Name of the current MEX function

mexGetVariable	C	F	Get a copy of a variable from another workspace
mexGetVariablePtr	C	F	Get a read-only pointer to a variable from another workspace
mexPutVariable	C	F	Copy an mxArray into a MATLAB workspace
mexEvalString	C	F	Execute a MATLAB command in the caller workspace
mexErrMsgTxt	C	F	Issue an error message and returns to MATLAB
mexErrMsgIdAndTxt	C	F	Issue an error message and identifier, and returns to MATLAB
mexWarnMsgTxt	C	F	Issue a warning message
mexWarnMsgIdAndTxt	C	F	Issue a warning message and identifier
mexLock	C	F	Lock a MEX-file so it cannot be cleared from memory
mexUnlock	C	F	Unlock a MEX-file so it can be cleared from memory
mexIsLocked	C	F	True if the MEX-file is locked
mexIsGlobal	C	F	True if the mxArray has global scope
mexPrintf	C		ANSI C printf-style output routine
mexPrintf		F	Output routine. Optional arguments, such as format strings, are not supported in FORTRAN.
mexSetTrapFlag	C	F	Control the response of mexCallMATLAB to errors
mexMakeArrayPersistent	C	F	Make an mxArray persist after the MEX-file completes
mexMakeMemoryPersistent	C	F	Make memory allocated by mxMalloc and mxCalloc persist
mexGet	C		Get the value of a Handle Graphics property
mexSet	C		Set the value of a Handle Graphics property

MATLAB version 7 has replaced the older mexGetArray, mexGetMatrix, and mexGetFull functions with a single mexGetVariable function. Similarly, the mexPutArray, mexPutMatrix, and mexPutFull functions have been replaced by the mexPutVariable function. Pointer functions have had a similar name change.

The mxGetName function, which returned the MATLAB name of an mxArray (similar to the MATLAB built-in function inputname), has been discontinued in MATLAB 7 and will no longer be supported. For backward compatibility, you can still use the -V5 option in the mex script, but that capability will be removed in the future. The only way to pass a MATLAB variable name into your program without using mxGetName is to pass the arguments as strings containing the variable names, rather than the variables themselves, and use mexGetVariable to get the data, using the variable name.

MEX Example 1: fact

This basic example creates a MEX-file that computes the factorial function (imbedded comments and error checking have been omitted):

```
/*
 * fact.c - returns the factorial of a nonnegative integer.
 *
 *   MATLAB usage:    p=fact(n)
 *
 *   Mastering MATLAB 7 C MEX Example 1
 */

#include "mex.h"

void mexFunction( int nlhs,       mxArray *plhs[],
                  int nrhs, const mxArray *prhs[] )
{
    double n, j, *p;
    int i;

    n=mxGetScalar(prhs[0]);
    plhs[0]=mxCreateDoubleMatrix(1,1,mxREAL);
    p=mxGetPr(plhs[0]);

    j=1.0;
    for (i=n;i>1;i--)
      j=j*i;
    *p=j;
}
```

First, the mex.h header file is included. Then, mexFunction is declared, along with the required parameters. A few variables are declared, and the value of the input argument is obtained by using the `mxGetScalar` function. An output matrix is created to be a 1-by-1 matrix of real doubles by using the `mxCreateDoubleMatrix` function, and a C pointer is obtained by using `mxGetPr`. The calculation is performed, and the result is assigned to the output matrix.

Issuing » `mex fact.c` creates a MEX-file that can be called from MATLAB. The name of the resulting MEX-file depends on the computer platform: `fact.dll` on a Windows PC or `fact.mexglx` on a Linux platform. If the source C file contains an error, compiler error messages are displayed that provide sufficient information to make the necessary corrections to the source code. No MEX output file is created if errors are found.

Running the compiled MEX-file produces output similar to the following:

```
>> x = 5

x =

      5


>> y = fact(x)

y =

    120


>> which fact

/home/work/matlab/fact.mexglx
```

Some of the more common MX and MEX functions are used in these examples, but many more are available. A complete list of MATLAB interface functions, along with descriptions of their arguments and return values (called the MATLAB application programming interface) is available in both HTML and PDF formats.

MEX Example 2: `mycalc`

This example introduces more elements. The function `mycalc` includes error checking and uses a subroutine to perform a calculation on the elements of an array. This style can be used to turn an existing C subroutine into a MEX-file by adding an interface section to existing source code, adding the `#include "mex.h"` directive, and compiling.

The `mycalc` MEX-file accepts a 2-D array of doubles and returns an array of the same size that contains the results of the calculation on each of the input elements. The interface section is used to do some error checking, determine the dimensions of the input array, create C pointers to the input and output arrays, and call the subroutine to do the calculation. Here is the code:

```
/*
 * mycalc.c - calculates x^2-x+1/x for each element of an array.
 *
 *   MATLAB usage:    p=mycalc(n)
 *
 *   Mastering MATLAB 7 C MEX Example 2:
 *         single 2-D real numeric array input,
 *         single array output.
 */

#include "mex.h"

/* This is the original subroutine that performs the calculation. */
static void mycalc( double p[], double n[], int r, int c)
{
    int i;
    for (i=0;i<r*c;i++)
        p[i]=n[i]*n[i]-n[i]+1.0/n[i];
}

/* This is the interface to MATLAB data types and arguments. */
void mexFunction( int nlhs,        mxArray *plhs[],
                  int nrhs, const mxArray *prhs[] )
{
    double *p, *n;
    int r, c;

    /* Do some error checking. */
    if (nrhs != 1)
      mexErrMsgTxt("One input argument required.");
    else if (nlhs > 1)
      mexErrMsgTxt("Too many output arguments.");
    else if (!mxIsNumeric(prhs[0]))
      mexErrMsgTxt("Input must be numeric.");
    else if (mxIsComplex(prhs[0]))
```

```
    mexErrMsgTxt("Input must be real.");
else if (mxGetNumberOfDimensions(prhs[0]) > 2)
    mexErrMsgTxt("N-Dimensional arrays are not supported.");

/* Get the input array dimensions. */
r=mxGetM(prhs[0]);
c=mxGetN(prhs[0]);

/* Create a matrix for the return argument */
plhs[0]=mxCreateDoubleMatrix(r,c,mxREAL);

/* Assign pointers to the parameters. */
p=mxGetPr(plhs[0]);
n=mxGetPr(prhs[0]);

/* Do the actual calculation in a subroutine. */
mycalc(p,n,r,c);
}
```

The subroutine appears first in the MEX C source file to avoid forward references and the necessity for a prototype statement. The mexErrMsgTxt function operates in the same way as the MATLAB error function. The error message text is printed in the MATLAB *Command* window, and the MEX function exits. Note that the standard C language double quotation marks (") are used to delimit a string argument. The mexWarnMsgTxt function can be used to emulate the MATLAB warning function, and it prints the text string in the *Command* window without exiting the MEX-file. The mexWarnMsgIdAndTxt and mexErrorMsgIdAndTxt functions could have been used to return a MATLAB message identifier string along with the text strings if desired. (See the MATLAB documentation for more information on using message identifiers to classify warnings and errors.)

The mxGetM and mxGetN functions return the number of rows and columns, respectively, of a 2-D mxArray. The statement

```
plhs[0]=mxCreateDoubleMatrix(r,c,mxREAL);
```

creates an r-by-c mxArray to contain real elements of MATLAB type double. The mxCOMPLEX flag, rather than the mxREAL flag, could be used to create an array containing complex elements. C-style pointers to the data stored in mxArrays are returned by the mxGetPr function. The function mxGetPr returns a pointer to the real elements of an mxArray. (Use mxGetPi to get a pointer to the imaginary elements of a complex mxArray.) The final step is to call the subroutine mycalc with the appropriate arguments.

MEX Example 3: count

The following example supports multidimensional arrays, uses #define statements to simplify the source code, and includes MATLAB help text for the count MEX-file:

```c
/*
 * count.c - count occurrences of values in an array.
 *
 *    MATLAB usage:    c=count(a,b,tol)
 *
 *    Mastering MATLAB 7 C MEX Example 3
 */

#include <math.h>
#include "mex.h"

/* Define some variables to make life easier. */
#define A   prhs[0] /* Pointer to first right-hand-side argument  */
#define B   prhs[1] /* Pointer to second right-hand-side argument */
#define TOL prhs[2] /* Pointer to third right-hand-side argument  */
#define C   plhs[0] /* Pointer to first left-hand-side argument   */

void  mexFunction( int nlhs,       mxArray *plhs[],
                   int nrhs, const mxArray *prhs[] )

{
    int i, j, sizea, sizeb;
    double tol, vcount;
    double *a, *b, *c;

    /* Do some error checking */
    if (nrhs < 2)
      mexErrMsgTxt("Missing input arguments.");
    else if (nrhs > 3)
      mexErrMsgTxt("Too many input arguments.");
```

```c
else if (nlhs > 1)
  mexErrMsgTxt("Too many output arguments.");

/* Get tolerance value if supplied, otherwise use EPS. */
if (nrhs == 3) {
  if (!mxIsNumeric(TOL) || mxIsComplex(TOL) ||
      mxGetNumberOfElements(TOL) != 1 )
    mexErrMsgTxt("TOL must be a real numeric scalar.");
  tol=mxGetScalar(TOL);
  if (tol < mxGetEps())
    mexErrMsgTxt("TOL must be a positive value.");
}
else
  tol=mxGetEps();

/* Make sure input arrays are noncomplex numeric arrays. */
if (!mxIsNumeric(A) || !mxIsNumeric(B) ||
    mxIsComplex(A) || mxIsComplex(B))
  mexErrMsgTxt("Input arguments must be real of type double.");

/* Create the output mxArray the same size as A */
C=mxCreateNumericArray(mxGetNumberOfDimensions(A),
      mxGetDimensions(A),mxDOUBLE_CLASS,mxREAL);
/* Get the number of elements in A and B and create pointers */
/* to the input and output arrays. */
sizea=mxGetNumberOfElements(A);
sizeb=mxGetNumberOfElements(B);
a=(double *) mxGetPr(A);
b=(double *) mxGetPr(B);
c=(double *) mxGetPr(C);


/* Cycle through the elements of the arrays and count values */
```

```
for (i=0;i<sizea;i++) {
  vcount=0.0;
  for (j=0;j<sizeb;j++)
  if ((fabs(a[i]-b[j])) <= tol)
          vcount++;
        c[i]=vcount;
      }
}
```

This example includes the standard `math.h` header file, in addition to the `mex.h` file, to support the `fabs` function used in the calculation section. Four `#define` statements are also used to reduce the amount of typing required and to improve the readability of the resulting code. The line

```
C=mxCreateNumericArray(mxGetNumberOfDimensions(A),
       mxGetDimensions(A),mxDOUBLE_CLASS,mxREAL);
```

is an example of an MX creation function for n-D mxArrays. Most mxArray creation functions have two distinct forms: a 2-D form and an n-D form. The 2-D version is

```
C=mxCreateNumericMatrix(mxGetM(A),mxGetN(A),mxDOUBLE_CLASS,mxREAL);
```

The 2-D form (`mxCreate...Matrix`) expects the number of rows and the number of columns as separate arguments. Any other arguments (the numerical class and the real/complex flag in this case) follow. The n-D form requires the number of dimensions and a vector of dimension sizes, followed by any other arguments. For example,

```
plhs[0]=mxCreateCellMatrix(3,4);
```

creates an empty 3-by-4 cell mxArray, while

```
plhs[0]=mxCreateCellArray(3,{2,3,4});
```

creates a 3-D 2-by-3-by-4 cell mxArray.

The `mxGetNumberOfElements` function returns the total number of elements in an mxArray. This is the number passed to the subfunction, rather than the dimensions of the arrays, since the subroutine uses single-subscript addressing to process the array elements.

Finally, we create an M-file to provide appropriate MATLAB help text for the `count` function. The file is named `count.m` and is saved in the same directory as the `count.mexglx` MEX-file:

```
function c=count(a,b,tol)
%COUNT Count Occurrences of Values in an Array.
% COUNT(A,B) returns an array the same size as A whose i-th element
% contains the number of times A(i) appears in the array B.
% A and B must be Real arrays.
%
% COUNT implements the following:
%          c = zeros(size(A)));
%          for i=1:prod(size(A))
%               c(i)=sum(A(i)==B);
%          end
```

When MATLAB finds a MEX-file and an M-file in the same directory, the MEX-file is called to execute the function, while the M-file provides help text.

MEX Example 4: mmcellstr

The next example leaves the realm of numbers and enters the world of strings and cells. The mmcellstr function is a C MEX-file implementation of the MATLAB cellstr function. The function mmcellstr creates a cell array of strings from a character array. Each row of the character array is placed in an individual cell of a cell array.

The following example demonstrates a number of new techniques, including allocating memory in a MEX-file, handling empty arrays, and duplicating an array:

```
/*
 *   mmcellstr.c - Create a cell array of strings from a 2-D character
array.
 *
 *   MATLAB usage: c=mmcellstr(s)
 *
 *   Mastering MATLAB 7 C MEX Example 4
 */

#include "mex.h"
```

```
void mexFunction( int nlhs,        mxArray *plhs[],
                  int nrhs, const mxArray *prhs[] )
{
  int m, n, i, j;
  char *buf;
  char **line;

  /* Do some error checking */
  if (nrhs < 1)
    mexErrMsgTxt("Missing input argument.");
  else if (nrhs > 1)
    mexErrMsgTxt("Too many input arguments.");
  else if (mxGetNumberOfDimensions(prhs[0]) != 2)
    mexErrMsgTxt("Input must be 2-D.");
  else if (nlhs > 1)
    mexErrMsgTxt("Too many output arguments.");

/****** Start of Region 1 ******/
  /* If the input is already a cell array, duplicate it. */
  if (mxIsCell(prhs[0]))
    if (mxIsChar(mxGetCell(prhs[0],0))) {
      plhs[0]=mxCreateCellMatrix(mxGetM(prhs[0]),mxGetN(prhs[0]));
      plhs[0]=mxDuplicateArray(prhs[0]);
      return;
    }
/******  End of Region 1  ******/

  /* Make sure the input is a character array. */
  if (!mxIsChar(prhs[0]))
    mexErrMsgTxt("Input must be a character array.");

  /* Handle the empty input case. */
  if (mxIsEmpty(prhs[0])) {
    plhs[0]=mxCreateCellMatrix(1, 1);
```

```
      mxSetCell(plhs[0],0,mxDuplicateArray(prhs[0]));
      return;
  }


  /* Determine the dimensions of the input structure array */
  m=mxGetM(prhs[0]);
  n=mxGetN(prhs[0]);

/****** Start of Region 2 ******/
  /* Stuff the input into a string buffer. */
  /* If mxArrayToString fails to allocate the buffer, the MFX file
terminates. */
  buf=mxArrayToString(prhs[0]);


  /* Create line buffers for the individual strings. */
  line=(char **)mxCalloc(m,sizeof(char*));
  for (i=0;i<m;i++)
    line[i]=mxCalloc(n+1,sizeof(char));
/******   End of Region 2   ******/


  /* Parse the buffer into individual lines. */
  for (j=0;j<n;j++)
    for (i=0;i<m;i++)
      line[i][j]=buf[i+m*j];


  /* Free the string buffer and create the output cell array. */
  mxFree(buf);
  plhs[0]=mxCreateCellMatrix(m, 1);


  for (i=0;i<m;i++) {
    /* For each line, remove trailing blanks... */
    j=n;
    while (--j >= 0)
```

```
    if (line[i][j] != ' ')
      break;
  line[i][j+1]='\0';

  /* insert the line into the output array... */
  mxSetCell(plhs[0],i,mxCreateString(line[i]));

  /* and free the line buffer memory. */
  mxFree(line[i]);
}
mxFree(line);
}
```

After the header section and the function definition, some variables are defined. The variable buf is a pointer to a character array (a string buffer), while line is a pointer to an array of character pointers. Memory is allocated for these buffers later in the function. Next comes some error checking. The next section deserves a closer look:

```
/****** Start of Region 1 ******/
  /* If the input is already a cell array, duplicate it. */
  if (mxIsCell(prhs[0]))
    if (mxIsChar(mxGetCell(prhs[0],0))) {

plhs[0]=mxCreateCellMatrix(mxGetM(prhs[0]),mxGetN(prhs[0]));
      plhs[0]=mxDuplicateArray(prhs[0]);
      return;
    }
/******  End of Region 1  ******/
```

This section handles the case where the input is already a cell array of strings. In this case, the array is duplicated and passed to the output. First, we check that the input is a cell array (mxIsCell). If it is, we check the class of the contents of a cell. The function mxGetCell(prhs[0],0) returns the contents of cell 0 of the input array; mxIsChar then determines whether the element contains character data. The line

```
plhs[0]=mxCreateCellMatrix(mxGetM(prhs[0]),mxGetN(prhs[0]));
```

creates an output array with the same dimensions as the input array. Then, mxDuplicateArray makes a ***deep*** copy of the input mxArray into the output mxArray. A deep copy duplicates all levels of an array; it is essentially a recursive copy. In this case, the line

```
plhs[0]=mxDuplicateArray(prhs[0]);
```

copies the entire contents of the input array into the output array. The function then returns to the calling workspace.

At this point, we have the correct number and type of arguments. In the second section, shown next, we obtain the dimensions of the input character array and then allocate a character buffer of the correct size and copy the character array into the new buffer as one long string by using mxArrayToString. The size is one more than the number of characters in the character array, because mxArrayToString converts the character array to a C-style null-terminated string. We then allocate a line buffer for each row in the input array. The size of each line buffer is the width of the character array, plus one space for the C string terminator '\0'. The code is as follows:

```
/****** Start of Region 2 ******/
  /* Stuff the input into a string buffer. */
  /* If mxArray To String fails to allocate the buffer, the MEX file
terminates.
  buf=mxArrayToString(prhs[0]);

  /* Create line buffers for the individual strings. */
  line=(char **)mxCalloc(m,sizeof(char*));
  for (i=0;i<m;i++)
    line[i]=mxCalloc(n+1,sizeof(char));
/****** End of Region 2  ******/
```

MEX-files should always use mxCalloc (or mxMalloc) to allocate memory and mxFree to return the allocated memory to the heap. These functions register the allocation and deallocation of memory with the MATLAB memory manager, which automatically frees all allocated memory when the function exits. The following code shows how mxFree is used:

```
/* Parse the buffer into individual lines. */
  for (j=0;j<n;j++)
    for (i=0;i<m;i++)
      line[i][j]=buf[i+m*j];

  /* Free the string buffer and create the output cell array. */
  mxFree(buf);
  plhs[0]=mxCreateCellMatrix(m, 1);

  for (i=0;i<m;i++) {
    /* For each line, remove trailing blanks... */
    j=n;
    while (--j >= 0)
      if (line[i][j] != ' ')
        break;
    line[i][j+1]='\0';

    /* insert the line into the output array... */
    mxSetCell(plhs[0],i,mxCreateString(line[i]));

    /* and free the line buffer memory. */
    mxFree(line[i]);
  }
  mxFree(line);
}
```

Since MATLAB arrays are stored columnwise, the buffer contains the input-array elements in column order, similar to the result of using the MATLAB colon operator (buf=chararray(:)). The remainder of the program extracts the correct elements of buff into the line buffers, frees the string buffer memory, and creates the output cell array. Then, each line is processed. A C-string terminator ('\0') is inserted into the buffer after the last nonblank character, and the line is copied and converted from a C string to a MATLAB string by using mxCreateString and is inserted into the output cell array by using mxSetCell. The line buffer memory is then freed by using mxFree, and the next line is processed. After all lines have been processed, mxFree is called one final time to free the line pointer array.

Windows PC Considerations

The previous examples were also compiled on a Windows PC, using the LCC compiler supplied with MATLAB. While this compiler does not offer many features, it does produce functional MEX-files. The command

```
» mex myprog.c
```

produces the MEX-file myprog.dll in the current directory. Adding the verbose (-v) option to the mex command line prints the compiler settings and shows the compile and link stages. The appropriate options file for this compiler was chosen previously by using the mex -setup command. A different options file for an individual compilation can be specified by using the -f option. The $MATLAB\bin\win32\mexopts directory contains options files for many different compilers. The MATLAB documentation offers more details about configuring compilers and debugging MEX programs.

FORTRAN Considerations

FORTRAN MEX-files can now create any MATLAB data type. Note that FORTRAN source code is not case sensitive, and so MXCREATEDOUBLEMATRIX, mxcreatedoublematrix, and mxCreateDoubleMatrix are the same function. FORTRAN examples in this chapter continue to use mixed case to maintain consistency and make the FORTRAN code easier to read.

Since few FORTRAN compilers support new data types, MATLAB passes a special identifier of type integer, called a ***pointer***, to the FORTRAN program for each input and output variable. These pointers can be used by MEX subroutines to obtain data in an appropriate FORTRAN native data type from the MATLAB arrays. Pointers must be declared as the default integer or integer*4 on all platforms, except those using a 64-bit processor, in which case pointers must be declared as integer*8.

Many FORTRAN compilers support the %val construct, which can be used to pass values from pointers obtained from functions such as p=mxGetPr() to subroutines. Otherwise, mxCopy... routines (such as mxCopyPtrToReal8 and mxCopyReal8ToPr) should be used to extract the data from the mxArray, pass it to the subroutine, and return the result to the mxArray.

In FORTRAN, the mexFunction subroutine definition is

```
subroutine mexFunction(nlhs, plhs, nrhs, prhs)
integer plhs(*), prhs(*)
integer nlhs, nrhs
```

where nlhs and nrhs contain the number of left- and right-hand-side arguments, respectively, and plhs and prhs are arrays of pointers to the arguments themselves.

The FORTRAN equivalent of the C MEX-file fact.c is as follows:

```
C-------------------------------------------------------------
C     fact.f - returns the factorial of a nonnegative integer.
C
C      MATLAB usage:    p=fact(n)
C
C     Mastering MATLAB 7 FORTRAN MEX Example 1

      subroutine mexFunction(nlhs, plhs, nrhs, prhs)
C-------------------------------------------------------------
C     These are pointers: integer*4 (integer*8 on 64-bit CPUs)
C
      integer plhs(*), prhs(*)
      integer mxGetPr, mxCreateDoubleMatrix
      integer y_pr
C-------------------------------------------------------------
      integer nlhs, nrhs
      integer i
      real*8  x, y, mxGetScalar
C-------------------------------------------------------------
      x = mxGetScalar(prhs(1))
      plhs(1) = mxCreateDoubleMatrix(1, 1, 0)
      y_pr = mxGetPr(plhs(1))
C
      y = 1.0
      do 10 i=x,1,-1
         y = y * i
 10   continue
C
      call mxCopyReal8ToPtr(y, y_pr, 1)
      return
      end
```

Note that FORTRAN, like MATLAB, uses one-based array and loop indexing, rather than the zero-based indexing used in C programs. Therefore, the pointer

to the first right-hand-side (input) argument (prhs[0] in a C program) is prhs(1) in a FORTRAN program.

34.3 CALLING MATLAB FROM C OR FORTRAN

Just as you are able to call C or FORTRAN routines from MATLAB, you can also call MATLAB to perform operations in the background as part of some larger C or FORTRAN program. In MATLAB, this is called the MATLAB Engine.

What Is It?

The MATLAB Engine consists of a communication library and a small collection of linkable routines that call MATLAB as a server process without linking in all of MATLAB. The MATLAB Engine allows you to start a MATLAB process, transfer data to MATLAB, execute MATLAB commands, capture the normal *Command* window output if desired, transfer data back to your program, and shut down the MATLAB process—all from within a C or FORTRAN program.

How It Works

The MATLAB Engine process runs in the background, separate from any interactive MATLAB session currently running. It does not interfere with any user running MATLAB. When an Engine process is started, a new MATLAB instance is created. This process is shared by all programs requesting access to a MATLAB Engine process on the specific host computer. An exclusive Engine process can be reserved in C programs, but the process is always shared in FORTRAN programs.

The MATLAB Engine communicates with your C or FORTRAN program by using pipes on Unix platforms and the Component Object Model (COM) interface on Windows PC platforms. You can specify a remote computer as the host of the MATLAB Engine on Unix platforms. See the engOpen function for specific syntax.

The eng prefix designates a MATLAB Engine function. MATLAB Engine functions available to C and FORTRAN programs are listed in the following table:

Function	C	F	Purpose
engOpen	C	F	Start up or shares an instance of the MATLAB Engine
engOpenSingleUse	C		Start an exclusive (nonshared) MATLAB Engine session
engPutVariable	C	F	Send a MATLAB array (mxArray) to the MATLAB Engine
engGetVariable	C	F	Get a MATLAB array (mxArray) from the MATLAB Engine
engOutputBuffer	C	F	Create a buffer to store MATLAB text output

Function	C	F	Purpose
engEvalString	C	F	Execute a MATLAB command within the MATLAB Engine
engGetVisible	C		Determine the visibility setting of a MATLAB Engine session. A visible session runs in a window on a PC desktop and is available for user interaction. An invisible session runs in the background and is hidden from the user.
engSetVisible	C		Set the visibility of a MATLAB Engine session
engClose	C	F	Shut down the MATLAB Engine

Engine Program Structure

The first step in using the MATLAB Engine is to open an Engine session by using engOpen or engOpenSingleUse. Then, create any mxArrays needed by using the appropriate MX functions, such as mxCreateDoubleMatrix. Populate the mxArrays and place them in the MATLAB environment by using engPutVariable. Create a buffer to capture *Command* window output, if desired, by using engOutputBuffer, and execute MATLAB commands by using engEvalString. Recover any output mxArrays from the MATLAB environment by using engGetVariable, and continue processing within the C or FORTRAN program. When you are done with the MATLAB environment, terminate the process by using engClose, and release allocated memory for mxArrays by using mxDestroyArray.

Engine Program Description

The engOpen function starts a MATLAB Engine session and returns a unique "Engine ID," which can be used to address this particular Engine. The function returns NULL if an error occurs. If the engOpen argument is NULL ('\0'), the Engine is started on the local computer, using the command matlab. On a UNIX or Linux computer, the argument can be a host name. In this case, the Engine is started on the remote host, using rsh. The DISPLAY environment variable is also set so that any graphics generated by the Engine (such as the result of a plot command) appear on the local computer display, rather than on the computer running the Engine process. Appropriate permissions must be set to allow remote execution and to allow the remote computer to display on your computer screen. If the argument to engOpen is anything other than a host name (such as a string containing spaces, tabs, or nonalphanumerical characters), the string is executed literally to start an Engine process. You can take advantage of this facility to customize your Engine session, such as by using ssh for encrypted communications with the remote host. The engOpen argument must be NULL on a Windows PC, since remote execution is not supported on this platform. The MATLAB application installed on the local PC is started to service any Engine requests.

You must create any MATLAB variables to be used within the Engine environment. One way to create a MATLAB variable is to use the engEvalString function to evaluate a MATLAB command such as y=22*pi:pi/25:2*pi; to create the variable y in the MATLAB workspace and assign values to y. Another method is to create a MATLAB array within your program and pass the resulting mxArray to the MATLAB Engine. For example, given a 1-by-10 array of double-precision data in the C variable dataset, the code

```
Engine *mat
mxArray *mydata = NULL;

mat = engOpen(NULL);
mydata = mxCreateDoubleMatrix(1, 10, mxREAL);
memcpy((void *)mxGetPr(mydata), (void *)dataset, sizeof(dataset));
engPutVariable(mat, "newdata", mydata);
engEvalString("sqdata=newdata.^2;");
```

opens a connection to MATLAB on the local computer, creates an mxArray called mydata, populates the array, assigns a MATLAB variable name newdata to the mydata array when sending the array to the Engine, squares the elements of the array, and assigns the output to the MATLAB variable sqdata. In this example, the C name and the MATLAB name are different, to illustrate how each is used. Normally, the same name is used to avoid confusion.

Once MATLAB performs the desired operations, the results must be passed back into your program. Once again, there are two ways to do this. The first method is to create a text buffer by using the engOutputBuffer function (to capture MATLAB text that is normally discarded), use the engEvalString function to generate the output text, and then parse the string buffer. For example, the code

```
char buf[256];
engOutputBuffer(mat, buf, 256);
engEvalString("disp(sum(sqdata))");
```

captures the output of the disp(sum(sqdata)) command, as it would be displayed in the MATLAB *Command* window, into the string buffer buf. The buffer can be parsed to extract the desired data.

```
double x;
x=atof(buf+2);
```

Each call to engEvalString replaces the data in the MATLAB output string buffer. Note that the first two characters of the output buffer always contain the MATLAB prompt characters ">>", and so the call to the atof function skips over these characters.

When the result is more substantial, you can retrieve one or more arrays from the MATLAB workspace. For example, the code

```
mxArray *sqdata = NULL;
double *sptr;

sqdata = engGetVariable(mat,"sqdata");
sptr = mxGetPr(sqdata);
```

makes the mxArray sqdata available to your program. Use mxGetPr to get a C pointer to the real part of the mxArray. The other MX functions are also available to access the sqdata array.

Be sure to release the memory that was allocated to mxArrays when you are finished with them, and close the MATLAB Engine when it is no longer needed.

```
mxDestroyArray(mydata);

mxDestroyArray(sqdata);

engClose(mat);
```

Compiling and Running Engine Programs

Programs that incorporate MATLAB Engine access functions must include the engine.h header file to support the MATLAB Engine functions, in addition to any other necessary header files. The engine.h header file also includes matrix.h for MX function support. The compilation step uses the mex command with the engopts.sh options file, in addition to any other compiler options. For example, if the MATLAB application is installed in /usr/local/matlab, the mex command to compile a C Engine program might be

```
mex -f /usr/local/matlab/bin/engopts.sh myprog.c
```

This command creates an executable program called myprog in the current directory, and can be run from an operating system command prompt or by double-clicking on a Windows PC. If the standard options file does not meet your needs, make a local copy of engopts.sh and modify it as appropriate.

When the compiled Engine program is run, one or more MATLAB shared libraries are loaded as well. The operating system must be told where to look for these libraries. The LD_LIBRARY_PATH environment variable (or the equivalent on your platform) lists additional directories to be searched when the operating system cannot find all of the required shared libraries in the standard system directories. You must add the appropriate system-specific MATLAB shared library directory to this list. If you have a Solaris operating system with MATLAB installed in the /opt/matlab directory and you are using a Bourne shell (sh, bsh, ksh), the commands are

```
LD_LIBRARY_PATH=/opt/matlab/extern/lib/sol2:$LD_LIBRARY_PATH
export LD_LIBRARY_PATH
```

If you are using a C shell (csh, tcsh) on a Linux platform with libc2 C libraries and have MATLAB installed in the /usr/local/matlab directory, the command is

```
setenv LD_LIBRARY_PATH /usr/local/matlab/extern/lib/glnx86:$LD_LIBRARY_PATH
```

The platform-specific library directories (sol2, glnx86, sgi, sgi64, win32, hpux, and others) are located in the $MATLAB/extern/lib directory. Some platforms use a different environment variable to list additional shared library locations. The SGI64 platform uses LD_LIBRARY64_PATH, rather than LD_LIBRARY_PATH. The HPUX platform uses SHLIB_PATH, and the Mac OS X platform uses DYLD_LIBRARY_PATH. The easiest way to set this variable is to add the appropriate command(s) to your shell start-up script (.profile, .cshrc, .bashrc, .tcshrc, and so on).

Engine Example

So far, we have covered the individual elements of MATLAB Engine programs. Now it is time to put it all together. This example builds on the preceding sections to create a complete C program that uses the MATLAB Engine as a back-end computational engine. This program creates an array of data, sends the array to MATLAB, calculates the squares of the elements and the sum of the squares, and returns the results to the C program for printing. Comments have been added to explain how the pieces are combined. The code is as follows:

```c
/*
 *  sos.c - Calculate the sum of the squares of the elements of a
vector.
 *
 *  Mastering MATLAB 7 C Engine Example 1
 *
 */

#include <stdio.h>
#include <string.h>
#include "engine.h"
#define  BUFSIZE 256

int main()
{
    Engine *mat;
```

```c
mxArray *mydata = NULL, *sqdata = NULL;
int i;
double x, *myptr, *sptr;
char buf[BUFSIZE];
double dataset[10] = { 0.0, 1.0, 2.0, 3.0, 4.0,
                       5.0, 6.0, 7.0, 8.0, 9.0 };

/* Start the MATLAB Engine on the local computer. */
if (!(mat = engOpen(NULL))) {
    fprintf(stderr, "\nCannot open connection to MATLAB!\n");
    return EXIT_FAILURE;
}

/* Create an mxArray and get a C pointer to the mxArray. */
mydata = mxCreateDoubleMatrix(1, 10, mxREAL);
myptr = mxGetPr(mydata);

/* Copy the dataset array to the new mxArray. */
memcpy((void *)myptr, (void *)dataset, sizeof(dataset));

/* Pass the mxArray to the Engine and square the elements. */
engPutVariable(mat, "newdata", mydata);
engEvalString(mat, "sqdata = newdata.^2");

/* Create an output buffer to capture MATLAB text output. */
engOutputBuffer(mat, buf, BUFSIZE);

/* Calculate the sum of the squares and save the result in x. */
engEvalString(mat,"disp(sum(sqdata))");
x=atof(buf+2);

/* Retrieve the array of squares from the Engine, */
if ((sqdata = engGetVariable(mat,"sqdata")) == NULL) {
    fprintf(stderr, "Cannot retrieve sqdata!\n\n");
    return EXIT_FAILURE;
}
```

```
/* and get a C pointer to the mxArray. */

sptr = mxGetPr(sqdata);

/* Print the results to stdout. */
printf("\nThe inputs are:\n");
for (i=0;i<10;i++)
  printf("%6.1f ",myptr[i]);
printf("\n\nThe squares are:\n");
for (i=0;i<10;i++)
  printf("%6.1f ",sptr[i]);
printf("\n\nThe sum of the squares is %6.1f \n\n",x);

/* Free the mxArray memory and quit MATLAB. */
mxDestroyArray(mydata);
mxDestroyArray(sqdata);
engClose(mat);

return EXIT_SUCCESS;
}
```

This example was compiled and tested on both Linux and Macintosh platforms.

Windows PC Considerations

The previous example also compiles on a Windows PC with the LCC compiler. The command

```
mex -f c:\matlab7\bin\win32\mexopts\lccengmatopts.bat sos.c
```

produces the executable file sos.exe in the current directory. If you are not using the LCC compiler, specify the appropriate options file on the command line. The $MATLAB\bin\win32\mexopts directory contains at least eight options files for different versions of the Borland, Microsoft, and Watcom compilers alone. If you have trouble, use the verbose (-v) option to the mex command to view the compiler settings and the various steps in the compilation process.

MATLAB Engine programs link with DLL libraries in the $MATLAB\bin\win32 directory. MATLAB adds this directory to your default path during installation so that Windows will be able to find the libraries. When executed from the command line or by double-clicking on the sos.exe file, a MATLAB session is run (minimized) to perform the calculations. It does not interfere with any interactive MATLAB session that may be open.

Consult the MATLAB documentation for more details about how to configure your specific programming software to compile MEX and Engine programs and for debugging these programs within your software development environment.

FORTRAN Considerations

The same MATLAB Engine program written in FORTRAN requires only a few changes. Most of the Engine functions return a status value which is tested for errors. Also, the sum of the squares is printed as a character array to avoid extracting the numerical value from the text buffer and then printing it. Here is the code:

```fortran
C
C     sos.f - Calculate the sum of the squares of the elements of a vector.
C
C     Mastering MATLAB 7 FORTRAN Engine Example 1
C
C================================================================
      program main
C----------------------------------------------------------------
C     Pointers
C
      integer engOpen, engGetVariable, mxCreateDoubleMatrix, mxGetPr
      integer mat, mydata, sqdata
C----------------------------------------------------------------
C     Other variable declarations
C
      double precision dataset(10), sqrs(10)
      integer engPutVariable, engEvalString, engClose, engOutputBuffer
      integer temp, status
      character*256 buf
      data dataset / 0.0, 1.0, 2.0, 3.0, 4.0, 5.0, 6.0, 7.0, 8.0, 9.0 /
C----------------------------------------------------------------
C     Start the MATLAB Engine on the local computer.
C
      mat = engOpen('matlab')
```

```fortran
      if (mat .eq. 0) then
         write(6,*) 'Cannot open connection to MATLAB!'
         stop
      endif
C
C     Create an mxArray, associate a MATLAB variable name with the
C        mxArray, and copy the data into the array.
C
      mydata = mxCreateDouble Matrix(1, 10, 0)
      call mxCopyReal8ToPtr(dataset, mxGetPr(mydata), 10)
C
C     Pass the variable mydata into the MATLAB workspace.
C
      status = engPutVariable(mat, 'newdata', mydata)
      if (status .ne. 0) then
         write(6,*) 'Cannot pass mydata to the Engine!'
         stop
      endif
C
C     Square the elements of the array.
C
      if (engEvalString(mat, 'sqdata = newdata.^2;') .ne. 0) then
         write(6,*) 'engEvalString failed'
         stop
      endif
C
C     Create an output buffer to capture MATLAB text output.
C
      if (engOutputBuffer(mat, buf) .ne. 0) then
         write(6,*) 'engOutputBuffer failed'
         stop
      endif
```

```
C
C     Calculate the sum of the squares and capture the result.
C

      if (engEvalString(mat, 'disp(sum(sqdata))') .ne. 0) then
         write(6,*) 'engEvalString failed'
         stop
      endif
C
C     Retrieve the mxArray of squares from the Engine,
C     copy the data into an array of doubles, and print.
C

      sqdata = engGetVariable(mat, 'sqdata')
      call mxCopyPtrToReal8(mxGetPr(sqdata), sqrs, 10)
C
 20   format(' ', G8.3, G8.3, G8.3, G8.3, G8.3, G8.3,
     & G8.3, G8.3, G8.3, G8.3)
      print *, 'The inputs are:'
      print 20, dataset
      print *, 'The squares are:'
      print 20, sqrs
      print *,  'The sum of the squares is ', buf(3:10)
C
C     Free the mxArray memory and quit MATLAB.
C

      call mxFreeMatrix(mydata)
      call mxFreeMatrix(sqdata)
      status = engClose(mat)
C
      if (status .ne. 0) then
         write(6,*) 'engClose failed'
         stop
      endif
```

```
C

     stop
     end
```

Note that, if your FORTRAN compiler uses shared libraries, you must add the appropriate MATLAB shared library directory to your LD_LIBRARY_PATH (or equivalent) before executing any compiled Engine program. If you are using a Windows PC, the $MATLAB\bin\win32 directory was added to your default path during MATLAB installation so that Windows can find the MATLAB shared libraries (DLLs).

34.4 EXCHANGING DATA WITH MAT-FILES

There are several ways to exchange data between MATLAB and other programs. Most methods involve creating a data file to serve as an exchange medium. Selecting the appropriate method depends on the amount and format of the data to be imported or exported. In this section, reading and writing standard MATLAB MAT-files from C and FORTRAN are considered.

MAT-files

MAT-files are platform-independent, in that platform differences (such as the native byte order) are stored in the MAT-file itself and MATLAB automatically translates the data format when loading data from the MAT-file. By supplying header files and libraries for use in your programs to read and write MAT-files, MATLAB provides these platform-independent features to C and FORTRAN programs, and enables easy data exchange with MATLAB.

MAT Functions

The mat prefix designates a MATLAB function that operates on MAT-files. MATLAB MAT functions available to C or FORTRAN programs are listed in the following table:

MAT Function	C	F	Purpose
matOpen	C	F	Open a MAT-file
matClose	C	F	Close a MAT-file
matGetDir	C	F	Get a list of MATLAB arrays from a MAT-file
matGetFp	C		Get an ANSI C file pointer to a MAT-file

MAT Function	C	F	Purpose
matGetVariable	C	F	Read a MATLAB array from a MAT-file
matGetVariableInfo	C	F	Load a MATLAB array header from a MAT-file
matGetNextVariable	C	F	Read the next MATLAB array from a MAT-file
matGetNextVariableInfo	C	F	Load the next MATLAB array header from a MAT-file
matPutVariable	C	F	Write a MATLAB array to a MAT-file
matPutVariableAsGlobal	C	F	Write a MATLAB array to a MAT-file as a global variable
matDeleteVariable	C	F	Delete a MATLAB array from a MAT-file

The matGetVariableInfo function creates an mxArray containing everything in the mxArray structure except the actual data itself. The matGetDir function creates a list of variable names and returns the number of MATLAB variables contained in the MAT-file. The matGetVariable function uses the variable name to access the contents of the variable, while matGetNextVariable and matGetNextVariableInfo functions access the variables sequentially. The matPutVariableAsGlobal function writes an array to the MAT-file as a global variable; that is, when the load command is used to load the MAT-file into MATLAB, the variable is loaded into the MATLAB workspace as a global variable.

MAT Program Structure

C or FORTRAN programs that read or write MAT-files use MX functions to create MATLAB data arrays (mxArrays) in the same general way that MX functions are used in MEX-files or Engine programs. When creating a MAT-file, you first open the MAT-file by using matOpen. Then you create, name, and populate mxArrays for each MATLAB variable; write the mxArrays to the MAT-file by using matPutVariable or matPutVaribleAsGlobal; and, finally, close the MAT-file by using matClose. When reading data from a MAT-file, you first open the MAT-file by using matOpen; get a list of the variables, if desired, by using matGetDir; create mxArrays by using matGetVariable or matGetNextVariable; and, finally, close the MAT-file by using matClose.

Compiling and Running MAT programs

Programs that incorporate MATLAB MAT-file access functions must include the mat.h header file to support the MATLAB MAT-file functions, in addition to any other necessary header files. The mat.h header file also includes matrix.h for MX function support. The compilation step uses the mex command with the matopts.sh options file, in addition to any other compiler options you may need. For example, if the MATLAB application is installed in /usr/local/matlab, the mex command to compile a C MAT program might be

```
mex -f /usr/local/matlab/bin/matopts.sh myprog.c
```

This command creates an executable program called `myprog` in the current directory and can be run from an operating system command prompt. The shared-library discussions in the Engine section relating to the UNIX and PC platforms apply here as well. An executable program created by the `mex` command uses shared libraries that must be found by your operating system when the program is run.

The Windows PC version uses the same options file that Engine programs do. Compiling the same `myprog.c` MAT program by using the LCC compiler might use the command

```
mex -f c:\matlab7\bin\win32\mexopts\lccengmatopts.bat myprog.c
```

The compilation produces the executable file `myprog.exe` in the current directory, which can be run from a command prompt or by double-clicking on the file. If you are not using the LCC compiler, specify the appropriate options file on the command line.

MAT Program Example 1: `writemat`

The following C program creates a MAT-file containing a string and an array of doubles:

```c
/*
 *  writemat.c - Create a binary MAT-file.
 *
 *  Mastering MATLAB 7 C MAT-file Example 1
 *
 */

#include "mat.h"

int makemat(const char *filename,
            double *data, int m, int n,
            char *mmstr)
{
  MATFile *mfile;
  mxArray *mdata, *mstr;
```

```
/* Open the MAT-file for writing. */
mfile = matOpen(filename, "w");
if (mfile == NULL) {
  printf("Cannot open %s for writing.\n", filename);
  return(EXIT_FAILURE);
}

/* Create the mxArray to hold the numeric data.    */
/* Note that the array dimensions are reversed.     */
/* C uses row order while MATLAB uses column order. */
/* The data array will be transposed in MATLAB.     */
mdata = mxCreateDoubleMatrix(n,m,mxREAL);

/* Copy the data to the mxArray. Note that mxGetData is */
/* similar to mxGetPr but returns a void pointer while  */
/* mxGetPr returns a pointer to a double.               */
memcpy((void *)(mxGetData(mdata)), (void *)data,
               m*n*sizeof(double));

/* Create the string array. */
mstr = mxCreateString(mmstr);

/* Write the mxArrays to the MAT-file. */
matPutVariable(mfile, "mydata", mdata);
matPutVariable(mfile, "mystr", mstr);

/* Free the mxArray memory. */
mxDestroyArray(mdata);
mxDestroyArray(mstr);

/* Close the MAT-file. */
if (matClose(mfile) != 0) {
  printf("Cannot close %s.\n",filename);
  return(EXIT_FAILURE);
}
```

```
  return(EXIT_SUCCESS);
}

int main()
{
  int status;
  char *mmstr = "Mastering MATLAB Rocks!";
  double data[3][4] = {{  1.0,   2.0,   3.0,   4.0 },
                       {  5.5,   6.6,   7.7,   8.8 },
                       { -4.0,  -3.0,  -2.0,  -1.0 }};

  status = makemat("mmtest.mat", *data, 3, 4, mmstr);
  return(status);
}
```

This program was compiled and tested on Linux, Macintosh, and Windows PC platforms. The resulting MAT-files were then loaded into different MATLAB sessions, with the following results:

```
>> clear all

>> load mmtest

>> whos
  Name          Size              Bytes  Class

  mydata        4x3                  96  double array
  mystr         1x23                 46  char array

Grand total is 35 elements using 142 bytes

>> mydata
mydata =
        1.0000    5.5000   -4.0000
        2.0000    6.6000   -3.0000
        3.0000    7.7000   -2.0000
        4.0000    8.8000   -1.0000
```

```
>> mystr

mystr =

Mastering MATLAB Rocks!
```

Note that the numerical array was transposed when loaded into MATLAB. This is because of the differences in the way that C and MATLAB store arrays—rowwise and columnwise, respectively. The FORTRAN version does not transpose the matrix, since both MATLAB and FORTRAN store arrays in columnwise format. The following is a FORTRAN version of the same program:

```
C
C    writemat.f - Create a binary MAT-file.
C
C    Mastering MATLAB 7 FORTRAN MAT-file Example 1
C
C

      program writemat
C---------------------------------------------------------------
C    Pointers.
C

      integer matOpen, mxCreateDoubleMatrix, mxCreateString
      integer matGetVariable, mxGetPr
      integer mfile, mdata, mstr
C---------------------------------------------------------------
C    Other variables
C

      integer status, matClose
      double precision dat(12)
      data dat / 1.0, 5.5, -4.0,
     &           2.0, 6.6, -3.0,
     &           3.0, 7.7, -2.0,
     &           4.0, 8.8, -1.0 /
C
```

```fortran
C     Open MAT-file for writing.
C
      mfile = matOpen('mmtest.mat', 'w')
      if (mfile .eq. 0) then
         write(6,*) 'Can''t open ''mmtest.mat'' for writing.'
         stop
      end if
C
C     Create the mxArray to hold the numeric data.
C
      mdata = mxCreateDoubleMatrix(4,3,0)
C
C     Copy the data to the mxArray.
C
      call mxCopyReal8ToPtr(dat, mxGetPr(mdata), 12)
C
C     Create the string array.
C
      mstr = mxCreateString('Mastering MATLAB Rocks!')
C
C     Write the mxArrays to the MAT-file.
C
      call matPutVariable(mfile, 'mydata', mdata)
      call matPutVariable(mfile, 'mystr', mstr)
C
C     Free the mxArray memory.
C
      call mxFreeMatrix(mdata)
      call mxFreeMatrix(mstr)
C
C     Close the MAT-file.
C
      status = matClose(mfile)
```

```
      if (status .ne. 0) then
         write(6,*) 'Cannot close mmtest.mat'
         stop
      end if
C

      stop
      end
```

MAT Program Example 2: whomat

The next example reads a MAT-file, examines its contents, and then prints variable lists in formats similar to the MATLAB who and whos commands:

```c
/*
 *  whomat.c - Examine a binary MAT-file and print a list
 *             of the contents (like "who" or "whos").
 *
 *  Mastering MATLAB 7 C MAT-file Example 2
 *
 */

#include "mat.h"
#include <string.h>

int whomat(const char *filename)
{
  MATFile *mfile;
  mxArray *marray;
  char **dir;
  char siz[25], buf[10];
  int i, j, num, nel, elsize, ndim, eltot, btot;
  const int *dims;
```

```c
/* Open the MAT-file for reading. */
mfile = matOpen(filename, "r");
if (mfile == NULL) {
  printf("Cannot open %s for reading.\n", filename);
  return(EXIT_FAILURE);
}

 /* Get the directory list and print in "who" format. */
dir = matGetDir(mfile, &num);
if (dir == NULL) {
  printf("Error reading the directory of %s.\n", filename);
  return(EXIT_FAILURE);
} else {
  printf("\n");
  printf("Variables in %s are:\n\n", filename);
  for (i=0; i<num; i++) {
    printf("%-10s",dir[i]);
    if (i>0 && i%4==0) printf("\n");
  }
}

/* Examine each variable and print a "whos" list. */
eltot=btot=0;
printf("\n\n  Name          Size        Bytes  Class\n\n");
for (i=0; i<num; i++) {
  marray=matGetVariable(mfile, dir[i]);
  if (marray == NULL) {
    printf("Cannot read file %s.\n\n", filename);
    return(EXIT_FAILURE);
  }

  /* If marray is a cell array or structure array, then  */
  /* mxGetElementSize returns the size of a pointer, not */
  /* the size of all the elements in each cell or field. */
```

```c
    /* To get the correct number of bytes would require    */
    /* traversing the array and summing leaf element sizes.*/
    /* Java arrays return 0x0 array dimensions and 0 size. */

    elsize=mxGetElementSize(marray);
    btot=btot+(nel*elsize);
    nel=mxGetNumberOfElements(marray);
    eltot=eltot+nel;
    ndim=mxGetNumberOfDimensions(marray);
    dims=mxGetDimensions(marray);
    siz[0]='\0';
    for (j=0; j<ndim; j++) {
      sprintf(buf,"%d",dims[j]);
      strcat(siz,buf);
      if (j<(ndim-1))
        strcat(siz,"x");
    }
    printf("  %-12s %-12s %5d  %s array\n", dir[i],
            siz,nel*elsize,mxGetClassName(marray));
    mxDestroyArray(marray);
  }
  printf("\nGrand total is %d elements using %d bytes\n\n",
          eltot,btot);

  /* Release the memory allocated for the directory. */
  mxFree(dir);

  /* Close the MAT-file. */
  if (matClose(mfile) != 0) {
    printf("Cannot close %s.\n",filename);
    return(EXIT_FAILURE);
  }
  return(EXIT_SUCCESS);
}
```

```
int main(int argc, char **argv)
{
  int status;

  if (argc > 1)
    status = whomat(argv[1]);
  else{
    status = EXIT_FAILURE;
    printf("Usage: whomat <matfile>");
  }
  return(status);
}
```

The output of this program, while very similar to the output of the MAT-LAB who and whos commands with the '-file' argument, reports incorrect results for cell arrays, structures, and Java arrays. If marray is a cell array or structure array, then mxGetElementSize(marray) returns the size of a pointer, not the size of all of the elements in each cell or field. To get the correct number of bytes would require traversing the array and summing leaf element sizes. If marray is a Java array, then mxGetElementSize(marray) returns zero and mxGetDimensions(marray) returns a pointer to a vector of zeros of length mxGetNumberOfDimensions(marray). Adding support for these data types is left as an exercise for the reader.

34.5 SHARED LIBRARIES

Libraries are collections of functions that are available to any program. These functions are precompiled into library files. There are two kinds of libraries. Static libraries are linked with your program at compile time, and the functions referenced in your program are included in your executable. Shared libraries are linked with your program at run time. Because they are not embedded in your executable, the shared libraries must be available at run time. Many different programs running on the same computer can share access to the same shared libraries at the same time. On Unix and Linux systems, these shared libraries are normally located in the standard system library directories (/lib, /usr/lib, /usr/shlib, or /usr/local/lib) and $MATLAB/bin/$ARCH. They are named by using the lib prefix and the .so suffix. (Macintosh systems use the .dylib suffix). On Windows PC systems, the shared libraries are called Dynamic Link Libraries and use the .DLL extension. At run time, the necessary functions from the appropriate shared libraries are loaded into memory and made available to the applications.

The MATLAB interface to shared libraries provides tools to load and unload a shared library, list the functions available in the library (including function name, arguments, and return values), and invoke these functions from within MATLAB. Any shared library can be accessed, as long as it has C-style arguments and types.

MATLAB shared library functions are listed in the following table:

Function	Purpose
loadlibrary	Load an external library into MATLAB
unloadlibrary	Unload an external library from memory
libisloaded	Determine if an external library is loaded
libfunctions	Return information on the functions in an external library
libfunctionsview	Create a window to display function information
calllib	Call a function in an external library
libpointer	Create a pointer object for use with external libraries
libstruct	Create a C-like structure that can be passed to external libraries

MATLAB automatically converts between MATLAB data types and C data types in most cases. For example, MATLAB structures passed as arguments to C library functions are converted to C structures automatically. MATLAB data types passed as values to C functions expecting pointers are converted into pointers automatically. The libpointer and libstruct functions permit explicit conversions under the control of the user. (For more details about shared library functions and data conversion issues, see the MATLAB documentation.)

34.6 SERIAL COMMUNICATIONS

Data are often gathered by using instruments connected to a serial port. MATLAB provides a built-in interface to manage serial ports and to communicate directly with devices connected to these serial ports. MATLAB supports the RS-232, RS-422, and RS-485 serial communications standards on Linux, Sun Solaris, and Windows platforms. Enhanced support for multifunction I/O boards, GPIB or VESA instrumentation, and additional serial-port functions are available in optional toolboxes. However, the built-in serial interface will suffice for most direct serial communications.

The MATLAB interface to serial ports is an object-based interface. The serial function creates an object connected to a specific serial port, with properties that can be queried and set by using the standard get and set functions. The serial function requires the name of a serial device and optional PropertyName and PropertyValue pairs. Serial devices have different names on different platforms. For example, the first serial port is /dev/term/a or /dev/ttya on a Solaris 7 platform, /dev/ttyS0 on an i386 Linux platform, and COM1 on a Windows PC.

Communicating with a serial device normally consists of a fixed sequence of operations. First, create the interface to the serial device with the `serial` function and `set` the appropriate properties. Then, open the device by using `fopen`, write and read by using `fprintf` and `fscanf` respectively, close the device by using `fclose`, and finally `delete` the serial object and `clear` the MATLAB reference to the variable.

Serial Communication Example

This is an interactive example of a communication session with a blood glucose meter. The meter requires serial port settings of 9600,8,N,1 with no handshaking. This means a baud rate of 9600 bits per second, 8-bit data, no parity checking, and one stop bit. The meter does not use hardware handshaking.

First, create a serial object attached to the first serial port, and check its properties:

```
>> s = serial('/dev/ttyS0');
>> get(s)
    ByteOrder = littleEndian
    BytesAvailable = 0
    BytesAvailableFcn =
    BytesAvailableFcnCount = 48
    BytesAvailableFcnMode = terminator
    BytesToOutput = 0
    ErrorFcn =
    InputBufferSize = 512
    Name = Serial-/dev/ttyS0
    ObjectVisibility = on
    OutputBufferSize = 512
    OutputEmptyFcn =
    RecordDetail = compact
    RecordMode = overwrite
    RecordName = record.txt
    RecordStatus = off
    Status = closed
    Tag =
    Timeout = 10
    TimerFcn =
```

```
TimerPeriod = 1
TransferStatus = idle
Type = serial
UserData = []
ValuesReceived = 0
ValuesSent = 0

SERIAL specific properties:
BaudRate = 9600
BreakInterruptFcn =
DataBits = 8
DataTerminalReady = on
FlowControl = none
Parity = none
PinStatus = [1x1 struct]
PinStatusFcn =
Port = /dev/ttyS0
ReadAsyncMode = continuous
RequestToSend = on
StopBits = 1
Terminator = LF
```

In this case, most of the default communications parameters are correct. However, the default line terminator is a line feed character, whereas the meter expects a carriage return/line feed sequence:

```
>> set(s,'Terminator','CR/LF');
>> get(s,'Terminator')
ans =
CR/LF
```

Now open the port and send a command to the device. In this case, the command instructs the meter to respond with its serial number and a communications checksum:

```
>> fopen(s);
>> fprintf(s,'DM@');
>> snum = fscanf(s)
```

```
snum =
    @ "RMF99CFBV" 031D
```

The next command zeros the meter's data. Write the command and check for a response:

```
>> fprintf(s,'DMZ')
>> resp = fscanf(s)
resp =
    Z 005A
```

No errors were encountered, so close the connection, delete the port object, and remove the variable from the MATLAB environment:

```
>> fclose(s)
>> delete(s)
>> clear s
```

Serial port objects are similar to graphics objects in that they support events and callbacks. Callbacks are functions to be executed when certain events occur. Functions can be specified either by using function handles or as cell arrays containing function names. Any object property with the string 'Fcn' in the property name can be configured to execute a callback function when the associated event occurs. Examples of serial events include timer expiration, errors, interrupts (break), output buffer empty, data available in the input buffer, or change of state of a serial line pin (CD, RI, DSR, and/or CTS).

Properties are available to monitor the status of control pins, control the DTR pin, implement data flow control using hardware (RTS/CTS) or software (Xon/Xoff), and automatically record details about the session to a data file. The number and types of data read and written and the event information, as well as all of the data values can be saved to one or more files if desired. (The MATLAB documentation contains more details and additional examples.)

34.7 SOURCE CODE CONTROL SYSTEMS

Large programming projects or projects involving many developers often use source code control systems (SCCS) to manage program source code and revisions. Files are collected into a central repository and are checked out by developers working on the project. Files are then checked back in after revisions. The SCCS ensures consistency and facilitates version rollbacks if needed.

Many existing source code control systems can be accessed from within the MATLAB environment. Supported systems include CVS, RCS, PVCS, CM Synergy, and ClearCase on Unix, Linux, or Macintosh systems, and any system conforming to the Microsoft Common Source Control standard on Windows PC platforms. Custom SCCS systems are also supported on all platforms except Windows PCs.

MATLAB supplies a graphical interface to your existing SCCS system when using the MATLAB Editor, as well as a function interface available from the *Command* window. Select your preferred SCCS in the Preferences window (select **Preferences** from the **File** menu of the *Command* window). A **Source Control** menu is then available from the **File** menu in the MATLAB Editor. Available actions are Check In, Check Out, and Undo Check-Out. SCCS functions are listed in the following table:

Function	Purpose
cmopts	Return the name of the selected source code control system
checkin	Check a file in to the source code control system
checkout	Check a file out of the source code control system
undocheckout	Undo the previous check-out
customverctl	Custom version control system template
rcs	Version control actions using RCS
cvs	Version control actions using CVS
pvcs	Version control actions using PVCS
sourcesafe	Version control actions using SourceSafe
clearcase	Version control actions using PVCS

34.8 WEB SERVICES

Now that the Internet is such a big part of computing, MATLAB has extended support for various Internet technologies. MATLAB now includes client support for making remote procedure calls over a network (local networks or the Internet) by using some of these technologies.

First, a little background and terminology are in order. The Standardized Generalized Markup Language (SGML), developed in 1985, is an international standard for defining descriptions of the structure of different types of electronic documents. Extensible Markup Language (XML) is a light-weight version of SGML without much of the overhead associated with the parent standard. Hypertext Markup Language (HTML) is a static subset of SGML describing a very simple class of report-style document and is widely used for Web pages. Simple Object Access Protocol (SOAP) is a lightweight protocol for exchange of information in a decentralized, distributed, networked environment that is based on XML. Web Services Description Language (WSDL) is a specification used to describe and locate networked XML-based services. WSDL provides a simple way for service providers to describe the basic format of requests to their systems, regardless of the underlying protocol (such as SOAP or XML) or encoding (such as Multipurpose Internet Messaging

Extensions or MIME). MATLAB provides support for WSDL and SOAP, but only supports remote procedure call (RPC) Web services.

MATLAB can act as a Web service client by sending requests to a server and handling the responses. The client, MATLAB in this case, initiates a connection with a server by using HTTP. When the server receives the request, which includes the operation to be performed and any necessary parameters, it sends back a response. A list of many of the available services is maintained at the XMethods website `<http://www.xmethods.net>`. The MATLAB documentation provides more information on these protocols, MATLAB Web services support, and examples.

34.9 SUMMARY

The MATLAB programming interfaces described in this chapter provide tools for maximizing MATLAB's productivity. MEX-files can be created to speed up loops that cannot be vectorized or to take advantage of previously written C functions or FORTRAN subroutines by adding a few lines of interface code. The resulting compiled MEX-files can then be called from MATLAB just like M-file functions. MATLAB can be used as a back-end computational engine by your own programs to take advantage of MATLAB's computational speed, efficiency, and visualization functions. You can read and write MATLAB MAT-files from your own programs, and transfer collections of variables and data into and out of MATLAB, by using simple `load` and `save` commands. Shared libraries can be accessed from within MATLAB. Direct control of serial ports and communication with serial devices can be accomplished from within MATLAB. Source code control systems can be accessed directly from MATLAB. XML-based Web services can be accessed by remote procedure calls, using standard protocols. All these *hooks* into the MATLAB environment add to the collection of tools you can use to solve problems or extend the capabilities of existing programs. (Much more extensive documentation of the MATLAB API—including a complete list of MX functions and details of the internal structure of MAT-files—and additional examples can be found in the MATLAB documentation.)

35

Extending MATLAB with Java

35.1 JAVA OVERVIEW

MATLAB 7 makes extensive use of the Java programming language. Every installation of MATLAB uses a Java virtual machine (JVM) supplied by the operating system or integrated into MATLAB. This Java interpreter is used extensively within MATLAB, as well as being the foundation of the MATLAB user interface. Because Java is so thoroughly integrated into the MATLAB environment, the Java virtual machine is also available to the MATLAB user. Java classes, objects, and methods can be manipulated within MATLAB, both from the command line and from MATLAB functions. Java integration provides opportunities to extend MATLAB in many different ways.

This chapter discusses the incorporation of Java into MATLAB. It is not an introduction to Java. The user is encouraged to consult a Java programming book to learn the details of the programming language. Only those features of Java that are useful in this chapter are covered here.

Java

Java is a programming language expressly designed for use in distributed environments on different kinds of computers under different operating systems. Java programs are compiled into Java platform-independent **bytecode** that can be run on any computer that has a Java virtual machine installed. The Java virtual machine is a program that interprets the bytecode into machine code that runs on the real computer hardware. This means that individual computer platform differences, such as instruction lengths and data storage differences, can be recognized and accommodated at the time the program is executed. Platform-specific versions of Java programs are not required. A single Java program can produce the same results on any computer with a JVM.

Java is an object-oriented programming language. If you have experience with Java, C++, or one of a number of other programming languages, or if you are comfortable with MATLAB classes and objects, the concepts and terms used to describe Java may be quite familiar.

The language of Java includes the following terms:

- **Class**–A Java class consists of data specification (variables) and a collection of operations (methods) available to objects of this class. A Java class is a template definition of a particular kind of object.

- **Object**–A Java object is a specific instance of a particular Java class or subclass in the same way that a MATLAB structure object is an instance of the MATLAB `struct` class. An object contains real values instead of variables, and all of the methods associated with the class can operate on any object instantiated from the class.

- **Method**–A method is a programmed procedure or operation that is defined as part of a class and is available for use by any object of that class. A Java method is analogous to a MATLAB class method—a function that operates only on objects of a specific class.

- **Variable** or **field**–A variable or field is the name associated with a value. Java variables are defined in Java classes. The terms *field* and *variable* are often used interchangeably.

- **Class library, toolkit, or package**–A package is a collection of related Java classes, such as the Abstract Windowing Toolkit (`java.awt.*`), providing windowing and GUI services, or the net class library (`java.net.*`), providing Internet and communications services.

- **Private/public**–*Private fields* are visible only within the class and can be changed only by public or private methods defined for the class. *Private methods* can be called only within the class and are not visible outside of the class. *Public fields* and *public methods* are visible outside of the class.

- **Static/nonstatic**–*Static fields* and *static methods* are associated with classes rather than with objects. Static fields are read-only; the contents cannot be changed. Static methods operate on classes. *Nonstatic fields* are associated with objects and often can be modified.

- **Final**–*Final classes* cannot have subclasses. *Final variables* or fields cannot be changed. *Final methods* cannot be overridden by subclasses. Private methods are effectively final. Variables labeled *public static final* are read-only class variables visible outside of the class. By convention, public static final variable names use all capital letters.

The ability to access arbitrary Java classes in the MATLAB environment opens up almost unlimited possibilities beyond the traditional MATLAB environment. Some of these possibilities are explored in examples later in this chapter.

Why Use Java?

Why would anyone want to use Java? There are a number of reasons. Anyone familiar with object-oriented programming languages or with MEX-file programming in MATLAB should be comfortable using Java. You can access existing Java classes and methods to add functionality to MATLAB. You can use special-purpose prewritten Java classes to add unique functionality. You can also create your own Java classes and access them from within MATLAB. *The MathWorks Inc.* has also made it easy to pass data between MATLAB data types and Java objects by automatically casting between Java and MATLAB data types. MATLAB support for Java arrays makes interaction between MATLAB and Java even easier.

35.2 JAVA CLASSES

Java classes are the foundation of Java. MATLAB makes a number of standard and MATLAB-specific class packages available automatically. Making additional classes or class libraries available entails two steps. First, you must tell MATLAB where to look for the `.class` or `.jar` files containing Java class definitions by adding entries to the `classpath.txt` file. Then, you must refer to a class by using the full class and package name, or `import` the class or package into the MATLAB workspace.

Using Java Classes in MATLAB

The systemwide `classpath.txt` file located in the `$toolbox/local` directory specifies the locations of Java class definitions. This text file can be edited to make additional classes available to all MATLAB users when MATLAB is launched. If this file is copied to your own startup directory, any changes made will affect only one user. To make individual `.class` files available, add the path to the directory containing the `.class` files to the `classpath.txt` file. Add the top directory of a class package directory tree to the `classpath.txt` file to add an entire package. Add the entire path to a `.jar` file (including the file name) to make an entire compressed Java archive available. The `classpath.txt` file is read only at startup, and so changes to the file are not recognized until MATLAB is restarted. MATLAB 7 provides the ability to dynamically change the Java class path during a MATLAB session by using the `javaclasspath`, `javaaddpath`, and `javarmpath` functions.

Once MATLAB knows where to look for Java classes, you can refer to any available class by using the full class name. For example, `java.lang.String` refers to the `String` class contained in the `java.lang` class library.

35.3 JAVA OBJECTS

Java objects are created by using either Java syntax or MATLAB syntax conventions. For example, the code

```
>> myFrameA = java.awt.Frame('A Cool Window');              % Java syntax
>> myFrameB = javaObject('java.awt.Frame','Another Window'); % MATLAB syntax
```

creates two objects, both instances of the `java.awt.Frame` class. The `javaObject` function is intended for unusual situations within functions and is rarely used. Java syntax is preferred.

The function `import` provides a shortcut method for refering to a given class within MATLAB. For instance, the code

```
>> import java.awt.Frame
>> myFrameA = Frame; myFrameB = Frame;
```

creates two objects of the `java.awt.Frame` class. Future references to `Frame` refer to `java.awt.Frame`. Entire class libraries can be imported in the same manner. For example, the code

```
>> import java.awt.* java.net.* com.mathworks.ide.help.HelpBrowser
```

makes all `java.awt` classes, all `java.net` classes, and the `com.mathworks.ide.help.HelpBrowser` class available in shorthand notation. For instance, the code

```
>> import java.awt.*
>> myButton = Button('Stop');
```

creates an instance of a `java.awt.Button` class. The `import` command with no arguments returns the current import list without adding to it. The `clear import` command clears the import list from the current workspace. Function import lists are cleared when the function returns.

Java Objects are References in MATLAB

It is important to understand that Java objects are **references** in MATLAB and are not copied on assignment or passed by value. A new reference is created whenever a Java object is assigned. When a Java object is passed to a function as an input argument, the variable name in the function is simply another reference to the original object. As an example, consider this short test function:

```
function javatest(obj)
% Test java object references

disp(obj.getLabel)                 % Display the object label
newRef=obj;                        % Create a new reference
set(newRef,'Label','Label One')    % Change the label using set()
disp(newRef.getLabel)              % Display the new label
setLabel(newRef,'Label Two')       % Use the setLabel method
disp(newRef.getLabel)              % Display the label again
newRef.setLabel('Label Three')     % Use Java object method syntax
disp(newRef.getLabel)              % Display the label
```

To illustrate use of the `javatest` function, first create an object of the `Button` class:

```
>> myBut = java.awt.Button('Label Zero')
 myBut =
 java.awt.Button[button1,0,0,0x0,invalid,label=Label Zero]
```

Then, pass the `Button` to the `javatest` function:

```
>> javatest(myBut)
Label Zero
Label One
Label Two
Label Three
```

The `Button` object in the MATLAB base workspace has changed!

```
>> myBut
myBut =
 java.awt.Button[button1,0,0,0x0,invalid,label=Label Three]
```

But the `newRef` reference remains local to the function workspace and is unknown in the base workspace:

```
>> newRef
??? Undefined function or variable 'newRef'.
```

This occurs because Java objects are passed by reference rather than by value, and assignments simply create another reference. Changes made to the object referenced by `newRef` within the function affect the object referenced by `myBut` in the base workspace, since they both refer to the *same* object. This is different from conventional MATLAB, where data is passed and assigned by value, not by reference.

> Java objects are created by object constructor methods and released by the `clear` function. Since every assignment creates a new object reference, Java objects cannot be duplicated. You cannot make a direct copy of any Java object.

Java arrays can be copied, but the elements of the new array are newly created references to the original objects.

35.4 JAVA METHODS

Java classes consist of data definitions and a collection of operations that can be performed on or by objects created from the class. The collection of operations, called methods, are the Java equivalent of operators and functions in MATLAB.

Invoking Methods on Java Objects

In the preceding example, the `Button`'s Label property was changed within the `javatest` function in three different ways:

```
set(newRef,'Label','Label One')    % Change the label using set()
setLabel(newRef,'Label Two')       % Use the setLabel method
newRef.setLabel('Label Three')     % Use Java object method syntax
```

The first assignment uses the MATLAB `set` function, overloaded for Java objects by using the functional syntax `set(object,property,value)`. The second assignment uses the `java.awt.Button` class `setLabel` method in a hybrid functional syntax: `method(object,value)`. The third assignment uses pure Java syntax: `object.method(value)`. All three variations are used in this example.
      There is one more way to invoke a method on a Java object. The `javaMethod` command uses pure functional notation to invoke a method. For example, the code

```
javaMethod('setLabel',newRef,'Label Four')
```

also invokes the `setLabel` method on the `Button` object `newRef`. The `javaMethod` function is available for unique situations, but is rarely used. Java or MATLAB syntax is preferred.

Getting Information about Classes and Objects

Classes are collections of variables and methods, and objects are instances of classes or subclasses. Each object inherits the methods and variables of the class. The `methods` command produces a list of the public methods available for a specific class:

```
>> methods java.lang.Double

Methods for class java.lang.Double:
```

Double	doubleToLongBits	intValue	parseDouble	strncmp
Number	doubleToRawLongBits	intersect	permute	strncmpi
Object	doubleValue	isInfinite	reshape	toString
byteValue	eq	isNaN	setdiff	transpose
char	equals	isletter	setxor	tril
compare	eval	ismember	shortValue	triu
compareTo	evalc	isspace	sort	unique
ctranspose	fieldnames	longBitsToDouble	str2double	valueOf
diag	findstr	longValue	str2num	wait

disp	floatValue	ne	strcmp
display	getClass	notify	strcmpi
double	hashCode	notifyAll	strmatch

Any object of the java.lang.Double class can use any of these methods. The argument to the methods command is the name of a Java class. A second argument '-full' produces a much more detailed list of methods, including information about their arguments and return values:

```
>> methods java.lang.Double -full

Methods for class java.lang.Double:

Double(double)
Double(java.lang.String) throws java.lang.NumberFormatException
Number()
Object()
byte byteValue()
byte byteValue()
char  % Inherited from opaque
static int compare(double,double)
abstract int compareTo(java.lang.Object)
int compareTo(java.lang.Object)
int compareTo(java.lang.Double)
ctranspose  % Inherited from opaque
diag  % Inherited from opaque
disp  % Inherited from opaque
display  % Inherited from opaque
double  % Inherited from opaque
static long doubleToLongBits(double)
static long doubleToRawLongBits(double)
abstract double doubleValue()
double doubleValue()
eq  % Inherited from opaque
boolean equals(java.lang.Object)
boolean equals(java.lang.Object)  % Inherited from java.lang.Object
(...)
```

The methodsview function presents the same information about the methods of a Java class by creating a new window displaying the methods and their attributes. When a method that does not exist for a Java class is invoked on an instance of that class, MATLAB will check for a built-in MATLAB function of the same name and try to invoke that function. If no built-in function is found, a message is returned indicating that a method by that name was not found for this class.

As the following code shows, the fieldnames command (introduced in the earlier discussion of structures) has been overloaded to return information about the public fields or variables of a Java class or object:

```
>> dObj = java.lang.Double(5.0);

>> fieldnames(dObj)
ans =

    'POSITIVE_INFINITY'
    'NEGATIVE_INFINITY'
    'NaN'
    'MAX_VALUE'
    'MIN_VALUE'
    'TYPE'

>> fieldnames(dObj,'-full')
ans =

    'static final double POSITIVE_INFINITY'
    'static final double NEGATIVE_INFINITY'
    'static final double NaN'
    'static final double MAX_VALUE'
    'static final double MIN_VALUE'
    'static final java.lang.Class TYPE'

>> fieldnames(java.awt.Dimension)
ans =

    'width'
    'height'

>> fieldnames(java.awt.Point,'-full')
ans =

    'int x'
    'int y'
```

Use the `object.field` syntax to access the public data fields of a Java object:

```
>> dObj.MAX_VALUE
ans =
    1.7977e+308
```

35.5 OBJECT PROPERTIES

Java objects in MATLAB have properties, just as Handle Graphics objects do. The MATLAB `get` and `set` methods can be used to access some of the standard Java class methods and to add common properties to Java objects. For example, the `java.awt.Frame` object has a MATLAB property called `'Background'`. The MATLAB `get` and `set` methods implement the Java `getBackground` and `setBackground` methods to change this property.

For example, create a `java.lang.Double` object with the code

```
>> dObj = java.lang.Double(6.0)
dObj. =
6.0
```

and examine its properties:

```
>> get(dObj)

        Class = [ (1 by 1) java.lang.Class array]
        Infinite = off
        NaN = off

        BeingDeleted = off
        ButtonDownFcn =
        Children = []
        Clipping = on
        CreateFcn =
        DeleteFcn =
        BusyAction = queue
        HandleVisibility = on
        HitTest = on
        Interruptible = on
        Parent = []
```

```
          Selected = off
          SelectionHighlight = on
          Tag =
          Type = java.lang.Double
          UIContextMenu = []
          UserData = []
          Visible = on
```

The common MATLAB object properties are available, including the 'Tag' and
'UserData' properties. The three object-specific properties, 'Class', 'Infinite',
and 'NaN', are all read-only properties. Object properties can be queried and/or
changed by using the get and set commands:

```
>> get(dObj,'Class')
ans =
class java.lang.Double
>> set(dObj,'Tag','MyDouble')
>> get(dObj,'Tag')
ans =
MyDouble
```

> Java objects do not have handles in the sense that Handle Graphics objects do. Java
> objects are not children of the *root* window and cannot be found by the findobj func-
> tion or any other MATLAB search technique.

35.6 DATA EXCHANGE

Java classes and objects are distinct from MATLAB classes and variables. Moving
data between different classes of objects or variables requires changing the form of
the data.

Automatic Data Type Conversions

MATLAB and Java, as in any typed programming language or environment, use
coercion or *casting*, a familiar concept to most programmers, to transfer data
between data types or classes. Each class contains at least one class constructor
method used to create an object from a different source data type. For example,
arithmetic operations on the MATLAB integer data types are performed internally by explicitly casting the integer values into the double data type, performing
the operation, and casting the result back into the appropriate integer data type.

The different integer data types cannot be mixed in an expression, but can be computed by explicit casting:

```
>> x = uint8(95)
x =
    95

>> y = uint16(43)
y =
    43

>> z = x+y
??? Error using ==> plus
Integers can only be combined with integers of the same class, or scalar
doubles.

>> z = y+uint16(x)
z =
   138

>> class(z)
ans =
uint16
```

MATLAB hides most of the complexity of Java data type conversion by automatically casting native MATLAB data types into standard Java classes as needed. In general, MATLAB numerical data types are converted to the most appropriate Java target data type. MATLAB strings and arrays of characters are converted to the java.lang.String class. Cell arrays of Java objects are converted to Java arrays of objects. Note that MATLAB arrays and Java arrays are not the same.

Java objects returned to MATLAB from Java methods are generally not automatically converted to MATLAB data types, so that they can continue to be used by Java methods. There is one general exception. Numerical values returned from Java methods are converted to MATLAB data types; scalars are converted to MATLAB doubles, and numerical arrays are converted to arrays of the MATLAB type most appropriate to save storage space. java.lang.String objects returned to MATLAB can be converted to character arrays, and Java String arrays can be converted to cell arrays of strings using the char function. Java objects explicitly created in MATLAB retain their original Java object form, including java.lang.Double and java.lang.String objects.

Explicit Data Type Conversions

Explicit data type conversions are also possible. Java objects belonging to any Java `Numeric` class or subclass, or any other Java class containing a `toDouble` method, can be converted to a MATLAB `double`. Any Java object of the `String` class or any other Java class containing a `toChar` method can be converted to a MATLAB `char`. The following example is illustrative:

```
>> wObj = java.lang.Double(3.12)
wObj =
3.12

>> wObj.intValue
ans =
    3

>> class(ans)
ans =
double

>> class(double(wObj))
ans =
double

>> sObj = java.lang.String('A Java String')
sObj =
A Java String

>> class(sObj)
ans =
java.lang.String
>> sObj.toUpperCase
ans =
A JAVA STRING

>> class(ans)
ans =
java.lang.String
```

```
>> butObj = java.awt.Button('OK')
but =
java.awt.Button[button1,0,0,0x0,invalid,label=OK]

>> butObj.getSize
ans =
java.awt.Dimension[width=0,height=0]

>> class(ans)
ans =
java.awt.Dimension
```

Java objects often have characteristics that are accessible by using field names defined as part of the object. This is very similar to the way MATLAB handles structures. As might be expected, the `struct` function can be used to convert Java objects into MATLAB structures. The MATLAB `cell` function can also be used to convert Java objects or arrays into cell arrays.

Displaying Java Objects

Whenever a Java object is listed by using the `disp` function or by omitting the semicolon on the command line, the `toString` method is used to display the object. The `toString` method for most Java objects returns a string containing the class name and information about some of the variables associated with the object:

```
>> butObj = java.awt.Button('OK')
butObj =
java.awt.Button[button0,0,0,0x0,invalid,label=OK]

>> f = java.awt.Frame('My Frame')
f =
java.awt.Frame[frame0,0,0,0x0,invalid,hidden,layout=java.awt.BorderLayout,
          title=My Frame,resizable,normal] % (line wrapped for printing)
```

Objects of the `Numeric` and `String` classes and subclasses are again exceptions to the rule. `Numeric` Java types are displayed as strings containing the value of the number, and `String` types are displayed as a character array containing the text of the Java `String`.

Java objects are very similar to MATLAB structures in some respects. Java class fields (the same fields listed by the `fieldnames` command) can be converted

to MATLAB structures and accessed by using standard MATLAB syntax, as in the
following example:

```
>> b = struct(butObj)
b =
          TOP_ALIGNMENT: 0
       CENTER_ALIGNMENT: 0.5000
       BOTTOM_ALIGNMENT: 1
         LEFT_ALIGNMENT: 0
        RIGHT_ALIGNMENT: 1
                  WIDTH: 1
                 HEIGHT: 2
             PROPERTIES: 4
               SOMEBITS: 8
              FRAMEBITS: 16
                ALLBITS: 32
                  ERROR: 64
                  ABORT: 128

>> b.PROPERTIES
ans =
     4

>> class(ans)
ans =
double
```

This feature provides a way to capture the names and values of all of the public
fields or variables of a Java class; however, the only way to change these values is by
using appropriate Java methods on an object of the class—if such methods are available. Note that these particular fields are likely to be public static final (the field
names are capitalized) and as such cannot be changed at all.

35.7 JAVA ARRAYS

So far, we have discussed passing singular Java objects between Java methods and
the MATLAB environment. Some methods can handle or even expect arrays of

objects, while others can return arrays of objects. Java arrays can be returned by Java methods or can be created by MATLAB commands.

Java arrays are based on the C language model and are significantly different from MATLAB arrays in their structure and in the way elements are accessed. Most of the time, the differences do not matter, as MATLAB handles the conversion transparently.

Java Array Structure

Java arrays are always 1-D vectors. 2-D Java arrays are constructed as arrays of arrays. 3-D Java arrays are arrays of arrays of arrays, and *n*-D Java arrays can be constructed in the same manner. This array structure is similar to that of nested cell arrays in MATLAB. Like cell arrays, nested Java arrays can be of different lengths. For example, 2-D MATLAB character arrays must have rectangular dimensions: Each row must have the same number of columns. Cell arrays of strings, however, can contain strings of different lengths. Java arrays can contain arrays of different lengths as well. These types of arrays are sometimes termed *ragged* arrays. When methods return ragged, nonrectangular arrays, MATLAB stores them in cell arrays.

Accessing Elements of Java Arrays

Java array indexing, like C, is zero-based, while MATLAB uses one-based indexing. Java programs access elements of an array of length N by using the indices 0 through N-1. MATLAB accesses the same elements by using the indices 1 through N. For example, myArray(4) is used to access the fourth element of a Java array in MATLAB, while a Java program accesses the same element of the array by using myArray[3]. Similarly, the element of a Java array that is accessed by myArray[2][4] in a Java program can be accessed by myArray(3,5) within MATLAB. However, MATLAB hides these differences. When a Java method is used or a Java array is accessed in the MATLAB environment, normal MATLAB syntax is used. MATLAB makes the necessary indexing conversions transparently, as the following example shows:

```
>> but1Obj = java.awt.Button('STOP')
but1Obj =
java.awt.Button[button1,0,0,0x0,invalid,label=STOP]

>> but2Obj = java.awt.Button('GO')
but2Obj =
java.awt.Button[button2,0,0,0x0,invalid,label=GO]

>> butArray = [but1Obj,but2Obj]
butArray =
java.awt.Button[]:
```

```
    [java.awt.Button]
    [java.awt.Button]
>> class(butArray)
ans =
java.awt.Button[]

>> size(butArray)
ans =
     2     1

>> getLabel(butArray(1))     % Functional syntax for the getLabel method
ans =
STOP

>> butArray(2).getLabel     % Java syntax for the getLabel method
ans =
GO

>> butArray(2).setLabel('YIELD')

>> but2Obj
but2Obj =
java.awt.Button[button2,0,0,0x0,invalid,label=YIELD]
```

Note that butArray is an array of **references** to the but1Obj and but2Obj objects. Changes made to these references change the original object.

Creating Java Arrays

Java arrays can be created as the output of the operation of a Java method, by concatenating objects or arrays, by extracting elements of existing Java arrays, by assignment to an array element, or by using the javaArray function.

As shown in the preceding example, the standard array operators [] (and the cat function) can be used to concatenate Java objects into a Java array. If the objects are of the same class, the resulting array is of the same class as well. If the objects are arrays of the same class, the arrays are stacked. The length of the resulting array is the sum of the lengths of the individual arrays. If the objects are 2-D arrays that differ in the lengths of the second dimension, the result is a ragged array. If the objects are of different classes, the resulting array is of the java.lang.Object class and its length is the number of arrays being combined.

As with normal MATLAB arrays, you can also create a Java array by assigning an object to a specific element of a nonexistent array:

```
>> bigButArray(3,4) = java.awt.Button('YES')
bigButArray =
 java.awt.Button[][]:
    []       []       []                        []
    []       []       []                        []
    []       []       []        [java.awt.Button]
```

Extending an array can produce unexpected results. Because of the structure of Java arrays and the fact that they can be ragged, extending an existing Java array by assignment does not necessarily result in a rectangular array. Consider an example using the bigButArray from the previous example:

```
>> bigButArray(4,5) = java.awt.Button('NO')
bigButArray =
java.awt.Button[][]:
    [4 element array]
    [4 element array]
    [4 element array]
    [5 element array]
```

This adds a fourth element to the top-level Java array containing a five-element Java array of Button objects. Each of the existing elements of the top-level array continue to contain four-element Java arrays of Button objects. Because the resulting array is nonrectangular or ragged, MATLAB stores the result as a cell array of Java arrays.

Similarly, when the empty matrix is assigned to an entire row or column of a MATLAB matrix, MATLAB removes the row or column and collapses the resulting matrix dimensions. Since Java arrays can be ragged, they cannot be handled in the same way. When an entire row or column of a Java array is assigned the empty matrix, the dimensions do not change. The elements are simply assigned the Java null character, which MATLAB represents as the empty matrix.

The javaArray function creates an unpopulated Java array (filled with null values) of a specific class by using the syntax javaArray('java_class',m,n,p,...), where java_class is a fully qualified Java class name (package and class name) and m, n, and p are array dimensions. For example, create a 3-by-4 empty Java array to contain Button objects with the following code:

```
>> buttonArray = javaArray('java.awt.Button',3,4)
buttonArray =
java.awt.Button[][]:
      []        []        []        []
      []        []        []        []
      []        []        []        []
```

Then, use standard MATLAB assignment syntax and a Button constructor method to populate the array with new Button objects:

```
>> buttonArray(2,2) = java.awt.Button('MAYBE')
buttonArray =
java.awt.Button[][]:
      []                          []        []        []
      []        [java.awt.Button]          []        []
      []                          []        []        []
```

Accessing Java Array Elements

The elements of a Java object array can be accessed by using the normal MATLAB array indexing syntax, as in the following example:

```
>> dArray = javaArray('java.lang.Double',3,4)
dArray =
java.lang.Double[][]:
      []        []        []        []
      []        []        []        []
      []        []        []        []

>> for i = 1:3
     for j = 1:4
       dArray(i,j) = java.lang.Double(i*10+j);
     end
   end

>> dArray
dArray =
java.lang.Double[][]:
     [11]      [12]      [13]      [14]
     [21]      [22]      [23]      [24]
     [31]      [32]      [33]      [34]
```

Notice that the result is a Java array consisting of three elements displayed as rows; each of these elements contains another Java array consisting of four elements displayed as columns. The element referenced by dArray[0][2] in Java is the element referenced by dArray(1,3) in MATLAB:

```
>> dArray(1,3)
ans =
13.0
```

The second element of the top-level dArray is a four-element Java array:

```
>> dArray(2)
ans =
java.lang.Double[]:
    [21]
    [22]
    [23]
    [24]
```

This is *not* the result you get when you use the same syntax on a MATLAB array. Given a 3-by-4 MATLAB array D of doubles, D(4) and D(1,2) refer to the same individual array element.

Subarrays

Subsets of Java arrays can be accessed by using familiar MATLAB array syntax. The following example uses the given Java array dArray and the corresponding MATLAB array D:

```
>> dArray
dArray =
java.lang.Double[][]:
    [11]    [12]    [13]    [14]
    [21]    [22]    [23]    [24]
    [31]    [32]    [33]    [34]

>> D = double(dArray)
D =
    11    12    13    14
    21    22    23    24
    31    32    33    34
```

```
>> dArray(2,2:3)
ans =
java.lang.Double[]:
    [22]
    [23]

>> D(2,2:3)
ans =
    22      23
```

On the other hand, MATLAB uses column-oriented storage for array elements, whereas Java uses row-oriented storage for array elements. A 2-by-2 MATLAB array is stored as ⌊11 21 12 22], while a 2-by-2 Java array (a two-element Java array containing two-element Java arrays) is stored as [11 12][21 22]. When using the colon operator (:) to create a vector from an array, the result differs because of these storage differences:

```
>> dArray(:)
ans =
java.lang.Double[]:
    [11]
    [12]
    [13]
    [14]
    [21]
    [22]
    [23]
    [24]
    [31]
    [32]
    [33]
    [34]
>> D(:)
ans =
    11
    21
    31
```

```
12

22

32

13

23

33

14

24

34
```

The colon operator allows assignment to every element of a Java array or sub-array with a single statement by treating the array as a single vector:

```
>> dblArray = javaArray('java.lang.Double',3,4)
dblArray =
java.lang.Double[][]:
    []      []      []      []
    []      []      []      []
    []      []      []      []

>> dblArray(:) = java.lang.Double(0)
dblArray =
java.lang.Double[][]:
    [0]     [0]     [0]     [0]
    [0]     [0]     [0]     [0]
    [0]     [0]     [0]     [0]

>> dblArray(2,:) = java.lang.Double(1)
dblArray =
java.lang.Double[][]:
    [0]     [0]     [0]     [0]
    [1]     [1]     [1]     [1]
    [0]     [0]     [0]     [0]
```

```
>> dblArray(:,3) = java.lang.Double(4)
dblArray =
java.lang.Double[][]:
    [0]      [0]      [4]      [0]
    [1]      [1]      [4]      [1]
    [0]      [0]      [4]      [0]
```

Remember that each of these elements is another reference to the same object. The original constructor assignment statement dblArray(:)=java.lang.Double(0) creates one java.lang.Double object containing the value 0 and a populated dblArray with 12 references to this object. Each subsequent assignment statement created one new object and multiple references to this object.

Duplicating Arrays

Assigning Java arrays creates new references as well. For example, the code

```
>> xArray = dArray
xArray =
java.lang.Double[][]:
    [11]      [12]      [13]      [14]
    [21]      [22]      [23]      [24]
    [31]      [32]      [33]      [34]
```

simply creates a new reference to dArray named xArray. You can also create a new object reference by assigning a name to a single element of an array:

```
>> xObj = dArray(2,3)
xObj =
23.0
```

In this example, xObj and dArray(2,3) reference the same object.

If you assign a top-level element of an array containing two or more levels, the result is another reference to the array contained in the top-level array:

```
>> yArray = dArray(2)
yArray =
java.lang.Double[]:
    [21]
    [22]
    [23]
    [24]
```

```
>> yArray(2)=java.lang.Double(0)
yArray =
java.lang.Double[]:
    [21]
    [ 0]
    [23]
    [24]


>> dArray
dArray =
java.lang.Double[][]:
    [11]    [12]    [13]    [14]
    [21]    [ 0]    [23]    [24]
    [31]    [32]    [33]    [34]
```

The preceding code creates yArray as a new reference to the second element of the top-level array in dArray.

You can create a new array from elements of a Java array (known as *cloning* in Java) by assigning a subset of the original Java array elements to a new variable:

```
>> zArray = dArray(1:2,2:3)
zArray =
java.lang.Double[][]:
    [12]    [13]
    [ 0]    [23]
```

In this example, zArray is a new array of Doubles containing values *which are new references to the original objects*. For example, zArray(1,1) and dArray(1,2) point to the same object. However, if you *replace* an element of zArray with a new object, dArray is not affected:

```
>> zArray(2,2) = java.lang.Double(-1)
zArray =
java.lang.Double[][]:
    [12]    [13]
    [ 0]    [-1]
```

```
>> dArray

dArray =

java.lang.Double[][]:

    [11]      [12]      [13]      [14]
    [21]      [ 0]      [23]      [24]
    [31]      [32]      [33]      [34]
```

Replacing an element of zArray replaces the existing object reference with a new reference to a new object, and therefore does not affect dArray. However, the other elements in zArray are still references. Changes to *attributes* of these elements are changes to the original elements.

An entire Java array can be copied by using the colon operator:

```
>> xArray = dArray(:,:)

xArray =

java.lang.Double[][]:

    [11]      [12]      [13]      [14]
    [21]      [ 0]      [23]      [24]
    [31]      [32]      [33]      [34]
```

```
>> xArray(2,3) = dArray(3,4)

xArray =

java.lang.Double[][]:

    [11]      [12]      [13]      [14]
    [21]      [ 0]      [34]      [24]
    [31]      [32]      [33]      [34]
```

```
>> dArray

dArray =

java.lang.Double[][]:

    [11]      [12]      [13]      [14]
    [21]      [ 0]      [23]      [24]
    [31]      [32]      [33]      [34]
```

The element at xArray(2,3) now contains a new reference to dArray(3,4), while dArray is unaffected.

A new array is also created when you concatenate Java arrays:

```
>> a = dArray(1,3)

a =

13.0
```

```
>> b = dArray(3,2)
b =
32.0

>> c = [a b]
c =
java.lang.Double[]:
    [13]
    [32]

>> c(2) = dArray(3,3)
c =
java.lang.Double[]:
    [13]
    [33]

>> a
a =
13.0

>> b
b =
32.0
```

> In general, a reference is created whenever you assign any Java object or array. A new
> array is created when you assign a subset of a Java array (or an entire array using the
> (:,:) notation) or concatenate Java arrays. The new array contains new references to
> the original array elements.

The statements

```
xArray = dArray;
yArray = dArray(2);
zObj = dArray(2,3);
```

create new references, while the statements

```
xArray = dArray(:.:);
yArray = dArray(2,:);
zArray = [dArray, dArray];
```

create new arrays containing additional references to the specified elements of the original array.

Java Array Size

Some MATLAB functions have been enhanced to support Java arrays, but full MATLAB functionality is not available because of the structure of Java arrays. For example, the `size` and `length` functions operating on a Java array return information about the top-level array only:

```
>> myArray = javaArray('java.lang.Double',3,4)
myArray =
java.lang.Double[][]:
        []       []       []        []
        []       []       []        []
        []       []       []        []

>> size(myArray)
ans =
      3     1

>> length(myArray)
ans =
      3
```

To get information on other dimensions, you must specifically address an element of the top-level array:

```
>> size(myArray(1))
ans =
      4     1

>> length(myArray(2))
ans =
      4
```

The total number of elements in a non-empty Java array can be obtained by using the `length` function:

```
>> length(dblArray(:))        % the dblArray created earlier
ans =
     12
```

```
>> length(myArray(:))            % the 0-by-1 (empty) myArray
ans =
    0

>> size(myArray(:))
ans =
    0    1
```

Since myArray is empty, size(myArray(:)) returns 0-by-1 and length (myArray(:)) returns 0. Similarly, you can use the end keyword when addressing Java arrays, but only in the top-level reference (the first dimension):

```
>> dblArray(2:end)
ans =
java.lang.Double[][]:
    [1]     [1]     [4]     [1]
    [0]     [0]     [4]     [0]
>> dblArray(2,2:end)
ans =
java.lang.Double[][]:
    [0 element array]
```

35.8 JAVA FUNCTIONS

In addition to those mentioned previously, other MATLAB functions have been created or modified to support Java classes, objects, and methods. As noted earlier, import can be used to add to the Java class and package import list and allow shorthand references to Java classes. Java objects can be removed and the import list cleared with the clear function. The inmem function has been enhanced to list the names of all of the Java classes that have been loaded into the MATLAB workspace in an optional third output argument. The exist function now recognizes and identifies Java objects and arrays. The class function recognizes the class of Java objects and arrays, and the isa function can test the class of a Java object or array. The new isjava function can be used to determine whether or not a variable is a Java object.

The fieldnames function returns a list of the public fields of a Java class or object. The methods command lists the names of the methods available to operate on objects of specific Java classes. The which function has been enhanced to return a list of the loaded Java classes that contain a given method or a combination of method and signature, as in the following example:

```
>> which -all resize
java.awt.Button.resize                              % Button method
```

```
java.awt.TextArea.resize                              % TextArea method
java.awt.Frame.resize                                 % Frame method
matlab6/toolbox/matlab/graph2d/@axisobj/resize.p % axisobj method
matlab6/toolbox/matlab/graph2d/@axisobj/resize.m  % Shadowed axisobj method
```

The save and load functions can store and recall Java objects and arrays in normal binary MAT-files. Java objects cannot be saved in ASCII formatted files, however. Java exceptions are automatically caught and converted to normal MATLAB errors. The text of the exception is captured and can be displayed by using lasterr.

The following table summarizes the MATLAB functions that support Java, some of which are standard MATLAB functions that have been overloaded for Java and others of which have been created specifically to support working with Java in MATLAB:

Function	Description
methods	List the public methods of a Java class
methodsview	List the public methods of a Java class in a new window
usejava	Determine if a Java feature is supported in MATLAB
fieldnames	List the public fields or variables of a Java class
import	List or add to the Java class and package import list
inmem	Can return a cell array of the names of all Java classes that have been loaded
which	Can return a list of matching methods in loaded Java classes
class	Determine the class of a Java object
clear	Clear Java objects from the workspace, or clear the Java class import list
double	Convert a Java object or array of class Numeric to MATLAB doubles
char	Convert a Java object or array of class String to a MATLAB character array or to a cell array of strings
struct	Create a structure from the public fields of a Java class or object
isjava	True if the argument is a Java object or Java array
isa	Determine if a variable is a Java object of a specific class
exist	Determine the nature of a variable; return 8 if the argument is a Java class
ismethod	Determine if the argument is a method of a Java object
isprop	Determine if the argument is a property of a Java object

Function	Description
inspect	Display a graphical interface to list and modify property values
[] and cat	Concatenate Java objects or arrays into a Java array
save	Store variables including Java objects to a binary MAT-file
load	Load variables including Java objects from a binary MAT-file
size	Return the size of the top-level array of a Java array
length	Determine the length of the top-level array of a Java array
javaArray	Create a new Java array and populate it with null values
javaMethod	Invoke a Java method on a Java object or Java array (not recommended)
javaObject	Create a Java object (not recommended)
javaaddpath	Add entries to a dynamic Java class path
javarmpath	Remove entries from a dynamic Java class path
javaclasspath	Set and get a dynamic Java class path
javachk	Generate an error message if a Java feature is not available

35.9 EXAMPLES

This section contains a number of examples that pull together many of the concepts and tools presented in this chapter to illustrate how Java can help to extend the capabilities of MATLAB. We start with few basic examples and proceed to more complete functions.

Example 1: whoami

This example uses elements of the java.net class library to return the host name and IP address of the computer running MATLAB:

```
function whoami
% Test function to illustrate the use of the java.net package.
%
% Mastering MATLAB 7 Java Example 1

% Use try and catch to avoid Java exception messages.
try
    me=java.net.InetAddress.getLocalHost;
```

```
catch
    error('Unable to get local host address.');
end

% Find my hostname and IP address
myname=me.getHostName;
myip=me.getHostAddress;

% and print the results.
disp(sprintf('My host name is %s',myname));
disp(sprintf('My IP address is %s',myip));
```

The whoami function produces results similar to the following:

```
>> whoami
My host name is www.phptr.com
My IP address is 204.179.152.74
```

Example 2: random

The next example uses the java.util package to generate a random number:

```
function num=random(varargin)
% Test function to illustrate the use of the java.util package.
%   Generate a random number between given limits (or 0:1).
%
% Mastering MATLAB 7 Java Example 2

% Check any input arguments.
if nargin == 0
  rmin=0; rmax=1;
elseif nargin < 3
  if nargin == 1
    lim = varargin{1};
```

```
  else
    lim = [varargin{1}, varargin{2}];
  end
  if isnumeric(lim) & length(lim) == 2
    rmin = min(lim); rmax = max(lim);
  else
    error('Invalid limits.');
  end
else
  error('Too many arguments.')
end

% Construct a Random object and generate a uniformly-distributed
% random number between the desired limits.
rNum = java.util.Random;
num = rNum.nextDouble * (rmax - rmin) + rmin;
```

The random function parses the input, creates a Random object, and uses the nextDouble method to obtain a uniformly distributed pseudorandom number. The nextGaussian method is available to generate a normally distributed random number. Other Random methods can be used to generate integer, long, or float values, or even a user-specified number of random bytes.

Example 3: hithere

This example creates a simple dialog box containing a line of text and a button. When the button is clicked, the dialog box exits:

```
% script hithere.m
% Example script to illustrate the use of the java.awt package.
%
% Mastering MATLAB 7 Java Example 3

% Create a Frame object and specify a 2x1 grid layout style.
dbox=java.awt.Frame('Hi There!');
dbox.setLayout(java.awt.GridLayout(2,1));
```

```
% Specify the window location, size and color. Use Java syntax
%    for the location, hybrid syntax for the size, and MATLAB
%    syntax to set the background color.
dbox.setLocation(50,50);
resize(dbox,200,100);
set(dbox,'Background',[.7,.8,.9]);

% Create a text label and a bright red button.
txt=java.awt.Label('Click the button to exit.',1);
but=java.awt.Button('Exit Button');
set(but,'Background',[1,0,0]);

% Define a callback for the button.
set(but,'MouseClickedCallback','dbox.dispose')

% Attach the label and button to the window.
dbox.add(txt);
dbox.add(but);

% The window is hidden by default. Make it visible.
dbox.show;
```

This example uses a script M-file, rather than a function, to avoid scoping issues with the Button callback. As with Handle Graphics objects, Java callbacks are either function handles or strings that are passed to the eval function in the MATLAB base workspace. The following figure is an example of the dialog box generated by the hithere.m script:

Example 4: netsearch

This example uses the java.net package to manage Internet communication, along with the java.io package to read a data stream. The netsearch function performs an Internet search and returns a cell array of the uniform resource locator (URL) strings found by the search engine. If no matches are found after contacting one search engine, another is tried. Here is the code:

```
function ulist=netsearch(varargin)
% MATLAB Java Demo function netsearch.m
%
%  Mastering MATLAB 7 Java Example 4
%    Open a connection to an internet search engine using
%    Java networking toolkit objects and return the URLs
%    found by the search in a cell array.

% Import the java.net and java.io toolkits to save typing.
import java.net.* java.io.*

% Define the url for a search command and a target string
%      for each host to try in order.
s_host={
        'http://www.google.com/search?q=',   '<p class=g><a href='
        'http://search.yahoo.com/search?p=', 'H=0/*-href='
        };

nhosts=size(s_host,1);       % Number of search hosts defined
ulist={};                    % Cell array to contain results

% Do some error checking.
if nargin < 1
  error('Nothing to search for.');
end
if nargout > 1
  error('Too many output arguments.');
end

% Create a search string from the input arguments.
for (idx=1:nargin)
  if ~ischar(varargin{idx})
    error('Search terms must be strings');
```

```
    else
      tmp_str=varargin{idx};

      % If this string argument contains spaces, quote the string
      % and replace the spaces with plus sign characters.
      if findstr(tmp_str,' ')
        tmp_str=['%22',strrep(tmp_str,' ','+'),'%22'];
      end


      % Build up the search string.
      if idx == 1
        s_str=tmp_str;
      else
        s_str=[s_str,'+',tmp_str];
      end
    end
end

% Start with the initial search host.
uidx=1;
hostidx=1;
while isempty(ulist) & (hostidx <= nhosts)

  % Construct a complete URL using the search string.
  s_url=[s_host{hostidx,1},s_str];

  % Create a URL connection to the search engine.
  s_con=jaca.net.URL(s_url);

  % Open an input stream on the connection.
  s_stream = openStream(s_con);

  % Open an input stream reader to read the stream.
  s_rdr = InputStreamReader(s_stream);
```

```matlab
    % Open a buffered reader to read one line at a time.
    shBuf = BufferedReader(s_rdr);


    % Read the lines of the page returned by the search engine
    %   and extract any target URLs until the the page ends.
    linebuf=shBuf.readLine;
    while ~isempty(linebuf)
       linebuf=char(linebuf);
       found = findstr(linebuf, s_host{hostidx,2});
       if ~isempty(found)
         tmp = strtok(linebuf(found(1)+length(s_host{hostidx,2}):end),'>');
          ulist{uidx}=strrep(tmp,'"','');   % Remove extra quote characters
          uidx=uidx+1;
       end
       linebuf=shBuf.readLine;
    end


    % We are done with this page so close the connection.
    shBuf.close;


    % If a target URL was not found, try the next host.
    if isempty(ulist)
       hostidx=hostidx+1;
    end
end

if isempty(ulist)
  ulist=[];
elseif length(ulist) == 1
  ulist=ulist{1};
end
```

The s_host array consists of two strings for each host. The first string contains the protocol, host, and search command portion of a search URL. The search terms are added to this string in an appropriate format to generate the complete URL string that is used to create a java.net.URL object. The openStream method is used on the searchHost object to open an Internet connection to the search host. An InputStreamReader object is created on this connection and passed to a BufferedReader to enable line-at-a-time processing of the input stream. The readLine method is then used to read one line at a time into a string buffer.

The second element of the s_host array is a target string. A Web page is composed of text containing formatting and other codes called *tags* delimited by '<' and '>' characters. Each target URL (or *hit*) that is found by a particular search engine is normally preceded by a specific character string in the page returned by the search URL. Each line of the Web page is examined for the presence of this target string. If the string is found, the target URL is extracted and copied into the output cell array. When the page has been completely processed, the connection is closed. If no results are found, the next search host is contacted, and the search continues. Note that Web pages change over time, and the strings in s_host may have to be modified appropriately.

Example 5: winfun

The next example is an expanded version of the hithere script shown in Example 3. This time, a window is created containing some buttons and menus with callbacks. The first time winfun is called, a window is constructed from java.awt toolkit objects, set to an initial state, and made visible. The function then exits. The Java objects still exist and are visible, but the object references (the variables) do not exist in the base workspace. If a button is pressed or a menu item is selected, the associated object callback is executed. The callback calls the winfun function again with a unique argument. When winfun is called with an argument, the argument is parsed to service the callback. Finally, if the winfun function is called with no arguments, but the Java objects still exist, the window is reset to the initial state and made visible again. The code is as follows:

```
function winfun(varargin)
% MATLAB Java Demo function winfun.m
%
%   Mastering MATLAB 7 Java Example 5
%     Create a window with buttons, menus and text objects using
```

```
%     Java windowing toolkit objects within a function. Use a
%     local function to implement callbacks. Use persistent variables
%     to maintain visibility of the Java objects between calls.
%

% Import the entire Java Abstract Window Toolkit.

import java.awt.*

% Make sure we can find the Java objects when servicing callbacks.

persistent win ta mi abt mq ma txt x y h w

if isempty(win)

  % Initial function call: create the necessary objects.
  % Start with some text for the About Box.
  txt=[' Cool Window version 1.4  ';...
       ' MM7 example function by  ';...
       'Mastering MATLAB authors  ';...
       'Hanselman and Littlefield '];
  % Reshape it for use in the textarea as well.
  tstr=reshape(txt',1,prod(size(txt)));

  % Create a window using the flow layout model (the simplest)
  %    and set the title. Set the layout using Java syntax.
  win=Frame('Cool Java Window');
  win.setLayout(FlowLayout);

  % Specify the window location, size and color. Use Java syntax
  %    for the location and size, and use MATLAB syntax to set
  %    the background color.
  x=150; y=150; w=430; h=160;
```

```
win.setLocation(x,y);
win.setSize(w,h);
set(win,'Background',[.8 .8 .8]);

% Create some menus: a 'File' menu and a 'Help' menu.
mf=Menu('File');
mh=Menu('Help');

% Create a 'Quit' menu item, define a callback to close the window,
%      and attach it to the File menu.
mq=MenuItem('Quit');
set(mq,'ActionPerformedCallback','winfun(''quit'')');
mf.add(mq);

% Create a 4x21 text area with no scroll bars (3).
ta=TextArea(tstr,4,21,3);

% Don't let the user change the text.
ta.setEditable(0);

% Set the text background color.
set(ta,'Background',[1 1 .9]);

% Create a 'Show Info' menu item and attach it to the Help menu.
%    Show or hide the textarea when the menu item is selected.
mi=MenuItem('Show Info');
set(mi,'ActionPerformedCallback','winfun(''info'')');
mh.add(mi);

% Create an 'About' menu item and attach it to the Help menu.
ma=MenuItem('About');
```

```
set(ma,'ActionPerformedCallback','winfun(''about'')');
mh.add(ma);

% Create a menubar object, add the menus, and attach
%    the menubar to the window.
mb=MenuBar;
mb.add(mf); mb.add(mh);
mb.setHelpMenu(mh);          % Move the Help menu to the right side.
win.setMenuBar(mb);

% Create some buttons.
bs=Button('Shrink');
be=Button('Expand');
bl=Button('Left');
bu=Button(' Up ');
bd=Button('Down');
br=Button('Right');
bq=Button('Quit');

% Define callbacks for the buttons.
set(bs,'MouseClickedCallback','winfun(''shrink'')')
set(be,'MouseClickedCallback','winfun(''expand'')')
set(bl,'MouseClickedCallback','winfun(''left'')');
set(bu,'MouseClickedCallback','winfun(''up'')');
set(bd,'MouseClickedCallback','winfun(''down'')');
set(br,'MouseClickedCallback','winfun(''right'')');
set(bq,'MouseClickedCallback','winfun(''quit'')');

% Specify some colors for the buttons.
set(bs,'Background',[.5,1,1]);
set(be,'Background',[.5,1,1]);
set(bq,'Background',[1,.4,.4]);
```

```
set(bl,'Background',[.9,.9,.9]);
set(bu,'Background',[.7,.7,.7]);
set(bd,'Background',[.7,.7,.7]);
set(br,'Background',[.9,.9,.9]);

% Add the first row of buttons to the window. Objects will be
%    positioned in the order in which they are added.
win.add(bs);
win.add(bl);
win.add(bu);
win.add(bd);
win.add(br);
win.add(be);

% Attach the text area to the window, but don't display it yet.
win.add(ta);
ta.setVisible(0);

% Now add the Quit button.
win.add(bq);

% Now that the main window has been created, it is time
%    to create a dialog box for the 'About' menu item.
abt = Dialog(win, 'About Winfun');
set(abt,'Background',[.95 .95 .95]);
abt.setLayout(GridLayout(5,1));

% Add some lines of text
abt.add(Label(txt(1,:)));
abt.add(Label(txt(2,:)));
abt.add(Label(txt(3,:)));
abt.add(Label(txt(4,:)));
```

```
  % and a button bar to close the About box.
  bok = Button('OK');
  set(bok,'ActionPerformedCallback','winfun(''ok'')');
  set(bok,'Background',[.6 .8 .8]);
  abt.add(bok);

  % Set the size of the dialog box but don't make it visible.
  abt.setSize(180,110);

  % All done. Now show the main window and exit.
  win.setVisible(1);

elseif nargin == 0
  % Reset the hidden window to the initial state and show it.
  abt.setVisible(0);
  ta.setVisible(0); mi.setLabel('Show Info');
  x=150; y=150; w=430; h=160;
  win.setLocation(x,y);
  win.setSize(w,h);
  win.setVisible(1);

elseif nargin == 1   % This is a callback.
    switch (varargin{1})
      case 'shrink'   % Shrink width by 4 pixels, height by 2
        w=w-4; h=h-2;
      case 'expand'   % Expand width by 4 pixels, height by 2
        w=w+4; h=h+2;
      case 'left'     % Move 4 pixels to the left
        x=x-4;
      case 'up'       % Move 2 pixels higher
        y=y-2;
      case 'down'     % Move 2 pixels lower
        y=y+2;
```

```
      case 'right'    % Move 4 pixels to the right
        x=x+4;
      case 'info'     % Toggle visibility of the text area
        if (ta.isVisible)
          ta.setVisible(0); mi.setLabel('Show Info');
        else
          ta.setVisible(1); mi.setLabel('Hide Info');
        end
      case 'about'    % Show the About Box to the right and lower
        abt.setLocation(x+w+20,y+10);
        abt.setVisible(1);
      case 'ok'       % Hide the About Box
        abt.setVisible(0);
      case 'quit'     % Hide the windows
        abt.setVisible(0);
        win.setVisible(0);
        return
      otherwise       % Bad argument
        error('Invalid callback');
    end
    win.setBounds(x,y,w,h); win.setVisible(1);

else    % Should not get here.
    error('Too many arguments.');
end
```

The callback argument determines which case statement is executed within the switch statement. When the argument is shrink, the size of the window object win is reduced by 4 pixels in the horizontal (x) direction and 2 pixels in the vertical (y) direction using the setBounds method. The setVisible method is then used to redraw the window with the changed characteristics. Similarly, the expand argument increases the window size. The up, down, left, and right arguments change the window location using the same methods.

When the `info` argument is invoked, the result depends on the state of the `ta` text area object. If `ta` is visible (as determined by the `isVisible` method), the text area is hidden and the menu item label is changed appropriately. If the text area is currently hidden, `ta` is made visible by using the `setVisible` method, and the menu item label is changed again. The function `win.setVisible` is then called to refresh the window.

When the `about` argument is received, the dialog box `abt` is positioned to the right of the `win` window and made visible. When the `ok` argument is received, the `abt` dialog box is hidden. When `quit` is received, both the `win` window and the `abt` dialog box are hidden.

The following figure is an example of the windows generated by the `winfun` function:

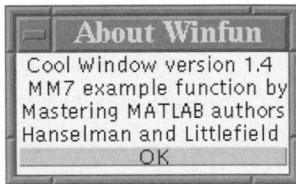

35.10 SUMMARY

Java is an object-oriented, platform-independent programming language that can be used to extend the scope of the MATLAB environment. Every installation of MATLAB 7 includes a Java virtual machine integrated into MATLAB and available to the MATLAB user. Java classes, objects, and methods can be manipulated within MATLAB, both from the command line and from MATLAB functions. Java integration provides new opportunities to extend MATLAB in many different ways. While MATLAB currently incorporates Java, it is not possible to call MATLAB functions from Java. It is expected that this feature will appear in a future MATLAB release.

The examples in this chapter explore only a few of the possibilities opened up by the availability of Java objects and methods within MATLAB. The MATLAB documentation contains additional Java examples and more details on some of the topics discussed in this chapter. Perhaps the most important property of Java objects is that they are passed by reference, not by value.

Important: Java objects are references in MATLAB. As a result, Java objects cannot be copied, only modified. In addition, Java objects never really disappear. The `clear` function removes references to these objects from the workspace, but the objects themselves may still exist. The Java `setVisible` method simply makes objects invisible. The Java `dispose` method may remove some attributes of objects, but the objects may still exist. Finally, if you lose all references to a Java object by exiting a workspace or using the `clear` command, they cannot be retreived. There is nothing similar to the `findobj` function for Java objects.

To end this chapter, we note that MATLAB 7 has greatly expanded Java support and that future versions of MATLAB are likely to continue to move toward full Java integration. Handle Graphics objects and Java graphics classes and objects are likely to converge. MATLAB appears to be moving toward a fully integrated development environment implemented in Java. This all suggests that Java will play an increasingly important role in the future development of MATLAB.

36

Windows Application Integration

It is not always necessary to write your own applications in a programming language such as C or FORTRAN to enable direct communication between applications. UNIX and Linux platforms use standard *pipes* (the standard output of one program is redirected to the standard input of another) to connect applications. Microsoft Windows operating systems implemented similar functionality with Dynamic Data Exchange (DDE). DDE permits applications to communicate directly with each other to send commands or exchange data. A component framework was built upon DDE and Visual Basic Extensions (VBX) and named Object Linking and Embedding (OLE). Later, some parts of OLE relating to graphical user interfaces (GUI) and the Internet were renamed ActiveX. Microsoft later renamed the entire component framework the *Component Object Model* or COM.

A COM object is an instance of a component object class that runs on a server application and is controlled by a client application. A COM object encapsulates all data and methods of the object and uses *interfaces* to access the methods of the object. An interface is a pointer to the methods of an object. COM objects can have multiple interfaces. (Refer to the documentation of the component vendor for details of the interfaces available for a specific component.)

Many commercial applications support COM objects, ActiveX, or DDE for interapplication communication on the Windows PC platform. MATLAB supports COM objects, ActiveX controls, and DDE in a limited capacity. It supports the creation of COM objects within MATLAB to control other applications and can also act as a COM server and respond to requests from other applications. COM and DDE are currently available only on the win32 platform under a *Microsoft Windows* operating system such as Windows 98, Windows 2000, or Windows XP.

36.1 COM OBJECTS: CLIENT/SERVER COMMUNICATION

The COM protocol specifies an object model and a collection of standard and custom interfaces that define each object's methods, properties, and events. Methods are actions that can be performed on or by an object (similar to MATLAB functions and object methods), properties are variables that determine the state of an object (similar to Handle Graphics properties), and events are notifications of changes in the state of an object that may trigger some action (similar to Handle Graphics or Java callbacks).

Applications that support COM can support server functions (the application responds to client requests), client functions (the application sends requests to a server), or both. ActiveX controls are objects that can be integrated into **control containers** (such as MATLAB *Figure* windows) by a COM client application and are used to initiate an action by a COM server application. COM servers are applications that can be controlled in some way by a COM client or by an ActiveX control.

MATLAB COM Support

MATLAB supports a limited set of COM interfaces. It can act as an Automation server by exchanging data with other applications and executing commands in the MATLAB workspace. MATLAB also can act as a COM client by exchanging data with other applications and controlling these Automation server applications (such as Microsoft Word, Excel, and PowerPoint) from the MATLAB command line or through M-files. MATLAB cannot be physically imbedded within other applications, but it can act as a control container by imbedding Automation server controls within MATLAB *Figure* windows, similar to the way *uicontrols* are used.

MATLAB as a COM Client

The `actxserver` function creates a COM object and opens a connection to a COM Automation server. Similarly, the `actxcontrol` function creates a control that can be imbedded in a MATLAB container (such as a *Figure* window) to control a COM Automation server. The functions (methods) that operate on COM objects are shown in the following table:

Function/Method	Description
`actxserver`	Create a COM Automation server
`actxcontrol`	Create an ActiveX control object to imbed within a *Figure* window
`actxcontrollist`	List all ActiveX controls that are currently installed
`fieldnames`	Return the property names of an object
`interfaces`	List the custom interfaces to a COM server
`methods`	List all of the public methods for the control or server

Function/Method	Description
events	Display a list of events the control can trigger
eventlisteners	Return the list of events attached to listeners
addproperty	Add a custom property to an object
deleteproperty	Remove a custom property from an object
invoke	Invoke a method on an interface or object, or displays a list of methods
registerevent	Register an event handler with a control to handle a specific event
unregisterevent	Unregister an event handler with a control
unregisterallevents	Unregister all event handlers with a control
isa	True if an argument is of a given MATLAB or Java class
iscom	True if the argument is a COM object
isevent	True if the argument is an event
isinterface	True if the argument is a COM interface
ismethod	True if the argument is an object method
isprop	True if the argument is an object property
inspect	Display a GUI to list and modify property values
methodsview	Display a GUI to list information about the methods of an object
actxcontrolselect	Display a GUI for creating an ActiveX control
class	Create an object or return the class of an object
set	Set a property value on an object or interface
get	Get a property value from an interface or display a list of properties
propedit	Display the built-in property page of a control if the page exists
release	Release an interface
delete	Delete a COM control object and all of its interfaces
load	Load the property state of a control from a file
save	Save the property state of a control to a file
move	Move or resize a control and return the new position

The `actxcontrol`, `actxserver`, `get`, and `invoke` functions can all return COM objects (or new *interfaces* to objects—essentially separate communication channels to objects), which can be controlled or queried by using other COM functions or methods. The `release` function should be used to release the resources used by a COM object or interface when it is no longer needed. The `delete` function closes all connections and releases all resources when the server or control object is no longer needed.

Help text is available by using the appropriate object method. For example, the code

```
>> help COM/delete
```

displays the help text for the `delete` method for COM objects.

A list of properties can be obtained by using the form `get(axhandle)` or `axhandle.get`, where `axhandle` is the COM object handle obtained from an `actxserver` or `actxcontrol` object creation method. The properties of a MS Power-Point COM server are as follows:

```
>> ppapp = actxserver('powerpoint.application');

ppapp =

    COM.powerpoint_application


>> get(ppapp)


      Presentations: [1x1 Interface.Microsoft_PowerPoint_9.0_Object_Library.Presentations]
            Windows: [1x1 Interface.Microsoft_PowerPoint_9.0_Object_Library.DocumentWindows]
       ActiveWindow: [1x203 char]
   ActivePresentation: [1x196 char]
    SlideShowWindows: [1x1 Interface.Microsoft_PowerPoint_9.0_Object_Library.SlideShowWindows]
        CommandBars: [1x1 Interface.Microsoft_Office_9.0_Object_Library._CommandBars]
               Path: 'C:\Program Files\Microsoft Office\Office'
               Name: 'Microsoft PowerPoint'
            Caption: 'Microsoft PowerPoint'
          Assistant: [1x1 Interface.Microsoft_Office_9.0_Object_Library.Assistant]
         FileSearch: [1x1 Interface.Microsoft_Office_9.0_Object_Library.FileSearch]
           FileFind: [1x51 char]
              Build: '6620'
            Version: '9.0'
    OperatingSystem: 'Windows (32-bit) 5.00'
      ActivePrinter: 'XColor'
            Creator: 1.3479e+009
             AddIns: [1x1 Interface.Microsoft_PowerPoint_9.0_Object_Library.AddIns]
                VBE: [1x1 Interface.Microsoft_Visual_Basic_for_Applications_Extensibility_5.3.VBE]
```

```
            Left: 99
             Top: 99
           Width: 720
          Height: 546.7500
     WindowState: 'ppWindowNormal'
         Visible: 'msoFalse'
          Active: 'msoFalse'
    AnswerWizard: [1x1 Interface.Microsoft_Office_9.0_Object_Library.AnswerWizard]
        COMAddIns: [1x1 Interface.Microsoft_Office_9.0_Object_Library.COMAddIns]
      ProductCode: '{00010409-78E1-11D2-B60F-0060774499C8E7}'
  DefaultWebOptions: [1x1 Interface.Microsoft_PowerPoint_9.0_Object_Library.DefaultWebOptions]
  LanguageSettings: [1x1 Interface.Microsoft_Office_9.0_Object_Library.LanguageSettings
ShowWindowsInTaskbar: 'msoTrue'
   FeatureInstall: 'msoFeatureInstallNone'
```

A list of methods for the `ppapp` object, along with the calling syntax, can be obtained by using the `invoke` function:

```
>> invoke(ppapp)
    Activate = void Activate(handle)
    Help = void Help(handle, Variant(Optional))
    Quit = void Quit(handle)
    Run = Variant Run(handle, string, SafeArray(Variant))
```

All of the methods available, including inherited methods for the `COM.powerpoint_application` class of objects, can be obtained by using the `methods` function:

```
>> ppapp.methods
```

Methods for class COM.powerpoint_application:

Activate	eq	load	str2double
Help	eval	move	str2num
Quit	evalc	ne	strcmp
Run	events	permute	strcmpi
addproperty	fieldnames	propedit	strmatch
char	findstr	release	strncmp

ctranspose	get	reshape	strncmpi
delete	interfaces	save	transpose
deleteproperty	intersect	send	tril
diag	invoke	set	triu
disp	isletter	setdiff	unique
display	ismember	setxor	
double	isspace	sort	

The fieldnames function provides a cell array containing the names of the public data fields available for the ppapp object:

```
>> ppapp.fieldnames
```

```
ans =

    'Presentations'
    'Windows'
    'ActiveWindow'
    'ActivePresentation'
    'SlideShowWindows'
    'CommandBars'
    'Path'
    'Name'
    'Caption'
    'Assistant'
    'FileSearch'
    'FileFind'
    'Build'
    'Version'
    'OperatingSystem'
    'ActivePrinter'
    'Creator'
    'AddIns'
    'VBE'
    'Left'
    'Top'
```

```
'Width'
'Height'
'WindowState'
'Visible'
'Active'
'AnswerWizard'
'COMAddIns'
'ProductCode'
'DefaultWebOptions'
'LanguageSettings'
'ShowWindowsInTaskbar'
'FeatureInstall'
```

Finally, the events function lists all of the events that the ppapp object can trigger (the list includes the function prototype used when calling the event handler and does not distinguish between registered and unregistered events):

```
>> ppapp.events
    WindowSelectionChange = void WindowSelectionChange(handle Sel)

    WindowBeforeRightClick = void WindowBeforeRightClick(handle Sel, bool Cancel)

    WindowBeforeDoubleClick = void WindowBeforeDoubleClick(handle Sel, bool Cancel)

    PresentationClose = void PresentationClose(handle Pres)

    PresentationSave = void PresentationSave(handle Pres)

    PresentationOpen = void PresentationOpen(handle Pres)

    NewPresentation = void NewPresentation(handle Pres)

    PresentationNewSlide = void PresentationNewSlide(handle Sld)

    WindowActivate = void WindowActivate(handle Pres, handle Wn)

    WindowDeactivate = void WindowDeactivate(handle Pres, handle Wn)

    SlideShowBegin = void SlideShowBegin(handle Wn)

    SlideShowNextBuild = void SlideShowNextBuild(handle Wn)

    SlideShowNextSlide = void SlideShowNextSlide(handle Wn)

    SlideShowEnd = void SlideShowEnd(handle Pres)

    PresentationPrint = void PresentationPrint(handle Pres)

    AfterNewPresentation = void AfterNewPresentation(handle Pres)

    AfterPresentationOpen = void AfterPresentationOpen(handle Pres)
```

With this introduction, several examples are presented to illustrate the key features of COM and ActiveX in MATLAB.

MATLAB Client Examples

This example uses MATLAB as a COM client and Microsoft Word as a COM Automation server. The MATLAB M-file function wordfig copies the contents of the current (or a specified) *Figure* window to the clipboard, starts a server instance of the Microsoft Word application, opens a selected document (or creates a new document) by using the uiputfile dialog box to specify the filename, pastes the graphic at the end of the document, closes the document, quits the Word application, and deletes the COM server object:

```
function wordfig(filespec,popt)
%WORDFIG Open a MSWord document and paste the current figure into it.
% WORDFIG Paste the current Figure window into a word document.
% WORDFIG(FILESPEC) Paste the current Figure window into the document
%   named FILESPEC. Use the complete path to the document if necessary.
% WORDFIG(FILESPEC,POPT) Paste a Figure window into the document FILESPEC
%   using the print options POPT (e.g. -f2 to select Figure window #2).
% If the FILESPEC argument is missing or empty, the uiputfile dialog box
%   is used to select a file name.

% Create or verify a valid file name.
if nargin < 1 | isempty(filespec) | ~ischar(filespec)
  [fname,dname]=uiputfile('*.doc','Modify or create the file:');
  if isequal(fname,0), return, end
  filespec=fullfile(dname,fname);
end
[dname,fname,fext]=fileparts(filespec);
if isempty(dname), dname=pwd; end
if isempty(fext), fext='.doc'; end
filespec=fullfile(dname,[fname,fext]);

% Copy the current Figure window onto the clipboard.
if nargin < 2
  print('-dmeta');
else
  print('-dmeta',popt);
end
```

```
% Start a session with MSWord.
wrd=actxserver('word.application');
wrd.Visible=1;  % Watch the action...

% Open or create a document.
if ~exist(filespec,'file')
  doc=invoke(wrd.Documents,'Add');
else
  doc=invoke(wrd.Documents,'Open',filespec);
end

% Insert some text at the end of the document.
myrange=doc.Content;
myrange.InsertParagraphAfter;
invoke(myrange,'InsertAfter','---Figure Top Caption Goes Here---');
myrange.InsertParagraphAfter;
% Paste AFTER the existing text.
invoke(myrange,'Collapse',0);
% Paste with "Picture Format" (1) and "Float Over Text" (3) options.
invoke(myrange,'PasteSpecial',0,0,1,0,3);
mmyrange.InsertParagraphAfter;
invoke(myrange,'InsertAfter','---Figure Bottom Caption Goes Here---');
myrange.InsertParagraphAfter;

% Save and close the document.
if ~exist(filespec,'file')
  invoke(doc,'SaveAs',filespec,1);
else
  doc.Save;
end
doc.Close;

% Quit Word and close the server connection.
wrd.Quit;
delete(wrd);
return
```

The next example uses MATLAB as a COM client and Microsoft PowerPoint as a COM Automation server. The MATLAB M-file function `addslide` starts a server instance of the Microsoft PowerPoint application and opens a selected presentation (or creates a new presentation) by using the `uiputfile` dialog box to specify the filename. A new slide is added to the end of the presentation for each valid variable argument. Valid arguments are character strings, cell arrays of strings, 2-D numeric arrays, and figure handles. After the new slides have been added, the presentation is saved to a file and closed, the application is terminated, and the COM server object is deleted. The code is as follows:

```
function addslide(filespec,varargin)
%ADDSLIDE Insert variables into a PowerPoint presentation.
%
% ADDSLIDE(filename,var1,var2,...) Insert variables into
%    new slides at the end of a PowerPoint presentation.
%
%    An invalid or empty filename or a new filename will
%    bring up a dialog box to enable the user to select an
%    existing file or create a new file.
%
%    Inputs can be strings, cell arrays of strings, 2D numeric
%    arrays, and figure handles. Variable sizes should be limited
%    to avoid overfilling the slides.

% Check for missing arguments and initialize variables
narg=nargin;
if narg < 2, error('Missing input arguments.'); end
nl=char(13);

% Make sure we have a valid file name
if ischar(filespec) & exist(filespec,'file')    % a real file
  fullname=filespec;
else
  if ischar(filespec) & ~isempty(filespec)       % use as default file
    [dname,fname,fext]=fileparts(filespec);
    if isempty(fext), fext='.ppt'; end
    defname=fullfile(dname,[fname fext]);
```

```
  else
    defname='*.ppt';                              % use file filter
  end
  [fname,dname]=uiputfile(defname,'Modify or create the file:');
  if isequal(fname,0), return, end
  fullname=fullfile(dname,fname);
end

% Start a session with PowerPoint
ppapp=actxserver('powerpoint.application');
ppapp.Visible=1;                              % Watch the action...

% Open or create a presentation
if ~exist(fullname,'file')
  ppres=invoke(ppapp.Presentations,'Add');
else
  ppres=invoke(ppapp.Presentations,'Open',fullname);
end

% Process the rest of the arguments
for idx=2:narg
  arg=varargin{idx-1};

  if ishandle(arg) & (floor(arg)==arg)     % figure handle
    ppslide=ppres.Slides.Add(ppres.Slides.Count+1,'ppLayoutBlank');
    print('-dmeta',sprintf('-f%d',arg));   % copy plot to clipboard
    ppfig=ppslide.Shapes.Paste;            % paste the clip into the slide
    ppfig.Align(1,-1);                     % center the plot
    ppfig.IncrementTop(200);               % lower the plot
    ppfig.ScaleWidth(1.5,0,1);             % scale width from middle

  elseif isstr(arg)                        % text string
    if length(arg) > 275, warning('possible overfull slide'); end
    ppslide=ppres.Slides.Add(ppres.Slides.Count+1,'ppLayoutTitleOnly');
```

```matlab
    ppslide.Shapes.Title.TextFrame.AutoSize=1;
    ppslide.Shapes.Title.TextFrame.VerticalAnchor=1;
    ppslide.Shapes.Title.TextFrame.TextRange.Text=arg;


  elseif iscellstr(arg)                      % cell array of strings
    ppslide=ppres.Slides.Add(ppres.Slides.Count+1,'ppLayoutTitleOnly');
    str='';
    for idx2=1:length(arg)
      str=[str,arg{idx2},nl];
    end
    if (length(arg) > 5) | (length(str) > 275)
      warning('possible overfull slide');
    end
    ppslide.Shapes.Title.TextFrame.AutoSize=1;
    ppslide.Shapes.Title.TextFrame.VerticalAnchor=1;
    ppslide.Shapes.Title.TextFrame.TextRange.Text=str(1:end-1);


  elseif isnumeric(arg) & ~all(ishandle(arg(:))) & (ndims(arg) <= 2)
    ppslide=ppres.Slides.Add(ppres.Slides.Count+1,'ppLayoutTitleOnly');
    [r,c]=size(arg);
    if numel(arg) > 20, warning('possible overfull slide'); end
    rs=int2str(r); cs=int2str(c);
    str=[sprintf('Array %s', inputname(idx)),nl,nl];
    for idx2=1:r
        str=[str, sprintf('    %5.5f',arg(idx2,:)), nl];
    end
    ppslide.Shapes.Title.TextFrame.AutoSize=1;
    ppslide.Shapes.Title.TextFrame.VerticalAnchor=1;
    ppslide.Shapes.Title.TextFrame.TextRange.Text=str;
  else
    warning([inputname(idx), ' is not an accepted input and will be ignored.'])
  end %if
end %for
```

```
% Save the file and exit
ppres.SaveAs(fullname,1,0);        % save as presentation
ppres.Close;                       % close the presentation
ppapp.Quit;                        % quit PPT
delete(ppapp);                     % done with the COM server
return
```

The final client example illustrates the use of COM events and event handlers. The cdemo function embeds an ActiveX control in a *Figure* window and performs certain actions when events are triggered. MATLAB ships with two very simple ActiveX controls (mwsamp.ocx and mwsamp2.ocx) and associated type libraries (mwsamp.tlb and mwsamp2.tlb) stored in the $MATLAB\bin\win32 directory. The Mwsamp class contains one event (Click) and three variables (Label, Radius, and MMPropertyContainer). The MMPropertyContainer is a Java object containing various properties of the control. The Mwsamp2 class is an expanded version of the Mwsamp class and adds two events (DblClick and MouseDown) and one variable (Ret_IDispatch) containing an interface function.

When an event is triggered by an ActiveX control, MATLAB passes a number of arguments to the event handler, including the name of the object, a numerical event identifier, zero or more event arguments, an event structure containing additional information about the event, and the event name as the final argument. The number and names of the event arguments are specific to the ActiveX control object. (Refer to the MATLAB documentation for more details about these arguments.)

The cdemo function creates a new *Figure* window containing a surf plot, embeds an MWSAMP2 control in the corner of the *Figure* window, sets some control properties, and exits. When a mouse-click event is detected within the control, the cdemo function is called again to service the callback. The control is deleted when the *Figure* window is closed. Following is the code:

```
function cdemo(varargin)
%CDEMO Sample function to manage an ActiveX object.
%   Function to create a sample ActiveX control. The function creates
%   a figure window, adds a nice plot, creates an MWSAMP control,
%   embeds the control in the Figure window, sets the 'Label' and
%   'Radius' properties of the control, and invokes the 'Redraw'
%   method on the control.
```

```
%
%   CDEMO is also the event handler for this control. The three events
%   fired by the control are 'Click', 'DblClick', and 'MouseDown'.
%   The event handler changes the text message in the control when
%   a valid event is fired.
%   The control is deleted when the figure window is closed.


% Keep track of a few things between calls.
persistent numclicks h


if nargin == 0        % Initial call -- do the setup.

   % Create a new figure window and draw a nice plot.
   f = figure;
   surf(peaks);
   numclicks=0;


   % Embed an MWSAMP2 ActiveX control in the lower left corner
   % of the Figure window and set the callback to recall this
   % function (cdemo).
   h = actxcontrol('MWSAMP.MwsampCtrl.2',[0 0 90 90],f,'cdemo');


   % Set the initial label and circle radius in the control
   % showing two methods of setting the property values.
   set(h, 'Label', 'Click Here');
   h.Radius=28;


   % Tell the control to redraw itself by invoking the Redraw method.
   invoke(h, 'Redraw');

else     % This part handles the callback. For each valid event
         % detected, the last argument will be a string that
         % resolves to the event name.
```

```matlab
if strcmp(varargin{end},'Click')
  % Increment the click total.
  numclicks = numclicks + 1;
  h.Label=['Click #',num2str(numclicks)];

elseif strcmp(varargin{end},'DblClick')
  % Decrement the click total by 2. The first of the pair
  % generated a Click event and incremented by 1. The second
  % click within the time limit generated this DblClick event.
  numclicks = numclicks - 2;
  h.Label=['Click #',num2str(numclicks)];

elseif strcmp(varargin{end},'MouseDown')
  % Display the x,y coordinates of the mouse pointer.
  h.Label=['(x,y)=(',num2str(varargin{5}),',',num2str(varargin{6}),')'];

else
  error('Invalid input.');
end
% Redraw the control.
h.Redraw;
end
```

MATLAB as COM Server

MATLAB can act as a COM Automation server when called by a COM client such as Visual Basic, Visual Basic for Applications, Visual C++, or even MATLAB itself. Microsoft Excel, Microsoft Access, or another COM client application can be an Automation controller as well. The client application can start and stop an instance of MATLAB, pass arrays to and from the MATLAB workspace, and execute MATLAB commands in the MATLAB workspace. In many ways, this functionality is identical to that provided by the MATLAB Engine. In fact, any application you build by using C or FORTRAN should use the MATLAB Engine interface rather than the COM interface to MATLAB. The COM interface should be reserved for use by applications that cannot access MATLAB as a server in any other way.

The registered MATLAB COM object name is **matlab.application**. MATLAB provides the following Automation methods to COM client applications:

Function	Description
Execute	Execute a MATLAB command contained in a string argument in the server
Feval	Evaluate a MATLAB command that cannot be contained in a single string in the server; examples include commands that use values contained in local variables unknown in the MATLAB server workspace
GetCharArray	Get a character array from the server
GetFullMatrix	Get a matrix from the server; returns a SAFEARRAY data type
GetWorkspaceData	Get data from the server workspace; returns a variant data type
MaximizeCommandWindow	Display the server window on the Windows desktop
MinimizeCommandWindow	Minimize the size of the server window
PutCharArray	Store a character array in the server
PutFullMatrix	Store a matrix in the server
PutWorkspaceData	Store data in the server workspace
Quit	Terminate the MATLAB server

This list of methods, along with the calling syntax, can be obtained by using the invoke function on a COM.matlab_application object handle. Properties of a MATLAB COM server can be obtained by using MATLAB as a COM client. Both of these uses are illustrated in the following code:

```
>> mlapp=actxserver('matlab.application')

mlapp =

    COM.matlab_application

>> get(mlapp)
    Visible: 1

>> mlapp.methods
```

```
Methods for class COM.matlab_application:
```

Execute	evalc	send
GetCharArray	events	set
GetFullMatrix	fieldnames	setdiff
MaximizeCommandWindow	findstr	setxor
MinimizeCommandWindow	get	sort
PutCharArray	interfaces	str2double
PutFullMatrix	intersect	str2num
Quit	invoke	strcmp
addproperty	isletter	strcmpi
char	ismember	strmatch
ctranspose	isspace	strncmp
delete	load	strncmpi
deleteproperty	move	transpose
diag	ne	tril
disp	permute	triu
display	propedit	unique
double	release	
eq	reshape	
eval	save	

```
>> mlapp.fieldnames

ans =

    'Visible'
```

A MATLAB COM server does not support any events:

```
>> mlapp.events
```

Example

This is an example of an Excel macro that starts MATLAB as an Automation server, passes the contents of cells B3:E8 to a MATLAB array, squares the contents of the array, passes back the result, and inserts the result in cells B12:E17 in the Excel spreadsheet. It also passes a string value obtained from cell B1 to MATLAB, appends " squared (in MATLAB)" to the string, passes back the result, and inserts this result in cell B10. The MATLAB server is then closed. The code is as follows:

```
Sub Square()
'

' Square Macro
' Square the contents of cells B3:E8 and place the result in B12:E17.
' Also append the phrase " squared (in MATLAB)" to a string from B1
' and place the result in B10.


' First define the variables.
Dim MatLab As Object
Dim Result, NewString As String
Dim MReal(6, 4) As Double
Dim MImag() As Double
Dim RealValue As Double
Dim i, j As Integer


' Invoke MATLAB.
Set MatLab = CreateObject("Matlab.Application")


' Fill the Mreal array with data from the sheet.
For i = 0 To 5
  For j = 0 To 3
    Real(i,j)=ActiveSheet.Range(Cells(i+3,j+2),Cells(i+3,j+2)).Value
  Next j
Next i
```

```
' Send the string and data from the spreadsheet to MATLAB.
Call MatLab.PutCharArray("instr", "base", ActiveSheet.Range("B1:B1").Value)
Call MatLab.PutFullMatrix("a", "base", MReal, Mimag)


' Send MATLAB some commands to execute.
Result = MatLab.Execute("b=a.^2;")
Result = MatLab.Execute("outstr=[instr, ' squared (in MATLAB)']")


' Retrieve the results and stuff them into spreadsheet cells.
Call MatLab.GetFullMatrix("b", "base", MReal, MImag)
ActiveSheet.Range("B12:E17").Value = MReal
ActiveSheet.Range("B10:B10").Value = MatLab.GetCharArray("outstr", "base")
'
End Sub
```

Suppose the Sub Square() macro is applied to the following spreadsheet:

	Input array of numeric values			
	1	2	3	4
	2	3	4	5
	3	4	5	6
	4	5	6	7
	5	6	7	8
	6	7	8	9

The results after the macro is run are as follows:

Input array of numeric values			
1	2	3	4
2	3	4	5
3	4	5	6
4	5	6	7
5	6	7	8
6	7	8	9
Input array of numeric values squared (in MATLAB)			
1	4	9	16
4	9	16	25
9	16	25	36
16	25	36	49
25	36	49	64
36	49	64	81

Of course, this example is of very limited practical value. However, it provides a solid foundation for further development.

An easy way to create a VBA macro is to record a new macro and edit the resulting VBA code. For example, open Excel and select the menu item **Tools/Macro/Record New Macro . . .** Give your macro a name in the dialog box (the previous example used the name "Square") and click on the OK button. Then, click on the Stop Recording button in the next dialog box. This creates an empty macro. Next, select the **Tools/Macro/Macros . . .** menu item and click on the **Edit** button. Now you can edit the macro, save the result, and run the macro whenever you wish.

The Visual Basic environment provides some tools that help determine the appropriate objects and methods for communicating with other COM applications in addition to the MATLAB `get` and `invoke` methods previously illustrated. The Object Browser can be used to determine the objects available in each of the object libraries (such as the Excel library or the MSForms library). The objects and members (properties, events, and functions) are searchable, and context-sensitive help is available.

36.2 DYNAMIC DATA EXCHANGE

Prior to the client/server capabilities of COM, Microsoft created a mostly peer-to-peer interapplication communication protocol called Dynamic Data Exchange. DDE enables two cooperating applications to exchange data and execution strings (commands) by using the Windows clipboard. One application can also register a request for notification of updates when certain data change in the other application. This DDE functionality has been incorporated into the more flexible COM protocol. DDE is still supported for existing code, but COM is normally used for new implementations.

DDE connections between applications are called **conversations**. Each application has a unique **service name** to identify the application. Each conversation is identified by a service name and a **topic** known to both applications. Each application supports a System topic, and most support one or more additional topics. The application that initiates a DDE request is designated the **client**, and the application that answers the request is designated the **server** for the duration of the conversation.

A typical DDE conversation starts with application A requesting a conversation with an instance of application B. The conversation consists of a service name/topic pair. If application B (identified by service name) recognizes the topic, then it establishes a conversation with application A. The elements of the conversation are items that are passed between the applications by using the Windows clipboard. All applications support the Text format for data exchange. Some applications support additional data formats, such as Bitmap and MetaFilePict for graphics and XLTable format for Excel spreadsheet data. MATLAB supports only the Text format when operating as a DDE client, but supports Text, XLTable, and MetaFilePict formats when operating as a DDE server.

A COM connection request can invoke an instance of an application (launch the application) to service a client request, but a DDE conversation can be initiated only between applications that are already running.

MATLAB as a DDE Client

The DDE functions supported by MATLAB are listed in the following table:

Function	Description
ddeadv	Set up an advisory link between MATLAB and a DDE server application
ddeexec	Send an execution string to a DDE server application
ddeinit	Initiate a DDE conversation between MATLAB and another application
ddepoke	Send data from MATLAB to a DDE server application

ddereq	Request data from a DDE server application
ddeterm	Terminate a DDE conversation between MATLAB and a server application
ddeunadv	Release an advisory link between MATLAB and a DDE server application

These functions are used by MATLAB to manage DDE connections to other applications acting as DDE servers. Examples of DDE servers include Microsoft Excel, Word, Access, PowerPoint, and even another instance of MATLAB.

As a server, Microsoft Excel supports two kinds of topics: the System topic and a topic consisting of the name of a workbook that is open in Excel or the name of a spreadsheet in the open workbook. A Microsoft Excel item is any cell reference—an individual cell or a range of cells. Microsoft Word supports the System topic and the name of an open document in Word. A Microsoft Word item is any bookmark in the open document. (Consult the on-line help or the printed documentation for these and other DDE applications for details about the topics and the items they support.)

The following is an example of a DDE conversation between MATLAB (the client) and an open copy of Excel (the server); first, we request a connection to the System topic (every DDE application supports the System topic):

```
>> xls = ddeinit('excel','system')
>> xls =
     4.7836e-299
```

The return value is the handle assigned to the conversation. A failed request returns a handle value of 0. Next, we request a list of the items available under the System topic by using the standard SysItems item. The optional third argument is a format vector. The first element indicates the clipboard format (1 for Text—the only format supported by MATLAB as a client). The second element determines how the result is stored: as 0 to interpret the data as numerical values (the default) or as 1 to interpret the data as a character string. These requests are for items that return tab-delimited character strings. Here is the code:

```
>> s = ddereq(xls,'SysItems',[1 1])
s =
SysItems   Topics  Status  Formats Selection   Protocols   EditEnvItems
```

Query each of them for more detail:

```
>> t = ddereq(xls,'Topics',[1 1])
t =
[Book1]Sheet1   [Book1]Sheet2   [Book1]Sheet3   System
```

```
>> t = ddereq(xls,'Status',[1 1])
t =
Ready

>> t = ddereq(xls,'Formats',[1 1])
t =
XlTable Microsoft Excel 8.0 Format  BIFF4 Biff3 SYLK Wk1 Csv Unicode Text
Text    Rich Text Format    DIF Bitmap  Picture (Enhanced Metafile)
Printer_Picture Screen Picture EMF

>> t = ddereq(xls,'Selection',[1 1])
t =
[Book1]Sheet1!R1C1

>> t = ddereq(xls,'Protocols',[1 1])
t =
StdFileEditing   Embedding

>> t = ddereq(xls,'EditenvItems',[1 1])
t =
StdHostNames    StdTargetDevice StdDocDimensions
```

Close the connection:

```
>> status = ddeterm(xls)
status =
     1
```

This shows that the open workbook named Book1 contains three worksheets, named Sheet1, Sheet2, and Sheet3; that Excel is ready to converse; that Excel supports many clipboard formats; that the currently selected cell is row 1, column 1; and other details. Excel supports the System topic and one topic for each of the available worksheets. The worksheet topics support only one item type—a range of cells.

The next example opens a conversation with one of the worksheets. First, create a magic square and plot the resulting surface:

```
>> m = magic(10);
>> h = surf(m);
```

Open a connection to a worksheet and insert the data into a range of cells:

```
>> xl = ddeinit('excel','Sheet1');
>> range = 'r1c1:r10c10';
>> status = ddepoke(xl,range,m);
```

Now set up an advisory link between the spreadsheet data and MATLAB to trigger a callback if any data in the range of cells change:

```
>> status = ddeadv(xl,range,'disp(''Spreadsheet data change alert!'')');
```

The third argument is the callback passed to eval. Now, anytime a change is made to the contents of one or more of these cells in the spreadsheet, a message is displayed in the *Command* window.

Next, close this link and open another. This time, when a change is made to the spreadsheet, the link passes the range of data back to MATLAB in the variable z and the callback updates the plot with the new data. The presence of the fourth argument notifies the server to send back data, in addition to sending a signal to MATLAB to execute the callback. The code is as follows:

```
>> status = ddeunadv(xl,range);
>> status = ddeadv(xl,range,'set(h,''ZData'',z);
          set(h,''CData'',z);','z');
```

Make some extensive changes to the data to see the effect on the plot:

```
>> range1 = 'r1c1:r9c9';
>> status = ddepoke(xl,range1,magic(9));
```

Change some data yourself to see how it affects the plot.

The ddeexec function passes a command to the DDE server to be evaluated. For example, the code

```
>> status = ddeexec(xl,'[formula.goto(''r3c5'')]'');
```

makes the cell in row 3, column 5, the current cell. (Consult the documentation for your DDE server application for a list of supported commands.)

When you are finished, close the link and terminate the conversation:

```
>> status = ddeunadv(xl,range);
>> status = ddeterm(xl);
```

MATLAB as a DDE Server

The DDE service name for MATLAB is, naturally enough, **MATLAB**. MATLAB supports two topics, System and Engine, when operating as a DDE server. The valid items for each of these topics are listed in the following table:

Topic	Item	Description
System	SysItems	A tab-delimited list of items supported under the System topic (SysItems, Format, and Topics)
	Format	A tab-delimited list of supported formats (Text, MetaFilePict, and XLTable)
	Topics	A tab-delimited list of supported topics (System and Engine)
Engine	EngEvalString	Item name (if required by the client calling syntax) when a command is sent for evaluation
	EngStringResult	String result of a DDE execute command
	EngFigureResult	Graphical result of a DDE execute command
	<matrixname>	The name of the matrix to be created, updated, or returned

DDE client applications open a conversation with MATLAB by using the service name **MATLAB** and either the System or Engine topic. The MATLAB Engine topic supports the following three client request operations: sending data to MATLAB, receiving data from MATLAB, and sending a command to MATLAB to be executed.

The client can send data by using a DDE poke operation, where the item is the name of the matrix to create or update and the format of the poke data is Text or XLTable. Text format data should be in tab-delimited ASCII format with CR/LF row delimiters, while XLTable format accepts any Excel binary data format. The client can request data from MATLAB by using the DDE request operation in three forms. To request the contents of an array, the item is the name of the matrix to send, and the data format is Text or XLTable. To request the text output from a previous MATLAB command, the item is EngStringResult, using the Text format. To request graphical output from a previous MATLAB command, the item is EngFigureResult and the format is MetaFilePict. Some clients are able to receive only Text from a request operation. For these clients, MATLAB provides the ability to specify item EngFigureResult with the Text format. In this case, the result is a status string. If the string is "Yes", the figure is available on the clipboard for the client to retrieve. If the operation fails, the text returned is "No".

The client can send a command to MATLAB by using the DDE `execute` operation. The item is `EngEvalString` and the format of the command is Text. Some clients omit the item and just send the command. MATLAB accepts either form. (Consult the documentation for your DDE client application or programming language for the appropriate syntax for writing DDE programs or functions to connect to an instance of MATLAB.)

36.3 MATLAB NOTEBOOK

Tha MATLAB Notebook is an easy way to create a report or other document and embed MATLAB commands, command output, and graphics. The MATLAB Notebook provides direct access to MATLAB as a COM Automation server from within a Microsoft Word document without requiring that you actually write your own C or FORTRAN programs. Microsoft Word is the only word processor supported by the MATLAB Notebook and must be purchased and installed separately on your Windows PC or Macintosh OS X computer. Using the Notebook, you can create a report with MATLAB commands and the resulting MATLAB output imbedded within the document itself. The Notebook is implemented by using a predefined collection of Microsoft Word macros and COM controls that allow you to connect to a MATLAB session from a Word document called an M-book. This functionality is automatically available when you install MATLAB on a `win32` PC platform along with Microsoft Word or Office.

The Notebook uses a Microsoft Word template to embed controls within your document. The command `notebook -setup` initializes the Notebook. You are prompted to select the version of Word you are using (97, 2000, or 2002). MATLAB will then attempt to determine the location of the installation and install the correct version of a template file. If the executable cannot be found, the installation script requests the location of the Microsoft Word program (by using a file requester) and the location of a template file (such as `normal.dot`) by using another file requester. After verification, the correct template file (called `m-book.dot`) is copied into the appropriate directory. The next time you use the `notebook` command, the Microsoft Word application is opened using the `m-book.dot` template. If a filename argument is supplied, Word attempts to open the requested file. If no argument is supplied, a new M-book document is opened.

Notebook support is not enabled in MATLAB 7 on the Macintosh platform; the `notebook` command is PC-specific in this version. The capability is there, however, and we expect that Macintosh support will be added in a future release.

When an M-book is opened by using the `notebook` command from MATLAB or by opening an existing M-book from Microsoft Word, a MATLAB server session is started and a new **Notebook** menu appears next to the **Table** menu in the Word application menus. The **Notebook** menu items enable such operations as defining or undefining MATLAB input statements or *cells*, grouping and ungrouping these cells, defining calculation zones within the document, and evaluating cells, calculation zones, or the entire M-book. When an input cell is evaluated, the statement is sent to the MATLAB server for evaluation, and the results (either the command-line results, a plot, or both) are inserted into the M-book document following the input statement. Notebook preferences can also be set from the **Notebook** menu.

> To create a new M-book without using the `notebook` command from MATLAB, start Word and select **Templates and Add-Ins . . .** from the **Tools** menu. (If `mbook.dot` is not listed in the **Global templates** area, click on the **Add . . .** button and select the m-book file.) Select the `m-book.dot` template (a checkmark will appear) and close the dialog box. The **Notebook** menu will appear and the document becomes an M-book. Save the file and close Word.

For example, consider the following illustrative M-book:

MATLAB Notebook Example 1

Assignment #2

```
Create a 3-by-3 magic square


    m=magic(3)



square the elements of the magic square


    m2=m.^2



and plot the result


    plot(1:9,m2(:))
```

Select one of the MATLAB statements, and then select the **Notebook/Define Input Cell** menu item. This defines the selected text as an *input cell*—a MATLAB statement to be evaluated. After each of the MATLAB statements has been defined as an input cell, the resulting M-book appears as follows:

MATLAB Notebook Example 1

Assignment #2

```
Create a 3-by-3 magic square


    [m=magic(3) ]


square the elements of the magic square


    [m2=m.^2 ]


and plot the result


    [plot(1:9,m2(:)) ]
```

The command text in the input cells changes to a fixed-point font, and the cells are delimited by special bold square brackets. (The brackets do not appear when the document is printed.) If the **Notebook/Evaluate M-book** menu item is then selected, the application connects to an open MATLAB session (or starts a new MATLAB session), evaluates the MATLAB statements, and inserts the results into the M-book:

MATLAB Notebook Example 1

Assignment #2

```
Create a 3-by-3 magic square


    [m=magic(3) ]
```

```
[m =

    8       1       6
    3       5       7
    4       9       2 ]
```

square the elements of the magic square

```
    [m2=m.^2 ]
```

```
[m2 =

    64       1      36
     9      25      49
    16      81       4 ]
```

and plot the result

```
    [plot(1:9,m2(:)) ]
```

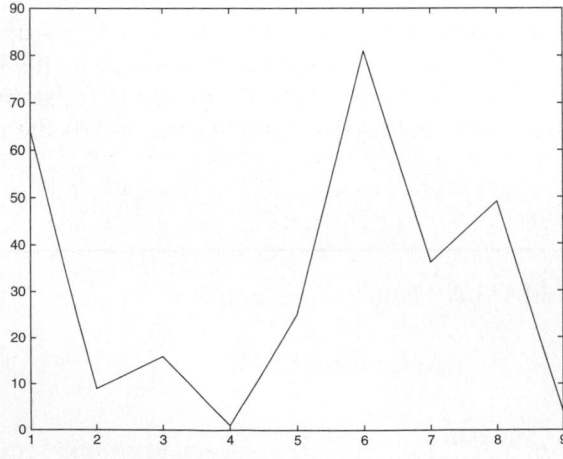

The Notebook can be used to create an annotated record of a MATLAB session (think of it as a *very* fancy diary command) or to insert MATLAB examples into a report. Any changes made to an input cell are reflected in the output cell whenever the input cell is evaluated. In that respect, an M-book is a dynamic document.

All M-books in a Word session share a single server instance of MATLAB, and all share the same MATLAB workspace. Thus, if you have multiple documents open, all variables are shared among all M-books; a change to a variable in one M-book is a change to variables of the same name in other M-books. Also, since you can define and undefine input cells and calculation zones, data dependencies can be easily broken by these changes. The **Purge Output Cells** and **Evaluate M-book** menu items should be selected periodically to ensure data consistency. When you evaluate any input cell or an entire M-book, text output is replaced, but graphical output is replicated. For example, if the example M-book document is evaluated a second time without first purging the output cells, the document will contain two graphs at the end.

36.4 MATLAB COM-RELATED TOOLBOXES

The MathWorks Inc. sells a product called *Excel Link* as an extra-cost toolbox that provides functionality similar to that of the MATLAB Notebook for the Microsoft Excel spreadsheet application. Data exchange and MATLAB commands can be easily controlled from within Excel. If your work requires moving large amounts of data between Excel and MATLAB or using MATLAB to perform complex calculations on Excel data, the time saved by purchasing *Excel Link* may be well worth the cost of the toolbox.

Other COM-related toolboxes are available as extra-cost additions to MATLAB. MATLAB COM Builder can be used to convert MATLAB algorithms to Common Object Model (COM) objects that are accessible from any COM-based application. MATLAB Excel Builder converts complex MATLAB algorithms into independent Excel add-ins. The MATLAB Report Generator lets you create standard and customized reports from your MATLAB models and organize data in multiple output formats, including HTML, RTF, XML, and SGML. All of these toolboxes consist of collections of functions built upon the techniques and functions previously described.

36.5 SUMMARY

The simplicity of using pipes to connect the standard input and standard output of applications on UNIX and Linux platforms has enabled widespread interapplication communication for many years. Standardized protocols for communication between applications on the Windows PC platform are now available as well.

MATLAB supports the Microsoft Dynamic Data Exchange and COM protocols to enable MATLAB to operate as a client or as a server for other DDE or COM-enabled applications, such as Microsoft Word, Excel, and PowerPoint. Extra-cost toolboxes are available to enhance your ability to easily integrate MATLAB with other COM-enabled applications.

37

Getting Help

With each major release of the software, the help capabilities of MATLAB have improved significantly. MATLAB 7 is no exception. Historically, help text was simply displayed in the *Command* window by using the function `help` and `lookfor`. As versions have changed, old help functions have changed, too. Some have become obsolete, and others that have been grandfathered now do different things. This chapter introduces the primary help capabilities available in MATLAB, including resources available via the Internet.

37.1 *COMMAND* WINDOW HELP

Before MATLAB included GUI features, help was available in the *Command* window by using the functions `help` and `lookfor`. These basic functions remain available today. For example, the code

```
>> help sqrt

 SQRT   Square root.
    SQRT(X) is the square root of the elements of X. Complex
    results are produced if X is not positive.

    See also SQRTM.
```

shows the help text for the function `sqrt`. In general, the function `help` is useful if you know the name of a particular function, but are unsure of its input or output arguments. In the previous example, function names are capitalized only to give them visual distinction. All functions in MATLAB are case sensitive. For example, the code

```
>> SQRT(2)
??? Undefined command/function 'SQRT'
```

shows that MATLAB does not recognize SQRT as the square root function.

If you are unsure which function performs a given task, but you know a keyword that is related to the function, the function lookfor is helpful:

```
>> lookfor inverse
INVHILB Inverse Hilbert matrix.
ACOS    Inverse cosine.
ACOSH   Inverse hyperbolic cosine.
ACOT    Inverse cotangent.
ACOTH   Inverse hyperbolic cotangent.
ACSC    Inverse cosecant.
ACSCH   Inverse hyperbolic cosecant.
ASEC    Inverse secant.
ASECH   Inverse hyperbolic secant.
ASIN    Inverse sine.
ASINH   Inverse hyperbolic sine.
ATAN    Inverse tangent.
ATAN2   Four quadrant inverse tangent.
ATANH   Inverse hyperbolic tangent.
ERFCINV Inverse complementary error function.
ERFINV Inverse error function.
INV     Matrix inverse.
PINV    Pseudoinverse.
IFFT Inverse discrete Fourier transform.
IFFT2 Two-dimensional inverse discrete Fourier transform.
IFFTN N-dimensional inverse discrete Fourier transform.
IFFTSHIFT Inverse FFT shift.
IPERMUTE Inverse permute array dimensions.
UPDHESS Performs the Inverse Hessian Update.
INVHESS Inverse of an upper Hessenberg matrix.
```

The function lookfor opens all of the function M-files on the MATLAB search path, looks for the given keyword on the first comment line (i.e., the H1 help line), and returns the lines where a match is found.

37.2 THE HELP BROWSER

In newer versions of MATLAB, help functions beyond help and lookfor have been consolidated in the Help Browser or simply in the *Help* window. This window can be opened by choosing **Help** from the **Window** menu on the MATLAB desktop or by typing helpwin, helpdesk, or doc at the MATLAB prompt. In addition to supplying a text area for displaying help text, the *Help* window offers a help navigator that contains user-selectable tabs for *Contents, Index, Search,* and *Demos.* The *Contents* tab provides a table of contents to all on-line documentation for MATLAB and all toolboxes. The *Index* tab provides an index to all on-line help entries. The *Search* tab allows you to search on-line documentation. Finally, the *Demos* tab provides an interface to MATLAB demo functions. These features are easier to explore firsthand than they are to illustrate in a text, so users are encouraged to explore the Help Browser for themselves.

The *Help* window equivalent to the function help is helpwin. For example, the code

```
>> helpwin mmpadd
```

displays the same help text as help mmpadd, but it does so in the *Help* window. Behind the scenes, MATLAB opens the file mmpadd.m, reads the help text found there, converts the file to HTML format, and displays the HTML text in the *Help* window. In the process, capitalized function names are converted to lowercase and functions listed on the See also line are converted to HTML links to the named functions. In addition, if the function has more extensive on-line documentation, a link to that documentation appears at the top of the displayed help text. If you create your own toolbox or collection of function M-files, helpwin will display help text (as described here) for these functions as well. If the argument to helpwin is the name of a toolbox directory, the Contents.m file in that directory is opened and displayed in the *Help* window.

The function doc bypasses M-file help text and goes directly to on-line documentation. For example, the code

```
>> doc print
```

displays the on-line documentation for the function print, which contains much more information than is available from helpwin print.

The functions whatsnew and whatsnew *toolbox* display release notes and last-minute changes for MATLAB or a selected toolbox in the *Help* window. Behind the scenes, whatsnew *toolbox* opens and displays the Readme.m file for the toolbox in the *Help* window.

37.3 INTERNET RESOURCES

The Mathworks Inc., the makers of MATLAB, was one of the first 100 commercial sites on the Internet. Their website, `http://www.mathworks.com`, provides a wealth of information covering all aspects of MATLAB. There is simply too much on their website to discuss here. In addition, as with most sites, features and links come and go, making it difficult to accurately describe what is available at any given time. Two of the most useful features of *The Mathworks Inc.* website are the solution search engine, which provides answers to common technical support questions, and the repository of user-contributed and Mathworks-contributed function M-files, called *MATLAB Central*. You should be able to find *Mastering MATLAB 7* listed on their website as well.

If you are in the middle of a MATLAB session, you need not leave MATLAB to go to their website. The MATLAB Help Browser accepts URLs and serves as a simple web browser. Alternatively, any other web browser will do.

In addition to the wealth of information on *The Mathworks Inc.* website, the Internet newsgroup `comp.soft-sys.matlab` is an unmoderated forum for discussion of MATLAB topics. Access to this newsgroup requires software and access to newsgroups. Alternatively, the newsgroup can be accessed through *MATLAB Central*. Many very knowledgeable people monitor the MATLAB newsgroup regularly and contribute answers to posted questions. The MATLAB newsgroup is a good place to ask questions that are not easily answered through on-line help. Before posting to the newsgroup, it is beneficial to follow conventional newsgroup etiquette. In addition, many common questions can be answered by consulting the newsgroup Frequently-Asked-Questions (FAQ) list, which can be found at the URL `www.mit.edu/~pwb/cssm/matlab-faq.html`.

37.4 *MASTERING MATLAB 7* HELP

In addition to the help capabilities available in MATLAB, the authors of this text provide two services. First, we maintain the *Mastering MATLAB* website at `http://www.eece.maine.edu/mm`. At this website, you will find a downloadable collection of script M-files to generate all of the figures in the text. These M-files spare you from having to type all of the code found in the text into the *Command* window. In addition, the Java, MEX, and function M-file code that appears in the text is available for downloading. You can also find a listing of errata for the text, as well as links to other MATLAB-related areas on the Internet. Second, the authors provide an email address, `mm@eece.maine.edu`, for your comments and feedback. We encourage you to submit errata, to give us feedback that will help us make the next edition of the text better, and to ask questions about the examples and content of *Mastering MATLAB 7*. You may submit general questions about MATLAB as well, but we might not provide an answer if that answer can be found easily by using the documentation that comes with MATLAB, in the MATLAB FAQ, or in the MATLAB newsgroup archive.

37.5 SUMMARY

The following table summarizes the important aspects of this chapter:

Item	Description
help *functionname*	Display help for the function functionname in the *Command* window
lookfor *keyword*	Keyword search of M-files, using the string keyword
helpwin *functionname*	Display help for the function functionname in the *Help* window
doc functionname	Display online help for the function functionname in the *Help* window
helpbrowser, helpdesk	Open the *Help* window to the documentation homepage
whatsnew	Display Release Notes or Readme.m toolbox file
http://www.mathworks.com	*The Mathworks Inc.* website
comp.soft-sys.matlab	The MATLAB newsgroup
www.mit.edu/~pwb/cssm/matlab-faq.html	The MATLAB newsgroup FAQ
http://www.eece.maine.edu/mm	*Mastering MATLAB 7* website
mm@eece.maine.edu	E-mail to *Mastering MATLAB 7* authors

38

Examples, Examples, Examples

It's been said that a picture is worth a thousand words. Likewise, when it comes to software, an example is worth a thousand words. This chapter is devoted to extensive examples. Some of the examples demonstrate *vectorization*, which is the process of writing code that maximizes the use of array operations. Other examples illustrate creating code that maximizes the benefits achieved from MATLAB's *JIT-Accelerator* features. Still other examples demonstrate the solution of typical problems. Before considering the examples, it is beneficial to introduce vectorization and JIT-acceleration.

38.1 VECTORIZATION

Vectorization means to write or rewrite code so that scalar operations on array elements are replaced by array operations. For example, the code

```
>> for i=1:n
      y(i) = sin(2*pi*i/m);
   end
```

is replaced by the *vectorized* code

```
>> i = 1:n;
>> y = sin(2*pi*i/m);
```

The preceding For Loop represents poor programming practice. Not only is the For Loop unnecessary (as shown by its vectorized equivalent), but it's also very slow, since memory is reallocated for the variable y each time through the loop.

Contrary to what some may believe, vectorization does not mean eliminating *all* For Loops. For Loops serve a very useful purpose; after all, MATLAB wouldn't include them as a control flow structure if they didn't. For Loops are a good choice when the loop makes full use of JIT-acceleration features. When that is not possible, For Loops are often a good choice when (1) the amount of code to be interpreted within the loop is small, especially if this code requires substantial floating-point operations; (2) the code within the loop makes minimum calls to M-file functions; (3) no memory is allocated or reallocated after the first pass through the loop; or (4) using a For Loop eliminates the need to create arrays larger than the computer can access with minimum delay. Obviously, the last case is not only platform-dependent, but also dependent on the hardware attributes of an individual platform.

While vectorization leads to efficient MATLAB programming, it does have a down side. That is, vectorized code is often more difficult to read or follow. The previous example is clearly an exception to this fact—the vectorized code is much easier to read than the nonvectorized code. The example represents a simple, easy-to-learn vectorization. Most often, it is the more difficult or less obvious vectorization challenges that lead to code that is more difficult to follow.

Vectorizing code makes use of a small number of MATLAB operators and functions. These operators and functions generally involve the manipulation of indices or the replication of arrays and can be divided into three categories, as shown in the following tables (the first two categories are basic, internal MATLAB capabilities and are therefore fast; the last category consists of optimized M-file code for implementing common array manipulation functions):

Operator	Description
:	Colon notation. n:m creates a row array that starts with n and ends with m. n:inc:m creates a row array that starts with n, counts by inc, and ends at or before m. As an array index, : means take all elements. Also, A(:) on the right-hand side of an equal sign means reshape A as a column vector. On the left-hand side of an equal sign, A(:) means fill contents of A with results of the right-hand side without reallocating memory for A.
.'	Nonconjugate transpose. Exchanges rows for columns.
[ ]	Brackets. Array concatenation. Slower than using cat(...).

Built-in Function	Description
all(x)	True if all elements of x are nonzero
any(x)	True if any elements of x are nonzero
cat(Dim,A,B, ...)	Concatenate A, B, ... along dimension Dim (faster than [ ])

Built-in Function	Description
`cumsum(x)`	Cumulative sum of elements of vector x
`diff(x)`	Difference between elements in x
`end`	Last index; inside array index identifies the last element along given dimension

`find(x),`
`find(x,n)`
`find(x,n, 'last')`
	Find indices where x is nonzero (usually slower than using the logical argument x directly for array addressing)

`logical(x)`	Convert x to logical data type to enable logical array addressing
`permute(A,Order)`	Generalized transpose; rearrange the dimensions of A so that they are in the order specified by the vector Order
`prod(x)`	Product of elements in x
`reshape(A,r,c)`	Reshape array A to be r-by-c
`sort(x)` `sort(x, 'descend')`	Sort array x in ascending or descending order
`sum(x)`	Sum of elements in x

M-file Functions	Description
`ind2sub(Size,idx)`	Convert single indices in idx to array subscripts of an array having dimensions Size
`ipermute(A,Order)`	Generalized transpose; inverse of `permute(A,Order)`
`kron(A,B)`	Kronecker tensor product of A and B
`meshgrid(x,y)`	Mesh domain generation from vectors x and y
`repmat(A,r,c)`	Replicate array A creating an r-by-c block array
`shiftdim(A,n)`	Shift dimensions of A by integer n
`squeeze(x)`	Remove singleton dimensions from array A
`sub2ind(Size,r,c)`	Convert array subscripts r and c of an array having dimensions Size to single indices

Using the above operators, built-in functions, and M-file functions, vectorization involves the substitution of scalar operations with equivalent array operations when such operations increase execution speed.

38.2 JIT-ACCELERATION

Starting with MATLAB 6.5, improvements were made to the MATLAB interpreter to minimize the processing overhead involved in executing loops. These improvements are collectively known as the *JIT-Accelerator*. As stated in the documentation, these improvements will appear over a series of MATLAB releases. At this time, the JIT-Accelerator offers significant improvements for a subset of MATLAB syntax, data types, and array sizes. MATLAB code containing loops benefits from JIT-acceleration if it has the following features and properties:

1. The loop structure is a For Loop.
2. The loop contains only logical data, character string data, double-precision real data, and less than 64-bit integer data.
3. The loop uses arrays that are three-dimensional or less.
4. All variables within a loop are defined prior to loop execution.
5. Memory for all variables within the loop are preallocated, and maintain constant size and data type for all loop iterations.
6. Loop indices are scalar quantities (e.g., the index i in for i=1:N).
7. Only built-in MATLAB functions are called within the loop.
8. Conditional statements using *if-then-else* or *switch-case* constructions involve scalar comparisons.
9. All lines within the block contain no more than one assignment statement.

JIT-acceleration provides the greatest benefit when the arrays addressed within the loop are relatively small. As array sizes increase, computational time increases and the percentage of time spent on processing overhead decreases, thereby leading to less dramatic improvements in overall execution time.

38.3 UP-DOWN SEQUENCE

As a first example, consider the following simple algorithm: Let N be some positive integer. If N is even, divide it by 2. On the other hand, if it is odd, multiply it by 3 and add 1. Repeat until N becomes 1. This algorithm has some interesting properties. It appears to converge to 1 for all numbers N. Some numbers require many iterations to converge. Others, such as $N = 2^m$, converge very quickly. While it is interesting to study the sequence of values generated by different values of N, let's just compute the number of iterations required to achieve convergence.

First, let N be a scalar. That is, let's write the algorithm for a single number. The following script M-file implements algorithm:

```
% updown1.m
% up-down algorithm

N = 25;      % number to test
count = 0;   % iteration count
```

```
while N>1
    if rem(N,2)==0 % even, since division by 2 gives zero remainder
        N = N/2;
        count = count+1;
    else            % odd
        N = (3*N+1)/2;
        count = count+2;
    end
end
count  % display iteration count
```

This code directly implements the algorithm, with one exception. When an odd number is multiplied by 3 and has 1 added, the resulting number is automatically even. As a result, the next pass through the algorithm always divides by 2. Since this always occurs, the divide-by-2 step is included, and the count is incremented by 2 to reflect the fact that two steps are taken.

Next, consider letting N be an array of numbers, for each of which we wish to find the iteration count. The most direct approach is to use a For Loop, as shown in the following script M-file:

```
% updown2.m
% up-down algorithm

Nums = 25:50;  % numbers to test

for i=1:length(Nums)
    N = Nums(i);  % number to test
    count = 0; % iteration count
        while N>1
        if rem(N,2)==0 % even
            N = N/2;
            count = count+1;
        else            % odd
            N = (3*N+1)/2;
```

```
            count = count+2;
        end
    end
    Counts(i) = count;
end
results = [Nums 'Counts']
```

Here, the earlier scalar algorithm appears within a For Loop. At the beginning of the loop, the *i*th element of the vector Nums is copied into N. The algorithm then runs to completion, and the iteration count is copied into the *i*th element of Counts. Finally, all results are displayed. This code segment violates a key memory allocation guideline—namely, the variable Counts is reallocated to a larger size at every pass through the For Loop. As a result, the previous code does not make use of JIT-acceleration.

To fix this problem, Counts must be preallocated as follows:

```
% updown3.m
% up-down algorithm

Nums = 25:50;                     % numbers to test
Counts = zeros(size(Nums));       % preallocate array
N = Nums(1);                      % predefine N data type and dimension
count = 0;                        % predefine count data type and dimension
for i=1:length(Nums)
    N = Nums(i);   % number to test
    count = 0;   % iteration count
        while N>1
        if rem(N,2)==0 % even
            N = N/2;
            count = count+1;
        else               % odd
            N = (3*N+1)/2;
            count = count+2;
```

```
      end
    end
    Counts(i) = count;
end
results = [Nums' Counts']
```

Now, every time through, the For Loop simply inserts the current count into a pre-existing location in Counts. Preallocation is always a first and most important step in any code optimization process. In addition, since N and count are used within the For Loop, their data types and dimensions are predefined. With these changes, updown3.m makes full use of JIT-acceleration.

For comparison purposes, it is beneficial to develop a vectorized version of this algorithm. In a vectorized version, all input data must be processed simultaneously. This eliminates the For Loop. The function rem returns an array the same size as its input, and so all numbers can be tested simultaneously at every iteration of the While Loop. Using this fact leads to the following script M-file:

```
% updown4.m
% up-down algorithm


Nums = 25:50; % numbers to test
N = Nums;                   % duplicate numbers
Counts = zeros(size(N)); % preallocate array


not1 = N>1;                       % True for numbers greater than one


while any(not1)

   odd = rem(N,2)~=0;          % True for odd values

   odd_not1 = odd & not1;      % True for odd values greater than one
   even_not1 = ~odd & not1;    % True for even values greater than one

   N(even_not1) = N(even_not1)/2;        % Process evens
   Counts(even_not1) = Counts(even_not1)+1;
```

```
    N(odd_not1) = (3*N(odd_not1)+1)/2;          % Process odds
    Counts(odd_not1) = Counts(odd_not1)+2;


    not1 = N>1;                      % Find remaining numbers not converged
end
results = [Nums' Counts']
```

When you consider all elements simultaneously, you must find a way to operate only on array elements that have not converged to 1. The statement not1 = N>1; logically identifies all elements of N that haven't converged. Inside of the While Loop, the nonconverged odd and even values are identified by the odd_not1 and even_not1 logical variables, respectively. With the use of these variables, the corresponding elements of N are processed, and the Counts values are updated.

This vectorized solution makes use of logical array addressing. Addressing by using numerical indices could have been done as well by using the find function; for example, the odd indices are given by find(odd_not1). However, use of this function adds more statements to the solution, which makes the code run slower.

Running the profiler on this solution shows that the greatest percentage of time is consumed by the odd = rem(N,2)~=0; statement. There are two reasons for this. First, this statement finds the remainder for all elements of N every time, not just the remaining of nonconverged numbers. For example, if there are 1000 data points in N and only 10 of them are nonconverged, the remainder is computed for 990 elements in N that are not required. Second, the remainder function incurs overhead because it is a function call and because it performs internal error checking. Both of these reasons for the time consumed by the call to the rem function can be addressed. Eliminating the cause of the first reason requires removing the converged values from N as they appear. Implementing this requires substantial complexity. However, the the rem function overhead can be minimized by replacing it with its definition, as shown in the following code:

```
% updown5.m
% up-down algorithm

Nums = 25:50;        % numbers to test

N = Nums;                    % duplicate numbers
```

```
Counts = zeros(size(N)); % preallocate array

not1 = N>1;                      % True for numbers greater than one

while any(not1)

    odd = (N-2*fix(N/2))~=0;     % True for odd values

    odd_not1 = odd & not1;       % True for odd values greater than one
    even_not1 = ~odd & not1;     % True for even values greater than one

    N(even_not1) = N(even_not1)/2;       % Process evens
    Counts(even_not1) = Counts(even_not1)+1;

    N(odd_not1) = (3*N(odd_not1)+1)/2;       % Process odds
    Counts(odd_not1) = Counts(odd_not1)+2;

    not1 = N>1;                      % Find remaining numbers
end
results = [Nums' Counts']
```

Running the profiler on this example shows that $(N-2*fix(N/2))$ is generally faster rem$(N,2)$.

As a final test, it is worth considering the use of integer arithmetic. Since N and Counts are always integers, it is possible to cast them into integer arrays. In this case, $(N-2*fix(N/2))$ no longer works as a replacement for rem$(N,2)$. However, odd = $2*(N/2)$~=N; is an alternative test for odd integers. The code shown next illustrates this integer implementation:

```
% updown6.m
% up-down algorithm

Nums = 25:50; % numbers to test

N = uint32(Nums);                    % duplicate numbers as uint32
```

```
Counts = zeros(size(N),'uint32'); % preallocate array as uint32

not1 = N>1;                       % True for numbers greater than one

while any(not1)

   odd = 2*(N/2)~=N;              % True for odd values

   odd_not1 = odd & not1;         % True for odd values greater than one
   even_not1 = ~odd & not1;       % True for even values greater than one


   N(even_not1) = N(even_not1)/2;          % Process evens
   Counts(even_not1) = Counts(even_not1)+1;


   N(odd_not1) = (3*N(odd_not1)+1)/2;      % Process odds
   Counts(odd_not1) = Counts(odd_not1)+2;


   not1=N>1;                      % Find remaining numbers
end
results = [Nums' Counts']
```

As a comparison, the last four implementations were timed with the profiler by using Nums = 1:2049; and with the last line results = [Nums' Counts'] removed. The following table documents the timing results:

Code	Relative Execution Time	Comments
updown3	1.0	Scalar operations using JIT-acceleration
updown5	1.2	Vectorized operations **without** the function rem
updown4	1.5	Vectorized operations with the function rem
updown6	2.3	Vectorized operations, using uint32 integers and without the function rem

These results clearly show the power of the JIT-accelerator. Without it, updown3 would have been up to an order of magnitude slower than the vectorized cases. Now it is the fastest. More surprising is the fact that the integer arithmetic case updown6 was slowest, even though it is otherwise identical to updown5.

It is important to note that the preceding timing performance is representative of data computed on the author's computer using the prerelease version of MATLAB 7. Every profiler run produces slightly different results. To generate your own timing information, you can download this code to your computer from the *Mastering MATLAB* website at *http://www.eece.maine.edu/mm*.

38.4 VANDERMONDE MATRIX

There are a number of numerical linear algebra problems that require the generation of a Vandermonde matrix. For a vector x having n elements, a Vandermonde matrix has the following form:

$$
V = \begin{bmatrix}
x_1^m & x_1^{m-1} & \cdots & x_1 & 1 \\
x_2^m & x_2^{m-1} & \cdots & x_2 & 1 \\
\vdots & \vdots & \ddots & \vdots & \vdots \\
x_n^m & x_n^{m-1} & \cdots & x_n & 1
\end{bmatrix}
$$

As shown, the columns of V are element-by-element powers of the components of x. Let's consider a variety of approaches to constructing this matrix.

The first approach that comes to mind is the straightforward application of a For Loop, as in the following script M-file:

```
% vander1.m
% construct a Vandermonde matrix.

x = (1:6)';     % column vector for input data
m = 5;          % highest power to compute
V = [];

for i = 1:m+1   % build V column by column
    V = [V x.^(m+1-i)];
end
```

This approach builds V column by column, starting from an empty matrix. There are a number of weaknesses in this implementation, the most obvious being that memory is reallocated for V each time through the loop. So the first vectorization step is to preallocate V, as shown in the following M-file:

```
% vander2.m
% construct a Vandermonde matrix.

x = (1:6)';         % column vector for input data
m = 5;              % highest power to compute
n = length(x);      % number of elements in x
V = ones(n,m+1);    % preallocate memory for result

for i=0:m-1         % build V column by column
    V(:,i+1) = x.^(m-i);
end
```

Here, V is initialized as a matrix containing all ones. Then, the individual columns of V are assigned within the For Loop. The last column is not assigned in the For Loop, since it already contains ones and there is no use in computing x.^0. There are still two problems with the preceding code. First, the columns of V are explicitly computed without making use of prior columns; second, the For Loop should be able to be eliminated. The following script M-file solves the first problem:

```
% vander3.m
% construct a Vandermonde matrix.

x = (1:6)';         % column vector for input data
m = 5;              % highest power to compute
n = length(x);      % number of elements in x
V = ones(n,m+1);    % preallocate memory for result

for i=m:-1:1        % build V column by column
    V(:,i) = x.*V(:,i+1);
end
```

Now the columns of V are assigned, starting with the second-to-last column and proceeding backward to the first. This is done because the ith column of V is equal to the $(i + 1)$th column multiplied elementwise by x. This is the implementation found in the MATLAB functions polyfit and vander.

At this point, the preceding implementation cannot be optimized further without eliminating the For Loop. Eliminating the For Loop requires some ingenuity and a lot of familiarity with the functions in MATLAB. Using the array manipulation tables found earlier in this chapter, the functions `repmat` and `cumprod` offer some promise. The following script M-file demonstrates an approach that uses `repmat`:

```
% vander4.m
% construct a Vandermonde matrix.

x = (1:6)';        % column vector for input data
m = 5;             % highest power to compute
n = length(x);     % number of elements in x

p = m:-1:0;        % column powers

V = repmat(x,1,m+1).^repmat(p,n,1);
```

This implementation uses `repmat` twice, once to replicate x, creating a matrix of m+1 columns each containing x, and the second time to create a matrix containing the powers to be applied to each element of the matrix containing x. Given these two matrices, element-by-element exponentiation is used to create the desired result. As with `vander2.m`, this implementation explicitly computes each column without using information from other columns. The function `cumprod` solves this problem, as shown in the following script M-file:

```
% vander5.m
% construct a Vandermonde matrix.

x = (1:6)';        % column vector for input data
m = 5;             % highest power to compute
n = length(x);     % number of elements in x

V = ones(n,m+1);   % preallocate memory for result

V(:,2:end) = cumprod(repmat(x,1,m),2);

V = V(:,m+1:-1:1); % reverse column order
```

Here, the function cumprod is used to compute the columns of V after using repmat to duplicate x. Since cumprod proceeds from left to right, the final result is found by reversing the columns of V. This implementation uses only one M-file function, repmat. Eliminating this function by array addressing should lead to the fastest possible implementation. Doing so leads to the following script M-file:

```
% vander6.m
% construct a Vandermonde matrix.

x = (1:6)';        % column vector for input data
m = 5;             % highest power to compute
n = length(x);     % number of elements in x

V = ones(n,m+1); % preallocate memory for result

V(:,2:end) = cumprod(x(:,ones(1,m)),2); % avoid call to repmat

V = V(:,m+1:-1:1); % reverse column order
```

Among these six implementations, vander3 and vander6 are the fastest. The MATLAB functions polyfit and vander employ the same algrorithm as vander3, which is easier to read and requires less memory than vander6. All of the other algorithms are not as efficient, from either a memory allocation or computational point of view.

38.5 REPEATED VALUE CREATION AND COUNTING

This section considers the following problem: Given a vector x containing data and a vector n of equal length containing nonnegative integers, construct a vector where $x(i)$ is repeated $n(i)$ times for every ith element in the two vectors. For example, $x = [3\,2\,0\,5\,6]$ and $n = [2\,0\,3\,1\,2]$ would produce the result $y = [3\,3\,0\,0\,0\,5\,6\,6]$. Note that since $n(2)$ is zero, $x(2)$ does not appear in the result y.

In addition to repeated value creation, the inverse problem of identifying and counting repeated values is also of interest. Thus, given the previous example y, find the vectors x and n that describe it.

Consider repeated value creation first. A scalar approach is straightforward, as shown in the following script M-file:

```
% repeat1.m
% repeated value creation and counting

x = [3 2 0 5 6];  % data to repeat
n = [2 0 3 1 2];  % repeat counts

y = [];
for i=1:length(x)

    y = [y repmat(x(i),1,n(i))];
end
```

In this implementation, the result is built by using brackets to concatenate the repeated values next to each other. JIT-acceleration does not work with this example, because memory is reallocated for y every loop iteration and the M-file function `repmat` appears in the loop. These issues can be resolved by preallocating y and by using indexing, as shown in the following script M-file:

```
% repeat2.m
% repeated value creation and counting

x = [3 2 0 5 6];  % data to repeat
n = [2 0 3 1 2];  % repeat counts

nz = n==0;              % locations of zero elements
n(nz) = [];             % eliminate zero counts
x(nz) = [];             % eliminate corresponding data
y = zeros(1,sum(n));    % preallocate output array

idx = 1;                % pointer into y
for i=1:length(x)

    y(idx:idx+n(i)-1) = x(i); % fill y using scalar expansion
    idx = idx+n(i);           % next pointer location
end
```

In the preceding code, zero counts are eliminated prior to the For Loop, since they do not contribute to the result. In this algorithm, sum(n) is the total number of elements in the result, and the variable idx is used to identify where the next data is to be placed in y.

Once again, it is worthwhile to consider a vectorized solution to this problem. To determine how to proceed, it is beneficial to look at how x is related to y. After eliminating zero counts in the example being considered, x, n, and y are as follows:

```
>> x

x =

      3      0      5      6
>> n

n =

      2      3      1      2
>> y

y =

      3      3      0      0      0      5      6      6
```

If we can create an index vector idx = [1 1 2 2 2 3 4 4], then x is related to y as

```
>> idx = [1 1 2 2 2 3 4 4];
>> y = x(idx)

y =

      3      3      0      0      0      5      6      6
```

So, rather than concentrate on getting the values of x into the correct places in y, if we can generate an index vector for x with the desired values, finding y simplifies to just one statement, y = x(idx). This is a common situation in vectorizing. The indices are often more important than the data.

The relationship between n and idx is straightforward. In addition, since idx looks like a cumulative sum, there's a chance that the cumsum function will be useful. If we can generate an array of ones and zeros at the indices where idx changes value, then idx is indeed a cumulative sum, as in the following example:

```
>> tmp = [1 0 1 0 0 1 1 0]

tmp =

      1      0      1      0      0      1      1      0
>> idx = cumsum(tmp)

idx =

      1      1      2      2      2      3      4      4
```

The nonzero values in tmp are related to n. To discover this relationship, look at its cumulative sum:

```
>> csn = cumsum(n)
csn =
     2    5    6    8
```

If the last value in csn is discarded and 1 is added to the remaining values, then the indices of all 1s in tmp are known, except for the first, which is always 1. Therefore, the indices of all 1s can be computed:

```
>> tmp2 = [1 csn(1:end-1)+1]
tmp2 =
     1    3    6    7
```

That's it. The values in tmp2 identify the 1s to be placed in tmp. All other values in tmp are 0. Now, all that remains is creating tmp. This is easily done, as in the following example:

```
>> tmp = zeros(1,csn(end)) % preallocate with all zeros
tmp =
     0    0    0    0    0    0    0    0
>> tmp(tmp2) = 1            % poke in ones with scalar expansion
tmp =
     1    0    1    0    0    1    1    0
>> idx = cumsum(tmp)        % form the desired cumsum
idx =
     1    1    2    2    2    3    4    4
>> y = x(idx)              % idx does the rest!
y =
     3    3    0    0    0    5    6    6
```

Using this approach, a vectorized implementation of creating repeated values is shown in the following script M-file:

```
% repeat3.m
% repeated value creation and counting

x = [3 2 0 5 6];  % data to repeat
```

```
n = [2 0 3 1 2];   % repeat counts

nz = n==0;                  % locations of zero elements
n(nz) = [];                 % eliminate zero counts
x(nz) = [];                 % eliminate corresponding data
csn = cumsum(n);                % cumulative sum of counts
tmp = zeros(1,csn(end));        % preallocate memory
tmp([1 csn(1:end-1)+1]) = 1;    % poke in ones
idx = cumsum(tmp);              % index vector
y = x(idx);                     % let array indexing do the work
```

As a comparison, these implementations were timed with the profiler, using larger x and n data sets than shown in the implementations. The following table documents the timing results:

Code	Relative Execution Time	Comments
repeat3	1.0	Vectorized solution
repeat2	4.0	Scalar operations using JIT-acceleration
repeat1	55	Nonoptimized solution without variable preallocation

In this case, the vectorized solution is much faster than the JIT-accelerated solution. This is the opposite conclusion as that reached in the Vandermonde matrix construction example illustrated earlier in this chapter. The downside of the vectorized solution is that the code is not transparent. It is almost impossible to tell what the vectorized code does without actually executing it and monitoring the intermediate variables.

Let's move on to the inverse of the above algorithm, namely, repeated value identification and counting. That is, starting with y, find x and n (except for any zero-count values, of course). Again, the most obvious solution is nonvectorized and uses a For Loop, as shown in the following script M-file:

```
% repeat4.m
% repeated value creation and counting
% inverse operation
```

```
y = [3 3 0 0 0 5 6 6]; % data to examine

x = y(1);                  % beginning data
n = 1;                     % beginning count
idx = 1;                   % index value
for i=2:length(y)
   if y(i)==x(idx) % value matches current x

      n(idx) = n(idx)+1; % increment current count

   else % new value found

      idx=idx+1;           % increment index
      x(idx) = y(i);       % poke in new x
      n(idx) = 1;          % start new count
   end
end
```

Here, a simple If-Else-End construction is used to decide if a particular element of y is a member of the current repeated value. If it is, the count is incremented. If it's not, a new repeat value is created. Although it is not obvious, each time the Else section is executed, memory is reallocated for x and n. As a result, this solution does not benefit from JIT-acceleration.

The next step is to use preallocation, as shown in the following script M-file:

```
% repeat5.m
% repeated value creation and counting
% inverse operation

y = [3 3 0 0 0 5 6 6]; % data to examine

x = zeros(size(y));      % preallocate results
n = zeros(size(y));

x(1) = y(1);                  % beginning data
```

```
n(1) = 1;                   % beginning count
idx = 1;                    % index value

for i=2:length(y)

    if y(i)==x(idx) % value matches current x

        n(idx) = n(idx)+1; % increment current count

    else % new value found
        idx = idx+1;        % increment index
        x(idx) = y(i);      % poke in new x
        n(idx) = 1;         % start new count
    end
end
nz = (n==0);  % find elements not used
x(nz) = [];   % delete excess allocations
n(nz) = [];
```

Since the lengths of x and n are unknown (but they cannot be any longer than y) x and n are preallocated to have the same size as y. At the end, the excess memory allocations are discarded.

Eliminating the For Loop to vectorize this algorithm requires study of an example. Consider the y vector that was the result of applying the creation algorithm to the original x and n data:

```
>> y
y =
      3     3     0     0     0     5     6     6
```

Because of the structure of this vector, the MATLAB function diff must be useful:

```
>> diff(y)
ans =
      0    -3     0     0     5     1     0
```

If this vector is shifted one element to the right, the nonzero elements line up with places in y that represent new repeated values. In addition, the first element is always a new repeated value. So, using logical operations, the repeated values can be identified, as in the following code:

```
>> y
y =
     3     3     0     0     0     5     6     6
>> tmp = [1 diff(y)]~=0
tmp =
     1     0     1     0     0     1     1     0
```

Given the logical variable tmp, the vector x is found by logical addressing:

```
>> x = y(tmp)
x =
     3     0     5     6
```

Finding the repeat counts n associated with each value in x takes further study. The repeat counts are equal to the distance between the ones in tmp. Therefore, finding the indices associated with tmp is useful, as the following example shows:

```
>> find(tmp)
ans =
     1     3     6     7
>> diff(ans)
ans =
     2     3     1
```

The difference in indices gives the repeat count for all but the last repeated element. The function diff misses this count because there is no marker identifying the end of the array, which marks the end of the last repeated value. Appending one True value to tmp solves this problem:

```
>> find([tmp 1])
ans =
     1     3     6     7     9
>> n = diff(ans)
n =
     2     3     1     2
```

The inverse algorithm is now known. Implementing it leads to the following script file:

```
% repeat6.m
% repeated value creation and counting
% inverse operation

y = [3 3 0 0 0 5 6 6]; % data to examine

tmp = ([1 diff(y)]~=0);
x = y(tmp);
n = diff(find([tmp 1]));
```

The preceding implementation demonstrates the compactness of vectorization. This solution requires 3 lines of code, compared to the 17 used in repeat5.m. This implementation also demonstrates the difficulty often encountered in reading vectorized code. These 3 lines are essentially meaningless, unless you execute and view the results of each line with a simple example.

A comparison of these three algorithms by using the profiler shows that repeat4 is generally the slowest and that repeat5 and repeat6 are essentially equal in performance. If there are few repeated values, the lack of preallocation does not slow repeat4 much. On the other hand, the deletion of the extra elements in the last three lines of repeat5 can be time consuming if the size of y is much greater than the size of x.

Before concluding, it is important to note that the preceding implementations fail if y contains any Inf or NaN elements. The function diff returns NaN for differences between Inf elements, as well as for differences containing NaN elements. Because of the utility of creating and counting repeated values, these algorithm encapsulated in the function mmrepeat shown as follows:

```
function [y,m]=mmrepeat(x,n)
%MMREPEAT Repeat or Count Repeated Values in a Vector.
% MMREPEAT(X,N) returns a vector formed from X where X(i) is repeated
% N(i) times. If N is a scalar it is applied to all elements of X.
% N must contain nonnegative integers. N must be a scalar or have the same
% length as X.
%
% For example, MMREPEAT([1 2 3 4],[2 3 1 0]) returns the vector
% [1 1  2 2 2  3]    (extra spaces added for clarity)
%
```

```
% [X,N]=MMREPEAT(Y) counts the consecutive repeated values in Y returning
% the values in X and the counts in N. Y=MMREPEAT(X,N) and [X,N]=MMREPEAT(Y)
% are inverses of each other if N contains no zeros and X contains unique
% elements.

if nargin==2 % MMREPEAT(X,N) MMREPEAT(X,N) MMREPEAT(X,N) MMREPEAT(X,N)
    xlen=length(x);
    nlen=length(n);
    if ndims(x)~=2 | numel(x)~=xlen
       error('X Must be a Vector.')
    else
       [r,c]=size(x);
    end
    if any(n<0) | any(fix(n)~=n)
       error('N Must Contain NonNegative Integers.')
    end
    if ndims(n)~=2 | numel(n)~=nlen | (nlen>1 & nlen~=xlen)
        error('N Must be a Scalar or Vector the Same Size as X.')
    end
    x=reshape(x,1,xlen); % make x a row vector


    if nlen==1 % scalar n case, repeat all elements the same amount
       if n==0 % quick exit for special case
          y=[];
          return
       end
       y=x(ones(1,n),:); % duplicate x to make n rows each containing x
       y=y(:);            % stack each column into a single column
       if r==1            % input was a row so return a row
          y=y.';
       end
    else % vector n case
```

```
        iz=find(n~=0);        % take out elements to be repeated zero times
        x=x(iz);
        n=n(iz);
        csn=cumsum(n);
        y=zeros(1,csn(end)); % preallocate temp/output variable
        y(csn(1:end-1)+1)=1; % mark indices where values increment
        y(1)=1;               % poke in first index
        y=x(cumsum(y));       % use cumsum to set indices
        if c==1               % input was a column so return a column
            y-y.';
        end
    end
elseif nargin==1 % MMREPEAT(Y) MMREPEAT(Y) MMREPEAT(Y) MMREPEAT(Y)
    xlen=length(x);
    if ndims(x).-=2 | prod(size(x)).-=xlen
        error('Y Must be a Vector.')
    else
        [r,c]=size(x);
    end
    x=reshape(x,1,xlen); % make x a row vector
    xnan=isnan(x);
    xinf=isinf(x);
    if any(xnan|xinf) % handle case with exceptions
        ntmp=sum(rand(1,4))*sqrt(realmax); % replacement for nan's
        itmp=1/ntmp;                       % replacement for inf's
        x(xnan)=ntmp;
        x(xinf)=itmp.*sign(x(xinf));
        y=[1 diff(x)]~=0;         % places where distinct values begin
        m=diff([find(y) xlen+1]); % counts
        x(xnan)=nan;              % poke nan's and inf's back in
        x(xinf)=inf*x(xinf);
```

```
        else % x contains only algebraic numbers
            y=[1 diff(x)]~=0;            % places where distinct values begin
            m=diff([find(y) xlen+1]); % counts
        end
        y=x(y); % the unique values
        if c==1
            y=y.';
        end
    else
        error('Incorrect Number of Input Arguments.')
    end
```

38.6 DIFFERENTIAL SUMS

This section considers computing the differential sum between elements in an array. That is, if $x = [3\,1\,2\,6\,3\,1\,-1]$, then the differential sum of x is $y = [4\,3\,8\,9\,4\,0]$. This is the dual, or complement, of the MATLAB function diff. Just as with diff, we wish to create a function that works along any dimension of its input, no matter how many dimensions the input has. Before generalizing to the n-D case, let's consider the vector and matrix cases. The vector case is straightforward:

```
>> x = [3 1 2 6 3 1 -1]
x =
      3     1     2     6     3     1    -1
>> x(1:end-1)
ans =
      3     1     2     6     3     1
>> x(2:end)
ans =
      1     2     6     3     1    -1
>> y = x(1:end-1)+x(2:end)
y =
      4     3     8     9     4     0
```

Simple array addressing is all that is required, and the preceding approach works if x is either a row or a column vector. When the input is a matrix, the default action is to perform a differential sum down the columns, which is along the row dimension, as in the following example:

```
>> x = magic(4) % 2-D data
x =
    16     2     3    13
     5    11    10     8
     9     7     6    12
     4    14    15     1
>> y = x(1:end-1,:)+x(2:end,:)
y =
    21    13    13    21
    14    18    16    20
    13    21    21    13
```

It is also desirable to be able to specify an operation across the columns, which is along the column dimension:

```
>> y = x(:,1:end-1)+x(:,2:end)
y =
    18     5    16
    16    21    18
    16    13    18
    18    29    16
```

Here, the row and column indices are reversed from the preceding row operation. This operation, along the column dimension, can be performed by the preceding row operation if x is transposed first, and then transposed again after the operation:

```
>> tmp = x'; % transpose data
>> y = tmp(1:end-1,:)+tmp(2:end,:);
>> y = y'    % transpose result
y =
    18     5    16
    16    21    18
    16    13    18
    18    29    16
```

This can also be accomplished with the *n*-D functions `permute` and `ipermute`, which are generalizations of the transpose operator, as in the following example:

```
>> tmp = permute(x,[2 1])
tmp =
    16     5     9     4
     2    11     7    14
     3    10     6    15
    13     8    12     1
>> y = tmp(1:end-1,:)+tmp(2:end,:);
>> y = ipermute(y,[2 1])
y =
    18     5    16
    16    21    18
    16    13    18
    18    29    16
```

Before extending this to the *n*-D case, consider the 3-D case, since it is relatively easy to visualize:

```
>> x = cat(3,hankel([3 1 6 -1]),pascal(4))
x(:,:,1) =
     3     1     6    -1
     1     6    -1     0
     6    -1     0     0
    -1     0     0     0
x(:,:,2) =
     1     1     1     1
     1     2     3     4
     1     3     6    10
     1     4    10    20
>> y = x(1:end-1,:,:)+x(2:end,:,:) % diff sum along row dimension
y(:,:,1) =
     4     7     5    -1
     7     5    -1     0
     5    -1     0     0
```

```
y(:,:,2) =
```

2	3	4	5
2	5	9	14
2	7	16	30

Note that the same process occurs here, but that added colons are required to reach both pages of x. If the 1:end-1 and 2:end indices are moved to other dimensions, the differential sum moves to that dimension, as in the following example:

```
>> y = x(:,:,1:end-1)+x(:,:,2:end)
y =
```

4	2	7	0
2	8	2	4
7	2	6	10
0	4	10	20

This is the differential sum between the two pages of x.

The previous example points to one way to generalize this algorithm to n-dimensions. The indices into x on the right-hand side are comma-separated lists. Therefore, if we create cell arrays containing the desired indices, the differential sum can be computed by using comma-separated list syntax:

```
>> y = x(1:end-1,:,:)+x(2:end,:,:) % duplicate this case
y(:,:,1) =
```

4	7	5	-1
7	5	-1	0
5	-1	0	0

```
y(:,:,2) =
```

2	3	4	5
2	5	9	14
2	7	16	30

```
>> c1 = {(1:3) ':' ':'}   % first set of indices
c1 =
    [1x3 double]     ':'      ':'
>> c2 = {(2:4) ':' ':'}   % second set of indices
c2 =
    [1x3 double]     ':'      ':'
```

```
>> y = x(c1{:})+x(c2{:})  % use comma separated list syntax
y(:,:,1) =
       4       7       5      -1
       7       5      -1       0
       5      -1       0       0
y(:,:,2) =
       2       3       4       5
       2       5       9      14
       2       7      16      30
```

The preceding example demonstrates the power of comma-separated list syntax. In the example, the keyword **end** could not be used because it has meaning only when used directly as an index to a variable. As a result, c1 and c2 contain the actual numerical indices.

The following script file generalizes this implementation for computing differential sums:

```
% diffsum1.m
% compute differential sum along a given dimension


x = cat(3,hankel([3 1 6 -1]),pascal(4)) % data to test
dim = 1  % dimension to work along


xsiz = size(x)
xdim = ndims(x)


tmp = repmat({':'},1,xdim) % cells of ':'
c1 = tmp;
c1{dim} = 1:xsiz(dim)-1     % poke in 1:end-1
c2 = tmp;
c2(dim) = {2:xsiz(dim)}     % poke in 2:end


y = x(c1{:})+x(c2{:})        % comma separated list syntax
```

With no semicolons at the end of the statements, diffsum1 produces the following output:

```
x(:,:,1) =
     3      1      6     -1
     1      6     -1      0
     6     -1      0      0
    -1      0      0      0
x(:,:,2) =
     1      1      1      1
     1      2      3      4
     1      3      6     10
     1      4     10     20
dim =
     1
xsiz =
     4      4      2
xdim =
     3
tmp =
    ':'     ':'     ':'
c1 =
    [1x3 double]    ':'     ':'
c2 =
    [1x3 double]    ':'     ':'
y(:,:,1) =
     4      7      5     -1
     7      5     -1      0
     5     -1      0      0
y(:,:,2) =
     2      3      4      5
     2      5      9     14
     2      7     16     30
```

Here, x is the input data and dim is the dimension chosen for computing the differential sum. Using information about the size and dimensions of x, repmat can produce a cell array for addressing all elements in all dimensions of x. Then, indices are inserted into the proper cells to create c1 and c2, as shown earlier. Finally, comma-separated list syntax is used to generate the final result.

　　The preceding algorithm is not the only way to compute differential sums for an arbitrary *n*-D array. The functions permute and ipermute can be used to transpose x so that the desired dimension for computing the sum is the row dimension. Applying this procedure to the previous 3-D example gives the following script M-file:

```
% diffsum2.m
% compute differential sum along a given dimension


x = cat(3,hankel([3 1 6 -1]),pascal(4)) % data to test
dim = 3  % dimension to work along


xsiz = size(x);
n = xsiz(dim);                  % size along desired dim
xdim = ndims(x);                % # of dimensions


perm = [dim:xdim 1:dim-1]       % put dim first
x = permute(x,perm)             % permute so dim is row dimension
x = reshape(x,n,[])             % reshape into a 2D array


y = x(1:n-1,:)+x(2:n,:)         % Differential sum along row dimension


xsiz(dim) = n-1                 % new size of dim dimension
y = reshape(y,xsiz(perm))       % put result back in original form
y = ipermute(y,perm)            % inverse permute dimensions
```

Here, the variable perm forms a permutation vector for transposing x so that the differential sum is computed along the row dimension. After permuting, x is reshaped into a 2-D array. For 3-D x, this means block-stacking the pages of x as additional columns. Then, the differential sum is computed, and the array result is reshaped and inverse-permuted to its original shape by using the fact that its size along the chosen dimension has decreased by 1. With no semicolons at the end of the statements, diffsum2 produces the following output:

```
x(:,:,1) =
    3     1     6    -1
    1     6    -1     0
    6    -1     0     0
   -1     0     0     0
x(:,:,2) =
    1     1     1     1
    1     2     3     4
    1     3     6    10
    1     4    10    20
dim =
    3
perm =
    3     1     2
x(:,:,1) =
    3     1     6    -1
    1     1     1     1
x(:,:,2) =
    1     6    -1     0
    1     2     3     4
x(:,:,3) =
    6    -1     0     0
    1     3     6    10
x(:,:,4) =
   -1     0     0     0
    1     4    10    20
x =
  Columns 1 through 12
    3     1     6    -1     1     6    -1     0     6    -1     0     0
    1     1     1     1     1     2     3     4     1     3     6    10
  Columns 13 through 16
   -1     0     0     0
    1     4    10    20
```

```
y =
  Columns 1 through 12
    4    2    7    0    2    8    2    4    7    2    6   10
  Columns 13 through 16
    0    4   10   20
xsiz =
    4    4    1
y(:,:,1) =
    4    2    7    0
y(:,:,2) =
    2    8    2    4
y(:,:,3) =
    7    2    6   10
y(:,:,4) =
    0    4   10   20
y =
    4    2    7    0
    2    8    2    4
    7    2    6   10
    0    4   10   20
```

Of these two approaches to generalizing functions, the *n*-D case (the approach taken in `diffsum2`) appears more often in M-File functions distributed with MATLAB. There is no significant difference in the speed of the approaches.

38.7 STRUCTURE MANIPULATION

Structures are a convenient data structure in MATLAB. They allow you to group associated data into a single variable and use descriptive field names to identify different data contained within the structure. Given the utility and convenience of structures, it is often convenient to pack a group of variables into a single structure, and then later extract them back out. To illustrate this, first consider the process of gathering variables and storing them as fields within a single structure, with field names matching variable names. Performing this task in the *Command* window is straightforward; you simply assign fields to like-named variables, as in the following example:

```
>> a = eye(2) % test data
a =
     1     0
     0     1
>> b = 'String'
b =
String
>> c = cell(2)
c =
     []     []
     []     []
>> y.a = a; % store variables in a structure
>> y.b = b;
>> y.c = c
y =
    a: [2x2 double]
    b: 'String'
    c: {2x2 cell}
```

The inverse of this process is also straightforward. Using the structure y as an example leads to the following statements:

```
>> a = y.a
a =
     1     0
     0     1
>> b = y.b
b =
String
>> c = y.c
c =
     []     []
     []     []
```

Here, the field names of the structure become the variable names.

This process of packing and unpacking variables is encapsulated in the M-file function mmv2struct:

```
function varargout=mmv2struct(varargin)
%MMV2STRUCT Pack/Unpack Variables to/from a Scalar Structure.
% MMV2STRUCT(X,Y,Z,...) returns a structure having fields X,Y,Z,...
% containing the corresponding data stored in X,Y,Z,...
% Inputs that are not variables are stored in fields named ansN
% where N is an integer identifying the Nth unnamed input.
%
% MMV2STRUCT(S)assigns the contents of the fields of the scalar structure
% S to variables in the calling workspace having names equal to the
% corresponding field names.
%
% [A,B,C,...]=MMV2STRUCT(S) assigns the contents of the fields of the
% scalar structure S to the variables A,B,C,... rather than overwriting
% variables in the caller. If there are fewer output variables than
% there are fields in S, the remaining fields are not extracted. Variables
% are assigned in the order given by fieldnames(S).

if nargin==0
    error('Input Arguments Required.')

elseif nargin==1 % Unpack Unpack Unpack Unpack Unpack Unpack Unpack Unpack

    argin = varargin{1};
    if ~isstruct(argin)||length(argin)~=1
        error('Single Input Must be a Scalar Structure.')
    end
    names = fieldnames(argin);

    if nargout==0 % assign in caller

        for i=1:length(names)
            assignin('caller',names{i},argin.(names{i}))
        end
    else % deal fields into variables in caller
```

```
            varargout = cell(1,nargout); % preallocate output
            for i=1:nargout
                varargout{i} = argin.(names{i});
            end
        end
    else % Pack Pack Pack Pack Pack Pack Pack Pack Pack Pack Pack Pack Pack Pack

        args = cell(2,nargin);
        num = 1;

        for i=1:nargin % build cells for call to struct

            args(:,i) = {inputname(i); varargin{i}};

            if isempty(args{1,i})
                args{1,i} = sprintf('ans%d',num);
                num = num+1;
            end
        end
        varargout{1}=struct(args{:}); % create struct using comma-separated list
end
```

Consider the code that unpacks a structure into variables. When called with no output arguments as mmv2struct(structvar), the function uses a For Loop to repeatedly call the function assignin to assign values to variables in the workspace that called mmv2struct. During each For Loop iteration, variables having names that match the corresponding field names of the structure are created. The statement fragment arg.(names{i}) addresses the field name string contained in names{i} of the structure argin, which is equal to structvar. The content of arg.(names{i}) is assigned to a variable named names{i} in the workspace of the calling program.

The code that packs variables into a structure creates a cell array containing the arguments required for the struct function (e.g., struct('field1', values1, 'field2', values2, ...)). For convenience, the arguments to the struct function are created with 'field1', 'field2', etc., as a first row in the variable args, and values1, values2, etc., as a second row in the variable args. Then, args{:} in the struct(arg{:}) statement creates a single row of arguments, as needed by the struct function, by using comma-separated list syntax.

38.8 INVERSE INTERPOLATION

In Chapter 19, Data Interpolation, 1-D interpolation was demonstrated by using the function `interp1(x,y,xi)`. While this function offers a variety of beneficial features, it assumes that there is one y value for each value in xi. If there is not, the function terminates with an error.

In some situations, it is necessary to interpolate in the reverse direction (i.e., to inverse interpolate data), whereby given a value of *y*, the problem is to find all values of *x* where $y = f(x)$. To illustrate this situation, consider the following example code and associated plot:

```
>> x = (1:10).';    % sample data
>> y = cos(pi*x);


>> yo = -0.2;       % inverse interpolation point
>> xol = [x(1); x(end)];
>> yol = [yo; yo];


>> plot(x,y,xol,yol)
>> xlabel X
>> ylabel Y
>> set(gca,'Ytick',[-1 yo 0 1])
>> title('Figure 38.1: Inverse Interpolation')
```

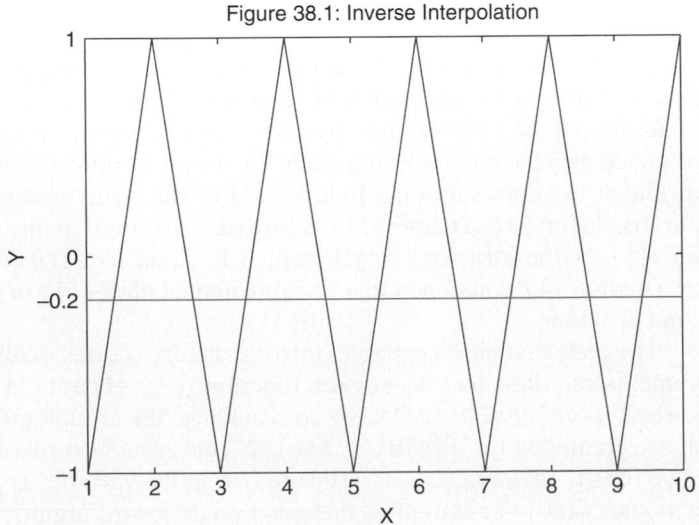

Figure 38.1: Inverse Interpolation

As shown by the triangle waveform in the figure, it is straightforward to use `interp1` to find y values associated with any set of x values, since, for each x value, there is one and only one value of y. On the other hand, in the inverse direction, there are numerous values of x for each value of y between −1 and 1. For example, for the horizontal line drawn yo = −0.2, there are nine corresponding values of y. The function `interp1` cannot be used to find those points:

```
>> interp1(y,x,yo)
??? Error using ==> interp1
The data abscissae should be distinct.
```

Since `interp1` does not work in this case, let's consider the development of an M-file function that performs this task.

To solve this problem, we must find all of the data points that stagger the desired point. As shown in the figure below, in some cases, a data point will appear below the interpolating point y_o and the next point will be above it. In other cases, a data point will appear above y_0 and the next point will be below it. Once the pairs of data points (x_k, y_k) and (x_{k+1}, y_{k+1}) are found, it is simply a matter of linearly interpolating between the data point pairs to find each x_o that corresponds with y_o.

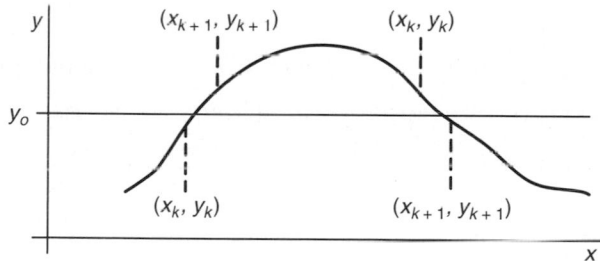

A straightforward, nonvectorized solution is shown in the following code segment:

```
% invinterp1

x = (1:10).';    % sample data
y = cos(pi*x);

yo = -0.2;        % chosen inverse interpolation point

if yo<min(y) || yo>max(y) % quick exit if no values exist
```

```
    xo = [];
else                          % search for the desired points

    n = length(y);
    xo = zeros(size(y))+nan; % preallocate space for found points
    alpha = 0;

    for k=1:n-1 % look through all data pairs

        if ( y(k)<yo && y(k+1)>=yo ) ||... % below then above
           ( y(k)>yo && y(k+1)<=yo )        % above then below

            alpha = (yo-y(k))/(y(k+1)-y(k));    % distance between x(k+1) and x(k)
            xo(k) = alpha*(x(k+1)-x(k)) + x(k); % linearly interpolate using alpha

        end
    end
    xo = xo(~isnan(xo));        % get rid of unneeded preallocated space
    yo = repmat(yo,size(xo)); % duplicate yo to match xo points found
end
```

Note that this code preallocates the interpolation points xo with an array of NaNs. An array of zeros or ones is not appropriate, since zero and one are potential valid points in xo. In some sense, this preallocation is wasteful, because xo typically has many fewer points than the input data. In any case, the unused points are eliminated in the second-to-last statement, and yo is replicated to be the same size as xo in the last statement. This replication makes it simple to plot the interpolated points, as in plot(xo,yo,'o').

It appears that this function can be vectorized, since logical comparisons can be done on an array basis. To investigate this possibility, consider the following example code:

```
>> x = (1:10);     % sample data
>> y = cos(pi*x);

>> yo = -0.2;      % chosen inverse interpolation point
```

```
>> below = y<yo    % True where below yo
below =
     1    0    1    0    1    0    1    0    1    0

>> above = y>=yo   % True where at or above yo
above =
     0    1    0    1    0    1    0    1    0    1
```

Because of the choice of x and y in the foregoing example, alternating elements are below and alternating elements are above. The value yo is bracketed by consecutive data points when below(k) and above(k+1) are True or when above(k) and below(k+1) are True. By shifting the below and above arrays by one, these tests are accomplished:

```
>> n = length(y);
>> below(1:n-1) & above(2:n)   % below(k) and above(k+1)
ans =
     1    0    1    0    1    0    1    0    1

>> above(1:n-1) & below(2:n)   % above(k) and below(k+1)
ans =
     1    0    1    0    1    0    1    0    1
```

Combining these gives the k points. The points immediately after these are the (k+1) points. These are found by the code

```
>> kth = (below(1:n-1)&above(2:n))|(above(1:n-1)&below(2:n)); % True at k points
>> kp1 = [false kth];                                         % True at k+1 points
```

These two logical arrays address the points in y that require interpolation. Using the same interpolation algorithm used earlier, a vectorized inverse interpolation solution is shown in the following code segment:

```
% invinterp2

x = (1:10).';   % sample data
y = cos(pi*x);

yo = -0.2;      % chosen inverse interpolation point
```

```
n = length(y);

if yo<min(y) || yo>max(y) % quick exit if no values exist
   xo = [];
else                        % find the desired points

   below = y<yo;    % True where below yo
   above = y>=yo;   % True where at or above yo

   kth = (below(1:n-1)&above(2:n))|(above(1:n-1)&below(2:n)); % k points
   kp1 = [false; kth];                                  % k+1 points

   alpha = (yo - y(kth))./(y(kp1)-y(kth));% distance between x(k+1) and x(k)
   xo = alpha.*(x(kp1)-x(kth)) + x(kth);  % linearly interpolate using alpha

   yo = repmat(yo,size(xo)); % duplicate yo to match xo points found
end
```

Searching with a For Loop is replaced with logical comparisons and logical manipulation. Again, the find function has been avoided, since it requires additional computations, which would slow the implementation down. All linear interpolations are performed with array mathematics. Here, the size of xo is determined as the code executes, so there is no need to preallocate it, as with the JIT-accelerated case earlier. The downside of this vectorized case is that it creates a number of potentially large arrays.

Testing both of these algorithms with the profiler shows that they have essentially identical performance. In some cases, the JIT-accelerated algorithm is faster; in others, the vectorized case is faster. Since neither implementation dominates, the following code illustrates M-file function creation, implementing the vectorized solution:

```
function [xo,yo]=mminvinterp(x,y,yo)
%MMINVINTERP 1-D Inverse Interpolation.
% [Xo,Yo]=MMINVINTERP(X,Y,Yo) linearly interpolates the vector Y to find
% the scalar value Yo and returns all corresponding values Xo interpolated
% from the X vector. Xo is empty if no crossings are found. For
% convenience, the output Yo is simply the scalar input Yo replicated so
```

```
% that size(Xo)=size(Yo).
% If Y maps uniquely into X, use INTERP1(Y,X,Yo) instead.
%
% See also INTERP1.

if nargin~=3
   error('Three Input Arguments Required.')
end
n = numel(y);
if ~isequal(n,numel(x))
   error('X and Y Must have the Same Number of Elements.')
end
if ~isscalar(yo)
   error('Yo Must be a Scalar.')
end

x = x(:); % stretch input vectors into column vectors
y = y(:);

if yo<min(y) || yo>max(y) % quick exit if no values exist
   xo = [];
   yo = [];
else                      % find the desired points

   below = y<yo;          % True where below yo
   above = y>=yo;         % True where at or above yo

   kth = (below(1:n-1)&above(2:n))|(above(1:n-1)&below(2:n)); % point k
   kp1 = [false; kth];                                        % point k+1

   alpha = (yo - y(kth))./(y(kp1)-y(kth));% distance between x(k+1) and x(k)
```

```
xo = alpha.*(x(kp1)-x(kth)) + x(kth);  % linearly interpolate using alpha

yo = repmat(yo,size(xo)); % duplicate yo to match xo points found
end
```

The preceding function is useful for finding the intersection of two plotted curves. For example, consider the problem of finding the intersection points of the two curves created by the following code:

```
>> x = linspace(0,10);
>> y = sin(x);
>> z = 2*cos(2*x);
>> plot(x,y,x,z)
>> xlabel X
>> ylabel Y
>> title 'Figure 38.2: Intersection of Two Curves'
```

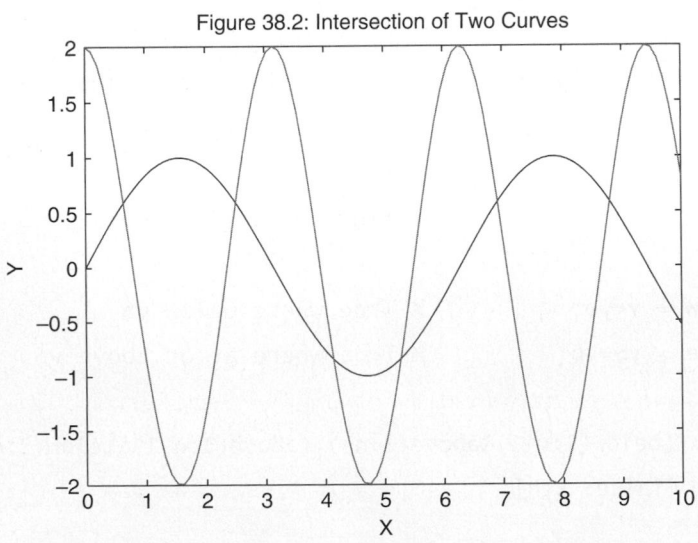

Figure 38.2: Intersection of Two Curves

The intersection of these two curves is given by the zero crossings of their difference. Given the preceding data and using mminvinterp, these points are found to be

```
>> xo = mminvinterp(x,y-z,0); % find zero crossings of difference


>> yo = interp1(x,y,xo);       % find corresponding y values


>> plot(x,y,x,z,xo,yo,'o')     % regenerate plot showing intersection points
>> xlabel X
>> ylabel Y
>> title 'Figure 38.3: Intersection Points'
```

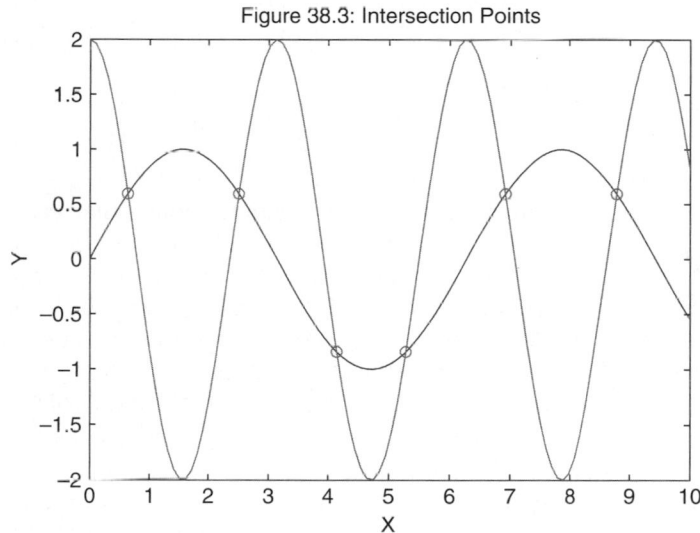

Figure 38.3: Intersection Points

38.9 POLYNOMIAL CURVE FITTING

As discussed in Chapter 20, polynomial curve fitting is performed by using the function `polyfit`. Because polynomial curve fitting is such a basic numerical analysis topic, it is worth exploring more fully. Consider a general polynomial, written as

$$y = p_1 x^n + p_2 x^{n-1} + \cdots + p_n x + p_{n+1}$$

Here, there are $n + 1$ coefficients for an nth order polynomial. For convenience, the coefficients have subscripts that are numbered in increasing order as the power of x decreases. Written in this way, the polynomial can be written as the matrix product

$$y = [x^n \quad x^{n-1} \quad \cdots \quad x \quad 1] \cdot \begin{bmatrix} p_1 \\ p_2 \\ \cdots \\ p_n \\ p_{n+1} \end{bmatrix}$$

In this format, the polynomial coefficients are grouped into a single column vector.

Common polynomial curve fitting uses this form to compute the polynomial coefficients, given a set of data points $\{x_i, y_i\}$ for $i = 1, 2, \ldots N$. Substituting each of these data points into the preceding relationship and grouping the results leads to the matrix equation

$$\begin{bmatrix} x_1^n & x_1^{n-1} & \cdots & x_1 & 1 \\ x_2^n & x_2^{n-1} & \cdots & x_2 & 1 \\ \vdots & \vdots & \cdots & \vdots & \vdots \\ x_N^n & x_N^{n-1} & \cdots & x_N & 1 \end{bmatrix} \cdot \begin{bmatrix} p_1 \\ p_2 \\ \cdots \\ p_n \\ p_{n+1} \end{bmatrix} = \begin{bmatrix} y_1 \\ y_2 \\ \vdots \\ y_N \end{bmatrix}$$

As written, the matrix on the left is a Vandermonde matrix, as discussed earlier in Section 38.4. If $N = n + 1$, the Vandermonde matrix is square. Furthermore, if the data points are distinct, the matrix has full rank, and the unique polynomial vector $p = [p_1 \, p_2 \cdots p_n \, p_{n+1}]'$ is found in MATLAB by using the backslash or left-division operator as p = V\y, where V is the Vandermonde matrix and y is the right-hand side vector.

On the other hand, when $N \geq n + 1$ and the data points are distinct, the Vandermonde matrix has more rows than columns, and no exact solution exists. In this case, p = V\y in MATLAB computes the polynomial coefficient vector that minimizes the least squared error in the set of equations. For example, the following code segment duplicates the example shown in Figure 20.2, without the use of polyfit:

```
% polyfit1.m
% find polynomial coefficients without polyfit

x = [0 .1 .2 .3 .4 .5 .6 .7 .8 .9 1]';   % column vector data
y = [-.447 1.978 3.28 6.16 7.08 7.34 7.66 9.56 9.48 9.30 11.2]';

n = 2; % desired polynomial order

pm = polyfit(x,y,n) % MATLAB polyfit result
```

```
% create Vandermonde matrix using code from vander3.m
m = length(x);   % number of elements in x
V = ones(m,n+1); % preallocate memory for result

for i=n:-1:1      % build V column by column
   V(:,i) = x.*V(:,i+1);
end


p = V\y; % find least squares solution

p = p'   % convert to row vector to match MATLAB's convention for polynomials
```

Running this code shows that both polynomial vectors are equal:

```
>> polyfit1
pm =
        -9.8108          20.129      -0.031671
p =
        -9.8108          20.129      -0.031671
```

From this basic understanding, it is possible to consider potential numerical problems. In particular, because the Vandermonde matrix contains elements ranging from 1 to x^n, it can suffer accuracy problems if the x data points differ a great deal from the number 1. For example, if the desired polynomial order is increased to 4, and the x data is scaled by 10^4, the Vandermonde matrix becomes

```
>> x = 1e4*[0 .1 .2 .3 .4 .5 .6 .7 .8 .9 1]'; % scale x data by 1e4

>> n = 4;                                      % change order to 4
>> m = length(x);    % number of elements in x
>> V = ones(m,n+1);  % preallocate memory for result
>> for i=n:-1:1      % build V column by column
      V(:,i) = x.*V(:,i+1);
   end
```

```
>> V
V =
```

0	0	0	0	1
1e+012	1e+009	1e+006	1000	1
1.6e+013	8e+009	4e+006	2000	1
8.1e+013	2.7e+010	9e+006	3000	1
2.56e+014	6.4e+010	1.6e+007	4000	1
6.25e+014	1.25e+011	2.5e+007	5000	1
1.296e+015	2.16e+011	3.6e+007	6000	1
2.401e+015	3.43e+011	4.9e+007	7000	1
4.096e+015	5.12e+011	6.4e+007	8000	1
6.561e+015	7.29e+011	8.1e+007	9000	1
1e+016	1e+012	1e+008	10000	1

Now the values in V vary in value from 1 to 10^{16}. This disparity causes trouble for the function polyfit:

```
>> p = polyfit(x,y,n)
  Warning: Rank deficient, rank = 4,  tol =   3.1633e+001.
  (Type "warning off MATLAB:rankDeficientMatrix" to suppress this warning.)
p =
  3.0675e-015 -4.8147e-011  9.5904e-008    0.0018927            0
```

Since we have reached the limits of double-precision mathematics, to eliminate this problem we must somehow scale the data so that the Vandermonde matrix does not exhibit this disparity in values. Of the numerous ways to scale the data, use of the mean and standard deviation often leads to good results. That is, instead of fitting a polynomial to the data in x, the fit is done with respect to a new independent variable z, given by

$$z = \frac{x - x_m}{s}$$

where x_m and s are the mean and standard deviation, respectively, of the x data. Subtracting by the mean shifts the data to the origin, and dividing by the standard deviation reduces the spread in the data values. Applying this to the previous data that were scaled by 10^4 gives

```
>> xm = mean(x)
>> s = std(x)
>> z = (x - xm)/s;
```

```
>> m = length(z);    % number of elements in x
>> V = ones(m,n+1); % preallocate memory for result
>> for i=n:-1:1      % build V column by column
   V(:,i) = z.*V(:,i+1);
end
>>V
V =
```

5.1653	-3.4263	2.2727	-1.5076	1
2.1157	-1.7542	1.4545	-1.206	1
0.66942	-0.74007	0.81818	-0.90453	1
0.13223	-0.21928	0.36364	-0.60302	1
0.0082645	-0.02741	0.090909	-0.30151	1
0	0	0	0	1
0.0082645	0.02741	0.090909	0.30151	1
0.13223	0.21928	0.36364	0.60302	1
0.66942	0.74007	0.81818	0.90453	1
2.1157	1.7542	1.4545	1.206	1
5.1653	3.4263	2.2727	1.5076	1

Now the Vandermonde matrix is numerically well conditioned, and the polynomial curve fit proceeds without difficulty. For this data, the resulting polynomial is

```
>> p = (V\y)'
p =
```

| 0.26689 | 0.58649 | -1.6858 | 2.4732 | 7.7391 |

Comparing these polynomial coefficients to those computed before scaling shows that these are much better scaled as well.

While the previous result works well, it changes the original problem to

$$y = p_1 z^n + p_2 z^{n-1} + \cdots + p_n z + p_{n+1}$$

That is, the polynomial is now a function of the variable z. Because of this, evaluation of the original polynomial requires a two-step process. First, using the values of x_m and s used to find z, x data points for polynomial evaluation must be converted to their z data equivalents by using $z = (x - x_m)/s$. Then, the polynomial can be evaluated, as in the following code:

```
>> xi = 1e4*[.25 .35 .45]; % sample data for polynomial evaluation

>> zi = (xi-xm)/s;           % convert to z data


>> yi = polyval(p,zi)        % evaluate using polyval
yi =

        4.752        6.2326           7.326
```

Rather than performing this step manually when data scaling is used, the MATLAB functions polyfit and polyval implement this data scaling through the use of an additional variable. For example, the previous results are duplicated by the following code:

```
>> [p,Es,mu] = polyfit(x,y,n);

>> p

p =

     0.26689        0.58649        -1.6858        2.4732        7.7391
>> yi = polyval(p,xi,Es,mu)
yi =

        4.752        6.2326           7.326
```

The optional polyfit output variable mu contains the mean and the standard deviation of the data used to compute the polynomial. Providing this variable to polyfit directs it to perform the conversion to z before evaluating the polynomial. As shown here, the polynomial p and the interpolated points yi are equal to those just computed by using the Vandermonde matrix.

The structure variable Es (which has not been discussed here) is used to compute error estimates in the solution. (See the documentation for further information regarding this variable.)

Beyond a consideration of data scaling, it is sometimes important to perform weighted polynomial curve fitting. That is, in some situations, there may be more confidence in some of the data points than in others. When this is true, the curve fits procedure should take this confidence into account and return a polynomial that reflects the weight given to each data point.

While polyfit does not provide this capability, it is easy to implement in a number of ways. Perhaps the simplest way is to weigh the data before the Vandermonde matrix is formed. For example, consider the case of a third-order polynomial being fit to five data points. Assuming that the confidence in the third data point is α times that of all other data points, the matrix equation to be solved becomes

$$
\begin{bmatrix}
x_1^3 & x_1^2 & x_1 & 1 \\
x_2^3 & x_2^2 & x_2 & 1 \\
\alpha \cdot x_3^3 & \alpha \cdot x_3^2 & \alpha \cdot x_3 & \alpha \\
x_4^3 & x_4^2 & x_4 & 1 \\
x_5^3 & x_5^2 & x_5 & 1
\end{bmatrix}
\cdot
\begin{bmatrix}
p_1 \\
p_2 \\
p_3 \\
p_4
\end{bmatrix}
=
\begin{bmatrix}
y_1 \\
y_2 \\
\alpha \cdot y_3 \\
y_4 \\
y_5
\end{bmatrix}
$$

By multiplying the third row of the matrix equation by α, the error in this equation is weighted by α. The process of minimizing the least squared error forces the error at this data point to decrease relative to the others. As α increases, the error at the data point decreases.

In general, it is possible to give a weight to each data point, not just to one of them, as previously shown. Implementing this approach in MATLAB is illustrated in the following code segment:

```
% polyfit2.m
% find weighted polynomial coefficients

x = [0 .1 .2 .3 .4 .5 .6 .7 .8 .9 1]';
y = [-.447 1.978 3.28 6.16 7.08 7.34 7.66 9.56 9.48 9.30 11.2]';

n = 3;

pm = polyfit(x,y,n) % MATLAB polyfit result

% create Vandermonde matrix using code from vander3.m

m = length(x);    % number of elements in x

V = ones(m,n+1); % preallocate memory for result

for i=n:-1:1      % build V column by column
    V(:,i) = x.*V(:,i+1);
end

w = ones(size(x)); % default weights of one
w(4) = 2;            % weigh 4th point by 2
w(7) = 10;           % weigh 7th point by 10
```

```
V = V.*repmat(w,1,n+1); % multiply rows by weights
y = y.*w;               % multiply y by weights

p = (V\y)' % find polynomial
```

Running this code shows that the unweighted and weighted polynomials are different:

```
>> polyfit2
pm =
        16.076       -33.924       29.325       -0.6104
p =
        28.576       -50.761       34.104       -0.67441
```

An alternative to this approach uses the MATLAB function lscov, which specifically computes weighted least squares solutions. Repeating the previous example by using this function is straightforward, as shown in the following code segment and its associated output:

```
% polyfit3
% find weighted polynomial coefficients using lscov

x = [0 .1 .2 .3 .4 .5 .6 .7 .8 .9 1]';
y = [-.447 1.978 3.28 6.16 7.08 7.34 7.66 9.56 9.48 9.30 11.2]';

n = 3;

pm = polyfit(x,y,n) % MATLAB polyfit result

% create Vandermonde matrix using code from vander3.m

m = length(x);   % number of elements in x
V = ones(m,n+1); % preallocate memory for result

for i=n:-1:1      % build V column by column
   V(:,i) = x.*V(:,i+1);
end
```

```
w = ones(size(x)); % default weights of one
w(4) = 2^2;          % here weights are the square of those used in polyfit2
w(7) = 10^2;

p = lscov(V,y,w)'
```

```
>> polyfit3
pm =
        16.076        -33.924        29.325        -0.6104
p =
        28.576        -50.761        34.104        -0.67441
```

To see how these polynomials differ, they can be plotted over the range of the data:

```
% mm3804.m
% find weighted polynomial coefficients using lscov

x = [0 .1 .2 .3 .4 .5 .6 .7 .8 .9 1]';
y = [-.447 1.978 3.28 6.16 7.08 7.34 7.66 9.56 9.48 9.30 11.2]';
n = 3;
pm = polyfit(x,y,n); % MATLAB polyfit result

% create Vandermonde matrix using code from vander3.m

m = length(x);    % number of elements in x
V = ones(m,n+1); % preallocate memory for result

for i=n:-1:1        % build V column by column
   V(:,i) = x.*V(:,i+1);
end

w = ones(size(x)); % default weights of one
w(4) = 2^2;
w(7) = 10^2;
```

```
p = lscov(V,y,w)';

xi = linspace(0,1,100);
ym = polyval(pm,xi);
yw = polyval(p,xi);

pt = false(size(x)); % logical array pointing to weighted points
pt([4 7]) = true;

plot(x(~pt),y(~pt),'x',x(pt),y(pt),'o',xi,ym,'--',xi,yw)
xlabel('x'), ylabel('y=f(x)')
title('Figure 38.4: Weighted Curve Fitting')
```

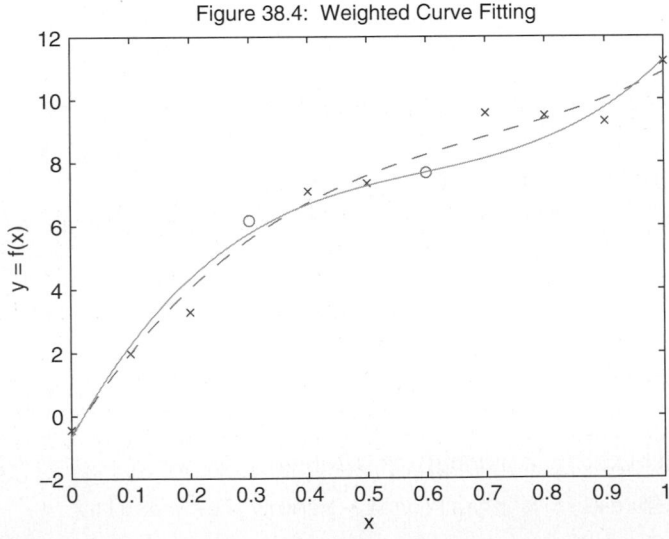

Figure 38.4: Weighted Curve Fitting

The weighted points are marked with circles in the figure, while the unweighted points are marked with an x. Clearly, the weighted polynomial is closer to the fourth and seventh data points than the original `polyfit` results. In addition, the seventh data point is closer than the fourth because its weight is much higher.

38.10 NONLINEAR CURVE FITTING

The polynomial curve fitting discussed in the previous section is popular in part because the problem can be written in matrix form and solved by using linear least-squares techniques. Perhaps most important, or most convenient, is that no initial guess is required. The optimum solution is found without searching an n-dimension-al solution space. In the more general case, where the unknown parameters do not appear linearly (i.e., nonlinear curve fitting), the solution space must be searched, starting with an initial guess. For example, consider fitting data to the function

$$f(t) = a + be^{\alpha t} + ce^{\beta t}$$

If α and β are known constants and a, b, and c are the unknowns to be found, this function can be written as

$$f(t) = [1 \quad e^{\alpha t} \quad e^{\beta t}] \cdot \begin{bmatrix} a \\ b \\ c \end{bmatrix}$$

Substituting a given a set of data points, $\{t_i, y_i = f(t_i)\}$ for $i = 1, 2, \ldots N$, into this expression and gathering the results in matrix format results in the equation

$$\begin{bmatrix} 1 & e^{\alpha t_1} & e^{\beta t_1} \\ 1 & e^{\alpha t_2} & e^{\beta t_2} \\ \vdots & \vdots & \vdots \\ 1 & e^{\alpha t_N} & e^{\beta t_N} \end{bmatrix} \cdot \begin{bmatrix} a \\ b \\ c \end{bmatrix} = \begin{bmatrix} y_1 \\ y_2 \\ \vdots \\ y_N \end{bmatrix}$$

Written in this way, a, b, and c appear linearly, just as the polynomial coefficients did in the previous section. As a result, the backslash operation in MATLAB produces the least-squares solution for the unknown variables. That is, if E is the N-by-3 matrix on the left, p is the vector $[a\,b\,c]'$, and y is the right-hand side vector, then p = E\y gives the least-squares solution for p.

When α and β are not known, the linear least-squares solution does not apply, because α and β cannot be separated as a, b, and c are in the preceding equation. In this case, it is necessary to use a minimization algorithm such as fminsearch in MATLAB. There are two ways to set up this problem for solution with fminsearch. The first is to write $f(t)$ as

$$f(t) = x_3 + x_4 e^{x_1 t} + x_5 e^{x_2 t}$$

where the unknown variables have been combined into a vector $x = [x_1\ x_2\ x_3\ x_4\ x_5]'$. Given this form and a set of data points $\{t_i, y_i\}$ for $i = 1, 2, \ldots N$, the least-squares solution minimizes the norm of error between the data points y_i and the function eval-uated at t_i, $f(t_i)$. That is, if y is a vector containing the y_i data points and if f is a vector containing the previous function evaluated at the time points t_i, then the least-squares solution is given by

$$\min_x \|y - f\|$$

To use `fminsearch`, this norm must be computed in a function as follows:

```
function enorm=fitfun1(x,tdata,ydata)
%ENORM Norm of fit to example nonlinear function
% f(t) = x(3)+x(4)*exp(x(1)*t)+x(5)*exp(x(2)*t)
%
% ENORM(X,Tdata,Ydata) returns norm(Ydata-f(Tdata))

f = x(3)+x(4)*exp(x(1)*tdata)+x(5)*exp(x(2)*tdata);

enorm = norm(f-ydata);
```

The following code segment tests this approach by creating some data, making an initial guess, and calling `fminsearch` to find a solution:

```
% testfitfun1
% script file to test nonlinear least squares problem

% create test data
x1 = -2; % alpha
x2 = -5; % beta
x3 = 10;
x4 = -4;
x5 = -6;

tdata = linspace(0,4,30)';
ydata = x3+x4*exp(x1*tdata)+x5*exp(x2*tdata);

% create an initial guess

x0 = zeros(5,1); % not a good one, but a common first guess

% call fminsearch

fitfun = @fitfun1; % create handle

options = []; % take default options
```

```
x = fminsearch(fitfun,x0,options,tdata,ydata)

% compute error norm at returned solution

enorm = fitfun(x,tdata,ydata)
```

Running this code produces the following output:

```
>> testfitfun1
Exiting: Maximum number of function evaluations has been exceeded
          - increase MaxFunEvals option.
          Current function value: 3.234558
x =
        0.17368
        -1.5111
         13.972
        -2.2059
        -9.4466


enorm =
          3.2346
```

For this initial guess, the algorithm has not converged. Increasing the number of function evaluations and algorithm iterations permitted to 2000 leads to

```
>> options = optimset('MaxFunEvals',2e3,'MaxIter',2e3);

x = fminsearch(fitfun,x0,options,tdata,ydata)
x =
       -0.78361
        -3.7185
         10.061
       -0.80533
        -9.1907
```

```
enorm =
   0.23727
```

The algorithm now converges, but not to the values $x = [-2 \; -5 \; 10 \; -4 \; -6]'$ used to create the data. A plot of the data, the actual function, and the fitted solution provides further information, computed and displayed as follows:

```
% testfitfun2
% script file to test nonlinear least squares problem

% create test data
x1 = -2; % alpha
x2 = -5; % beta
x3 = 10;
x4 = -4;
x5 = -6;
tdata = linspace(0,4,30)';
ydata = x3+x4*exp(x1*tdata)+x5*exp(x2*tdata);

% create an initial guess

x0 = zeros(5,1); % not a good one, but a common first guess

% call fminsearch

fitfun = @fitfun1; % create handle

options = optimset('MaxFunEvals',2e3,'MaxIter',2e3);

x = fminsearch(fitfun,x0,options,tdata,ydata);

ti = linspace(0,4); % evaluation points
actual = x3+x4*exp(x1*ti)+x5*exp(x2*ti); % actual function

fitted = x(3)+x(4)*exp(x(1)*ti)+x(5)*exp(x(2)*ti); % fitted solution
subplot(2,1,1)
plot(tdata,ydata,'o',ti,actual,ti,fitted)
xlabel t
title 'Figure 38.5: Nonlinear Curve Fit'
```

```
subplot(2,1,2)
plot(ti,actual-fitted)
xlabel t
ylabel Error
```

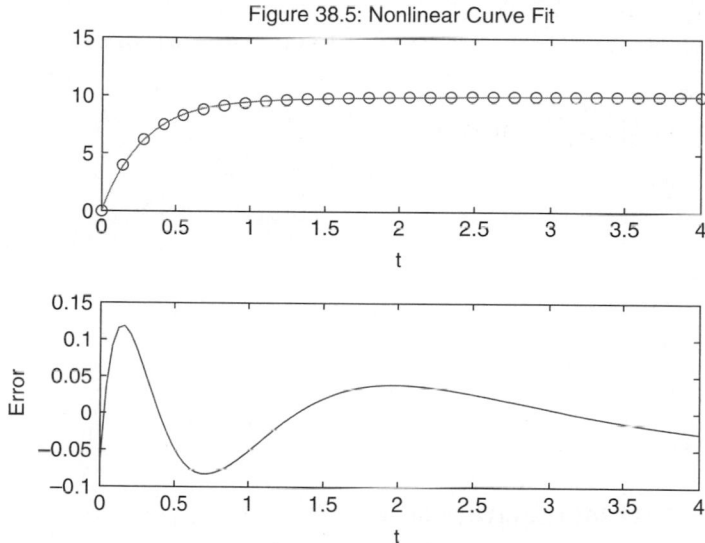

Figure 38.5: Nonlinear Curve Fit

The fitted solution in the upper plot looks good visually. However, the lower error plot shows that there is significant error. If the overall goal of this problem is to minimize the error between the actual coefficients represented by the data and the results returned by fminsearch, then this algorithm essentially failed, even though the fitted solution plot looks good visually.

At this point, you must decide whether this solution is satisfactory. Certainly, you can try setting tighter convergence criteria, as in the following code:

```
>> options=optimset('TolX',1e-10,'TolFun',1e-10,'MaxFunEvals',2e3,'MaxIter',2e3);

>> x = fminsearch(fitfun,x0,options,tdata,ydata)
x =
      -0.78359
       -3.7185
        10.061
```

```
        -0.80533
        -9.1907
>> enorm = fitfun(x,tdata,ydata)
enorm =
        0.23727
```

In this case, the solution has not changed. You can also try better initial guesses, such as the following:

```
>> x0 = [-1 -4 8 -3 -7]'; % much closer initial guess


>> options = []; % default options


>> x = fminsearch(fitfun,x0,options,tdata,ydata)
x =
              -2
              -5
              10
              -4
              -6
>> enorm = fitfun(x,tdata,ydata)
enorm =
      3.2652e-006
```

In this case, fminsearch returns coefficients that very closely match those used to create the data. This example illustrates an ambiguity inherent in most nonlinear optimization algorithms. That is, other than having the error norm be zero, there is no set way of knowing when a solution is the best that can be expected.

When a nonlinear curve fitting problem contains a mixture of linear and nonlinear terms, as is true in this example, it is possible to use linear least squares to compute the linear terms and use a function such as fminsearch to compute the nonlinear terms. This leads to better results in almost all cases.

To understand how this is done, consider the problem formulation stated earlier, where α and β were known constants:

$$\begin{bmatrix} 1 & e^{\alpha t_1} & e^{\beta t_1} \\ 1 & e^{\alpha t_2} & e^{\beta t_2} \\ \vdots & \vdots & \vdots \\ 1 & e^{\alpha t_N} & e^{\beta t_N} \end{bmatrix} \cdot \begin{bmatrix} a \\ b \\ c \end{bmatrix} = \begin{bmatrix} y_1 \\ y_2 \\ \vdots \\ y_N \end{bmatrix}$$

In this case, α and β become the variables manipulated by fminsearch. Within the function that evaluates the error norm, this linear least-squares problem is solved by using the current estimates for α and β. That is, if $x_1 = \alpha$, $x_2 = \beta$, and $p = [a\ b\ c]'$, then fitfun1 is rewritten as follows:

```
function [enorm,p]=fitfun2(x,tdata,ydata)
%ENORM Norm of fit to example nonlinear function
% f(t) = p(1)+p(2)*exp(x(1)*t)+p(3)*exp(x(2)*t)
%
% ENORM(X,Tdata,Ydata) returns norm(Ydata-f(Tdata))
%
% [e,p]=ENORM(...) returns the linear least squares
%                       parameter vector p

% solve linear least squares problem given x input supplied by fminsearch

E = [ones(size(tdata)) exp(x(1)*tdata) exp(x(2)*tdata)];
p = E\ydata; % least squares solution for p=[a b c]'

% use p vector to compute error norm
f = p(1)+p(2)*exp(x(1)*tdata)+p(3)*exp(x(2)*tdata);

enorm = norm(f-ydata);
```

Now there are only two parameters for fminsearch to manipulate. The following code segment tests this alternative approach:

```
% testfitfun3
% script file to test nonlinear least squares problem

% create test data
x1 = -2; % alpha
x2 = -5; % beta
p1 = 10;
p2 = -4;
p3 = -6;
```

```
tdata = linspace(0,4,30)';
ydata = p1+p2*exp(x1*tdata)+p3*exp(x2*tdata);

% create an initial guess

x0 = zeros(2,1); % only two parameters to guess now!

% call fminsearch

fitfun = @fitfun2; % create handle to new function

options = []; % take default options

x = fminsearch(fitfun,x0,options,tdata,ydata)

% find p and compute error norm at returned solution

[enorm,p] = fitfun(x,tdata,ydata)
```

Running this code produces the following output:

```
>> testfitfun3
x =
              -2
              -5
enorm =
    7.1427e-006
p =
              10
              -4
              -6
```

Without any difficulty, fminsearch finds a minimum very close to the actual values. The incorporation of an internal linear least-squares algorithm within a nonlinear problem reduces dimension of the parameter space to be searched. This almost always improves the speed and convergence of the nonlinear algorithm. In the preceding example, the problem was reduced from five dimensions to two.

Finally, note how the parameter p was included as a second output argument to `fitfun2`. The minimization algorithm `fminsearch` ignores this argument, so it does not influence the search process. However, after the algorithm terminates, calling the `fitfun2` again with two output arguments returns the values of the linear parameters at the solution point. Global variables should not be used to obtain p, since there is no way of knowing if `fminsearch` evaluates `fitfun2` at its solution on its last iteration. Calling `fitfun2` after `fminsearch` finishes execution guarantees that the parameter vector p is evaluated at the solution returned by `fminsearch`.

38.11 PICTURE-IN-A-PICTURE ZOOM

The final example in this chapter demonstrates use of the Handle Graphics features in MATLAB. As illustrated in the next figure, this example implements a picture-in-a-picture zoom. That is, the user calls the function, then goes to the current axes and, by using the mouse, drags a selection rectangle. Once created, a dotted outline of the selection rectangle remains, and a new, but smaller, axis is created showing the graphical contents inside of the drawn selection rectangle. This function allows the user to zoom into a portion of an axis without hiding the original plot. Although not shown, the smaller axis can be selected, dragged, and resized.

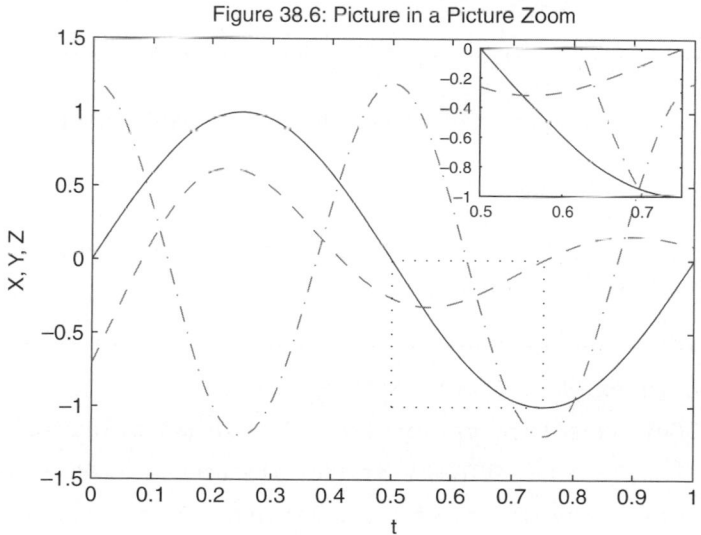

Figure 38.6: Picture in a Picture Zoom

There are a number of steps involved in the creation of this function. Rather than implement the function as one large function, it is convenient to create supporting functions first that could be subfunctions of the primary functions or simply functions on the MATLAB search path. The first function `getn` simplifies the assignment of the output of the `get` function into individual variables:

```
function varargout=getn(H,varargin)
%GETN Get Multiple Object Properties.
% [Prop1,Prop2,...] = GETN(H,PName1,PName2,...) returns the requested
% properties of the scalar handle H in the corresponding output arguments.
%
% For example, [Xlim,Ylim,Xlabel] = GETN(gca,'Xlim','Ylim','Xlabel')
% returns the requested axes properties in like-named output variables.
%
% This simplifies the construct
% [Xlim,Ylim,Xlabel] = deal(get(gca,{'Xlim','Ylim','Xlabel'}))

if max(size(H))~=1 || ~ishandle(H)
    error('Scalar Object Handle Required.')
end
varargout = get(H,varargin);
```

This function is not absolutely needed, but it is convenient and demonstrates use of `varargin` and `varargout`.

The next function to consider is implementation of the selection rectangle and capturing its *x*- and *y*-axis limits. As shown in the following example, this requires use of the MATLAB functions `rbbox`, `waitforbuttonpress`, and the `CurrentPoint` properties of the *axes* and *figure* objects:

```
function [xbox,ybox,prect]=getbox
%GETBOX Get axes information from user-drawn selection rectangle.
% [Xbox,Ybox,Prect] = GETBOX waits for the user to drag a selection box The
% x and y axis data limits of the selection box are returned in Xbox, Ybox,
% and Prect.
% Xbox is a two element vector containing the minimum and maximum limits
% along the x axis, i.e., Xbox = [min(x) max(x)]
% Ybox is a two element vector containing the minimum and maximum limits
% along the y axis, i.e., Ybox = [min(y) max(y)]
```

```
% Prect is a four element vector containing the selection box position
% in standard position vector format, Prect = [left bottom width height]
% Data returned is in the axis data units.
%
% The selection box is limited to the x and y axis limits of the axes where
% the selection rectangle was drawn.

% waitforbuttonpress waits until user presses a mouse button over a figure
% waitforbuttonpress return False if that happens. Alternatively, it
% returns True if the user presses a key on the keyboard.

if waitforbuttonpress % Returns True if a key is pressed, abort
    xlim = [];
    ylim = [];
    prect = [];
    return
end
% Function only gets here if user presses a mouse button in a figure

Hf = gcf;       % get current figure where button was pressed
Ha = gca(Hf);   % get current axes where button was pressed

AxesPt = get(Ha,'CurrentPoint'); % get first axes data point clicked
FigPt = get(Hf,'CurrentPoint');   % get first figure point clicked

% call the function rbbox, i.e., rubberband box, to create the selection
% rectangle. This function needs to know where to start from. It does not
% automatically start at the mouse click unless told to do so.

% drag selection rectangle starting at first figure point
rbbox([FigPt 0 0],FigPt) % function returns as soon as mouse button is up

% get point on opposite corner of selection rectangle; add to first point
AxesPt = [AxesPt; get(Ha,'CurrentPoint')];
```

```
% get axis limits of axes where selection rectangle was drawn
[Xlim,Ylim] = getn(Ha,'Xlim','Ylim');


% convert AxesPt data into usable output vectors.
xbox = [min(AxesPt(:,1)) max(AxesPt(:,1))]; % x axis limits of selection
xbox = [max(xbox(1),Xlim(1)) min(xbox(2),Xlim(2))]; % limit to axes size


ybox = [min(AxesPt(:,2)) max(AxesPt(:,2))];
ybox = [max(ybox(1),Ylim(1)) min(ybox(2),Ylim(2))]; % limit to axes size


prect = [xbox(1) ybox(1) diff(xbox) diff(ybox)]; % position rectangle
```

The comments contained in getbox describe its operation. The CurrentPoint property of an *axes* object returns a two-by-three matrix containing the *x*, *y*, and *z* data points of the back and front of the current point in 3-D space. Since this is a 2-D axes, only the *x* and *y* data points in the first two columns are relevant. Manipulation of this data produces the desired output variables.

Using getn and getbox, the function mmzoom implements the picture-in-a-picture zoom:

```
function [Hza,Hzr]=mmzoom(arg)
%MMZOOM Picture in a Picture Zoom.
% MMZOOM creates a new axes containing the data inside a box formed by a
% click and drag with the mouse in the current axes. The new zoomed axes
% is placed in the upper right of the current axes, but can be moved with
% the mouse. Clicking in the figure border disables dragging.
%
% Previous axes created by MMZOOM are deleted if MMZOOM is called again.
%
% [Hza,Hzr] = MMZOOM returns handles to the created axes and rectangle
% marking the zoomed area respectively.
%
% MMZOOM DRAG enables dragging of a zoomed axes.
```

```matlab
% MMZOOM RESET disables dragging of a zoomed axes.
% MMZOOM OFF removes the zoomed axes and rectangle marking the zoomed area.

if nargin == 0
   arg = [];
end
if isempty(arg) % zoom zoom zoom zoom zoom zoom zoom zoom zoom zoom zoom


   Hzoom = findobj(0,'Tag','MMZOOM'); % find previous zoomed axes
   if ~isempty(Hzoom)    % delete prior zoomed axes if it exists
      delete(Hzoom)
   end

   [xlim,ylim,prect] = getbox;  % get selection box for zoom

   if ~isempty(prect) % act only if rectangle exists
      Haxes = gca; % handle of axes where selection box was drawn
      Hzr = rectangle('Position',prect,... % place rectangle object
         'Linestyle',':',...                % to mark selection box
         'Tag','MMZOOM');

      Hfig = gcf; % handle of Figure where selection box was drawn
      Hzoom = copyobj(Haxes,Hfig); % copy original axes and its children


      OldUnits = get(Haxes,'Units');  % get position vector of original
      set(Haxes,'Units','normalized') % axes in normalized units
      Pvect = get(Haxes,'Position');
      set(Haxes,'Units',OldUnits)


      % scale and shift zoomed axes relative to original axes
      alpha = 1/3;   % position scaling for zoomed axes
      beta = 98/100; % position shift for zoomed axes
```

```
% compute position vector for zoomed axes
Zwidth = alpha*Pvect(3);                  % zoomed axes width
Zheight = alpha*Pvect(4);                 % zoomed axes height
Zleft = Pvect(1)+beta*Pvect(3)-Zwidth;    % zoomed axes left
Zbottom = Pvect(2)+beta*Pvect(4)-Zheight; % zoomed axes bottom


% modify zoomed axes as required
set(Hzoom,'units','Normalized',...              % make units normalized
    'Position',[Zleft Zbottom Zwidth Zheight],...% axes position
    'Xlim',xlim,'Ylim',ylim,...                 % axis data limits
    'Box','on',...                              % axis box on
    'Xgrid','off','Ygrid','off',...             % grid lines off
    'FontUnits','points',...
    'FontSize',8,...                            % shrink font size
    'ButtonDownFcn',@selectmoveresize,...       % enable drag
    'Tag','MMZOOM',...                          % tag zoomed axes
    'UserData',Haxes)                           % store original axes


[Htx,Hty,Htt] = getn(Hzoom,'Xlabel','Ylabel','Title');
set([Htx,Hty,Htt],'String','')          % delete labels on zoomed axes


set(Haxes,'DeleteFcn',...                   % delete both axes together
    'delete(findobj(0,''Type'',''axes'',''Tag'',''MMZOOM''))')


% place zoomed axes at top of object stack
Hchild = findobj(Hfig,'type','axes'); % get all axes in figure
Hchild(Hchild==Hzoom) = [];               % remove zoomed axes from list
```

```
        set(Hfig,'Children',[Hzoom;Hchild],... % put zoom axes at top of stack
            'CurrentAxes',Haxes,...                % make original axes current
            'ButtonDownFcn','mmzoom reset')  % enable reset

        if nargout>=1  % provide output only if requested
            Hza = Hzoom;
        end
    end


elseif strncmpi(arg,'d',1) % drag zoom axes drag zoom axes drag zoom axes

    Hzoom = findobj(0,'Type','axes','Tag','MMZOOM');
    if ~isempty(Hzoom)
        set(Hzoom,'ButtonDownFcn',@selectmoveresize)
    end


elseif strncmpi(arg,'r',1) % reset reset reset reset reset reset reset

    Hzoom = findobj(0,'Type','axes','Tag','MMZOOM');
    if ~isempty(Hzoom)
        [Hfig,Haxes] = getn(Hzoom,'Parent','UserData');
        set(Hzoom,'ButtonDownFcn','','Selected','off')% turn off selection
        set(Hfig,'CurrentAxes',Haxes)                 % make Haxes current
    end

elseif strncmpi(arg,'o',1)  % off off off off off off off off off off off

    Hzoom = findobj(0,'Tag','MMZOOM');
    if ~isempty(Hzoom)
        delete(Hzoom)
```

```
    end
else
    error('Unknown Input Argument.')
end
```

The structure of mmzoom uses a *switchyard* approach that evaluates code on the basis of the input argument. When no input argument is used, mmzoom creates the zoomed *axes*. The function copyobj is used to make a copy of the original *axes* and all of its children. This new *axes* is then modified to create the zoomed *axes*. Finally, the zoomed *axes* is placed on top of the original axes, which is made the current *axes*. The ButtonDownFcn callback of the zoomed axes executes the function selectmoveresize, which performs exactly what its name implies. The ButtonDownFcn callback of the *figure* calls mmzoom('reset'), which deselects the zoomed axes and deletes its callback. Since this function is fairly simple, the function findobj is used to find the zoomed *axes* and the *rectangle* marking the selection box. Using findobj makes it difficult to allow multiple simultaneous instances of this function to exist. As a result, only one is permitted. If multiple simultaneous instances were desired, function handle callbacks with added arguments as described in Chapter 32 would be required.

Appendix

MATLAB Release Information

This appendix documents the new and revised functionality and functions available in individual releases of MATLAB, starting with MATLAB 5. This material makes it possible to identify what functionality and functions appeared in which version of MATLAB. With worldwide MATLAB users running a variety of MATLAB releases, and with the proliferation and ease of file sharing, it is important to be able to identify the backward and forward compatibility of MATLAB M-files. The information contained in this appendix facilitates this process by documenting MATLAB release information that is otherwise not centrally available.

Each MATLAB release cited here identifies the public release date, the functionality and functions introduced in the release, as well as the revised functionality and functions that became effective with the release. Obsolete functions are also identified. Changed functionality and functions have (Change) in their description. When appropriate, a reference to the *Mastering MATLAB 7* chapter covering the cited functionality is included.

The material presented is limited to the base MATLAB product only and is based on information contained in the release notes for each release. Because the release notes are not always comprehensive, some information may be missing. Furthermore, because they constitute too much information, Handle Graphics objects and their properties are not included here.

MATLAB 5.0 (December 1996)

Functionality	Description	Chapter
Multidimensional arrays	Arrays can have any number of dimensions (e.g., `A(i,j,k,...)`).	6
Structures	A container variable class identified by a variable name and fields, with fields denoted by a preceding period (e.g., `varname.field1`)	8
Cell arrays	A container variable class identified by a name, with contents enclosed with { and }	8
Object-oriented programming	User-defined variable classes with function and operator overloading, data encapsulation, methods, inheritance, etc.	33
Subfunctions	Function M-files can contain multiple functions, with those appearing after the first or primary function being subfunctions.	12
Private functions	Function M-files in `private` subdirectories just off of the MATLAB path	12
Switch/case statements	Alternative control flow construction	11
Variable length function input and output argument lists	Cell arrays `varargin` and `varargout` supported in function input and output argument lists	12
Pseudocode creation	Ability to precompile M-files into an encrypted format	12
Single byte data type for images	Image object supports 8-bit integer data, using the `uint8` data type.	7, 29
Truecolor	Images in Truecolor are supported.	28
end as last array element	Last element along a dimension can be addressed by using the keyword end (e.g., `A(1:2:end)`	5
Scalar expansion assignment	(Change) `A(...) = x`, performs scalar expansion to fill all addressed elements of A with the scalar x.	5
Character string storage	(Change) Each character occupies 2 bytes of storage rather than 8.	9
Dimension specification in data analysis functions	(Change) Data analysis functions `sum`, `prod`, `cumprod`, and `cumsum` now accept a last input argument that specifies the dimension along which the function operates.	18

Empty arrays	(Change) Empty arrays may have some nonzero dimensions.	5
Marker Style Enhancements	(Change) New line markers are available, and markers can be specified independently from line styles.	26
Camera viewing model	Axes now have camera properties to set viewing properties.	27
TeX support	(Change)Text objects can contain TeX commands.	31
Z-buffering	Z-buffer graphics rendering supported	28
Light object	New graphics object	31

Math Function	**Description**
airy	Airy functions
besselh	Bessel functions of the third kind (Hankel)
condeig	Condition number with respect to eigenvalues
condest	1-norm matrix condition number estimate
dblquad	Numerical 2-D integration
mod	Modulus
normest	2-norm estimate

n-D Function	**Description**
cat	Concatenate arrays
flipdim	Flip array along specified dimension
ndgrid	Generate arrays for *n*-D functions
ndims	Number of array dimensions
permute, ipermute	Permute and inverse permute dimensions of *n*-D array
reshape	Change shape of an array
shiftdim	Shift dimensions
squeeze	Eliminate singleton dimensions
sub2ind, ind2sub	Single index from subscripts and subscripts from single linear index

Cell Array and Structure Function	Description
cell	Create cell array
cell2struct	Convert cell to structure
celldisp	Display cell structure
cellplot	Graphically display cell structure
num2cell	Convert matrix to cell array
fieldnames	Field names of structure
getfield	Get field from structure
rmfield	Remove field from structure
setfield	Set field in structure
struct	Create structure array
struct2cell	Convert structure to cell array

Character Function	Description
char	Convert to string array
mat2str	Convert matrix to string
strcat	String concatenation
strmatch	Find matches for a string
strncmp	Compare first n characters
strvcat	Vertical string concatenation

Logical Function	Description
iscell	True for cell array
isequal	True if arrays are equal
isfinite	True for finite elements
islogical	True for logical arrays
isnumeric	True for numeric array
isstruct	True for structure
logical	Convert to logical array

M-file Function	Description
assignin	Assign variable in workspace
evalin	Evaluate expression in workspace
inmem	Functions in memory
inputname	Input argument name
mfilename	Name of currently running M-file
mexext	MEX-file extension
pcode	Create pseudocode
profile	Profile M-file execution
varargin, varargout	Pass or return a variable number of function arguments
warning	Display warning message

File and Directory Function	Description
addpath	Append directory to MATLAB search path
edit	Edit M-file
editpath	Modify MATLAB search path
fullfile	Build full filename from parts

Set, Bit, and Base Function	Description
intersect	Set intersection
ismember	Detect members of a set
setdiff	Set difference
setxor	Set exclusive OR
union	Union of two sets
unique	Unique elements of a vector
bitand	Bitwise AND
bitcmp	Compare bits
bitget	Get bit
bitmax	Maximum floating-point integer
bitor	Bitwise OR

Set, Bit, and Base Function	Description
bitset	Set bit
bitshift	Bitwise shift
bitxor	Bitwise XOR
base2dec	Convert base to decimal
bin2dec	Convert binary to decimal
dec2base	Convert decimal to base
dec2bin	Convert decimal to binary

Time and Date Function	Description
calendar	Produce monthly calendar
datenum	Serial date number
datestr	Create date string
datetick	Create date-formatted tick labels
datevec	Date components
eomday	End of month
now	Current date and time
weekday	Day of week

Matrix Function	Description
cholinc	Incomplete Cholesky factorization
gallery	More than 50 text matrices
luinc	Incomplete LU factorization
repmat	Replicate and tile an array
sprand	Random uniformly distributed sparse arrays

Sparse Matrix Function	Description
bicg	Biconjugate gradients method
bicgstab	Biconjugate gradients stabilized method
cgs	Conjugate gradients squared method
eigs	Find several eigenvalues and eigenvectors

gmres	Generalized minimal residual method
pcg	Preconditioned conjugate gradients method
qmr	Quasi-minimal residual method
svds	A few singular values

Data Analysis Function	**Description**
convhull	Convex hull
cumtrapz	Cumulative trapezoidal numerical integration
delaunay	Delaunay triangularization
dsearch	Search for nearest point
factor	Primer factors
inpolygon	Detect points inside a polygonal region
isprime	True for prime numbers
nchoosek	All possible combinations of n elements taken k at a time
perms	All possible permutations
polyarea	Area of a polygon
primes	Generate a list of prime numbers
sortrows	Sort rows in ascending order
tsearch	Search for enclosing Delaunay triangle
voronoi	Voronoi diagram
sum([]) = 0, prod([]) = 1, max([]) = [], min([]) =[]	(Change) Defined output for empty inputs

Interpolation Function	**Description**
interp3	3-D data interpolation
interpn	n-D data interpolation
ndgrid	Generate arrays for n-D functions and interpolation
griddata	(Change) Uses triangle-based interpolation

ODE Function	Description
ode45, ode23, ode113, ode23s, ode15s	Solve differential equations by using various methods
odefile	Problem definition file for ODE solvers
odeget	Extract options from ODE options structure
odeset	Create or edit options structure for ODE solvers

Plot Function	Description
area	Filled area plot
bar3	3-D bar chart
bar3h	3-D horizontal bar chart
barh	2-D horizontal bar chart
box	Turn On or Off axes box
countourf	Filled contour plot
pie	Pie chart
pie3	3-D pie chart
plotyy	Plot with y-axis labels on both the left and right
quiver3	3-D quiver plot
ribbon	Draw lines as 3-D strips
stem	(Change) stem tips can be filled or unfilled
stem3	3-D stem plot
trimesh	Triangular mesh plot
trisurf	Triangular surface plot

Color and Light Function	Description
autumn	Colormap of red and yellow
colorcube	Colormap of regularly spaced colors
colordef	Select figure color scheme
lines	Colormap that follows axes colororder property
spring	Colormap of magenta and yellow
summer	Colormap of green and yellow
winter	Colormap of blue and green

Image Function	**Description**
imread	Read image data
imwrite	Write image data

Handle Graphics and GUI Function	**Description**
dragrect	Drag rectangle
inputdlg	Display input dialog
msgbox	Display message dialog
questdlg	Display question dialog
rbbox	Rubberband box
selectmoveresize	Interactively select, move, or resize objects
uiresume	Resume suspended M-file execution
uiwait	Block M-file execution
waitfor	Block execution until condition is satisfied

MATLAB 5.1 (June 1997)

Functionality	**Description**	**Chapter**
find function returns empty	(Change) find(...) returns [] if no indices where argument is True are found.	5
Multibyte characters	(Change) Two byte characters are supported.	9

Function	**Description**
ismember	Now calls the MEX function ismemc to maximize speed
pagedlg	Open Page Layout dialog
printdlg	Open Print dialog
scatter	2-D scatter plot
scatter3	3-D scatter plot

MATLAB 5.2, Release 10 (March 1998)

Functionality	Description	Chapter
`try/catch` blocks	Error handling using `try/catch` blocks	11
M-file locking	M-files can now be locked so that `clear` does not purge that M-file from memory.	12
Persistent variables	Variables can be declared persistent so that they persist from one call to a function to the next.	12
`S = load(...)`	(Change) Load file contents to structure	14
HDF file support	Files in HDF format can be read and written to.	14
OpenGL support	OpenGL graphics rendering supported	28
Tooltips	UI controls have a `tooltip` string property.	32
Toggle Buttons	New *Uicontrol* object	32
Uicontextmenu	New *Uimenu* object	32
Support for cell arrays of strings	The functions `intersect`, `ismember`, `lower`, `setdiff`, `setxor`, `sort`, `union`, `unique`, and `upper` now handle inputs that are cell arrays of strings.	9

Function	Description
`camdolly`	Translate camera position and target
`camlight`	Create or move light object in camera coordinate system
`camorbit`	Rotate camera position around camera target
`campan`	Rotate camera target around camera position
`camroll`	Rotate camera about camera viewing axis
`camzoom`	Zoom camera in or out
`campos`	Set or get camera position and position mode
`camproj`	Set or get camera projection type
`camtarget`	Set or get camera target and camera target mode
`camup`	Set or get camera up vector and up vector mode
`camva`	Set or get camera view angle and view angle mode

`cholinc`	Sparse incomplete Cholesky and Cholesky-infinity factorization
`cholupdate`	Rank 1 update to Cholesky factorization
`daspect`	Set or get data aspect ratio and aspect ratio mode
`ifftshift`	Inverse FFT shift
`lastwarn`	Return last warning string
`lightangle`	Create or move light object in spherical coordinates
`mislocked`	True if M-file cannot be cleared
`mlock`	Prevent M-file clearing
`munlock`	Permit M-file clearing
`ode23t`	Solve moderately stiff differential equations
`ode23tb`	Solve stiff differential equations by using crude error tolerances
`pbaspect`	Set or get plot box aspect ratio and plot box aspect mode
`persistent`	Declare variable persistent
`strcmpi`	Compare strings, ignoring case
`strncmpi`	Compare first n characters in strings, ignoring case
`strjust`	(Change) Now does right, left, and center justification
`xlim`	Set or get x-axis limits and limit mode
`ylim`	Set or get y-axis limits and limit mode
`zlim`	Set or get z-axis limits and limit mode

MATLAB 5.3, Release 11 (March 1999)

Functionality	Description	Chapter
Integer data types	Support for `int8`, `int16`, `unint16`, `int32`, and `uint32` added	7
Single precision data	Storage support for `single` added.	7
Execute function when quitting	When MATLAB is quit, MATLAB executes the function `finish`.	4
Mass matrix support	All functions in the ODE suite support the use of a mass matrix.	25
Optimization parameters	Options for optimization functions `fminbnd`, `fminsearch`, and `lsqnonneg` are set, using an options structure variable rather than an options vector.	23

Functionality	Description	Chapter
Rectangle object	New handle graphics object for drawing rectangle, ovals, and circle	31
Double buffering	*Figure* windows now support double buffering to reduce flickering.	31
Portable Network Graphics	Support for images in PNG format	29

Function	Description
blkdiag	Create block diagonal matrix
cellfun	Perform common operations on cell arrays
complex	Create complex array from real and imaginary parts
coneplot	Velocity vector cone plot in a 3-D vector field
contourslice	Draw contours in volume slices
datenum, datestr, datevec	(Change) Now accept pivotyear argument
evalc	String evaluation with output converted to a string
ezcontour	Easy contour plotter
ezcontourf	Easy filled contour plotter
ezmesh	Easy mesh plotter
ezmeshc	Easy mesh with contour plotter
ezplot	Easy 2-D line plotter
ezplot3	Easy 3-D line plotter
ezpolar	Easy polar plotter
ezsurf	Easy surface plotter
ezsurfc	Easy surface with contour plotter
findfigs	Find all visible *Figure* windows
fminbnd	(Change) New name and calling sequence for function fmin
fminsearch	(Change) New name and calling sequence for function fmins
hist	(Change) Now calls MEX function histc
histc	Histogram binning given bin edges

`int8, int16, uint16, int32, unit32`	Conversion to integer data types
`isocaps`	Compute isosurface end-cap geometry
`isonormals`	Compute normals at isosurface vertices
`isosurface`	Extract isosurface data from volume data
`lsqnonneg`	(Change) New name and calling sequence for function `nnls`
`optimget, optimset`	Get, set, and modify optimization options
`pause`	(Change) Fractional second accepted
`reducepatch`	Reduce number of patch faces
`reducevolume`	Reduce number of volume data elements
`shrinkfaces`	Reduce size of patch faces
`single`	Conversion to single-precision data type
`smooth3`	Smooth 3-D data
`str2double`	Convert character string to double-precision value
`stream2`	Compute 2-D stream lines
`stream3`	Compute 3-D stream lines
`streamline`	Draw stream lines
`subvolume`	Extract subset of volume data set
`sum`	(Change) Can now be used with all integer data types
`surf2patch`	Convert surface data to patch data
`texlabel`	Create the TeX format from a character string

MATLAB 6.0, Release 12 (November, 2000)

Functionality	**Description**	**Chapter**
MATLAB Desktop	User interface organized and named the MATLAB desktop, which is implemented in the Java language	3
Java	MATLAB now supports an interface with the Java language	
Matrix math	(Change) MATLAB now uses LAPACK with BLAS for matrix computations	17

Functionality	Description	Chapter
Fast Fourier Transform	(Change) MATLAB now uses the MIT FFTW library for computing Fast Fourier Transforms	22
Function handles	New data type for capturing function information for evaluation	12
Transparency	(Change) Surfaces, patches, and images now support transparency	29
Operator precedence	(Change) Logical AND now has higher precedence than logical OR	10

Function	Description
beep	Make computer beep
numel	Number of elements in an array
continue	Skip rest of For Loop or While Loop
genpath	Generate path string that includes all directories below a specified directory
sort	(Change) Now works on data types other than double precision
std	(Change) std([]) now returns NaN rather than empty.
polyfit, polyval	(Change) Now support centering and scaling of data

Logical Function	Description
iskeyword	Generate or test if argument is a MATLAB keyword
isvarname	True if input string is a valid variable name

M-file Function	Description
check_syntactic_warnings	Run syntax check on M-files
func2str	Construct function name string from function handle
functions	Display information about a function handle
nargoutchk	Validate number of output arguments
rehash	Refresh function and file system caches
str2func	Construct function handle from function name string

Sparse Matrix Function	Description
colamd	Compute approximate column minimum degree permutation
lsqr	LSQR implementation of Conjugate Gradients on Normal Equations
minres	Solve system of equations by using Minimal Residual Method
symamd	Symmetric approximate minimum degree permutation
symmlq	Solve system of equations by using symmetric LQ method

Interpolation Function	Description
convhulln	n-D convex hull
delaunay3	3-D Delaunay tessellation
delaunayn	n-D Delaunay tessellation
desearchn	n-D nearest point search
griddata3	3-D data gridding
griddatan	n-D data gridding
interp1	(Change) 'cubic' option now calls pchip
pchip	Piecewise cubic Hermite interpolating polynomial interpolation
voronoin	n-D Voronoi diagram

Optimization or Integration Function	Description
quadl	Numerically compute integral by using Lobatto quadrature; replaces quad8
dblquad	(Change) Can now pass extra arguments to integrand function
fzero	(Change) Calling sequence changed

Differential Equation Function	Description
bvp4c	Solve two-point boundary value problem by collocation
bvpget	Get BVP option from option structure
bvpinit	Form initial guess for bvp4c
bvpset	Create or change BVP option structure
bvpval	Evaluate solution from bvp4c
ode*	(Change) ODE solvers can now solve problems without using an ODE file
pdepe	Solve partial differential equations in one dimension
pdeval	Evaluate solution computed by pdepe

3-D Visualization Function	Description
coneplot	Create 3-D coneplot
curl	Compute curl and angular velocity perpendicular to flow
divergence	Compute divergence of vector field
interpstreamspeed	Interpolate streamline vertices from speed
isocolors	Compute colors of isosurface vertices
isosurface	Extract isosurface
streamparticles	Draw stream particles
streamribbon	Draw stream ribbons
streamslice	Draw stream lines
streamtube	Draw stream tubes
volumebounds	Get coordinate and color limits for volume data

Handle Graphics and GUI Function	Description
alpha	Set or get transparency properties
alphamap	Specify figure alphamap
alim	Set or get axes alpha limits

Java Function	Description
import	Add to the current Java packages import list
isjava	True for Java Object
javaArray	Create Java array
javaMethod	Invoke Java method
javaObject	Construct Java object
methodsview	Display information on all methods implemented by Java or MATLAB class

Obsolete Function	Description
errortrap	Replaced by try-catch block
flops	Floating-point operation count
fmin	Replaced by fminbnd
fmins	Replaced by fminsearch
foptions	Replaced by optimget and optimset
interp4, interp5, interp6,	Replaced by interp2
isdir	Replaced with exist
isiee	All MATLAB platforms use IEEE arithmetic
isstr	Replaced by ischar
meshdom	Replaced by meshgrid
nnls	Replaced by lsnonneg
quad8	Replaced by quadl
saxis	No longer used
setstr	Replaced by char
str2mat	Replaced by char
table1	Replaced by interp1
table2	Replaced by interp2

MATLAB 6.1, Release 12.1 (June 2001)

Functionality	Description	Chapter
Transparent legend	Can now make axes legend box transparent	26

Function	Description
audioplayer	Create audio object to play audio data on Windows platforms
audiorecorder	Create audio object to record audio data on Windows platforms
bvpval	(Obsolete) Replaced by deval
convhull, delaunay, griddata, voronoi	(Change) Now makes use of Qhull
cdfinfo	Get information about a CDF file
cdfread	Read CDF file
convhull	(Change) [K,a]=convhull(x,y) now returns the area a of the convex hull; in addition, it now ignores the third input argument
convhulln	(Change) [K,v]=convhulln(x,y) now returns the volume v of the convex hull
datenum, datestr	(Change) Now accept date vector as an input argument
delaunay	(Change) Now ignores third input argument
deval	Evaluate ODE solution; also replaces obsolete bvpval function
erfcinv	Inverse complementary error function
fitsinfo	Get information about a FITS file
fitsread	Read FITS file
hdfinfo	Get information about a HDF file
hdfread	Read HDF file
histc	(Change) Bug fixed
interp1	(Change) Now enables data extrapolation
numel	(Change) numel(A,varargin) returns the number of subscripted elements in A(varargin{:})

ode*	(Change) Now optionally return solution structure for use by `deval`
`polyeig`	(Change) Can now return only the eigenvalues
`ppval`	(Change) Now supports `ppval(xx,pp)` to permit use of `ppval` with function functions
`quad`	(Change) Function sampling bug fixed
`reshape`	(Change) `reshape(A,..., [],...)` now calculates the size required for the empty dimension
`sortrows`	(Change) Now calls MEX function `sortrowsc` to maximize speed. With cell array of strings input, now calls MEX function `sortcellchar` to maximize speed
`strfind`	Search for occurrence of second string argument in first string argument
`svd`	(Change) Can now return only the first two outputs, U and S
`tetramesh`	Tetrahedron mesh plot for use with `delaunayn`
`triplot`	2-D triangular plot for use with `delaunay`

MATLAB 6.5, Release 13 (August 2002)

Functionality	Description	Chapter
JIT-Acceleration	For Loops with specific properties now execute at maximum speed, thereby avoiding the need to vectorize code under some circumstances.	11
Dynamic structure field names	Need for `getfield` and `setfield` eliminated; `var.(fstr)`, where `fstr` is a character string, addresses the field identified by `fstr` in the variable `var`	8
Maximum name length for variables and functions	Variable names and function names can now have up to 63 characters.	2
Integer array subscripts	Array subscripts must be real positive integer values or logical values (e.g., `x(1.3)` reports an error).	5
Relational operators and 64 bit integer arrays	(MATLAB 6.5.1) All relational operators now support `int64` and `unit64` data types.	7
Delay differential equations	Support for solving delay differential equations provided	25

Functionality	Description	Chapter
BLAS for Pentium 4	Specific BLAS provided for the Pentium 4 processor	
UMFPACK support	When appropriate, sparse matrix solutions now use UMFPACK library functions.	17
Regular Expressions	MATLAB now supports regular expressions.	9
Logical class	Logical arrays are now a separate MATLAB class that use one byte of storage per array element.	10
Logical True and False	MATLAB now supports direct creation of arrays containing logical True and logical False by using the functions `true` and `false`.	10
Empty comparisons	`[]==[]` and `[]==scalar` now return empty (`[]`) result to coincide with other operators.	10
Sparse class	Sparse is now an attribute of the underlying variable class, as opposed to a distinct or separate variable class. In MATLAB 6.5, logical and double data types can be sparse.	17
Formatted error and warning strings	Functions `error` and `warning` now accept data formatting inputs similar to `sprintf`.	12
Message identifiers	Error and warning messages can now contain identifier tags to make the message uniquely identifiable.	12
Warning control	Individual warnings can be suppressed.	12
Logical operators for short-circuiting	New operators && and || short circuit comparisons when an early exit is possible.	10
Scheduled execution	New timer object permits scheduled execution of MATLAB code.	4
Text object properties	Text objects now have properties for their background box.	31

Function	Description
`audiodevinfo`	Get information about installed audio devices on Windows platforms
`cdfepoch`	Convert MATLAB date number or string to CDF format
`cdfwrite`	Write data to CDF file

cell2mat	Combine cell array of matrices into one matrix; previously part of the *Neural Networks Toolbox*
colormapeditor	Interactive colormap editor
copyfile	(Change) Now also copies directories
corrcoef	(Change) Three new syntaxes
dde23	Solve delay differential equations with constant delays
ddeget	Get properties from DDE options structure
ddeget	Create or modify properties in DDE options structure
deval	(Change) Now accepts output from dde23
false	Create array of logical False
fileattrib	Set or get file attributes
gallery	(Change) New test matrices available
imformats	Eases the task of adding read and write support for new file formats
int64, unit64	Create signed or unsigned 64-bit integer array
isequal	(Change) When used to compare structures, input argument field creation order no longer has an impact on equality
isequalwithequalnans	True if arrays are equal with NaNs considered equal
ismember	(Change) [tf, idx]=ismember(...) now returns indices idx of located members; in this case, MEX function ismemc2 is called to maximize speed
issorted	True if array is sorted
lasterror	Return last error message and related information
lu	(Change) Now uses UMFPACK for sparse matrices
mat2cell	Break matrix up into a cell array of matrices (previously part of the *Neural Networks Toolbox*)
movefile	(Change) Now also rename a file or directory
multibandread	Supports reading data from raw files
multibandwrite	Supports writing data to raw files
namelengthmax	Return maximum variable and function name lengths
orderfields	Order fields of a structure
perl	Call Perl script using platform executable
profview	Produce graphical profile report

Function	Description
psi	Evaluate Digamma function
qrdelete, qrinsert	(Change) Can now insert or delete rows as well as columns
regexp	Match regular expression
regexpi	Match regular expression, ignoring case
regexprep	Replace string, using regular expression
rethrow	Reissue error
rmdir	Remove directory and, optionally, contents as well
sendmail	Send e-mail
timer	Create and control timer objects to schedule execution of MATLAB code
triplequad	Evaluate triple integral
true	Create array of logical True
urlread, urlwrite	Read and write content, using URL
ver	(Change) Now returns more detailed information, and hostid information is no longer provided
winopen	On Windows platforms, opens a file in its appropriate application
xmlread, xmlwrite	Read or write XML document
xslt	Transform XML document, using XSLT engine
zip, unzip	Compress and uncompress files and directories

MATLAB 7.0, Release 14 (June 2004)

Functionality	Description	Chapter
Case sensitivity	Function and directory names are now case-sensitive; in prior versions, they were case-sensitive only on UNIX platforms.	4,12,14
Function precedence	Built-in and M-file functions now share the same calling precedence; in prior versions, built-in functions had higher precedence.	12
Nondouble arithmetic	Mathematical operations on integer and single-precision data are now supported; many built-in functions support these data types as well.	8

Anonymous functions	Single line function specification as a replacement for in-line functions	12
Nested functions	M-file functions can be nested, allowing shared workspaces among the primary and all nested functions.	12
Unicode character storage	MATLAB now encodes character strings in Unicode format.	9
Java 1.4	MATLAB now uses Java version 1.4.	
Block comments	Now supports block comments using %{ and %} syntax	4
Code cells	Editor permits creation, execution, etc., on sections of code called code cells.	4
Results publishing	M-files and figures can be published to HTML, XML, LaTeX, Word, and Powerpoint.	
Regular expressions	Expanded regular expression capabilities	9
Freeform date and time	Date and time functions now support user-specified date and time specifications.	16
File ftp support	Access files using ftp commands in MATLAB	14
Annotation layer	*Figure* windows now have an annotation layer where annotations appear. Annotations include rectangles, ellipses, arrows, double arrows, text arrows, text boxes, lines, colorbars, and legends.	31
Plot objects	High-level and specialized plotting functions now create plot objects that have properties specific to the type of plot created. In the past, these functions returned handles to the core graphics objects used to create the graphics.	31
Group objects.	Graphics objects can now be grouped or linked to each other.	31
Toolbars	New user interface object	32
Panels and button groups	New user interface container objects	32
Generic dynamic linked libraries	MATLAB supports interaction with Generic DLLs on the Windows platform.	36
TeX support	Text objects now optionally support complete TeX capabilities.	31
Code checking	M-file code can be assessed for potential problems and improvement by using the functions `mlint` or `mlintrpt`.	13

Mathematical Function	Description
acosd, acotd, acscd, asecd, asind, atand	Inverse trigonometric functions returning angles in degrees
balance	(Change) Now returns different outputs and offers balancing without permuting rows and columns
bvp4c	(Change) Can now solve multipoint boundary value problems
convhulln, delaunayn, voronoin	(Change) Now supports user-settable options
cosd, cotd, cscd, secd, sind, tand	Trigonometric functions with arguments in degrees
decic	Compute consistent initial conditions for ODE15I
deval	(Change) Now optionally returns derivative at points as well
eps	(Change) Now accepts arguments to specify single- or double-precision values and to return eps relative to any value, not just 1
expm1	Compute $\exp(x)$-1 accurately
eye, ones, zeros	(Change) Now accepts a final argument specifying numeric data type of result
fftw	Tune or set options in FFTW library for FFT computations
fminbnd, fminsearch, fzero	(Change) Now support calling an output function at each iteration
interp1, ppval, spline	(Change) Now support multidimensional arrays for Y
interp1	(Change) Now optionally returns a pp-form for evaluation with ppval
linsolve	Solve $Ax = y$, given specific structure of A
ltitr	(Change, previously undocumented) Linear time-invariant time response kernel
log1p	Compute $\log(1+x)$ accurately
mimofr	(Change, previously undocumented) Linear time-invariant frequency response
nthroot	n^{th} real root
ode15i	Ordinary differential equation solver for implicit equations

odextend	Extend solution of ordinary differential equations
ordqz	Reorder QZ factorization
ordschur	Reorder Shur factorization
pwch	Piecewise cubic Hermite interpolation
quadv	Vectorized quad function
sort	(Change) Now supports an optional last argument that specifies the sort direction

Logical Function	**Description**
iscom	True for COM/ActiveX objects
isevent	True if event of object
isfloat	True for floating point data
ishghandle	True for Handle Graphics Object handle
isinteger	True for integer data
isinterface	True for COM interface
ispuma	True for computers running Mac OS X 10.1.x
isscalar	True if argument is a scalar
isstrprop	True for string elements matching a variety of specifications
isvector	True if argument is a row or column vector

M-file Function	**Description**
auditcontents	Audit Contents.m file for a given directory
deleteconfirm	Confirm the deletion of a file with a dialog box
deprpt	Scan a file or directory for dependencies
diff2asv	Compare file to autosaved version, if one exists
diffrpt	Visual directory browser
dofixrpt	Scan a file or directory for all TODO, FIXME, or NOTE messages
helprpt	Scan a file or directory for help
makemcode	Make M-file for regenerating object and its children

M-file Function	Description
makecontentsfile	Make a new `Contents.m` file
mlint, mlintrpt	Examine M-file or a directory of M-files for potential problems and make suggestions for possible improvements
path2rc	(Obsolete) Replaced by `savepath`
publish	Run a script M-file and save the results
recycle	Determine if deleted files go to Recycle Bin
restoredefaultpath	Restore default MATLAB path
savepath	Save current MATLAB path; replaces `path2rc` function

Graphics and GUI Function	Description
addsubplot	Add subplot to figure in given location
ancestor	Get ancestor of graphics object
annotation	Add *annotation* object
axes	(Change) `ActivePositionProperty`, `OuterPosition`, `TightInset` properties added
axescheck	Process leading *axes* object from input list
axis	(Change) Now accepts *axes* handle as first argument
commandhistory	Open *Command History* window or select it
commandwindow	Open *Command* window or select it
datacursormode	Interactively create data cursors on plot
exportsetupdlg	Show figure export style dialog
figure	(Change) `DockControls` property added, `KeyPressFcn` property modified
figureheaderdlg	Show figure header dialog
figurepalette	Show or hide the palette for a figure
getpixelposition	Get position of object in pixels
hasbehavior	Sets or gets behaviors of Handle Graphics objects
hgexport	Export a figure
hggroup	Create a Handle Graphics group object
hgtransform	Create graphics transformation object

hold	(Change) Now supports all option, which holds the plot so that subsequent plots do not reset the color and linestyle order
linkaxes	Synchronize limits of specified *axes* objects
linkprop	Maintain same value for corresponding Handle Graphics properties
makehgtransform	Create graphical transformation matrix
pan	Interactively pan the plot view
plottools	Show or hide the plot-editing tools for a figure
printdlg	Print dialog box
printpreview	Display preview of figure to be printed
propertyeditor	Show or hide the property editor for a figure
refreshdata	Refresh data in plot
setpixelposition	Set position of object in pixels
showplottool	Show or hide one of the plot-editing components for a figure
title	(Change) Now accepts *axes* handle as first argument
uibuttongroup	Create *buttongroup* object
uicontainer	Create *container* object
uicontrol	(Change) uicontrol(h) now transfers focus to the *uicontrol* having handle h. Multiline 'edit' style *uicontrol* objects now have a vertical scroll bar. *uicontrol* objects now have a 'KeyPressFcn' callback
uigetfile	(Change) Now permits selection of multiple files
uipanel	Create *uipanel* container object
uipushtool	Create pushbutton in *uitoolbar* object
uitable	Create a *uitable* object
uitoggletool	Create togglebutton in *uitoolbar* object
uitoolbar	Create *uitoolbar* object
uitree	Create *uitree* object
uitreenode	Creates a *node* object in a *uitree* component
uiwait	(Change) uiwait(handle,t) now times out after time t has elapsed
xlabel,ylabel,zlabel	(Change) Now accepts *axes* handle as first argument

External Inteface Function	Description

`calllib,`
`libfunctions,`
`libfunctionsview,`
`libisloaded,`
`libpointer,`
`libstruct,`
`loadlibrary`
`unloadlibrary`
	Generic DLL Interface functions

`callsoapservice`	Send a SOAP message off to an endpoint
`eventlisteners`	Lists all events that are registered
`ftp`	Create *ftp* object
`instrfindall`	Find all serial port objects with specified property values
`javaaddpath`	Add directories to the dynamic Java path
`javaclasspath`	Get and set Java path
`javacomponent`	Create a Java AWT Component and put in a figure
`javarmpath`	Remove directory from dynamic Java path
`registerevent`	Registers events for a specified control at runtime
`unregisterallevents`	Unregister all events for a specified control at runtime
`unregisterevent`	Unregister events for a specified control at runtime

Other Function	Description
`accumarray`	Construct array with accumulation; that is, if any element is specified more than once, later elements add to the current value rather than overwrite it
`addtodate`	Modify a particular field of a date number
`audiorecorder,` `audioplayer`	(Change) Now supported on both UNIX and Windows platforms
`cast`	Cast a variable to a different data type
`docsearch`	Search HTML documentation in the Help browser
`find`	(Change) Now supports optional arguments specifying an upper limit on the number of indices returned, and whether the search begins at the start or end of the array

`fixquote`	Double up single quotes in a string
`genvarname`	Generate variable name from candidate name

`hd5info,`
`hd5read,`
`hdf5write`
	HDF5 file information, read, and write

`hex2num,` `num2hex`	Convert number to and from IEEE hexadecimal format
`intmax, intmin`	Maximum and minimum integer values given integer data type
`intwarning`	Controls state of the integer data type warnings
`mmcompinfo`	Multimedia compressor information
`mmfileinfo`	Get information about a multimedia file
`save`	(Change) Now supports compressing MAT-files
`strfind, strtok`	(Change) Now support cell array of strings as input
`strtrim`	Remove leading and trailing white space from a string
`textscan`	Read text file into a cell array; has more features than `textread`
`timerfindall`	Find all timer objects with specified property values
`xlswrite`	Write Matrix to Excel spreadsheet

Index